SCHAUM'S OUTLINE OF

THEORY AND PROBLEMS

of

COLLEGE ALGEBRA

.

BY

MURRAY R. SPIEGEL, Ph.D.

Professor of Mathematics
Rensselaer Polytechnic Institute

.

SCHAUM'S OUTLINE SERIES

McGRAW-HILL BOOK COMPANY

New York, St. Louis, San Francisco, Toronto, Sydney

ISBN 07-060226-3

29 30 SH SH 8 7

Preface

Algebra, which has a glorious history of more than three thousand years, might very well be called a universal language of civilization. It provides a foundation upon which higher mathematics is built and it is the language of modern science and technology. Many problems which would prove difficult to one armed only with arithmetic become far easier when translated into algebra.

Like so many languages, however, algebra requires much study before one can become proficient in it. The old adage that there is no royal road to learning is no exception here. One must develop a clear understanding of its basic principles, and much practice and drill may be necessary before one can "speak it" fluently.

This book is designed primarily to assist students in acquiring a more thorough knowledge and proficiency in this basic field. In addition to the use of the book by students taking a formal course in algebra, it should also be of considerable value to those who wish to review the fundamental principles and applications in anticipation of further work in mathematics, science and engineering.

The contents are divided into chapters covering duly-recognized areas of theory and study. Each chapter begins with a clear summary of the pertinent definitions, principles and theorems, together with illustrative and other descriptive material. This is followed by graded sets of solved and supplementary problems. The solved problems have been chosen and solutions arranged so that the principles are clearly established. They serve to illustrate and amplify the theory, provide the repetition of basic principles so vital to effective teaching, and bring into sharp focus those fine points without which the student continually feels himself on unsafe ground. Derivations of formulas and proofs of theorems are included among the solved problems. The treatment of many topics is considerably more extensive and thorough than in most texts; such topics include complex numbers, theory of equations, permutations and combinations, probability, determinants, and infinite series. This is done to meet any choice of topics which the instructor may make, to provide a book of reference, and to stimulate further interest in the subject.

M. R. SPIEGEL

Rensselaer Polytechnic Institute
September, 1956

Contents

CHAPTER 1

Fundamental Operations with Numbers

FOUR OPERATIONS are fundamental in algebra, as in arithmetic. These are addition, subtraction, multiplication, and division.

ADDITION. When two numbers a and b are added, their sum is indicated by $a + b$. Thus $3 + 2 = 5$.

SUBTRACTION. When a number b is subtracted from a number a, the difference is indicated by $a - b$. Thus $6 - 2 = 4$.

Subtraction may be defined in terms of addition. That is, we may define $a - b$ to represent that number x such that x added to b yields a, or $x + b = a$. For example, $8 - 3$ is that number x which when added to 3 yields 8, i.e. $x + 3 = 8$; thus $8 - 3 = 5$.

MULTIPLICATION. The product of two numbers a and b is a number c such that $a \times b = c$. The operation of multiplication may be indicated by a cross or a dot or parentheses. Thus $5 \times 3 = 5 \cdot 3 = 5(3) = (5)(3) = 15$, where the factors are 5 and 3 and the product is 15. When letters are used as in algebra, the notation $p \times q$ is usually avoided since $\times$ may be confused with a letter representing a number.

DIVISION. When a number a is divided by a number b, the quotient obtained is written $a \div b$ or $\frac{a}{b}$ or a/b, where a is called the dividend and b the divisor. The expression a/b is also called a fraction having numerator a and denominator b.

Division by zero is not defined. See Prob. 1(b), 1(e).

Division may be defined in terms of multiplication. That is, we may consider a/b as that number x which upon multiplication by b yields a, or $bx = a$. For example, $6/3$ is that number x such that 3 multiplied by x yields 6, or $3x = 6$; thus $6/3 = 2$.

THE SYSTEM OF REAL NUMBERS as we know it today is a result of gradual progress, as the following indicates.

1) *Natural Numbers* $1, 2, 3, 4, \ldots$ (three dots mean 'and so on') used in counting are also known as the positive integers. If two such numbers are added or multiplied, the result is always a natural number.

2) *Positive Rational Numbers* or positive fractions are the quotients of two positive integers, such as 2/3, 8/5, 121/17. The positive rational numbers include the set of natural numbers. Thus the rational number 3/1 is the natural number 3.

3) *Positive Irrational Numbers* are numbers which are not rational, such as $\sqrt{2}$, π.

1

4) *Zero*, written 0, arose in order to enlarge the number system so as to permit such operations as $6-6$, $10-10$, etc. Zero has the property that any number multiplied by zero is zero. Zero divided by any number $\neq 0$ (i.e., not equal to zero) is zero.

5) *Negative* integers, negative rational numbers and negative irrational numbers such as -3, $-2/3$, and $-\sqrt{2}$, arose in order to enlarge the number system so as to permit such operations as $2-8$, $\pi-3\pi$, $2-2\sqrt{2}$, etc.

When no sign is placed before a number, a plus sign is understood. Thus 5 is +5, $\sqrt{2}$ is $+\sqrt{2}$. Zero is considered a rational number without sign.

THE REAL NUMBER SYSTEM consists of the collection of positive and negative rational and irrational numbers and zero.

Note. The word real is used in contradiction to still other numbers involving $\sqrt{-1}$ which will be taken up later and which are known as *imaginary* although they are very useful in mathematics and the sciences. Unless otherwise specified we shall deal with real numbers.

GRAPHICAL REPRESENTATION OF REAL NUMBERS. It is often useful to represent real numbers by points on a line. To do this, we choose a point on the line to represent the real number zero and call this point the origin. The positive integers +1, +2, +3, ... are then associated with points on the line at distances 1, 2, 3, ... units respectively to the *right* of the origin (see figure), while the negative integers -1, -2, -3, ... are associated with points on the line at distances 1, 2, 3, ... units respectively to the *left* of the origin.

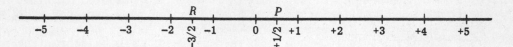

The rational number 1/2 is represented on this scale by a point P halfway between 0 and +1. The negative number $-3/2$ or $-1\frac{1}{2}$ is represented by a point R $1\frac{1}{2}$ units to the left of the origin.

It can be proved that corresponding to each real number there is one and only one point on the line; and conversely, to every point on the line there corresponds one and only one real number.

THE POSITION OF REAL NUMBERS on a line establishes an order to the real number system. If a point A lies to the right of another point B on the line we say that the number corresponding to A is *greater* or *larger* than the number corresponding to B, or that the number corresponding to B is *less* or *smaller* than the number corresponding to A. The symbols for 'greater than' and 'less than' are > and < respectively. These symbols are called 'inequality signs'.

Thus since 5 is to the right of 3, 5 is greater than 3 or $5 > 3$; we may also say 3 is less than 5 and write $3 < 5$. Similarly since -6 is to the left of -4, -6 is smaller than -4, i.e. $-6 < -4$; we may also write $-4 > -6$.

BY THE ABSOLUTE VALUE or numerical value of a number is meant the number regardless of sign. Absolute value is indicated by two vertical lines surrounding the number. Thus $|-6| = 6$, $|+4| = 4$, $|-3/4| = 3/4$.

LAWS OF ADDITION AND MULTIPLICATION.

1) Commutative Law for Addition. The order of addition of two numbers does not affect the result.

Thus $a + b = b + a$, $5 + 3 = 3 + 5 = 8$.

2) **Associative Law for Addition.** The terms of a sum may be grouped in any manner without affecting the result.

$$a + b + c = a + (b + c) = (a + b) + c, \qquad 3 + 4 + 1 = 3 + (4 + 1) = (3 + 4) + 1 = 8$$

3) **Commutative Law for Multiplication.** The order of the factors of a product does not affect the result.

$$a \cdot b = b \cdot a, \qquad 2 \cdot 5 = 5 \cdot 2 = 10$$

4) **Associative Law for Multiplication.** The factors of a product may be grouped in any manner without affecting the result.

$$abc = a(bc) = ab(c), \qquad 3 \cdot 4 \cdot 6 = 3(4 \cdot 6) = (3 \cdot 4)6 = 72$$

5) **Distributive Law for Multiplication.** The product of a number a by the sum of two numbers $(b + c)$ is equal to the sum of the products ab and ac.

$$a(b + c) = ab + ac, \qquad 4(3 + 2) = 4 \cdot 3 + 4 \cdot 2 = 20$$

Extensions of these laws may be made. Thus we may add the numbers a, b, c, d, e by grouping in any order, as $(a + b) + c + (d + e)$, $a + (b + c) + (d + e)$, etc. Similarly, in multiplication we may write $(ab)c(de)$ or $a(bc)(de)$, the result being independent of order or grouping.

RULES OF SIGNS.

1) To add two numbers with like signs, add their absolute values and prefix the common sign. Thus $3 + 4 = 7$, $(-3) + (-4) = -7$.

2) To add two numbers with unlike signs, find the difference between their absolute values and prefix the sign of the number with greater absolute value.

Examples. $17 + (-8) = 9$, $(-6) + 4 = -2$, $(-18) + 15 = -3$

3) To subtract one number b from another number a, change the sign of b and add to a.

Examples. $12 - (7) = 12 + (-7) = 5$, $(-9) - (4) = -9 + (-4) = -13$, $2 - (-8) = 2 + 8 = 10$

4) To multiply (or divide) two numbers having like signs, multiply (or divide) their absolute values and prefix a plus sign (or no sign).

Examples. $(5)(3) = 15$, $(-5)(-3) = 15$, $\dfrac{-6}{-3} = 2$

5) To multiply (or divide) two numbers having unlike signs, multiply (or divide) their absolute values and prefix a minus sign.

Examples. $(-3)(6) = -18$, $(3)(-6) = -18$, $\dfrac{-12}{4} = -3$

EXPONENTS AND POWERS. When a number a is multiplied by itself n times, the product $a \cdot a \cdot a \cdots a$ (n times) is indicated by the symbol a^n which is referred to as 'the nth power of a' or 'a to the nth power' or 'a to the nth'.

Examples. $2 \cdot 2 \cdot 2 \cdot 2 \cdot 2 = 2^5 = 32$, $(-5)^3 = (-5)(-5)(-5) = -125$

$2 \cdot x \cdot x \cdot x = 2x^3$, $a \cdot a \cdot a \cdot b \cdot b = a^3 b^2$, $(a - b)(a - b)(a - b) = (a - b)^3$

In a^n, the number a is called the *base* and the positive integer n is the *exponent*.

LAWS OF EXPONENTS. If p and q are positive integers, then:

1) $a^p \cdot a^q = a^{p+q}$ Examples. $2^3 \cdot 2^4 = 2^{3+4} = 2^7$

2) $\dfrac{a^p}{a^q} = a^{p-q} = \dfrac{1}{a^{q-p}}$ if $a \neq 0$ $\dfrac{3^5}{3^2} = 3^{5-2} = 3^3$, $\dfrac{3^4}{3^6} = \dfrac{1}{3^{6-4}} = \dfrac{1}{3^2}$

3) $(a^p)^q = a^{pq}$ $(4^2)^3 = 4^6$, $(3^4)^2 = 3^8$

4) $(ab)^p = a^p b^p$, $\left(\dfrac{a}{b}\right)^p = \dfrac{a^p}{b^p}$ if $b \neq 0$ $(4 \cdot 5)^2 = 4^2 \cdot 5^2$, $\left(\dfrac{5}{2}\right)^3 = \dfrac{5^3}{2^3}$

OPERATIONS WITH FRACTIONS may be performed according to the following rules.

     1) The value of a fraction remains the same if its numerator and denominator are both multiplied or divided by the same number provided the number is not zero.

Examples. $\dfrac{3}{4} = \dfrac{3 \cdot 2}{4 \cdot 2} = \dfrac{6}{8}$, $\dfrac{15}{18} = \dfrac{15 \div 3}{18 \div 3} = \dfrac{5}{6}$

     2) Changing the sign of the numerator or denominator of a fraction changes the sign of the fraction.

Example. $\dfrac{-3}{5} = -\dfrac{3}{5} = \dfrac{3}{-5}$

     3) Adding two fractions with a common denominator yields a fraction whose numerator is the sum of the numerators of the given fractions and whose denominator is the common denominator.

Example. $\dfrac{3}{5} + \dfrac{4}{5} = \dfrac{3+4}{5} = \dfrac{7}{5}$

     4) The sum or difference of two fractions having different denominators may be found by writing the fractions with a common denominator.

Example. $\dfrac{1}{4} + \dfrac{2}{3} = \dfrac{3}{12} + \dfrac{8}{12} = \dfrac{11}{12}$

     5) The product of two fractions is a fraction whose numerator is the product of the numerators of the given fractions and whose denominator is the product of the denominators of the fractions.

Examples. $\dfrac{2}{3} \cdot \dfrac{4}{5} = \dfrac{2 \cdot 4}{3 \cdot 5} = \dfrac{8}{15}$, $\dfrac{3}{4} \cdot \dfrac{8}{9} = \dfrac{3 \cdot 8}{4 \cdot 9} = \dfrac{24}{36} = \dfrac{2}{3}$

     6) The reciprocal of a fraction is a fraction whose numerator is the denominator of the given fraction and whose denominator is the numerator of the given fraction. Thus the reciprocal of 3 (i.e. 3/1) is 1/3. Similarly the reciprocals of 5/8 and −4/3 are 8/5 and 3/−4 or −3/4 respectively.

     7) To divide two fractions, multiply the first by the reciprocal of the second.

Examples. $\dfrac{a}{b} \div \dfrac{c}{d} = \dfrac{a}{b} \cdot \dfrac{d}{c} = \dfrac{ad}{bc}$, $\dfrac{2}{3} \div \dfrac{4}{5} = \dfrac{2}{3} \cdot \dfrac{5}{4} = \dfrac{10}{12} = \dfrac{5}{6}$

This result may be established as follows: $\dfrac{a}{b} \div \dfrac{c}{d} = \dfrac{a/b}{c/d} = \dfrac{a/b \cdot bd}{c/d \cdot bd} = \dfrac{ad}{bc}$.

SOLVED PROBLEMS

1. Write the sum S, difference D, product P and quotient Q of each of the following pairs of numbers: a) 48, 12; b) 8, 0; c) 0, 12; d) 10, 20; e) 0, 0.

   a) S = 48 + 12 = 60,   D = 48 − 12 = 36,   P = 48(12) = 576,   Q = 48 ÷ 12 = $\frac{48}{12}$ = 4

   b) S = 8 + 0 = 8,   D = 8 − 0 = 8,   P = 8(0) = 0,   Q = 8 ÷ 0 or $\frac{8}{0}$.   But by definition $\frac{8}{0}$ is that number x (if it exists) such that $x(0)$ = 8.   Clearly there is no such number, since any number multiplied by 0 must yield 0.

   c) S = 0 + 12 = 12,   D = 0 − 12 = −12,   P = 0(12) = 0,   Q = $\frac{0}{12}$ = 0

   d) S = 10 + 20 = 30,   D = 10 − 20 = −10,   P = 10(20) = 200,   Q = 10 ÷ 20 = $\frac{10}{20}$ = $\frac{1}{2}$

   e) S = 0 + 0 = 0,   D = 0 − 0 = 0,   P = 0(0) = 0.   Q = 0 ÷ 0 or $\frac{0}{0}$ is by definition that number x (if it exists) such that $x(0)$ = 0. Since this is true for *all* numbers x there is no *one* number which $\frac{0}{0}$ represents.

   From b) and e) it is seen that division by zero is an undefined operation.

2. Perform each of the indicated operations.

   a) 42 + 23, 23 + 42
   b) 27 + (48 + 12), (27 + 48) + 12
   c) 125 − (38 + 27)
   d) 6·8, 8·6
   e) 4(7·6), (4·7)6

   f) 35·28
   g) 756 ÷ 21
   h) $\dfrac{(40 + 21)(72 − 38)}{(32 − 15)}$

   i) 72 ÷ 24 + 64 ÷ 16
   j) 4 ÷ 2 + 6 ÷ 3 − 2 ÷ 2 + 3·4
   k) 128 ÷ (2·4), (128 ÷ 2)·4

   a) 42 + 23 = 65, 23 + 42 = 65. Thus 42 + 23 = 23 + 42.
   This illustrates the commutative law for addition.

   b) 27 + (48 + 12) = 27 + 60 = 87, (27 + 48) + 12 = 75 + 12 = 87. Thus 27 + (48 + 12) = (27 + 48) + 12.
   This illustrates the associative law for addition.

   c) 125 − (38 + 27) = 125 − 65 = 60

   d) 6·8 = 48, 8·6 = 48. Thus 6·8 = 8·6, illustrating the commutative law for multiplication.

   e) 4(7·6) = 4(42) = 168, (4·7)6 = (28)6 = 168. Thus 4(7·6) = (4·7)6.
   This illustrates the associative law for multiplication.

   f) (35)(28) = 35(20 + 8) = 35(20) + 35(8) = 700 + 280 = 980 by the distributive law for multiplication.

   g) $\dfrac{756}{21}$ = 36   Check: 21·36 = 756

   h) $\dfrac{(40 + 21)(72 − 38)}{(32 − 15)}$ = $\dfrac{(61)(34)}{17}$ = $\dfrac{61 \cdot \overset{2}{\cancel{34}}}{\cancel{17}}$ = 61·2 = 122

   i) Computations in arithmetic, by convention, obey the following rule: Operations of multiplication and division precede operations of addition and subtraction.
   Thus 72 ÷ 24 + 64 ÷ 16 = 3 + 4 = 7.

   j) The rule of i) is applied here. Thus 4 ÷ 2 + 6 ÷ 3 − 2 ÷ 2 + 3·4 = 2 + 2 − 1 + 12 = 15.

   k) 128 ÷ (2·4) = 128 ÷ 8 = 16, (128 ÷ 2)·4 = 64·4 = 256
   Hence if one wrote 128 ÷ 2·4 without parentheses there would be ambiguity.

3. Classify each of the following numbers according to the categories: real number, positive integer, negative integer, rational number, irrational number, none of the foregoing.

$$-5, \quad 3/5, \quad 3\pi, \quad 2, \quad -1/4, \quad 6.3, \quad 0, \quad \sqrt{5}, \quad \sqrt{-1}, \quad .3782, \quad \sqrt{4}, \quad -18/7$$

If the number belongs to one or more categories this is indicated by a check mark.

	Real number	Positive integer	Negative integer	Rational number	Irrational number	None of foregoing
−5	√		√	√		
3/5	√			√		
3π	√				√	
2	√	√		√		
−1/4	√			√		
6.3	√			√		
0	√			√		
√5	√				√	
√−1						√
.3782	√			√		
√4	√	√		√		
−18/7	√			√		

4. Represent (approximately) by a point on a graphical scale each of the real numbers in Prob. 3.

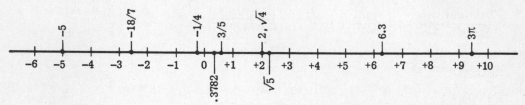

Note. 3π is approximately $3(3.14) = 9.42$, so that the corresponding point is between +9 and +10 as indicated. $\sqrt{5}$ is between 2 and 3, its value to three decimal places being 2.236.

5. Place an appropriate inequality symbol (< or >) between each pair of real numbers.
 a) 2, 5   c) 3, −1   e) −4, −3   g) $\sqrt{7}$, 3   i) −3/5, −1/2
 b) 0, 2   d) −4, +2   f) π, 3   h) $-\sqrt{2}$, −1

 a) $2 < 5$ (or $5 > 2$), i.e. 2 is *less than* 5 (or 5 is *greater than* 2)
 b) $0 < 2$ (or $2 > 0$)　　　　　 e) $-4 < -3$ (or $-3 > -4$)　　　 h) $-\sqrt{2} < -1$ ($-1 > -\sqrt{2}$)
 c) $3 > -1$ (or $-1 < 3$)　　　　 f) $\pi > 3$ (or $3 < \pi$)　　　　 i) $-3/5 < -1/2$ since $-.6 < -.5$
 d) $-4 < +2$ (or $+2 > -4$)　　　 g) $3 > \sqrt{7}$ (or $\sqrt{7} < 3$)

6. Arrange each of the following groups of real numbers in ascending order of magnitude.
 a) −3, 22/7, $\sqrt{5}$, −3.2, 0 ;   b) $-\sqrt{2}$, $-\sqrt{3}$, −1.6, −3/2.

 a) $-3.2 < -3 < 0 < \sqrt{5} < 22/7$　　　　　 b) $-\sqrt{3} < -1.6 < -3/2 < -\sqrt{2}$

7. Write the absolute value of each of the following real numbers.
　　　　　−1, +3, 2/5, $-\sqrt{2}$, −3.14, 2.83, −3/8, −π, +5/7

We may write the absolute values of these numbers as

$$|-1|, \quad |+3|, \quad |2/5|, \quad |-\sqrt{2}|, \quad |-3.14|, \quad |2.83|, \quad |-3/8|, \quad |-\pi|, \quad |+5/7|$$

which in turn may be written 1, 3, $2/5$, $\sqrt{2}$, 3.14, 2.83, $3/8$, π, $5/7$ respectively.

8. The following illustrate addition and subtraction of real numbers.

a) $(-3) + (-8) = -11$ d) $-2 + 5 = 3$ g) $50 - 23 - 27 = 0$

b) $(-2) + 3 = 1$ e) $-15 + 8 = -7$ h) $-3 - (-4) = -3 + 4 = 1$

c) $(-6) + 3 = -3$ f) $(-32) + 48 + (-10) = 6$ i) $-(-14) + (-2) = 14 - 2 = 12$

9. Write the sum S, difference D, product P and quotient Q of each of the following pairs of real numbers: a) $-2, 2$; b) $-3, 6$; c) $0, -5$; d) $-5, 0$.

a) $S = -2 + 2 = 0$, $D = (-2) - 2 = -4$, $P = (-2)(2) = -4$, $Q = -2/2 = -1$

b) $S = (-3) + 6 = 3$, $D = (-3) - 6 = -9$, $P = (-3)(6) = -18$, $Q = -3/6 = -1/2$

c) $S = 0 + (-5) = -5$, $D = 0 - (-5) = 5$, $P = (0)(-5) = 0$, $Q = 0/-5 = 0$

d) $S = (-5) + 0 = -5$, $D = (-5) - 0 = -5$, $P = (-5)(0) = 0$, $Q = -5/0$ an undefined operation

10. Perform the indicated operations.

a) $(5)(-3)(-2) = [(5)(-3)](-2) = (-15)(-2) = 30$
$$= (5)[(-3)(-2)] = (5)(6) = 30$$
The arrangement of the factors of a product does not affect the result.

b) $8(-3)(10) = -240$

c) $\dfrac{8(-2)}{-4} + \dfrac{(-4)(-2)}{2} = \dfrac{-16}{-4} + \dfrac{8}{2} = 4 + 4 = 8$

d) $\dfrac{12(-40)(-12)}{5(-3) - 3(-3)} = \dfrac{12(-40)(-12)}{-15 - (-9)} = \dfrac{12(-40)(-12)}{-6} = -960$

11. Evaluate.

a) $2^3 = 2 \cdot 2 \cdot 2 = 8$

b) $5(3)^2 = 5 \cdot 3 \cdot 3 = 45$

c) $2^4 \cdot 2^6 = 2^{4+6} = 2^{10} = 1024$

d) $2^5 \cdot 5^2 = (32)(25) = 800$

e) $\dfrac{3^4 \cdot 3^3}{3^2} = \dfrac{3^7}{3^2} = 3^{7-2} = 3^5 = 243$

f) $\dfrac{5^2 \cdot 5^3}{5^7} = \dfrac{5^5}{5^7} = \dfrac{1}{5^{7-5}} = \dfrac{1}{5^2} = \dfrac{1}{25}$

g) $(2^3)^2 = 2^{3 \cdot 2} = 2^6 = 64$

h) $\left(\dfrac{2}{3}\right)^4 = \dfrac{2^4}{3^4} = \dfrac{16}{81}$

i) $\dfrac{(3^4)^3 \cdot (3^2)^4}{(-3)^{15} \cdot 3^4} = \dfrac{3^{12} \cdot 3^8}{-3^{15} \cdot 3^4} = -\dfrac{3^{20}}{3^{19}} = -3^1 = -3$

j) $\dfrac{3^8}{3^5} - \dfrac{4^2 \cdot 2^4}{2^6} + 3(-2)^3 = 3^3 - \dfrac{4^2}{2^2} + 3(-8) = 27 - 4 - 24 = -1$

12. Convert each of the following fractions into an equivalent fraction having the indicated denominator.

a) $1/3$; 6 b) $3/4$; 20 c) $5/8$; 48 d) $-3/7$; 63 e) $-12/5$; 75

a) To obtain the denominator 6, multiply numerator and denominator of the fraction 1/3 by 2.

Then $\dfrac{1}{3} = \dfrac{1}{3} \cdot \dfrac{2}{2} = \dfrac{2}{6}$.

b) $\dfrac{3}{4} = \dfrac{3 \cdot 5}{4 \cdot 5} = \dfrac{15}{20}$

d) $-\dfrac{3}{7} = -\dfrac{3 \cdot 9}{7 \cdot 9} = -\dfrac{27}{63}$

c) $\dfrac{5}{8} = \dfrac{5 \cdot 6}{8 \cdot 6} = \dfrac{30}{48}$

e) $-\dfrac{12}{5} = -\dfrac{12 \cdot 15}{5 \cdot 15} = -\dfrac{180}{75}$

13. Find the sum S, difference D, product P and quotient Q of each of the following pairs of rational numbers: *a)* 1/3, 1/6; *b)* 2/5, 3/4; *c)* −4/15, −11/24 .

a) 1/3 may be written as the equivalent fraction 2/6.

$S = \dfrac{1}{3} + \dfrac{1}{6} = \dfrac{2}{6} + \dfrac{1}{6} = \dfrac{3}{6} = \dfrac{1}{2}$

$P = \left(\dfrac{1}{3}\right)\left(\dfrac{1}{6}\right) = \dfrac{1}{18}$

$D = \dfrac{1}{3} - \dfrac{1}{6} = \dfrac{2}{6} - \dfrac{1}{6} = \dfrac{1}{6}$

$Q = \dfrac{1/3}{1/6} = \dfrac{1}{3} \cdot \dfrac{6}{1} = \dfrac{6}{3} = 2$

b) 2/5 and 3/4 may be expressed with denominator 20: 2/5 = 8/20, 3/4 = 15/20.

$S = \dfrac{2}{5} + \dfrac{3}{4} = \dfrac{8}{20} + \dfrac{15}{20} = \dfrac{23}{20}$

$P = \left(\dfrac{2}{5}\right)\left(\dfrac{3}{4}\right) = \dfrac{6}{20} = \dfrac{3}{10}$

$D = \dfrac{2}{5} - \dfrac{3}{4} = \dfrac{8}{20} - \dfrac{15}{20} = -\dfrac{7}{20}$

$Q = \dfrac{2/5}{3/4} = \dfrac{2}{5} \cdot \dfrac{4}{3} = \dfrac{8}{15}$

c) −4/15 and −11/24 have a least common denominator 120: −4/15 = −32/120, −11/24 = −55/120.

$S = \left(-\dfrac{4}{15}\right) + \left(-\dfrac{11}{24}\right) = -\dfrac{32}{120} - \dfrac{55}{120} = -\dfrac{87}{120} = -\dfrac{29}{40}$

$P = \left(-\dfrac{4}{15}\right)\left(-\dfrac{11}{24}\right) = \dfrac{11}{90}$

$D = \left(-\dfrac{4}{15}\right) - \left(-\dfrac{11}{24}\right) = -\dfrac{32}{120} + \dfrac{55}{120} = \dfrac{23}{120}$

$Q = \dfrac{-4/15}{-11/24} = \left(-\dfrac{4}{15}\right)\left(-\dfrac{24}{11}\right) = \dfrac{32}{55}$

14. Evaluate the following expressions, given $x = 2$, $y = -3$, $z = 5$, $a = 1/2$, $b = -2/3$.

a) $2x + y = 2(2) + (-3) = 4 - 3 = 1$

b) $3x - 2y - 4z = 3(2) - 2(-3) - 4(5) = 6 + 6 - 20 = -8$

c) $4x^2 y = 4(2)^2(-3) = 4 \cdot 4 \cdot (-3) = -48$

d) $\dfrac{x^3 + 4y}{2a - 3b} = \dfrac{2^3 + 4(-3)}{2(1/2) - 3(-2/3)} = \dfrac{8 - 12}{1 + 2} = -\dfrac{4}{3}$

e) $\left(\dfrac{x}{y}\right)^2 - 3\left(\dfrac{b}{a}\right)^3 = \left(\dfrac{2}{-3}\right)^2 - 3\left(\dfrac{-2/3}{1/2}\right)^3 = \left(-\dfrac{2}{3}\right)^2 - 3\left(-\dfrac{4}{3}\right)^3 = \dfrac{4}{9} - 3\left(-\dfrac{64}{27}\right) = \dfrac{4}{9} + \dfrac{64}{9} = \dfrac{68}{9}$

SUPPLEMENTARY PROBLEMS

15. Write the sum S, difference D, product P and quotient Q of each of the following pairs of numbers: *a*) 54, 18; *b*) 4, 0; *c*) 0, 4; *d*) 12, 24; *e*) 50, 75.

16. Perform each of the indicated operations.

    a) $38 + 57$, $57 + 38$

    b) $15 + (33 + 8)$, $(15 + 33) + 8$

    c) $(23 + 64) - (41 + 12)$

    d) $12 \cdot 8$, $8 \cdot 12$

    e) $6(4 \cdot 8)$, $(6 \cdot 4)8$

    f) $42 \cdot 68$

    g) $1296 \div 36$

    h) $\dfrac{(35 - 23)(28 + 17)}{43 - 25}$

    i) $45 \div 15 + 84 \div 12$

    j) $10 \div 5 - 4 \div 2 + 15 \div 3 + 2 \cdot 5$

    k) $112 \div (4 \cdot 7)$, $(112 \div 4) \cdot 7$

    l) $\dfrac{15 + 3 \cdot 2}{9 - 4 \div 2}$

17. Place an appropriate inequality symbol (< or >) between each of the following pairs of real numbers.

    a) 4, 3

    b) −2, 0

    c) −1, 2

    d) 3, −2

    e) −8, −7

    f) 1, $\sqrt{2}$

    g) −3, $-\sqrt{11}$

    h) −1/3, −2/5

18. Arrange each of the following groups of real numbers in ascending order of magnitude.

    a) $-\sqrt{3}$, −2, $\sqrt{6}$, −2.8, 4, 7/2

    b) 2π, −6, $\sqrt{8}$, -3π, 4.8, 19/3

19. Write the absolute value of each of the following real numbers: 2, −3/2, $-\sqrt{6}$, +3.14, 0, 5/3, $\sqrt{4}$, −.001, $-\pi - 1$.

20. Evaluate.

    a) $6 + 5$

    b) $(-4) + (-6)$

    c) $(-4) + 3$

    d) $6 + (-4)$

    e) $-8 + 4$

    f) $-4 + 8$

    g) $(-18) + (-3) + 22$

    h) $40 - 12 + 4$

    i) $-12 - (-8)$

    j) $-(-16) - (-12) + (-5) - 15$

21. Write the sum S, difference D, product P and quotient Q of each of the following pairs of real numbers: *a*) 12, 4;   *b*) −6, −3;   *c*) −8, 4;   *d*) 0, −4;   *e*) 3, −2.

22. Perform the indicated operations.

    a) $(-3)(2)(-6)$

    b) $(6)(-8)(-2)$

    c) $4(-1)(5) + (-3)(2)(-4)$

    d) $\dfrac{(-4)(6)}{-3} + \dfrac{(-16)(-9)}{12}$

    e) $(-8) \div (-4) + (-3)(2)$

    f) $\dfrac{(-3)(8)(-2)}{(-4)(-6) - (2)(-12)}$

23. Evaluate.

    a) 3^3

    b) $3(4)^2$

    c) $2^4 \cdot 2^3$

    d) $4^2 \cdot 3^2$

    e) $\dfrac{5^6 \cdot 5^3}{5^5}$

    f) $\dfrac{3^4 \cdot 3^8}{3^6 \cdot 3^5}$

    g) $\dfrac{7^5}{7^3 \cdot 7^4}$

    h) $(3^2)^3$

    i) $(\tfrac{1}{2})^6 \cdot 2^5$

    j) $\dfrac{(-2)^3 \cdot (2)^3}{3(2^2)^2}$

    k) $\dfrac{3(-3)^2 + 4(-2)^3}{2^3 - 3^2}$

    l) $\dfrac{5^7}{5^4} + \dfrac{2^{10}}{8^2 \cdot (-2)^3} - 4(-3)^4$

24. Convert each of the following fractions into an equivalent fraction having the indicated denominator.

    a) 2/5 ; 15

    b) −4/7 ; 28

    c) 5/16 ; 64

    d) −10/3 ; 42

    e) 11/12 ; 132

    f) 17/18 ; 90

25. Find the sum S, difference D, product P and quotient Q of each of the following pairs of rational numbers: a) 1/4, 3/8; b) 1/3, 2/5; c) −4, 2/3; d) −2/3, −3/2.

26. Evaluate the following expressions, given $x = -2$, $y = 4$, $z = 1/3$, $a = -1$, $b = 1/2$.

   a) $3x - 2y + 6z$

   b) $2xy + 6az$ d) $\dfrac{3y^2 - 4x}{ax + by}$ e) $\dfrac{x^2 y (x + y)}{3x + 4y}$ f) $(\dfrac{y}{x})^3 - 4(\dfrac{a}{b})^2 - \dfrac{xy}{z^2}$

   c) $4b^2 x^3$

ANSWERS TO SUPPLEMENTARY PROBLEMS.

15. a) S = 72, D = 36, P = 972, Q = 3   d) S = 36, D = −12, P = 288, Q = 1/2
    b) S = 4, D = 4, P = 0, Q undefined   e) S = 125, D = −25, P = 3750, Q = 2/3
    c) S = 4, D = −4, P = 0, Q = 0

16. a) 95, 95   c) 34   e) 192, 192   g) 36   i) 10   k) 4, 196
    b) 56, 56   d) 96, 96   f) 2856   h) 30   j) 15   l) 3

17. a) 3 < 4 or 4 > 3   d) −2 < 3 or 3 > −2   g) $-\sqrt{11} < -3$ or $-3 > -\sqrt{11}$
    b) −2 < 0 or 0 > −2   e) −8 < −7 or −7 > −8   h) −2/5 < −1/3 or −1/3 > −2/5
    c) −1 < 2 or 2 > −1   f) $1 < \sqrt{2}$ or $\sqrt{2} > 1$

18. a) $-2.8 < -2 < -\sqrt{3} < \sqrt{6} < 7/2 < 4$ b) $-3\pi < -6 < \sqrt{8} < 4.8 < 2\pi < 19/3$

19. 2, 3/2, $\sqrt{6}$, 3.14, 0, 5/3, $\sqrt{4}$, .001, $\pi + 1$

20. a) 11   c) −1   e) −4   g) 1   i) −4
    b) −10   d) 2   f) 4   h) 32   j) 8

21. a) S = 16, D = 8, P = 48, Q = 3   d) S = −4, D = 4, P = 0, Q = 0
    b) S = −9, D = −3, P = 18, Q = 2   e) S = 1, D = 5, P = −6, Q = −3/2
    c) S = −4, D = −12, P = −32, Q = −2

22. a) 36   b) 96   c) 4   d) 20   e) −4   f) 1

23. a) 27   c) 128   e) 5^4 = 625   g) 1/49   i) 1/2   k) 5
    b) 48   d) 144   f) 3   h) 3^6 = 729   j) −4/3   l) −201

24. a) 6/15   b) −16/28   c) 20/64   d) −140/42   e) 121/132   f) 85/90

25. a) S = 5/8, D = −1/8, P = 3/32, Q = 2/3
    b) S = 11/15, D = −1/15, P = 2/15, Q = 5/6
    c) S = −10/3, D = −14/3, P = −8/3, Q = −6
    d) S = −13/6, D = 5/6, P = 1, Q = 4/9

26. a) −12   b) −18   c) −8   d) 14   e) 16/5   f) 48

CHAPTER 2

Fundamental Operations with Algebraic Expressions

AN ALGEBRAIC EXPRESSION is a combination of ordinary numbers and letters which represent numbers.

Thus $3x^2 - 5xy + 2y^4$, $2a^3 b^5$, $\dfrac{5xy + 3z}{2a^3 - c^2}$ are algebraic expressions.

A TERM consists of products and quotients of ordinary numbers and letters which represent numbers. Thus $6x^2 y^3$, $5x/3y^4$, $-3x^7$ are terms.

However $6x^2 + 7xy$ is an algebraic expression consisting of two terms.

A MONOMIAL is an algebraic expression consisting of only one term.

Thus $7x^3 y^4$, $3xyz^2$, $4x^2/y$ are monomials.

Because of this definition, monomials are sometimes simply called terms.

A BINOMIAL is an algebraic expression consisting of two terms.

Thus $2x + 4y$, $3x^4 - 4xyz^3$ are binomials.

A TRINOMIAL is an algebraic expression consisting of three terms.

Thus $3x^2 - 5x + 2$, $2x + 6y - 3z$, $x^3 - 3xy/z - 2x^3 z^7$ are trinomials.

A MULTINOMIAL is an algebraic expression consisting of more than one term.

Thus $7x + 6y$, $3x^3 + 6x^2 y - 7xy + 6$, $7x + 5x^2/y - 3x^3/16$ are multinomials.

COEFFICIENT. One factor of a term is said to be the coefficient of the rest of the term. Thus in the term $5x^3 y^2$, $5x^3$ is the coefficient of y^2, $5y^2$ is the coefficient of x^3, and 5 is the coefficient of $x^3 y^2$.

NUMERICAL COEFFICIENT. If a term consists of the product of an ordinary number and one or more letters, we call the number the numerical coefficient (or simply the coefficient) of the term.

Thus in $-5x^3 y^2$, -5 is the numerical coefficient or simply the coefficient.

LIKE TERMS or similar terms are terms which differ only in numerical coefficients.

For example, $7xy$ and $-2xy$ are like terms; $3x^2 y^4$ and $-\frac{1}{2}x^2 y^4$ are like terms; however, $-2a^2 b^3$ and $-3a^2 b^7$ are unlike terms.

Two or more like terms in an algebraic expression may be combined into one term. Thus $7x^2 y - 4x^2 y + 2x^2 y$ may be combined and written $5x^2 y$.

A TERM IS INTEGRAL AND RATIONAL in certain literals (letters which represent numbers) if the term consists of

11

a) positive integral powers of literal numbers multiplied by a factor not containing the letters or

b) no literal numbers at all.

For example, the terms $6x^2y^3$, $-5y^4$, 7, $-4x$, $\sqrt{3}\,x^3y^6$ are integral and rational in the letters present. However $3\sqrt{x}$ is not rational in x, $4/x$ is not integral in x.

A POLYNOMIAL is a monomial or multinomial in which every term is integral and rational in the literals.

For example, $3x^2y^3 - 5x^4y + 2$, $2x^4 - 7x^3 + 3x^2 - 5x + 2$, $4xy + z$, $3x^2$ are polynomials. However $3x^2 - 4/x$, $4\sqrt{y} + 3$ are not polynomials.

THE DEGREE OF A MONOMIAL is the sum of all the exponents in the literal part of the term. Thus the degree of $4x^3y^2z$ is $3 + 2 + 1 = 6$. The degree of a constant such as 6, 0, $-\sqrt{3}$, π is zero.

THE DEGREE OF A POLYNOMIAL is the same as that of the term having highest degree and non-zero coefficient.

Thus $7x^3y^2 - 4xz^5 + 2x^3y$ has terms of degree 5, 6, 4 respectively; hence the degree of the polynomial is 6.

A SYMBOL OF GROUPING such as parentheses ( ), brackets [ ], or braces { } is often used to show that the terms contained in them are considered as a single quantity.

For example, the sum of two algebraic expressions $5x^2 - 3x + y$ and $2x - 3y$ may be written $(5x^2 - 3x + y) + (2x - 3y)$. The difference of these may be written $(5x^2 - 3x + y) - (2x - 3y)$, and their product $(5x^2 - 3x + y)(2x - 3y)$.

Occasionally a vinculum, which is a bar over the terms, is used as a symbol of grouping. Thus $\overline{5x - 3y}$ has the same meaning as $(5x - 3y)$.

REMOVAL OF SYMBOLS OF GROUPING is governed by the following laws.

1) If a + sign precedes a symbol of grouping, this symbol of grouping may be removed without affecting the terms contained.

Thus $(3x + 7y) + (4xy - 3x^3) = 3x + 7y + 4xy - 3x^3$.

2) If a − sign precedes a symbol of grouping, this symbol of grouping may be removed if each sign of the terms contained is changed.

Thus $(3x + 7y) - (4xy - 3x^3) = 3x + 7y - 4xy + 3x^3$.

3) If more than one symbol of grouping is present, the inner ones are to be removed first.

Thus $2x - \{4x^3 - (3x^2 - 5y)\} = 2x - \{4x^3 - 3x^2 + 5y\} = 2x - 4x^3 + 3x^2 - 5y$.

ADDITION OF ALGEBRAIC EXPRESSIONS is achieved by combining like terms. In order to accomplish this addition, the expressions may be arranged in rows with like terms in the same column; these columns are then added.

Example. Add $7x + 3y^3 - 4xy$, $3x - 2y^3 + 7xy$ and $2xy - 5x - 6y^3$.

Write	$7x$	$3y^3$	$-4xy$
	$3x$	$-2y^3$	$7xy$
	$-5x$	$-6y^3$	$2xy$
Addition:	$5x$	$-5y^3$	$5xy$.

Hence the result is $5x - 5y^3 + 5xy$.

SUBTRACTION OF TWO ALGEBRAIC EXPRESSIONS is achieved by changing the sign of every term in the expression which is being subtracted (sometimes called the subtrahend) and adding this result to the other expression (called the minuend).

Example.   Subtract $2x^2 - 3xy + 5y^2$ from $10x^2 - 2xy - 3y^2$.

$$
\begin{array}{r}
10x^2 - 2xy - 3y^2 \\
2x^2 - 3xy + 5y^2 \\
\hline
\end{array}
$$

Subtraction: $\quad 8x^2 + xy - 8y^2$

We may also write $(10x^2 - 2xy - 3y^2) - (2x^2 - 3xy + 5y^2)$

$$= 10x^2 - 2xy - 3y^2 - 2x^2 + 3xy - 5y^2 \ = \ 8x^2 + xy - 8y^2.$$

MULTIPLICATION OF ALGEBRAIC EXPRESSIONS.

1) To Multiply Two or More Monomials: Use the laws of exponents, the rules of signs, and the commutative and associative laws of multiplication.

Example.   Multiply $-3x^2y^3z$, $2x^4y$ and $-4xy^4z^2$.

Write $(-3x^2y^3z)(2x^4y)(-4xy^4z^2)$.

Arranging according to the commutative and associative laws,

$$\{(-3)(2)(-4)\}\{(x^2)(x^4)(x)\}\{(y^3)(y)(y^4)\}\{(z)(z^2)\}. \qquad (1)$$

Combine using rules of signs and laws of exponents to obtain
$$24x^7y^8z^3.$$

Step (1) may be done mentally when experience is acquired.

2) To Multiply a Polynomial by a Monomial: Multiply each term of the polynomial by the monomial and combine results.

Example.   Multiply $3xy - 4x^3 + 2xy^2$ by $5x^2y^4$.

Write $(5x^2y^4)(3xy - 4x^3 + 2xy^2)$

$$= (5x^2y^4)(3xy) + (5x^2y^4)(-4x^3) + (5x^2y^4)(2xy^2)$$
$$= 15x^3y^5 - 20x^5y^4 + 10x^3y^6.$$

3) To Multiply a Polynomial by a Polynomial: Multiply each of the terms of one polynomial by each of the terms of the other polynomial and combine results.

It is very often useful to arrange the polynomials according to ascending (or descending) powers of one of the letters involved.

Example.   Multiply $-3x + 9 + x^2$ by $3 - x$.

Arranging in descending powers of x,

$$x^2 - 3x + 9 \qquad (2)$$
$$-x + 3$$

Multiplying (2) by $-x$, $-x^3 + 3x^2 - 9x$
Multiplying (2) by 3, $ 3x^2 - 9x + 27$

Adding, $-x^3 + 6x^2 - 18x + 27$

DIVISION OF ALGEBRAIC EXPRESSIONS.

1) To Divide a Monomial by a Monomial: Find the quotient of the numerical coefficients, find the quotients of the literal factors, and multiply these quotients.

Example. Divide $24x^4y^2z^3$ by $-3x^3y^4z$.

Write $\dfrac{24x^4y^2z^3}{-3x^3y^4z} = (\dfrac{24}{-3})(\dfrac{x^4}{x^3})(\dfrac{y^2}{y^4})(\dfrac{z^3}{z}) = (-8)(x)(\dfrac{1}{y^2})(z^2) = -\dfrac{8xz^2}{y^2}.$

2) To Divide a Polynomial by a Polynomial:

a) Arrange the terms of both polynomials in descending (or ascending) powers of one of the letters common to both polynomials.

b) Divide the first term in the dividend by the first term in the divisor. This gives the first term of the quotient.

c) Multiply the first term of the quotient by the divisor and subtract from the dividend, thus obtaining a new dividend.

d) Use the dividend obtained in c) to repeat steps b) and c) until a remainder is obtained which is either of degree lower than the degree of the divisor or zero.

e) The result is written $\dfrac{\text{dividend}}{\text{divisor}} = \text{quotient} + \dfrac{\text{remainder}}{\text{divisor}}.$

Example. Divide $x^2 + 2x^4 - 3x^3 + x - 2$ by $x^2 - 3x + 2$.

Write the polynomials in descending powers of x and arrange the work as follows.

$$
\begin{array}{r}
2x^2 + 3x + 6 \\
x^2 - 3x + 2 \overline{)\ 2x^4 - 3x^3 + x^2 + x - 2} \\
\underline{2x^4 - 6x^3 + 4x^2} \\
3x^3 - 3x^2 + x - 2 \\
\underline{3x^3 - 9x^2 + 6x} \\
6x^2 - 5x - 2 \\
\underline{6x^2 - 18x + 12} \\
13x - 14
\end{array}
$$

Hence $\dfrac{2x^4 - 3x^3 + x^2 + x - 2}{x^2 - 3x + 2} = 2x^2 + 3x + 6 + \dfrac{13x - 14}{x^2 - 3x + 2}.$

SOLVED PROBLEMS

1. Evaluate each of the following algebraic expressions, given that $x = 2$, $y = -1$, $z = 3$, $a = 0$, $b = 4$, $c = 1/3$.

a) $2x^2 - 3yz = 2(2)^2 - 3(-1)(3) = 8 + 9 = 17$

b) $2z^4 - 3z^3 + 4z^2 - 2z + 3 = 2(3)^4 - 3(3)^3 + 4(3)^2 - 2(3) + 3 = 162 - 81 + 36 - 6 + 3 = 114$

c) $4a^2 - 3ab + 6c = 4(0)^2 - 3(0)(4) + 6(1/3) = 0 - 0 + 2 = 2$

d) $\dfrac{5xy + 3z}{2a^3 - c^2} = \dfrac{5(2)(-1) + 3(3)}{2(0)^3 - (1/3)^2} = \dfrac{-10 + 9}{-1/9} = \dfrac{-1}{-1/9} = 9$

e) $\dfrac{3x^2 y}{z} - \dfrac{bc}{x+1} = \dfrac{3(2)^2(-1)}{3} - \dfrac{4(1/3)}{3} = -4 - 4/9 = -40/9$

f) $\dfrac{4x^2 y(z-1)}{a + b - 3c} = \dfrac{4(2)^2(-1)(3-1)}{0 + 4 - 3(1/3)} = \dfrac{4(4)(-1)(2)}{4 - 1} = -\dfrac{32}{3}$

2. Classify each of the following algebraic expressions according to the categories: term or monomial, binomial, trinomial, multinomial, polynomial.

a) $x^3 + 3y^2 z$ *d)* $y + 3$ *g)* $\sqrt{x^2 + y^2 + z^2}$

b) $2x^2 - 5x + 3$ *e)* $4z^2 + 3z - 2\sqrt{z}$ *h)* $\sqrt{y} + \sqrt{z}$

c) $4x^2 y/z$ *f)* $5x^3 + 4/y$ *i)* $a^3 + b^3 + c^3 - 3abc$

If the expression belongs to one or more categories, this is indicated by a check mark.

	Term or Monomial	Binomial	Trinomial	Multinomial	Polynomial
$x^3 + 3y^2 z$		✓		✓	✓
$2x^2 - 5x + 3$			✓	✓	✓
$4x^2 y/z$	✓				
$y + 3$		✓		✓	✓
$4z^2 + 3z - 2\sqrt{z}$			✓	✓	
$5x^3 + 4/y$		✓		✓	
$\sqrt{x^2 + y^2 + z^2}$	✓				
$\sqrt{y} + \sqrt{z}$		✓		✓	
$a^3 + b^3 + c^3 - 3abc$				✓	✓

3. Find the degree of each of the following polynomials.

a) $2x^3 y + 4xyz^4$. The degree of $2x^3 y$ is 4, of $4xyz^4$ is 6; hence the polynomial is of degree 6.

b) $x^2 + 3x^3 - 4$. The degree of x^2 is 2, of $3x^3$ is 3, of -4 is 0; hence the degree of the polynomial is 3.

c) $y^3 - 3y^2 + 4y - 2$ is of degree 3.

d) $xz^3 + 3x^2 z^2 - 4x^3 z + x^4$. Each term is of degree 4; hence the polynomial is of degree 4.

e) $x^2 - 10^5$ is of degree 2. (The degree of the constant 10^5 is zero.)

4. Remove the symbols of grouping in each of the following and simplify the resulting expressions by combining like terms.

a) $3x^2 + (y^2 - 4z) - (2x - 3y + 4z) = 3x^2 + y^2 - 4z - 2x + 3y - 4z = 3x^2 + y^2 - 2x + 3y - 8z$

b) $2(4xy + 3z) + 3(x - 2xy) - 4(z - 2xy) = 8xy + 6z + 3x - 6xy - 4z + 8xy = 10xy + 3x + 2z$

c) $x - 3 - 2\{2 - 3(x - y)\} = x - 3 - 2\{2 - 3x + 3y\} = x - 3 - 4 + 6x - 6y = 7x - 6y - 7$

d) $4x^2 - \{3x^2 - 2[y - 3(x^2 - y)] + 4\} = 4x^2 - \{3x^2 - 2[y - 3x^2 + 3y] + 4\}$

       $= 4x^2 - \{3x^2 - 2y + 6x^2 - 6y + 4\} = 4x^2 - \{9x^2 - 8y + 4\}$

       $= 4x^2 - 9x^2 + 8y - 4 = -5x^2 + 8y - 4$

5. Add the algebraic expressions in each of the following groups.

a) $x^2 + y^2 - z^2 + 2xy - 2yz$, $y^2 + z^2 - x^2 + 2yz - 2zx$, $z^2 + x^2 - y^2 + 2zx - 2xy$, $1 - x^2 - y^2 - z^2$

Arranging,

$$
\begin{array}{l}
x^2 + y^2 - z^2 + 2xy - 2yz \\
-x^2 + y^2 + z^2 \qquad\quad + 2yz - 2zx \\
x^2 - y^2 + z^2 - 2xy \qquad\quad + 2zx \\
-x^2 - y^2 - z^2 \qquad\qquad\qquad\qquad + 1
\end{array}
$$

Adding, $0 + 0 + 0 + 0 + 0 + 0 + 1$ The result of the addition is 1.

b) $5x^3y - 4ab + c^2$, $3c^2 + 2ab - 3x^2y$, $x^3y + x^2y - 4c^2 - 3ab$, $4c^2 - 2x^2y + ab^2 - 3ab$

Arranging,

$$
\begin{array}{l}
5x^3y - 4ab + \; c^2 \\
-3x^2y \qquad\quad + 2ab + 3c^2 \\
x^2y + \; x^3y - 3ab - 4c^2 \\
-2x^2y \qquad\qquad - 3ab + 4c^2 + ab^2
\end{array}
$$

Adding, $-4x^2y + 6x^3y - 8ab + 4c^2 + ab^2$

6. Subtract the second of each of the following expressions from the first.

a) $a - b + c - d$, $c - a + d - b$.   Write $\begin{array}{l} a - b + c - d \\ -a - b + c + d \end{array}$

Subtracting, $2a + 0 + 0 - 2d$ The result is $2a - 2d$.

Otherwise: $(a - b + c - d) - (c - a + d - b) = a - b + c - d - c + a - d + b = 2a - 2d$

b) $4x^2y - 3ab + 2a^2 - xy$, $4xy + ab^2 - 3a^2 + 2ab$

Write $\begin{array}{l} 4x^2y - 3ab + 2a^2 - \; xy \\ 2ab - 3a^2 + 4xy + ab^2 \end{array}$

Subtracting, $4x^2y - 5ab + 5a^2 - 5xy - ab^2$

Otherwise: $(4x^2y - 3ab + 2a^2 - xy) - (4xy + ab^2 - 3a^2 + 2ab)$

$$= 4x^2y - 3ab + 2a^2 - xy - 4xy - ab^2 + 3a^2 - 2ab$$

$$= 4x^2y - 5ab + 5a^2 - 5xy - ab^2$$

7. In each of the following find the indicated product of the algebraic expressions.

a) $(-2ab^3)(4a^2b^5)$

b) $(-3x^2y)(4xy^2)(-2x^3y^4)$

c) $(3ab^2)(2ab + b^2)$

d) $(x^2 - 3xy + y^2)(4xy^2)$

e) $(x^2 - 3x + 9)(x + 3)$

f) $(x^4 + x^3y + x^2y^2 + xy^3 + y^4)(x - y)$

g) $(x^2 - xy + y^2)(x^2 + xy + y^2)$

h) $(2x + y - z)(3x - z + y)$

a) $(-2ab^3)(4a^2b^5) = \{(-2)(4)\}\{(a)(a^2)\}\{(b^3)(b^5)\} = -8a^3b^8$

b) $(-3x^2y)(4xy^2)(-2x^3y^4) = \{(-3)(4)(-2)\}\{(x^2)(x)(x^3)\}\{(y)(y^2)(y^4)\} = 24x^6y^7$

c) $(3ab^2)(2ab + b^2) = (3ab^2)(2ab) + (3ab^2)(b^2) = 6a^2b^3 + 3ab^4$

d) $(x^2 - 3xy + y^2)(4xy^2) = (x^2)(4xy^2) + (-3xy)(4xy^2) + (y^2)(4xy^2)$

$$= 4x^3y^2 - 12x^2y^3 + 4xy^4$$

e) $x^2 - 3x + 9$

$\quad x + 3$
$\overline{x^3 - 3x^2 + 9x}$

$\qquad 3x^2 - 9x + 27$
$\overline{x^3 + 0 \;+\; 0 + 27}$

Ans. $x^3 + 27$

f) $x^4 + x^3y + x^2y^2 + xy^3 + y^4$

$\quad x - y$
$\overline{x^5 + x^4y + x^3y^2 + x^2y^3 + xy^4}$

$\qquad - x^4y - x^3y^2 - x^2y^3 - xy^4 - y^5$
$\overline{x^5 + 0 \;+\; 0 \;+\; 0 \;+\; 0 \; - y^5}$

Ans. $x^5 - y^5$

g) $x^2 - xy + y^2$

$\quad x^2 + xy + y^2$
$\overline{x^4 - x^3y + x^2y^2}$

$\qquad x^3y - x^2y^2 + xy^3$

$\qquad\quad x^2y^2 - xy^3 + y^4$
$\overline{x^4 + 0 \;+ x^2y^2 + 0 \;\; + y^4}$

Ans. $x^4 + x^2y^2 + y^4$

h) $2x + y - z$

$\quad 3x + y - z$
$\overline{6x^2 + 3xy - 3xz}$

$\qquad 2xy \qquad + y^2 - yz$

$\qquad\quad - 2xz \qquad - yz + z^2$
$\overline{6x^2 + 5xy - 5xz + y^2 - 2yz + z^2}$

8. Perform the indicated divisions.

a) $\dfrac{24x^3y^2z}{4xyz^2} = \left(\dfrac{24}{4}\right)\left(\dfrac{x^3}{x}\right)\left(\dfrac{y^2}{y}\right)\left(\dfrac{z}{z^2}\right) = (6)(x^2)(y)\left(\dfrac{1}{z}\right) = \dfrac{6x^2y}{z}$

b) $\dfrac{-16a^4b^6}{-8ab^2c} = \left(\dfrac{-16}{-8}\right)\left(\dfrac{a^4}{a}\right)\left(\dfrac{b^6}{b^2}\right)\left(\dfrac{1}{c}\right) = \dfrac{2a^3b^4}{c}$

c) $\dfrac{3x^3y + 16xy^2 - 12x^4yz^4}{2x^2yz} = \left(\dfrac{3x^3y}{2x^2yz}\right) + \left(\dfrac{16xy^2}{2x^2yz}\right) + \left(\dfrac{-12x^4yz^4}{2x^2yz}\right) = \dfrac{3x}{2z} + \dfrac{8y}{xz} - 6x^2z^3$

d) $\dfrac{4a^3b^2 + 16ab - 4a^2}{-2a^2b} = \left(\dfrac{4a^3b^2}{-2a^2b}\right) + \left(\dfrac{16ab}{-2a^2b}\right) + \left(\dfrac{-4a^2}{-2a^2b}\right) = -2ab - \dfrac{8}{a} + \dfrac{2}{b}$

e) $\dfrac{2x^4 + 3x^3 - x^2 - 1}{x - 2}$

f) $\dfrac{16y^4 - 1}{2y - 1}$

$$
\begin{array}{r}
2x^3 + 7x^2 + 13x + 26 \\
x - 2 \overline{\smash{\big)}\, 2x^4 + 3x^3 - x^2 \phantom{{}+13x} - 1} \\
\underline{2x^4 - 4x^3} \\
7x^3 - x^2 \phantom{{}+13x} - 1 \\
\underline{7x^3 - 14x^2} \\
13x^2 \phantom{{}+13x} - 1 \\
\underline{13x^2 - 26x} \\
26x - 1 \\
\underline{26x - 52} \\
51
\end{array}
$$

$$
\begin{array}{r}
8y^3 + 4y^2 + 2y + 1 \\
2y - 1 \overline{\smash{\big)}\, 16y^4 \phantom{{}+4y^2+2y} - 1} \\
\underline{16y^4 - 8y^3} \\
8y^3 \phantom{{}+4y^2+2y} - 1 \\
\underline{8y^3 - 4y^2} \\
4y^2 \phantom{{}+2y} - 1 \\
\underline{4y^2 - 2y} \\
2y - 1 \\
\underline{2y - 1} \\
0
\end{array}
$$

Thus $\dfrac{2x^4 + 3x^3 - x^2 - 1}{x - 2} = 2x^3 + 7x^2 + 13x + 26 + \dfrac{51}{x - 2}$ and $\dfrac{16y^4 - 1}{2y - 1} = 8y^3 + 4y^2 + 2y + 1.$

g) $\dfrac{2x^6 + 5x^4 - x^3 + 1}{-x^2 + x + 1}$.   Arrange in descending powers of x.

$$
\require{enclose}
\begin{array}{r}
-2x^4 - 2x^3 - 9x^2 - 10x - 19 \\
-x^2 + x + 1 \enclose{longdiv}{2x^6 \qquad\ + 5x^4 -\ \ x^3 \qquad\qquad\ + 1} \\
\underline{2x^6 - 2x^5 - 2x^4 \qquad\qquad\qquad\ } \\
2x^5 + 7x^4 -\ \ x^3 \qquad\qquad\ + 1 \\
\underline{2x^5 - 2x^4 -\ 2x^3 \qquad\qquad\ } \\
9x^4 +\ \ x^3 \qquad\qquad\ + 1 \\
\underline{9x^4 - 9x^3 -\ 9x^2 \qquad\qquad\ } \\
10x^3 + 9x^2 \qquad\ + 1 \\
\underline{10x^3 - 10x^2 - 10x \qquad\ } \\
19x^2 + 10x\ + 1 \\
\underline{19x^2 - 19x - 19} \\
29x + 20
\end{array}
$$

Thus $\dfrac{2x^6 + 5x^4 - x^3 + 1}{-x^2 + x + 1} = -2x^4 - 2x^3 - 9x^2 - 10x - 19 + \dfrac{29x + 20}{-x^2 + x + 1}$.

h) $\dfrac{x^4 - x^3y + x^2y^2 + 2x^2y - 2xy^2 + 2y^3}{x^2 - xy + y^2}$.   Arrange in descending powers of a letter, say x.

$$
\require{enclose}
\begin{array}{r}
x^2 + 2y \\
x^2 - xy + y^2 \enclose{longdiv}{x^4 - x^3y + x^2y^2 + 2x^2y - 2xy^2 + 2y^3} \\
\underline{x^4 - x^3y + x^2y^2 \qquad\qquad\qquad\qquad} \\
2x^2y - 2xy^2 + 2y^3 \\
\underline{2x^2y - 2xy^2 + 2y^3} \\
0
\end{array}
$$

Thus $\dfrac{x^4 - x^3y + x^2y^2 + 2x^2y - 2xy^2 + 2y^3}{x^2 - xy + y^2} = x^2 + 2y$.

9. Check the work in Problems 7h) and 8g) by using the values $x = 1$, $y = -1$, $z = 2$.

From Problem 7h), $(2x + y - z)(3x - z + y) = 6x^2 + 5xy - 5xz - 2yz + z^2 + y^2$.
Substitute $x = 1$, $y = -1$, $z = 2$ and obtain

$$[2(1) + (-1) - 2]\,[3(1) - (2) - 1] = 6(1)^2 + 5(1)(-1) - 5(1)(2) - 2(-1)(2) + (2)^2 + (-1)^2$$

or $[-1]\,[0] = 6 - 5 - 10 + 4 + 4 + 1$,   i.e. $0 = 0$.

From Problem 8g), $\dfrac{2x^6 + 5x^4 - x^3 + 1}{-x^2 + x + 1} = -2x^4 - 2x^3 - 9x^2 - 10x - 19 + \dfrac{29x + 20}{-x^2 + x + 1}$.

Put $x = 1$ and obtain $\dfrac{2 + 5 - 1 + 1}{-1 + 1 + 1} = -2 - 2 - 9 - 10 - 19 + \dfrac{29 + 20}{-1 + 1 + 1}$ or $7 = 7$.

Although a check by substitution of numbers for letters is not conclusive, it can be used to indicate possible errors.

SUPPLEMENTARY PROBLEMS

10. Evaluate each algebraic expression, given that $x = -1$, $y = 3$, $z = 2$, $a = 1/2$, $b = -2/3$.

a) $4x^3y^2 - 3xz^2$

b) $(x-y)(y-z)(z-x)$

c) $9ab^2 + 6ab - 4a^2$

d) $\dfrac{xy^2 - 3z}{a + b}$

e) $\dfrac{z(x+y)}{8a^2} - \dfrac{3ab}{y - x + 1}$

f) $\dfrac{(x-y)^2 + 2z}{ax + by}$

g) $\dfrac{1}{x} + \dfrac{1}{y} + \dfrac{1}{z}$

h) $\dfrac{(x-1)(y-1)(z-1)}{(a-1)(b-1)}$

11. Determine the degree of each of the following polynomials.

a) $3x^4 - 2x^3 + x^2 - 5$

b) $4xy^4 - 3x^3y^3$

c) $x^5 + y^5 + z^5 - 5xyz$

d) $\sqrt{3}\,xyz - 5$

e) -10^3

f) $y^2 - 3y^5 - y + 2y^3 - 4$

12. Remove the symbols of grouping and simplify the resulting expressions by combining like terms.

a) $(x + 3y - z) - (2y - x + 3z) + (4z - 3x + 2y)$

b) $3(x^2 - 2yz + y^2) - 4(x^2 - y^2 - 3yz) + x^2 + y^2$

c) $3x + 4y + 3\{x - 2(y - x) - y\}$

d) $3 - \{2x - [1 - (x + y)] + [x - 2y]\}$

13. Add the algebraic expressions in each of the following groups.

a) $2x^2 + y^2 - x + y$, $3y^2 + x - x^2$, $x - 2y + x^2 - 4y^2$

b) $a^2 - ab + 2bc + 3c^2$, $2ab + b^2 - 3bc - 4c^2$, $ab - 4bc + c^2 - a^2$, $a^2 + 2c^2 + 5bc - 2ab$

c) $2a^2bc - 2acb^2 + 5c^2ab$, $4b^2ac + 4bca^2 - 7ac^2b$, $4abc^2 - 3a^2bc - 3ab^2c$, $b^2ac - abc^2 - 3a^2bc$

14. Subtract the second of each of the following expressions from the first.

a) $3xy - 2yz + 4zx$, $3zx + yz - 2xy$

b) $4x^2 + 3y^2 - 6x + 4y - 2$, $2x - y^2 + 3x^2 - 4y + 3$

c) $r^3 - 3r^2s + 4rs^2 - s^3$, $2s^3 + 3s^2r - 2sr^2 - 3r^3$

15. Subtract $xy - 3yz + 4xz$ from twice the sum of the following expressions: $3xy - 4yz + 2xz$ and $3yz - 4zx - 2xy$.

16. Obtain the product of the algebraic expressions in each of the following groups.

a) $4x^2y^5$, $-3x^3y^2$

b) $3abc^2$, $-2a^3b^2c^4$, $6a^2b^2$

c) $-4x^2y$, $3xy^2 - 4xy$

d) $r^2s + 3rs^3 - 4rs + s^3$, $2r^2s^4$

e) $y - 4$, $y + 3$

f) $y^2 - 4y + 16$, $y + 4$

g) $x^3 + x^2y + xy^2 + y^3$, $x - y$

h) $x^2 + 4x + 8$, $x^2 - 4x + 8$

i) $3r - s - t^2$, $2s + r + 3t^2$

j) $3 - x - y$, $2x + y + 1$, $x - y$

17. Perform the indicated divisions.

a) $\dfrac{-12x^4yz^3}{3x^2y^4z}$

b) $\dfrac{-18r^3s^2t}{-4r^5st^2}$

c) $\dfrac{4ab^3 - 3a^2bc + 12a^3b^2c^4}{-2ab^2c^3}$

d) $\dfrac{4x^3 - 5x^2 + 3x - 2}{x + 1}$

18. Perform the indicated divisions.

a) $\dfrac{27s^3 - 64}{3s - 4}$ b) $\dfrac{1 - x^2 + x^4}{1 - x}$ c) $\dfrac{2y^3 + y^5 - 3y - 2}{y^2 - 3y + 1}$ d) $\dfrac{4x^3y + 5x^2y^2 + x^4 + 2xy^3}{x^2 + 2y^2 + 3xy}$

19. Perform the indicated operations and check by using the values $x = 1$, $y = 2$.

a) $(x^4 + x^2y^2 + y^4)(y^4 - x^2y^2 + x^4)$ b) $\dfrac{x^4 + xy^3 + x^3y + 2x^2y^2 + y^4}{xy + x^2 + y^2}$

ANSWERS TO SUPPLEMENTARY PROBLEMS.

10. a) -24 b) -12 c) -1 d) 90 e) $11/5$ f) -8 g) $-1/6$ h) $-24/5$

11. a) 4 b) 6 c) 5 d) 3 e) 0 f) 5

12. a) $3y - x$ b) $8y^2 + 6yz$ c) $12x - 5y$ d) $y - 4x + 4$

13. a) $2x^2 + x - y$ b) $a^2 + b^2 + 2c^2$ c) abc^2

14. a) $5xy - 3yz + zx$ b) $x^2 + 4y^2 - 8x + 8y - 5$ c) $4r^3 - r^2s + rs^2 - 3s^3$

15. $xy + yz - 8xz$

16. a) $-12x^5y^7$ f) $y^3 + 64$

      b) $-36a^6b^5c^6$ g) $x^4 - y^4$

      c) $-12x^3y^3 + 16x^3y^2$ h) $x^4 + 64$

      d) $2r^4s^5 + 6r^3s^7 - 8r^3s^5 + 2r^2s^7$ i) $3r^2 + 5rs + 8rt^2 - 2s^2 - 5st^2 - 3t^4$

      e) $y^2 - y - 12$ j) $y^3 - 2y^2 - 3y + 3x + 5x^2 - 3xy - 2x^3 - x^2y + 2xy^2$

17. a) $-\dfrac{4x^2z^2}{y^3}$ b) $\dfrac{9s}{2r^2t}$ c) $-\dfrac{2b}{c^3} + \dfrac{3a}{2bc^2} - 6a^2c$ d) $4x^2 - 9x + 12 + \dfrac{-14}{x + 1}$

18. a) $9s^2 + 12s + 16$ b) $-x^3 - x^2 + \dfrac{1}{1 - x}$ c) $y^3 + 3y^2 + 10y + 27 + \dfrac{68y - 29}{y^2 - 3y + 1}$ d) $x^2 + xy$

19. a) $x^8 + x^4y^4 + y^8$. Check: $21(13) = 273$.   b) $x^2 + y^2$. Check: $35/7 = 5$

CHAPTER 3

Special Products

SPECIAL PRODUCTS. The following are some of the products which occur frequently in mathematics, and the student should become familiar with them as soon as possible. Proofs of these results may be obtained by performing the multiplications.

$$\text{I} \quad a(c + d) = ac + ad$$

$$\text{II} \quad (a + b)(a - b) = a^2 - b^2$$

$$\text{III} \quad (a + b)(a + b) = (a + b)^2 = a^2 + 2ab + b^2$$

$$\text{IV} \quad (a - b)(a - b) = (a - b)^2 = a^2 - 2ab + b^2$$

$$\text{V} \quad (x + a)(x + b) = x^2 + (a + b)x + ab$$

$$\text{VI} \quad (ax + b)(cx + d) = acx^2 + (ad + bc)x + bd$$

$$\text{VII} \quad (a + b)(c + d) = ac + bc + ad + bd$$

The following products are also useful.

$$\text{VIII} \quad (a + b)(a + b)(a + b) = (a + b)^3 = a^3 + 3a^2b + 3ab^2 + b^3$$

$$\text{IX} \quad (a - b)(a - b)(a - b) = (a - b)^3 = a^3 - 3a^2b + 3ab^2 - b^3$$

$$\text{X} \quad (a - b)(a^2 + ab + b^2) = a^3 - b^3$$

$$\text{XI} \quad (a + b)(a^2 - ab + b^2) = a^3 + b^3$$

$$\text{XII} \quad (a + b + c)^2 = a^2 + b^2 + c^2 + 2ab + 2ac + 2bc$$

It may be verified by multiplication that

$$(a - b)(a^2 + ab + b^2) = a^3 - b^3 \qquad \text{(X)}$$

$$(a - b)(a^3 + a^2b + ab^2 + b^3) = a^4 - b^4$$

$$(a - b)(a^4 + a^3b + a^2b^2 + ab^3 + b^4) = a^5 - b^5$$

$$(a - b)(a^5 + a^4b + a^3b^2 + a^2b^3 + ab^4 + b^5) = a^6 - b^6$$

etc., the rule being clear. These may be summarized by

$$\text{XIII} \quad (a - b)(a^{n-1} + a^{n-2}b + a^{n-3}b^2 + \cdots + ab^{n-2} + b^{n-1}) = a^n - b^n$$

where n is *any positive integer* $(1, 2, 3, 4, \cdots)$.

Similarly, it may be verified that

$$(a + b)(a^2 - ab + b^2) = a^3 + b^3 \qquad \text{(XI)}$$

$$(a + b)(a^4 - a^3b + a^2b^2 - ab^3 + b^4) = a^5 + b^5$$

21

$$(a + b)(a^6 - a^5b + a^4b^2 - a^3b^3 + a^2b^4 - ab^5 + b^6) = a^7 + b^7$$

etc., the rule being clear. These may be summarized by

XIV $(a + b)(a^{n-1} - a^{n-2}b + a^{n-3}b^2 - \cdots - ab^{n-2} + b^{n-1}) = a^n + b^n$

where n is *any positive odd integer* $(1, 3, 5, 7, \cdots)$.

SOLVED PROBLEMS

Find each of the following products.

PRODUCTS I-VII

1. a) $3x(2x + 3y) = (3x)(2x) + (3x)(3y) = 6x^2 + 9xy$,   using I with $a = 3x$, $c = 2x$, $d = 3y$.

   b) $x^2y(3x^3 - 2y + 4) = (x^2y)(3x^3) + (x^2y)(-2y) + (x^2y)(4) = 3x^5y - 2x^2y^2 + 4x^2y$

   c) $(3x^3y^2 + 2xy - 5)(x^2y^3) = (3x^3y^2)(x^2y^3) + (2xy)(x^2y^3) + (-5)(x^2y^3)$
   $$= 3x^5y^5 + 2x^3y^4 - 5x^2y^3$$

   d) $(2x + 3y)(2x - 3y) = (2x)^2 - (3y)^2 = 4x^2 - 9y^2$, using II with $a = 2x$, $b = 3y$.

   e) $(1 - 5x^3)(1 + 5x^3) = (1)^2 - (5x^3)^2 = 1 - 25x^6$

   f) $(5x + x^3y^2)(5x - x^3y^2) = (5x)^2 - (x^3y^2)^2 = 25x^2 - x^6y^4$

   g) $(3x + 5y)^2 = (3x)^2 + 2(3x)(5y) + (5y)^2 = 9x^2 + 30xy + 25y^2$, using III with $a = 3x$, $b = 5y$.

   h) $(x + 2)^2 = x^2 + 2(x)(2) + 2^2 = x^2 + 4x + 4$

   i) $(7x^2 - 2xy)^2 = (7x^2)^2 - 2(7x^2)(2xy) + (2xy)^2$
   $$= 49x^4 - 28x^3y + 4x^2y^2,   \text{using IV with } a = 7x^2, \ b = 2xy.$$

   j) $(ax - 2by)^2 = (ax)^2 - 2(ax)(2by) + (2by)^2 = a^2x^2 - 4axby + 4b^2y^2$

   k) $(x^4 + 6)^2 = (x^4)^2 + 2(x^4)(6) + (6)^2 = x^8 + 12x^4 + 36$

   l) $(3y^2 - 2)^2 = (3y^2)^2 - 2(3y^2)(2) + (2)^2 = 9y^4 - 12y^2 + 4$

   m) $(x + 3)(x + 5) = x^2 + (3 + 5)x + (3)(5) = x^2 + 8x + 15$,   using V with $a = 3$, $b = 5$.

   n) $(x - 2)(x + 8) = x^2 + (-2 + 8)x + (-2)(8) = x^2 + 6x - 16$

   o) $(x + 2)(x - 8) = x^2 + (2 - 8)x + (2)(-8) = x^2 - 6x - 16$

   p) $(t^2 + 10)(t^2 - 12) = (t^2)^2 + (10 - 12)t^2 + (10)(-12) = t^4 - 2t^2 - 120$

   q) $(3x + 4)(2x - 3) = (3)(2)x^2 + [(3)(-3) + (4)(2)]x + (4)(-3)$
   $$= 6x^2 - x - 12,   \text{using VI with } a = 3, \ b = 4, \ c = 2, \ d = -3.$$

   r) $(2x + 5)(4x - 1) = (2)(4)x^2 + [(2)(-1) + (5)(4)]x + (5)(-1) = 8x^2 + 18x - 5$

   s) $(3x + y)(4x - 2y) = (3x)(4x) + (y)(4x) + (3x)(-2y) + (y)(-2y)$
   $$= 12x^2 - 2xy - 2y^2,   \text{using VII with } a = 3x, \ b = y, \ c = 4x, \ d = -2y.$$

   t) $(3t^2s - 2)(4t - 3s) = (3t^2s)(4t) + (-2)(4t) + (3t^2s)(-3s) + (-2)(-3s)$
   $$= 12t^3s - 8t - 9t^2s^2 + 6s$$

   u) $(3xy + 1)(2x^2 - 3y) = (3xy)(2x^2) + (3xy)(-3y) + (1)(2x^2) + (1)(-3y)$
   $$= 6x^3y - 9xy^2 + 2x^2 - 3y$$

   v) $(x + y + 3)(x + y - 3) = (x + y)^2 - 3^2 = x^2 + 2xy + y^2 - 9$

$w)$ $\quad (2x - y - 1)(2x - y + 1)$ $=$ $(2x - y)^2 - (1)^2$ $=$ $4x^2 - 4xy + y^2 - 1$

$x)$ $\quad (x^2 + 2xy + y^2)(x^2 - 2xy + y^2)$ $=$ $(x^2 + y^2 + 2xy)(x^2 + y^2 - 2xy)$

$\qquad = (x^2 + y^2)^2 - (2xy)^2$ $=$ $x^4 + 2x^2y^2 + y^4 - 4x^2y^2$ $=$ $x^4 - 2x^2y^2 + y^4$

$y)$ $\quad (x^3 + 2 + xy)(x^3 - 2 + xy)$ $=$ $(x^3 + xy + 2)(x^3 + xy - 2)$

$\qquad = (x^3 + xy)^2 - 2^2$ $=$ $x^6 + 2(x^3)(xy) + (xy)^2 - 4$ $=$ $x^6 + 2x^4y + x^2y^2 - 4$

PRODUCTS VIII - XII

2. a) $\quad (x + 2y)^3$ $=$ $x^3 + 3(x)^2(2y) + 3(x)(2y)^2 + (2y)^3$

$\qquad = x^3 + 6x^2y + 12xy^2 + 8y^3,$ $\quad$ using VIII with $a = x$, $b = 2y$.

b) $\quad (3x + 2)^3$ $=$ $(3x)^3 + 3(3x)^2(2) + 3(3x)(2)^2 + (2)^3$ $=$ $27x^3 + 54x^2 + 36x + 8$

c) $\quad (2y - 5)^3$ $=$ $(2y)^3 - 3(2y)^2(5) + 3(2y)(5)^2 - (5)^3$

$\qquad = 8y^3 - 60y^2 + 150y - 125,$ $\quad$ using IX with $a = 2y$, $b = 5$.

d) $\quad (xy - 2)^3$ $=$ $(xy)^3 - 3(xy)^2(2) + 3(xy)(2)^2 - (2)^3$ $=$ $x^3y^3 - 6x^2y^2 + 12xy - 8$

e) $\quad (x^2y - y^2)^3$ $=$ $(x^2y)^3 - 3(x^2y)^2(y^2) + 3(x^2y)(y^2)^2 - (y^2)^3$ $=$ $x^6y^3 - 3x^4y^4 + 3x^2y^5 - y^6$

f) $\quad (x - 1)(x^2 + x + 1)$ $=$ $x^3 - 1,$ $\quad$ using X with $a = x$, $b = 1$.

If the form is not recognized, multiply as follows.

$\qquad (x - 1)(x^2 + x + 1)$ $=$ $x(x^2 + x + 1) - 1(x^2 + x + 1)$ $=$ $x^3 + x^2 + x - x^2 - x - 1 = x^3 - 1$

g) $\quad (x - 2y)(x^2 + 2xy + 4y^2)$ $=$ $x^3 - (2y)^3$ $=$ $x^3 - 8y^3,$ $\quad$ using X with $a = x$, $b = 2y$.

h) $\quad (xy + 2)(x^2y^2 - 2xy + 4)$ $=$ $(xy)^3 + (2)^3$ $=$ $x^3y^3 + 8,$ $\quad$ using XI with $a = xy$, $b = 2$.

i) $\quad (2x + 1)(4x^2 - 2x + 1)$ $=$ $(2x)^3 + 1$ $=$ $8x^3 + 1$

j) $\quad (2x + 3y + z)^2$ $=$ $(2x)^2 + (3y)^2 + (z)^2 + 2(2x)(3y) + 2(2x)(z) + 2(3y)(z)$

$\qquad = 4x^2 + 9y^2 + z^2 + 12xy + 4xz + 6yz,$ using XII with $a = 2x$, $b = 3y$, $c = z$.

k) $\quad (u^3 - v^2 + 2w)^2$ $=$ $(u^3)^2 + (-v^2)^2 + (2w)^2 + 2(u^3)(-v^2) + 2(u^3)(2w) + 2(-v^2)(2w)$

$\qquad = u^6 + v^4 + 4w^2 - 2u^3v^2 + 4u^3w - 4v^2w$

PRODUCTS XIII - XIV

3. a) $\quad (x - 1)(x^5 + x^4 + x^3 + x^2 + x + 1)$ $=$ $x^6 - 1,$ $\quad$ using XIII with $a = x$, $b = 1$, $n = 6$.

b) $\quad (x - 2y)(x^4 + 2x^3y + 4x^2y^2 + 8xy^3 + 16y^4)$ $=$ $x^5 - (2y)^5$

$\qquad = x^5 - 32y^5,$ $\quad$ using XIII with $a = x$, $b = 2y$.

c) $\quad (3y + x)(81y^4 - 27y^3x + 9y^2x^2 - 3yx^3 + x^4)$ $=$ $(3y)^5 + x^5$

$\qquad = 243y^5 + x^5,$ $\quad$ using XIV with $a = 3y$, $b = x$.

MISCELLANEOUS PRODUCTS

4. a) $\quad (x + y + z)(x + y - z)(x - y + z)(x - y - z).$ $\quad$ The first two factors may be written as

$\qquad (x + y + z)(x + y - z)$ $=$ $(x + y)^2 - z^2$ $=$ $x^2 + 2xy + y^2 - z^2,$

and the second two factors as

$\qquad (x - y + z)(x - y - z)$ $=$ $(x - y)^2 - z^2$ $=$ $x^2 - 2xy + y^2 - z^2.$

The result may be written

$$(x^2 + y^2 - z^2 + 2xy)(x^2 + y^2 - z^2 - 2xy) \;=\; (x^2 + y^2 - z^2)^2 - (2xy)^2$$

$$= \;(x^2)^2 + (y^2)^2 + (-z^2)^2 + 2(x^2)(y^2) + 2(x^2)(-z^2) + 2(y^2)(-z^2) - 4x^2 y^2$$

$$= \;x^4 + y^4 + z^4 - 2x^2 y^2 - 2x^2 z^2 - 2y^2 z^2$$

b) $(x + y + z + 1)^2 \;=\; [(x + y) + (z + 1)]^2 \;=\; (x + y)^2 + 2(x + y)(z + 1) + (z + 1)^2$

$$= \;x^2 + 2xy + y^2 + 2xz + 2x + 2yz + 2y + z^2 + 2z + 1$$

c) $(u - v)^3(u + v)^3 \;=\; [(u - v)(u + v)]^3 \;=\; (u^2 - v^2)^3$

$$= \;(u^2)^3 - 3(u^2)^2 v^2 + 3(u^2)(v^2)^2 - (v^2)^3 \;=\; u^6 - 3u^4 v^2 + 3u^2 v^4 - v^6$$

d) $(x^2 - x + 1)^2(x^2 + x + 1)^2 \;=\; [(x^2 - x + 1)(x^2 + x + 1)]^2 = [(x^2 + 1 - x)(x^2 + 1 + x)]^2$

$$= \;[(x^2 + 1)^2 - x^2]^2 \;=\; [x^4 + 2x^2 + 1 - x^2]^2 \;=\; (x^4 + x^2 + 1)^2$$

$$= \;(x^4)^2 + (x^2)^2 + 1^2 + 2(x^4)(x^2) + 2(x^4)(1) + 2(x^2)(1)$$

$$= \;x^8 + x^4 + 1 + 2x^6 + 2x^4 + 2x^2 \;=\; x^8 + 2x^6 + 3x^4 + 2x^2 + 1$$

e) $(e^y + 1)(e^y - 1)(e^{2y} + 1)(e^{4y} + 1)(e^{8y} + 1) \;=\; (e^{2y} - 1)(e^{2y} + 1)(e^{4y} + 1)(e^{8y} + 1)$

$$= \;(e^{4y} - 1)(e^{4y} + 1)(e^{8y} + 1) \;=\; (e^{8y} - 1)(e^{8y} + 1) \;=\; e^{16y} - 1$$

SUPPLEMENTARY PROBLEMS

Find each of the following products.

5. a) $2xy(3x^2 y - 4y^3) \;=\; 6x^3 y^2 - 8xy^4$

b) $3x^2 y^3(2xy - x - 2y) \;=\; 6x^3 y^4 - 3x^3 y^3 - 6x^2 y^4$

c) $(2st^3 - 4rs^2 + 3s^3 t)(5rst^2) \;=\; 10rs^2 t^5 - 20r^2 s^3 t^2 + 15rs^4 t^3$

d) $(3a + 5b)(3a - 5b) \;=\; 9a^2 - 25b^2$

e) $(5xy + 4)(5xy - 4) \;=\; 25x^2 y^2 - 16$

f) $(2 - 5y^2)(2 + 5y^2) \;=\; 4 - 25y^4$

g) $(3a + 5a^2 b)(3a - 5a^2 b) \;=\; 9a^2 - 25a^4 b^2$

h) $(x + 6)^2 \;=\; x^2 + 12x + 36$

i) $(y + 3x)^2 \;=\; y^2 + 6xy + 9x^2$

j) $(z - 4)^2 \;=\; z^2 - 8z + 16$

k) $(3 - 2x^2)^2 \;=\; 9 - 12x^2 + 4x^4$

l) $(x^2 y - 2z)^2 \;=\; x^4 y^2 - 4x^2 yz + 4z^2$

m) $(x + 2)(x + 4) \;=\; x^2 + 6x + 8$

n) $(x - 4)(x + 7) \;=\; x^2 + 3x - 28$

o) $(y + 3)(y - 5) \;=\; y^2 - 2y - 15$

p) $(xy + 6)(xy - 4) \;=\; x^2 y^2 + 2xy - 24$

q) $(2x - 3)(4x + 1) \;=\; 8x^2 - 10x - 3$

r) $(4 + 3r)(2 - r) \;=\; 8 + 2r - 3r^2$

s) $(5x + 3y)(2x - 3y) \;=\; 10x^2 - 9xy - 9y^2$

t) $(2t^2 + s)(3t^2 + 4s) \;=\; 6t^4 + 11t^2 s + 4s^2$

$u)$ $(x^2 + 4y)(2x^2y - y^2)$ $=$ $2x^4y + 7x^2y^2 - 4y^3$

$v)$ $x(2x - 3)(3x + 4)$ $=$ $6x^3 - x^2 - 12x$

$w)$ $(r + s - 1)(r + s + 1)$ $=$ $r^2 + 2rs + s^2 - 1$

$x)$ $(x - 2y + z)(x - 2y - z)$ $=$ $x^2 - 4xy + 4y^2 - z^2$

$y)$ $(x^2 + 2x + 4)(x^2 - 2x + 4)$ $=$ $x^4 + 4x^2 + 16$

6. $a)$ $(2x + 1)^3$ $=$ $8x^3 + 12x^2 + 6x + 1$

   $b)$ $(3x + 2y)^3$ $=$ $27x^3 + 54x^2y + 36xy^2 + 8y^3$

   $c)$ $(r - 2s)^3$ $=$ $r^3 - 6r^2s + 12rs^2 - 8s^3$

   $d)$ $(x^2 - 1)^3$ $=$ $x^6 - 3x^4 + 3x^2 - 1$

   $e)$ $(ab^2 - 2b)^3$ $=$ $a^3b^6 - 6a^2b^5 + 12ab^4 - 8b^3$

   $f)$ $(t - 2)(t^2 + 2t + 4)$ $=$ $t^3 - 8$

   $g)$ $(z - x)(x^2 + xz + z^2)$ $=$ $z^3 - x^3$

   $h)$ $(x + 3y)(x^2 - 3xy + 9y^2)$ $=$ $x^3 + 27y^3$

7. $a)$ $(x - 2y + z)^2$ $=$ $x^2 - 4xy + 4y^2 + 2zx - 4zy + z^2$

   $b)$ $(s - 1)(s^3 + s^2 + s + 1)$ $=$ $s^4 - 1$

   $c)$ $(1 + t^2)(1 - t^2 + t^4 - t^6)$ $=$ $1 - t^8$

   $d)$ $(3x + 2y)^2(3x - 2y)^2$ $=$ $81x^4 - 72x^2y^2 + 16y^4$

   $e)$ $(x^2 + 2x + 1)^2(x^2 - 2x + 1)^2$ $=$ $x^8 - 4x^6 + 6x^4 - 4x^2 + 1$

   $f)$ $(y - 1)^3(y + 1)^3$ $=$ $y^6 - 3y^4 + 3y^2 - 1$

   $g)$ $(u + 2)(u - 2)(u^2 + 4)(u^4 + 16)$ $=$ $u^8 - 256$

CHAPTER 4

Factoring

THE FACTORS of a given algebraic expression consist of two or more algebraic expressions which when multiplied together produce the given expression.

For example, the algebraic expression $x^2 - 7x + 6$ may be written as the product of the two factors $(x-1)(x-6)$.

Similarly, $x^2 + 2xy - 8y^2 = (x + 4y)(x - 2y)$.

THE FACTORIZATION PROCESS is generally restricted to finding factors of polynomials with integer coefficients in each of its terms. In such cases it is required that the factors also be polynomials with integer coefficients. Unless otherwise stated we shall adhere to this limitation.

Thus we shall not consider $(x-1)$ as being factorable into $(\sqrt{x} + 1)(\sqrt{x} - 1)$ because these factors are not polynomials. Similarly, we shall not consider $(x^2 - 3y^2)$ as being factorable into $(x - \sqrt{3}y)(x + \sqrt{3}y)$ because these factors are not polynomials with integer coefficients. Also, even though $3x + 2y$ could be written $3(x + \frac{2}{3}y)$ we shall not consider this to be a factored form because $x + \frac{2}{3}y$ is not a polynomial with integer coefficients.

A given polynomial with integer coefficients is said to be *prime* if it cannot itself be factored in accordance with the above restrictions. Thus $x^2 - 7x + 6 = (x - 1)(x - 6)$ has been expressed as a product of the prime factors $x - 1$ and $x - 6$.

A polynomial is said to be factored completely when it is expressed as a product of prime factors.

Note 1. In factoring we shall allow trivial changes in sign. Thus $x^2 - 7x + 6$ can be factored either as $(x - 1)(x - 6)$ or $(1 - x)(6 - x)$. It can be shown that factorization into prime factors, apart from the trivial changes in sign and arrangement of factors, is possible in one and only one way. This is often referred to as the Unique Factorization Theorem.

Note 2. Sometimes the following definition of prime is used. A polynomial is said to be prime if it has no factors other than plus or minus itself and ±1. This is in analogy with the definition of a prime number or integer such as $2, 3, 5, 7, 11, \ldots$ and may be seen to be equivalent to the previous definition.

Note 3. Occasionally we may factor polynomials with rational coefficients, e.g. $x^2 - 9/4 = (x + 3/2)(x - 3/2)$. In such cases the factors should be polynomials with rational coefficients.

IN FACTORING, formulas I – XIV of Chapter 3 are very useful. Just as when read from left to right they helped to obtain *products*, so when read from right to left they help to find *factors*.

THE FOLLOWING PROCEDURES in factoring are very useful.

A) COMMON MONOMIAL FACTOR. Type: $ac + ad = a(c + d)$

Examples. 1) $6x^2y - 2x^3 = 2x^2(3y - x)$

2) $2x^3y - xy^2 + 3x^2y = xy(2x^2 - y + 3x)$

B) DIFFERENCE OF TWO SQUARES. Type: $a^2 - b^2 = (a + b)(a - b)$

Examples. 1) $x^2 - 25 = x^2 - 5^2 = (x + 5)(x - 5)$ where $a = x$, $b = 5$

2) $4x^2 - 9y^2 = (2x)^2 - (3y)^2 = (2x + 3y)(2x - 3y)$ where $a = 2x$, $b = 3y$

C) PERFECT SQUARE TRINOMIALS. Types: $\begin{aligned} a^2 + 2ab + b^2 &= (a + b)^2 \\ a^2 - 2ab + b^2 &= (a - b)^2 \end{aligned}$

It follows that a trinomial is a perfect square if two terms are perfect squares and the third term is numerically twice the product of the square roots of the other two terms.

Examples. 1) $x^2 + 6x + 9 = (x + 3)^2$

2) $9x^2 - 12xy + 4y^2 = (3x - 2y)^2$

D) OTHER TRINOMIALS. Types: $\begin{aligned} x^2 + (a + b)x + ab &= (x + a)(x + b) \\ acx^2 + (ad + bc)x + bd &= (ax + b)(cx + d) \end{aligned}$

Examples. 1) $x^2 - 5x + 4 = (x - 4)(x - 1)$ where $a = -4$, $b = -1$ so that their sum $(a + b) = -5$ and their product $ab = 4$.

2) $x^2 + xy - 12y^2 = (x - 3y)(x + 4y)$ where $a = -3y$, $b = 4y$

3) $3x^2 - 5x - 2 = (x - 2)(3x + 1)$. Here $ac = 3$, $bd = -2$, $ad + bc = -5$; and we find by trial that $a = 1$, $c = 3$, $b = -2$, $d = 1$ satisfies $ad + bc = -5$.

4) $6x^2 + x - 12 = (3x - 4)(2x + 3)$

5) $8 - 14x + 5x^2 = (4 - 5x)(2 - x)$

E) SUM, DIFFERENCE OF TWO CUBES. Types: $\begin{aligned} a^3 + b^3 &= (a + b)(a^2 - ab + b^2) \\ a^3 - b^3 &= (a - b)(a^2 + ab + b^2) \end{aligned}$

Examples. 1) $8x^3 + 27y^3 = (2x)^3 + (3y)^3$

$= (2x + 3y)[(2x)^2 - (2x)(3y) + (3y)^2]$

$= (2x + 3y)(4x^2 - 6xy + 9y^2)$

2) $8x^3y^3 - 1 = (2xy)^3 - 1^3 = (2xy - 1)(4x^2y^2 + 2xy + 1)$

F) GROUPING OF TERMS. Type: $ac + bc + ad + bd = c(a + b) + d(a + b) = (a + b)(c + d)$

Example. $2ax - 4bx + ay - 2by = 2x(a - 2b) + y(a - 2b) = (a - 2b)(2x + y)$

G) FACTORS OF $a^n \pm b^n$. Here we use formulas XIII and XIV of Chapter 3.

Examples. 1) $32x^5 + 1 = (2x)^5 + 1^5 = (2x + 1)[(2x)^4 - (2x)^3 + (2x)^2 - 2x + 1]$

$= (2x + 1)(16x^4 - 8x^3 + 4x^2 - 2x + 1)$

2) $x^7 - 1 = (x - 1)(x^6 + x^5 + x^4 + x^3 + x^2 + x + 1)$

H) ADDITION AND SUBTRACTION OF SUITABLE TERMS.

Example.   Factor $x^4 + 4$.

Adding and subtracting $4x^2$ (twice the product of the square roots of x^4 and 4), we have

$$x^4 + 4 = (x^4 + 4x^2 + 4) - 4x^2 = (x^2 + 2)^2 - (2x)^2$$
$$= (x^2 + 2 + 2x)(x^2 + 2 - 2x) = (x^2 + 2x + 2)(x^2 - 2x + 2)$$

I) MISCELLANEOUS COMBINATIONS OF PREVIOUS METHODS.

Example. $x^4 - xy^3 - x^3 y + y^4 = (x^4 - xy^3) - (x^3 y - y^4)$
$$= x(x^3 - y^3) - y(x^3 - y^3)$$
$$= (x^3 - y^3)(x - y) = (x - y)(x^2 + xy + y^2)(x - y)$$
$$= (x - y)^2 (x^2 + xy + y^2)$$

THE HIGHEST COMMON FACTOR (H.C.F.) of two or more given polynomials is the polynomial of highest degree and largest numerical coefficients (apart from trivial changes in sign) which is a factor of all the given polynomials.

The following method is suggested for finding the H.C.F. of several polynomials. *a*) Write each polynomial as a product of prime factors. *b*) The H.C.F. is the product obtained by taking each factor to the *lowest* power to which it occurs in any of the polynomials.

Example.   The H.C.F. of
$$2^3 3^2 (x - y)^3 (x + 2y)^2, \quad 2^2 3^3 (x - y)^2 (x + 2y)^3, \quad 3^2 (x - y)^2 (x + 2y)$$
is $3^2 (x - y)^2 (x + 2y)$.

Two or more polynomials are *relatively prime* if their H.C.F. is ±1.

THE LOWEST COMMON MULTIPLE (L.C.M.) of two or more given polynomials is the polynomial of lowest degree and smallest numerical coefficients (apart from trivial changes in sign) for which each of the given polynomials will be a factor.

The following procedure is suggested for determining the L.C.M. of several polynomials. *a*) Write each polynomial as a product of prime factors. *b*) The L.C.M. is the product obtained by taking each factor to the *highest* power to which it occurs.

Example.   The L.C.M. of
$$2^3 3^2 (x - y)^3 (x + 2y)^2, \quad 2^2 3^3 (x - y)^2 (x + 2y)^3, \quad 3^2 (x - y)^2 (x + 2y)$$
is $2^3 3^3 (x - y)^3 (x + 2y)^3$.

SOLVED PROBLEMS

COMMON MONOMIAL FACTOR.   Type: $ac + ad = a(c + d)$

1. *a*) $2x^2 - 3xy = x(2x - 3y)$ *c*) $3x^2 + 6x^3 + 12x^4 = 3x^2(1 + 2x + 4x^2)$

 b) $4x + 8y + 12z = 4(x + 2y + 3z)$ *d*) $9s^3 t + 15s^2 t^3 - 3s^2 t^2 = 3s^2 t(3s + 5t^2 - t)$

 e) $10a^2 b^3 c^4 - 15a^3 b^2 c^4 + 30a^4 b^3 c^2 = 5a^2 b^2 c^2 (2bc^2 - 3ac^2 + 6a^2 b)$

 f) $4a^{n+1} - 8a^{2n} = 4a^{n+1}(1 - 2a^{n-1})$

DIFFERENCE OF TWO SQUARES.   Type: $a^2 - b^2 = (a+b)(a-b)$

2. a) $x^2 - 9 = x^2 - 3^2 = (x+3)(x-3)$

b) $25x^2 - 4y^2 = (5x)^2 - (2y)^2 = (5x+2y)(5x-2y)$

c) $9x^2y^2 - 16a^2 = (3xy)^2 - (4a)^2 = (3xy+4a)(3xy-4a)$

d) $1 - m^2n^4 = 1^2 - (mn^2)^2 = (1+mn^2)(1-mn^2)$

e) $3x^2 - 12 = 3(x^2-4) = 3(x+2)(x-2)$

f) $x^2y^2 - 36y^4 = y^2[x^2 - (6y)^2] = y^2(x+6y)(x-6y)$

g) $x^4 - y^4 = (x^2)^2 - (y^2)^2 = (x^2+y^2)(x^2-y^2) = (x^2+y^2)(x+y)(x-y)$

h) $1 - x^8 = (1+x^4)(1-x^4) = (1+x^4)(1+x^2)(1-x^2) = (1+x^4)(1+x^2)(1+x)(1-x)$

i) $32a^4b - 162b^5 = 2b(16a^4 - 81b^4) = 2b(4a^2+9b^2)(4a^2-9b^2)$
$$= 2b(4a^2+9b^2)(2a+3b)(2a-3b)$$

j) $x^3y - y^3x = xy(x^2-y^2) = xy(x+y)(x-y)$

k) $(x+1)^2 - 36y^2 = [(x+1)+(6y)][(x+1)-(6y)] = (x+6y+1)(x-6y+1)$

l) $(5x+2y)^2 - (3x-7y)^2 = [(5x+2y)+(3x-7y)][(5x+2y)-(3x-7y)] = (8x-5y)(2x+9y)$

PERFECT SQUARE TRINOMIALS.   Types: $a^2 + 2ab + b^2 = (a+b)^2$
$$a^2 - 2ab + b^2 = (a-b)^2$$

3. a) $x^2 + 8x + 16 = x^2 + 2(x)(4) + 4^2 = (x+4)^2$ d) $x^2 - 16xy + 64y^2 = (x-8y)^2$

b) $1 + 4y + 4y^2 = (1+2y)^2$ e) $25x^2 + 60xy + 36y^2 = (5x+6y)^2$

c) $t^2 - 4t + 4 = t^2 - 2(t)(2) + 2^2 = (t-2)^2$ f) $16m^2 - 40mn + 25n^2 = (4m-5n)^2$

g) $9x^4 - 24x^2y + 16y^2 = (3x^2-4y)^2$

h) $2x^3y^3 + 16x^2y^4 + 32xy^5 = 2xy^3(x^2+8xy+16y^2) = 2xy^3(x+4y)^2$

i) $16a^4 - 72a^2b^2 + 81b^4 = (4a^2-9b^2)^2 = [(2a+3b)(2a-3b)]^2 = (2a+3b)^2(2a-3b)^2$

j) $(x+2y)^2 + 10(x+2y) + 25 = (x+2y+5)^2$ k) $a^2x^2 - 2abxy + b^2y^2 = (ax-by)^2$

l) $4m^6n^6 + 32m^4n^4 + 64m^2n^2 = 4m^2n^2(m^4n^4 + 8m^2n^2 + 16) = 4m^2n^2(m^2n^2+4)^2$

OTHER TRINOMIALS.   Types: $x^2 + (a+b)x + ab = (x+a)(x+b)$
$$acx^2 + (ad+bc)x + bd = (ax+b)(cx+d)$$

4. a) $x^2 + 6x + 8 = (x+4)(x+2)$ e) $x^2 - 7xy + 12y^2 = (x-3y)(x-4y)$

b) $x^2 - 6x + 8 = (x-4)(x-2)$ f) $x^2 + xy - 12y^2 = (x+4y)(x-3y)$

c) $x^2 + 2x - 8 = (x+4)(x-2)$ g) $16 - 10x + x^2 = (8-x)(2-x)$

d) $x^2 - 2x - 8 = (x-4)(x+2)$ h) $20 - x - x^2 = (5+x)(4-x)$

i) $3x^3 - 3x^2 - 18x = 3x(x^2-x-6) = 3x(x-3)(x+2)$

j) $y^4 + 7y^2 + 12 = (y^2+4)(y^2+3)$

k) $m^4 + m^2 - 2 = (m^2+2)(m^2-1) = (m^2+2)(m+1)(m-1)$

l) $(x+1)^2 + 3(x+1) + 2 = [(x+1)+2][(x+1)+1] = (x+3)(x+2)$

m) $s^2t^2 - 2st^3 - 63t^4 = t^2(s^2 - 2st - 63t^2) = t^2(s-9t)(s+7t)$

n) $z^4 - 10z^2 + 9 = (z^2-1)(z^2-9) = (z+1)(z-1)(z+3)(z-3)$

o) $2x^6y - 6x^4y^3 - 8x^2y^5 = 2x^2y(x^4 - 3x^2y^2 - 4y^4)$
$$= 2x^2y(x^2+y^2)(x^2-4y^2) = 2x^2y(x^2+y^2)(x+2y)(x-2y)$$

$p)$ $x^2 - 2xy + y^2 + 10(x-y) + 9$ $=$ $(x-y)^2 + 10(x-y) + 9$

$\qquad\qquad\qquad\qquad = [(x-y) + 1][(x-y) + 9]$ $=$ $(x - y + 1)(x - y + 9)$

$q)$ $4x^8y^{10} - 40x^5y^7 + 84x^2y^4$ $=$ $4x^2y^4(x^6y^6 - 10x^3y^3 + 21)$ $=$ $4x^2y^4(x^3y^3 - 7)(x^3y^3 - 3)$

$r)$ $x^{2a} - x^a - 30$ $=$ $(x^a - 6)(x^a + 5)$

$s)$ $x^{m+2n} + 7x^{m+n} + 10x^m$ $=$ $x^m(x^{2n} + 7x^n + 10)$ $=$ $x^m(x^n + 2)(x^n + 5)$

$t)$ $a^{2(y-1)} - 5a^{y-1} + 6$ $=$ $(a^{y-1} - 3)(a^{y-1} - 2)$

5. $a)$ $3x^2 + 10x + 3$ $=$ $(3x + 1)(x + 3)$ $\qquad$ $d)$ $10s^2 + 11s - 6$ $=$ $(5s - 2)(2s + 3)$

$b)$ $2x^2 - 7x + 3$ $=$ $(2x - 1)(x - 3)$ $\qquad$ $e)$ $6x^2 - xy - 12y^2 = (3x + 4y)(2x - 3y)$

$c)$ $2y^2 - y - 6$ $=$ $(2y + 3)(y - 2)$ $\qquad$ $f)$ $10 - x - 3x^2$ $=$ $(5 - 3x)(2 + x)$

$g)$ $4z^4 - 9z^2 + 2$ $=$ $(z^2 - 2)(4z^2 - 1)$ $=$ $(z^2 - 2)(2z + 1)(2z - 1)$

$h)$ $16x^3y + 28x^2y^2 - 30xy^3$ $=$ $2xy(8x^2 + 14xy - 15y^2)$ $=$ $2xy(4x - 3y)(2x + 5y)$

$i)$ $12(x+y)^2 + 8(x+y) - 15$ $=$ $[6(x+y) - 5][2(x+y) + 3]$ $=$ $(6x + 6y - 5)(2x + 2y + 3)$

$j)$ $6b^{2n+1} + 5b^{n+1} - 6b$ $=$ $b(6b^{2n} + 5b^n - 6)$ $=$ $b(2b^n + 3)(3b^n - 2)$

$k)$ $18x^{4p+m} - 66x^{2p+m}y^2 - 24x^my^4$ $=$ $6x^m(3x^{4p} - 11x^{2p}y^2 - 4y^4) = 6x^m(3x^{2p} + y^2)(x^{2p} - 4y^2)$

$\qquad\qquad\qquad\qquad = 6x^m(3x^{2p} + y^2)(x^p + 2y)(x^p - 2y)$

$l)$ $64x^{12}y^3 - 68x^8y^7 + 4x^4y^{11}$ $=$ $4x^4y^3(16x^8 - 17x^4y^4 + y^8)$ $=$ $4x^4y^3(16x^4 - y^4)(x^4 - y^4)$

$\qquad\qquad\qquad\qquad = 4x^4y^3(4x^2 + y^2)(4x^2 - y^2)(x^2 + y^2)(x^2 - y^2)$

$\qquad\qquad\qquad\qquad = 4x^4y^3(4x^2 + y^2)(2x + y)(2x - y)(x^2 + y^2)(x + y)(x - y)$

SUM OR DIFFERENCE OF TWO CUBES.　Types: $\begin{aligned} a^3 + b^3 &= (a + b)(a^2 - ab + b^2) \\ a^3 - b^3 &= (a - b)(a^2 + ab + b^2) \end{aligned}$

6. $a)$ $x^3 + 8$ $=$ $x^3 + 2^3$ $=$ $(x + 2)(x^2 - 2x + 2^2)$ $=$ $(x + 2)(x^2 - 2x + 4)$

$b)$ $a^3 - 27$ $=$ $a^3 - 3^3$ $=$ $(a - 3)(a^2 + 3a + 3^2)$ $=$ $(a - 3)(a^2 + 3a + 9)$

$c)$ $a^6 + b^6$ $=$ $(a^2)^3 + (b^2)^3$ $=$ $(a^2 + b^2)[(a^2)^2 - a^2b^2 + (b^2)^2]$

$\qquad\qquad\qquad = (a^2 + b^2)(a^4 - a^2b^2 + b^4)$

$d)$ $a^6 - b^6$ $=$ $(a^3 + b^3)(a^3 - b^3)$ $=$ $(a + b)(a^2 - ab + b^2)(a - b)(a^2 + ab + b^2)$

$e)$ $a^9 + b^9$ $=$ $(a^3)^3 + (b^3)^3$ $=$ $(a^3 + b^3)[(a^3)^2 - a^3b^3 + (b^3)^2]$

$\qquad\qquad\qquad = (a + b)(a^2 - ab + b^2)(a^6 - a^3b^3 + b^6)$

$f)$ $a^{12} + b^{12}$ $=$ $(a^4)^3 + (b^4)^3$ $=$ $(a^4 + b^4)(a^8 - a^4b^4 + b^8)$

$g)$ $64x^3 + 125y^3$ $=$ $(4x)^3 + (5y)^3$ $=$ $(4x + 5y)[(4x)^2 - (4x)(5y) + (5y)^2]$

$\qquad\qquad\qquad = (4x + 5y)(16x^2 - 20xy + 25y^2)$

$h)$ $(x+y)^3 - z^3$ $=$ $(x + y - z)[(x+y)^2 + (x+y)z + z^2]$

$\qquad\qquad\qquad = (x + y - z)(x^2 + 2xy + y^2 + xz + yz + z^2)$

$i)$ $(x-2)^3 + 8y^3$ $=$ $(x-2)^3 + (2y)^3$ $=$ $(x - 2 + 2y)[(x-2)^2 - (x-2)(2y) + (2y)^2]$

$\qquad\qquad\qquad = (x - 2 + 2y)(x^2 - 4x + 4 - 2xy + 4y + 4y^2)$

$j)$ $x^6 - 7x^3 - 8$ $=$ $(x^3 - 8)(x^3 + 1)$

$\qquad\qquad\qquad = (x^3 - 2^3)(x^3 + 1)$ $=$ $(x - 2)(x^2 + 2x + 4)(x + 1)(x^2 - x + 1)$

$k)$ $x^8y - 64x^2y^7 = x^2y(x^6 - 64y^6) = x^2y(x^3 + 8y^3)(x^3 - 8y^3) = x^2y[x^3 + (2y)^3][x^3 - (2y)^3]$
$$= x^2y(x + 2y)(x^2 - 2xy + 4y^2)(x - 2y)(x^2 + 2xy + 4y^2)$$

$l)$ $54x^6y^2 - 38x^3y^2 - 16y^2 = 2y^2(27x^6 - 19x^3 - 8) = 2y^2(27x^3 + 8)(x^3 - 1)$
$$= 2y^2[(3x)^3 + 2^3](x^3 - 1) = 2y^2(3x + 2)(9x^2 - 6x + 4)(x - 1)(x^2 + x + 1)$$

GROUPING OF TERMS.   Type: $ac + bc + ad + bd = c(a + b) + d(a + b) = (a + b)(c + d)$

7. $a)$ $bx - ab + x^2 - ax = b(x - a) + x(x - a) = (x - a)(b + x) = (x - a)(x + b)$

$b)$ $3ax - ay - 3bx + by = a(3x - y) - b(3x - y) = (3x - y)(a - b)$

$c)$ $6x^2 - 4ax - 9bx + 6ab = 2x(3x - 2a) - 3b(3x - 2a) = (3x - 2a)(2x - 3b)$

$d)$ $ax + ay + x + y = a(x + y) + (x + y) = (x + y)(a + 1)$

$e)$ $x^2 - 4y^2 + x + 2y = (x + 2y)(x - 2y) + (x + 2y) = (x + 2y)(x - 2y + 1)$

$f)$ $x^3 + x^2y + xy^2 + y^3 = x^2(x + y) + y^2(x + y) = (x + y)(x^2 + y^2)$

$g)$ $x^7 + 27x^4 - x^3 - 27 = x^4(x^3 + 27) - (x^3 + 27) = (x^3 + 27)(x^4 - 1)$
$$= (x^3 + 3^3)(x^2 + 1)(x^2 - 1) = (x + 3)(x^2 - 3x + 9)(x^2 + 1)(x + 1)(x - 1)$$

$h)$ $x^3y^3 - y^3 + 8x^3 - 8 = y^3(x^3 - 1) + 8(x^3 - 1) = (x^3 - 1)(y^3 + 8)$
$$= (x - 1)(x^2 + x + 1)(y + 2)(y^2 - 2y + 4)$$

$i)$ $a^6 + b^6 - a^2b^4 - a^4b^2 = a^6 - a^2b^4 + b^6 - a^4b^2 = a^2(a^4 - b^4) - b^2(a^4 - b^4)$
$$= (a^4 - b^4)(a^2 - b^2) = (a^2 + b^2)(a^2 - b^2)(a + b)(a - b)$$
$$= (a^2 + b^2)(a + b)(a - b)(a + b)(a - b) = (a^2 + b^2)(a + b)^2(a - b)^2$$

$j)$ $a^3 + 3a^2 - 5ab + 2b^2 - b^3 = (a^3 - b^3) + (3a^2 - 5ab + 2b^2)$
$$= (a - b)(a^2 + ab + b^2) + (a - b)(3a - 2b)$$
$$= (a - b)(a^2 + ab + b^2 + 3a - 2b)$$

FACTORS OF $a^n \pm b^n$.

8. $a^n + b^n$ has $a + b$ as a factor if and only if n is a positive odd integer. Then
$$a^n + b^n = (a + b)(a^{n-1} - a^{n-2}b + a^{n-3}b^2 - \ldots - ab^{n-2} + b^{n-1}).$$

$a)$ $a^3 + b^3 = (a + b)(a^2 - ab + b^2)$

$b)$ $64 + y^3 = 4^3 + y^3 = (4 + y)(4^2 - 4y + y^2) = (4 + y)(16 - 4y + y^2)$

$c)$ $x^3 + 8y^6 = x^3 + (2y^2)^3 = (x + 2y^2)[x^2 - x(2y^2) + (2y^2)^2]$
$$= (x + 2y^2)(x^2 - 2xy^2 + 4y^4)$$

$d)$ $a^5 + b^5 = (a + b)(a^4 - a^3b + a^2b^2 - ab^3 + b^4)$

$e)$ $1 + x^5y^5 = 1^5 + (xy)^5 = (1 + xy)(1 - xy + x^2y^2 - x^3y^3 + x^4y^4)$

$f)$ $z^5 + 32 = z^5 + 2^5 = (z + 2)(z^4 - 2z^3 + 2^2z^2 - 2^3z + 2^4)$
$$= (z + 2)(z^4 - 2z^3 + 4z^2 - 8z + 16)$$

$g)$ $a^{10} + x^{10} = (a^2)^5 + (x^2)^5 = (a^2 + x^2)[(a^2)^4 - (a^2)^3x^2 + (a^2)^2(x^2)^2 - (a^2)(x^2)^3 + (x^2)^4]$
$$= (a^2 + x^2)(a^8 - a^6x^2 + a^4x^4 - a^2x^6 + x^8)$$

$h)$ $u^7 + v^7 = (u + v)(u^6 - u^5v + u^4v^2 - u^3v^3 + u^2v^4 - uv^5 + v^6)$

$i)$ $x^9 + 1 = (x^3)^3 + 1^3 = (x^3 + 1)(x^6 - x^3 + 1) = (x + 1)(x^2 - x + 1)(x^6 - x^3 + 1)$

9. $a^n - b^n$ has $a - b$ as a factor if n is any positive integer. Then

$$a^n - b^n = (a - b)(a^{n-1} + a^{n-2}b + a^{n-3}b^2 + \ldots + ab^{n-2} + b^{n-1}).$$

If n is an even positive integer, $a^n - b^n$ also has $a + b$ as factor.

a) $a^2 - b^2 = (a - b)(a + b)$

b) $a^3 - b^3 = (a - b)(a^2 + ab + b^2)$

c) $27x^3 - y^3 = (3x)^3 - y^3 = (3x - y)[(3x)^2 + (3x)y + y^2] = (3x - y)(9x^2 + 3xy + y^2)$

d) $1 - x^3 = (1 - x)(1^2 + 1x + x^2) = (1 - x)(1 + x + x^2)$

e) $a^5 - 32 = a^5 - 2^5 = (a - 2)(a^4 + a^3 \cdot 2 + a^2 \cdot 2^2 + a \cdot 2^3 + 2^4)$

$$= (a - 2)(a^4 + 2a^3 + 4a^2 + 8a + 16)$$

f) $y^7 - z^7 = (y - z)(y^6 + y^5z + y^4z^2 + y^3z^3 + y^2z^4 + yz^5 + z^6)$

g) $x^6 - a^6 = (x^3 + a^3)(x^3 - a^3) = (x + a)(x^2 - ax + a^2)(x - a)(x^2 + ax + a^2)$

h) $u^8 - v^8 = (u^4 + v^4)(u^4 - v^4) = (u^4 + v^4)(u^2 + v^2)(u^2 - v^2)$

$$= (u^4 + v^4)(u^2 + v^2)(u + v)(u - v)$$

i) $x^9 - 1 = (x^3)^3 - 1 = (x^3 - 1)(x^6 + x^3 + 1) = (x - 1)(x^2 + x + 1)(x^6 + x^3 + 1)$

j) $x^{10} - y^{10} = (x^5 + y^5)(x^5 - y^5)$

$$= (x + y)(x^4 - x^3y + x^2y^2 - xy^3 + y^4)(x - y)(x^4 + x^3y + x^2y^2 + xy^3 + y^4)$$

ADDITION AND SUBTRACTION OF SUITABLE TERMS.

10. a) $a^4 + a^2b^2 + b^4$ (adding and subtracting a^2b^2)

$$= (a^4 + 2a^2b^2 + b^4) - a^2b^2 = (a^2 + b^2)^2 - (ab)^2$$

$$= (a^2 + b^2 + ab)(a^2 + b^2 - ab)$$

b) $36x^4 + 15x^2 + 4$ (adding and subtracting $9x^2$)

$$= (36x^4 + 24x^2 + 4) - 9x^2 = (6x^2 + 2)^2 - (3x)^2$$

$$= [(6x^2 + 2) + 3x][(6x^2 + 2) - 3x] = (6x^2 + 3x + 2)(6x^2 - 3x + 2)$$

c) $64x^4 + y^4$ (adding and subtracting $16x^2y^2$)

$$= (64x^4 + 16x^2y^2 + y^4) - 16x^2y^2 = (8x^2 + y^2)^2 - (4xy)^2$$

$$= (8x^2 + y^2 + 4xy)(8x^2 + y^2 - 4xy)$$

d) $u^8 - 14u^4 + 25$ (adding and subtracting $4u^4$)

$$= (u^8 - 10u^4 + 25) - 4u^4 = (u^4 - 5)^2 - (2u^2)^2$$

$$= (u^4 - 5 + 2u^2)(u^4 - 5 - 2u^2) = (u^4 + 2u^2 - 5)(u^4 - 2u^2 - 5)$$

MISCELLANEOUS PROBLEMS.

11. a) $x^2 - 4z^2 + 9y^2 - 6xy = (x^2 - 6xy + 9y^2) - 4z^2$

$$= (x - 3y)^2 - (2z)^2 = (x - 3y + 2z)(x - 3y - 2z)$$

b) $16a^2 + 10bc - 25c^2 - b^2 = 16a^2 - (b^2 - 10bc + 25c^2)$

$$= (4a)^2 - (b - 5c)^2 = (4a + b - 5c)(4a - b + 5c)$$

c) $x^2 + 7x + y^2 - 7y - 2xy - 8 = (x^2 - 2xy + y^2) + 7(x - y) - 8$

$$= (x - y)^2 + 7(x - y) - 8 = (x - y + 8)(x - y - 1)$$

d) $a^2 - 8ab - 2ac + 16b^2 + 8bc - 15c^2 = (a^2 - 8ab + 16b^2) - (2ac - 8bc) - 15c^2$

$$= (a - 4b)^2 - 2c(a - 4b) - 15c^2 = (a - 4b - 5c)(a - 4b + 3c)$$

e) $m^4 - n^4 + m^3 - mn^3 - n^3 + m^3n = (m^4 - mn^3) + (m^3n - n^4) + (m^3 - n^3)$

$$= m(m^3 - n^3) + n(m^3 - n^3) + (m^3 - n^3)$$

$$= (m^3 - n^3)(m + n + 1) = (m - n)(m^2 + mn + n^2)(m + n + 1)$$

HIGHEST COMMON FACTOR AND LOWEST COMMON MULTIPLE.

12. a) $9x^4y^2 = 3^2x^4y^2$, $12x^3y^3 = 2^2 \cdot 3x^3y^3$

$\qquad$ H.C.F. $= 3x^3y^2$, $\quad$ L.C.M. $= 2^2 \cdot 3^2x^4y^3 = 36x^4y^3$

b) $48r^3t^4 = 2^4 \cdot 3r^3t^4$, $54r^2t^6 = 2 \cdot 3^3r^2t^6$, $60r^4t^2 = 2^2 \cdot 3 \cdot 5r^4t^2$

$\qquad$ H.C.F. $= 2 \cdot 3r^2t^2 = 6r^2t^2$, $\quad$ L.C.M. $= 2^4 \cdot 3^3 \cdot 5r^4t^6 = 2160r^4t^6$

c) $6x - 6y = 2 \cdot 3(x - y)$, $4x^2 - 4y^2 = 2^2(x^2 - y^2) = 2^2(x + y)(x - y)$

$\qquad$ H.C.F. $= 2(x - y)$, $\quad$ L.C.M. $= 2^2 \cdot 3(x + y)(x - y)$

d) $y^4 - 16 = (y^2 + 4)(y + 2)(y - 2)$, $\quad y^2 - 4 = (y + 2)(y - 2)$, $\quad y^2 - 3y + 2 = (y - 1)(y - 2)$

$\qquad$ H.C.F. $= y - 2$, $\quad$ L.C.M. $= (y^2 + 4)(y + 2)(y - 2)(y - 1)$

e) $3 \cdot 5^2(x + 3y)^2(2x - y)^4$, $\quad 2^3 \cdot 3^2 \cdot 5(x + 3y)^3(2x - y)^2$, $\quad 2^2 \cdot 3 \cdot 5(x + 3y)^4(2x - y)^5$

$\qquad$ H.C.F. $= 3 \cdot 5(x + 3y)^2(2x - y)^2$, $\quad$ L.C.M. $= 2^3 \cdot 3^2 \cdot 5^2(x + 3y)^4(2x - y)^5$

SUPPLEMENTARY PROBLEMS

Factor each expression.

13. a) $3x^2y^4 + 6x^3y^3$

$\quad$ b) $12s^2t^2 - 6s^5t^4 + 4s^4t$

$\quad$ c) $2x^2yz - 4xyz^2 + 8xy^2z^3$

$\quad$ d) $4y^2 - 100$

$\quad$ e) $1 - a^4$

$\quad$ f) $64x - x^3$

$\quad$ g) $8x^4 - 128$

$\quad$ h) $18x^3y - 8xy^3$

$\quad$ i) $(2x + y)^2 - (3y - z)^2$

$\quad$ j) $4(x + 3y)^2 - 9(2x - y)^2$

$\quad$ k) $x^2 + 4x + 4$

$\quad$ l) $4 - 12y + 9y^2$

$\quad$ m) $x^2y^2 - 8xy + 16$

$\quad$ n) $4x^3y + 12x^2y^2 + 9xy^3$

$\quad$ o) $3a^4 + 6a^2b^2 + 3b^4$

$\quad$ p) $(m^2 - n^2)^2 + 8(m^2 - n^2) + 16$

$\quad$ q) $x^2 + 7x + 12$

$\quad$ r) $y^2 - 4y - 5$

$\quad$ s) $x^2 - 8xy + 15y^2$

$\quad$ t) $2z^3 + 10z^2 - 28z$

$\quad$ u) $15 + 2x - x^2$

14. a) $m^4 - 4m^2 - 21$

$\quad$ b) $a^4 - 20a^2 + 64$

$\quad$ c) $4s^4t - 4s^3t^2 - 24s^2t^3$

$\quad$ d) $x^{2m+4} + 5x^{m+4} - 50x^4$

$\quad$ e) $2x^2 + 3x + 1$

$\quad$ f) $3y^2 - 11y + 6$

$\quad$ g) $5m^3 - 3m^2 - 2m$

$\quad$ h) $6x^2 + 5xy - 6y^2$

$\quad$ i) $36z^6 - 13z^4 + z^2$

$\quad$ j) $12(x - y)^2 + 7(x - y) - 12$

$\quad$ k) $4x^{2n+2} - 4x^{n+2} - 3x^2$

15. a) $y^3 + 27$

$\quad$ b) $x^3 - 1$

$\quad$ c) $x^3y^3 + 8$

$\quad$ d) $8z^4 - 27z^7$

$\quad$ e) $8x^4y - 64xy^4$

$\quad$ f) $m^9 - n^9$

$\quad$ g) $y^6 + 1$

$\quad$ h) $(x - 2)^3 + (y + 1)^3$

$\quad$ i) $8x^6 + 7x^3 - 1$

16. a) $xy + 3y - 2x - 6$ c) $ax^2 + bx - ax - b$ e) $z^7 - 2z^6 + z^4 - 2z^3$

    b) $2pr - ps + 6qr - 3qs$ d) $x^3 - xy^2 - x^2y + y^3$ f) $m^3 - mn^2 + m^2n - n^3 + m^2 - n^2$

17. a) $z^5 + 1$ b) $x^5 + 32y^5$ c) $32 - u^5$ d) $m^{10} - 1$ e) $1 - z^7$

18. a) $z^4 + 64$ c) $x^8 - 12x^4 + 16$ e) $6ab + 4 - a^2 - 9b^2$

    b) $4x^4 + 3x^2y^2 + y^4$ d) $m^2 - 4p^2 + 4mn + 4n^2$ f) $9x^2 - x^2y^2 + 4y^2 + 12xy$

               g) $x^2 + y^2 - 4z^2 + 2xy + 3xz + 3yz$

19. Find the H.C.F. and L.C.M. of each group of polynomials.

    a) $16y^2z^4$, $24y^3z^2$

    b) $9r^3s^2t^5$, $12r^2s^4t^3$, $21r^5s^2$

    c) $x^2 - 3xy + 2y^2$, $4x^2 - 16xy + 16y^2$

    d) $6y^3 + 12y^2z$, $6y^2 - 24z^2$, $4y^2 - 4yz - 24z^2$

    e) $x^5 - x$, $x^5 - x^2$, $x^5 - x^3$

ANSWERS TO SUPPLEMENTARY PROBLEMS.

13. a) $3x^2y^3(y + 2x)$ b) $2s^2t(6t - 3s^3t^3 + 2s^2)$ c) $2xyz(r - 2z + 4yz^2)$ d) $4(y + 5)(y - 5)$
   e) $(1 + a^2)(1 + a)(1 - a)$ f) $x(8 + x)(8 - x)$ $8(x^2 + 4)(x + 2)(x - 2)$
   h) $2xy(3x + 2y)(3x - 2y)$ i) $(2x + 4y - z)(2x - 2y + z)$ j) $(8x + 3y)(9y - 4x)$
   k) $(x + 2)^2$ l) $(2 - 3y)^2$ m) $(xy - 4)^2$ n) $xy(2x + 3y)^2$ o) $3(a^2 + b^2)^2$
   p) $(m^2 - n^2 + 4)^2$ q) $(x + 3)(x + 4)$ r) $(y - 5)(y + 1)$ s) $(x - 3y)(x - 5y)$
   t) $2z(z + 7)(z - 2)$ u) $(5 - x)(3 + x)$

14. a) $(m^2 - 7)(m^2 + 3)$ b) $(a + 2)(a - 2)(a + 4)(a - 4)$ c) $4s^2t(s - 3t)(s + 2t)$
   d) $x^4(x^m - 5)(x^m + 10)$ e) $(2x + 1)(x + 1)$ f) $(3y - 2)(y - 3)$
   g) $m(5m + 2)(m - 1)$ h) $(2x + 3y)(3x - 2y)$ i) $z^2(2z + 1)(2z - 1)(3z + 1)(3z - 1)$
   j) $(4x - 4y - 3)(3x - 3y + 4)$ k) $x^2(2x^n + 1)(2x^n - 3)$

15. a) $(y + 3)(y^2 - 3y + 9)$ b) $(x - 1)(x^2 + x + 1)$ c) $(xy + 2)(x^2y^2 - 2xy + 4)$
   d) $z^4(2 - 3z)(4 + 6z + 9z^2)$ e) $8xy(x - 2y)(x^2 + 2xy + 4y^2)$
   f) $(m - n)(m^2 + mn + n^2)(m^6 + m^3n^3 + n^6)$ g) $(y^2 + 1)(y^4 - y^2 + 1)$
   h) $(x + y - 1)(x^2 - xy + y^2 - 5x + 4y + 7)$ i) $(2x - 1)(4x^2 + 2x + 1)(x + 1)(x^2 - x + 1)$

16. a) $(x + 3)(y - 2)$ b) $(2r - s)(p + 3q)$ c) $(ax + b)(x - 1)$ d) $(x - y)^2(x + y)$
   e) $z^3(z - 2)(z + 1)(z^2 - z + 1)$ f) $(m + n)(m - n)(m + n + 1)$

17. a) $(z + 1)(z^4 - z^3 + z^2 - z + 1)$ b) $(x + 2y)(x^4 - 2x^3y + 4x^2y^2 - 8xy^3 + 16y^4)$
   c) $(2 - u)(16 + 8u + 4u^2 + 2u^3 + u^4)$ d) $(m + 1)(m^4 - m^3 + m^2 - m + 1)(m - 1)(m^4 + m^3 + m^2 + m + 1)$
   e) $(1 - z)(1 + z + z^2 + z^3 + z^4 + z^5 + z^6)$

18. a) $(z^2 + 4z + 8)(z^2 - 4z + 8)$ b) $(2x^2 + xy + y^2)(2x^2 - xy + y^2)$ c) $(x^4 + 2x^2 - 4)(x^4 - 2x^2 - 4)$
   d) $(m + 2n + 2p)(m + 2n - 2p)$ e) $(2 + a - 3b)(2 - a + 3b)$ f) $(3x + xy + 2y)(3x - xy + 2y)$
   g) $(x + y + 4z)(x + y - z)$

19. a) H.C.F. $= 2^3y^2z^2 = 8y^2z^2$,   L.C.M. $= 2^4 \cdot 3y^3z^4 = 48y^3z^4$
   b) H.C.F. $= 3r^2s^2$,   L.C.M. $= 252r^5s^4t^5$
   c) H.C.F. $= x - 2y$,   L.C.M. $= 4(x - y)(x - 2y)^2$
   d) H.C.F. $= 2(y + 2z)$,   L.C.M. $= 12y^2(y + 2z)(y - 2z)(y - 3z)$
   e) H.C.F. $= x(x - 1)$,   L.C.M. $= x^3(x + 1)(x - 1)(x^2 + 1)(x^2 + x + 1)$

CHAPTER 5

Fractions

A RATIONAL ALGEBRAIC FRACTION is an expression which can be written as the quotient of two polynomials, P/Q. P is called the numerator and Q the denominator of the fraction.

Thus $\dfrac{3x-4}{x^2-6x+8}$ and $\dfrac{x^3+2y^2}{x^4-3xy+2y^3}$ are rational algebraic fractions.

RULES for manipulation of algebraic fractions are the same as for fractions in arithmetic. One such fundamental rule is: The value of a fraction is unchanged if its numerator and denominator are both multiplied by the same quantity or both divided by the same quantity, provided only that this quantity is not zero. In such case we call the fractions *equivalent*.

For example, if we multiply the numerator and denominator of $\dfrac{x+2}{x-3}$ by $(x-1)$ we obtain the equivalent fraction $\dfrac{(x+2)(x-1)}{(x-3)(x-1)} = \dfrac{x^2+x-2}{x^2-4x+3}$ provided $(x-1)$ is not zero, i.e. provided $x \neq 1$.

Similarly, given the fraction $\dfrac{x^2+3x+2}{x^2+4x+3}$ we may write it as $\dfrac{(x+2)(x+1)}{(x+3)(x+1)}$ and divide numerator and denominator by $(x+1)$ to obtain $\dfrac{x+2}{x+3}$ provided $(x+1)$ is not zero, i.e. provided $x \neq -1$. The operation of dividing out common factors of the numerator and denominator is called *cancellation* and may be indicated by a sloped line thus: $\dfrac{(x+2)(x+1)}{(x+3)(x+1)} = \dfrac{x+2}{x+3}$.

TO SIMPLIFY a given fraction is to convert it into an equivalent form in which numerator and denominator have no common factor (except ± 1). In such case we say that the fraction is *reduced to lowest terms*. This reduction is achieved by factoring numerator and denominator and cancelling common factors assuming they are not equal to zero.

Thus $\dfrac{x^2-4xy+3y^2}{x^2-y^2} = \dfrac{(x-3y)(x-y)}{(x+y)(x-y)} = \dfrac{x-3y}{x+y}$ provided $(x-y) \neq 0$.

THREE SIGNS are associated with a fraction: the sign of the numerator, of the denominator, and of the entire fraction. Any two of these signs may be changed without changing the value of the fraction. If there is no sign before a fraction, a plus sign is implied.

Examples. $\dfrac{-a}{b} = \dfrac{a}{-b} = -\dfrac{a}{b}$, $\dfrac{-a}{-b} = \dfrac{a}{b}$, $-\left(\dfrac{-a}{-b}\right) = -\dfrac{a}{b}$

Change of sign may often be of use in simplification.   Thus

$$\frac{x^2 - 3x + 2}{2 - x} = \frac{(x-2)(x-1)}{2-x} = \frac{(x-2)(x-1)}{-(x-2)} = \frac{x-1}{-1} = 1 - x.$$

THE ALGEBRAIC SUM OF FRACTIONS having a *common denominator* is a fraction whose numerator is the algebraic sum of the numerators of the given fractions and whose denominator is the common denominator.

Examples. $\dfrac{3}{5} - \dfrac{4}{5} - \dfrac{2}{5} + \dfrac{1}{5} = \dfrac{3 - 4 - 2 + 1}{5} = \dfrac{-2}{5} = -\dfrac{2}{5}$

$$\frac{2}{x-3} - \frac{3x+4}{x-3} + \frac{x^2+5}{x-3} = \frac{2 - (3x+4) + (x^2+5)}{x-3} = \frac{x^2 - 3x + 3}{x-3}$$

TO ADD AND SUBTRACT FRACTIONS having *different denominators*, write each of the given fractions as equivalent fractions all having a common denominator.

The *least common denominator* (L.C.D.) of a given set of fractions is the L.C.M. of the denominators of the fractions.

Thus the L.C.D. of $\dfrac{3}{4}, \dfrac{4}{5}, \dfrac{7}{10}$ is the L.C.M. of 4, 5, 10 which is 20, and

the L.C.D. of $\dfrac{2}{x^2}, \dfrac{3}{2x}, \dfrac{x}{7}$ is $14x^2$.

Examples. $\dfrac{3}{4} - \dfrac{4}{5} + \dfrac{7}{10} = \dfrac{15}{20} - \dfrac{16}{20} + \dfrac{14}{20} = \dfrac{15 - 16 + 14}{20} = \dfrac{13}{20}$

$$\frac{2}{x^2} - \frac{3}{2x} - \frac{x}{7} = \frac{2(14) - 3(7x) - x(2x^2)}{14x^2} = \frac{28 - 21x - 2x^3}{14x^2}$$

$$\frac{2x+1}{x(x+2)} - \frac{3}{(x+2)(x-1)} = \frac{(2x+1)(x-1) - 3x}{x(x+2)(x-1)} = \frac{2x^2 - 4x - 1}{x(x+2)(x-1)}$$

THE PRODUCT of two or more given fractions produces a fraction whose numerator is the product of the numerators of the given fractions and whose denominator is the product of the denominators of the given fractions.

Examples. $\dfrac{2}{3} \cdot \dfrac{4}{5} \cdot \dfrac{15}{16} = \dfrac{2 \cdot 4 \cdot 15}{3 \cdot 5 \cdot 16} = \dfrac{1}{2}$

$$\frac{x^2 - 9}{x^2 - 6x + 5} \cdot \frac{x-5}{x+3} = \frac{(x+3)(x-3)}{(x-5)(x-1)} \cdot \frac{x-5}{x+3}$$

$$= \frac{(x+3)(x-3)(x-5)}{(x-5)(x-1)(x+3)} = \frac{x-3}{x-1}$$

THE QUOTIENT of two given fractions is obtained by inverting the divisor and then multiplying.

Examples.

$$\frac{3}{8} \div \frac{5}{4} \quad \text{or} \quad \frac{3/8}{5/4} = \frac{3}{8} \cdot \frac{4}{5} = \frac{3}{10}$$

$$\frac{7}{x^2 - 4} \div \frac{xy}{x+2} = \frac{7}{(x+2)(x-2)} \cdot \frac{x+2}{xy} = \frac{7}{xy(x-2)}$$

A COMPLEX FRACTION is one which has one or more fractions in the numerator or denominator, or in both. To simplify a complex fraction:
1) Reduce the numerator and denominator to simple fractions.
2) Divide the two resulting fractions.

Example.

$$\frac{x - \dfrac{1}{x}}{1 + \dfrac{1}{x}} = \frac{\dfrac{x^2 - 1}{x}}{\dfrac{x+1}{x}} = \frac{x^2-1}{x} \cdot \frac{x}{x+1} = \frac{x^2-1}{x+1} = x-1$$

SOLVED PROBLEMS

REDUCTION OF FRACTIONS TO LOWEST TERMS

1. *a)* $\dfrac{15x^2}{12xy} = \dfrac{3 \cdot 5 \cdot x \cdot x}{3 \cdot 4 \cdot x \cdot y} = \dfrac{5x}{4y}$

c) $\dfrac{14a^3b^3c^2}{-7a^2b^4c^2} = -\dfrac{2a}{b}$

b) $\dfrac{4x^2y}{18xy^3} = \dfrac{2 \cdot 2 \cdot x \cdot x \cdot y}{2 \cdot 9 \cdot x \cdot y \cdot y^2} = \dfrac{2x}{9y^2}$

d) $\dfrac{8x - 8y}{16x - 16y} = \dfrac{8(x-y)}{16(x-y)} = \dfrac{1}{2}$ (where $x-y \neq 0$)

e) $\dfrac{x^3y - y^3x}{x^2y - xy^2} = \dfrac{xy(x^2 - y^2)}{xy(x-y)} = \dfrac{xy(x-y)(x+y)}{xy(x-y)} = x + y$

f) $\dfrac{x^2 - 4xy + 3y^2}{y^2 - x^2} = \dfrac{(x-3y)(x-y)}{(y-x)(y+x)} = -\dfrac{(x-3y)(x-y)}{(x-y)(y+x)} = -\dfrac{x-3y}{y+x} = \dfrac{3y-x}{y+x}$

g) $\dfrac{6x^2 - 3xy}{-4x^2y + 2xy^2} = \dfrac{3x(2x-y)}{2xy(y-2x)} = -\dfrac{3x(2x-y)}{2xy(2x-y)} = -\dfrac{3}{2y}$

h) $\dfrac{r^3s + 3r^2s + 9rs}{r^3 - 27} = \dfrac{rs(r^2 + 3r + 9)}{r^3 - 3^3} = \dfrac{rs(r^2 + 3r + 9)}{(r-3)(r^2 + 3r + 9)} = \dfrac{rs}{r-3}$

i) $\dfrac{(8xy + 4y^2)^2}{8x^3y + y^4} = \dfrac{(4y[2x+y])^2}{y(8x^3 + y^3)} = \dfrac{16y^2(2x+y)^2}{y(2x+y)(4x^2 - 2xy + y^2)} = \dfrac{16y(2x+y)}{4x^2 - 2xy + y^2}$

j) $\dfrac{x^{2n+1} - x^{2n}y}{x^{n+3} - x^ny^3} = \dfrac{x^{2n}(x-y)}{x^n(x^3 - y^3)} = \dfrac{x^{2n}(x-y)}{x^n(x-y)(x^2 + xy + y^2)} = \dfrac{x^n}{x^2 + xy + y^2}$

MULTIPLICATION OF FRACTIONS

2. a) $\dfrac{2x}{3y^2} \cdot \dfrac{6y}{x^2} = \dfrac{12xy}{3x^2y^2} = \dfrac{4}{xy}$

 b) $\dfrac{9}{3x+3} \cdot \dfrac{x^2-1}{6} = \dfrac{9}{3(x+1)} \cdot \dfrac{(x+1)(x-1)}{6} = \dfrac{x-1}{2}$

 c) $\dfrac{x^2-4}{xy^2} \cdot \dfrac{2xy}{x^2-4x+4} = \dfrac{(x+2)(x-2)}{xy^2} \cdot \dfrac{2xy}{(x-2)^2} = \dfrac{2(x+2)}{y(x-2)}$

 d) $\dfrac{6x-12}{4xy+4x} \cdot \dfrac{y^2-1}{2-3x+x^2} = \dfrac{6(x-2)}{4x(y+1)} \cdot \dfrac{(y+1)(y-1)}{(2-x)(1-x)}$

 $= -\dfrac{6(x-2)(y+1)(y-1)}{4x(y+1)(x-2)(1-x)} = -\dfrac{3(y-1)}{2x(1-x)} = \dfrac{3(y-1)}{2x(x-1)}$

 e) $\left(\dfrac{ax+ab+cx+bc}{a^2-x^2}\right)\left(\dfrac{x^2-2ax+a^2}{x^2+(b+a)x+ab}\right) = \dfrac{(a+c)(x+b)}{(a-x)(a+x)} \cdot \dfrac{(x-a)(x-a)}{(x+a)(x+b)}$

 $= -\dfrac{(a+c)(x+b)}{(x-a)(a+x)} \cdot \dfrac{(x-a)(x-a)}{(x+a)(x+b)} = -\dfrac{(a+c)(x-a)}{(x+a)^2} = \dfrac{(a+c)(a-x)}{(x+a)^2}$

DIVISION OF FRACTIONS

3. a) $\dfrac{5}{4} \div \dfrac{3}{11} = \dfrac{5}{4} \cdot \dfrac{11}{3} = \dfrac{55}{12}$

 c) $\dfrac{3x}{2} \div \dfrac{6x^2}{4} = \dfrac{3x}{2} \cdot \dfrac{4}{6x^2} = \dfrac{1}{x}$

 b) $\dfrac{9}{7} \div \dfrac{4}{7} = \dfrac{9}{7} \cdot \dfrac{7}{4} = \dfrac{9}{4}$

 d) $\dfrac{10xy^2}{3z} \div \dfrac{5xy}{6z^3} = \dfrac{10xy^2}{3z} \cdot \dfrac{6z^3}{5xy} = 4yz^2$

 e) $\dfrac{x+2xy}{3x^2} \div \dfrac{2y+1}{6x} = \dfrac{x+2xy}{3x^2} \cdot \dfrac{6x}{2y+1} = \dfrac{x(1+2y)}{3x^2} \cdot \dfrac{6x}{(2y+1)} = 2$

 f) $\dfrac{9-x^2}{x^4+6x^3} \div \dfrac{x^3-2x^2-3x}{x^2+7x+6} = \dfrac{9-x^2}{x^4+6x^3} \cdot \dfrac{x^2+7x+6}{x^3-2x^2-3x}$

 $= \dfrac{(3-x)(3+x)}{x^3(x+6)} \cdot \dfrac{(x+1)(x+6)}{x(x-3)(x+1)} = -\dfrac{3+x}{x^4}$

 g) $\dfrac{2x^2-5x+2}{\dfrac{2x-1}{3}} = (2x^2-5x+2) \cdot \dfrac{3}{2x-1} = (2x-1)(x-2) \cdot \dfrac{3}{2x-1} = 3(x-2)$

 h) $\dfrac{\dfrac{x^2-5x+6}{x^2+7x-8}}{\dfrac{9-x^2}{64-x^2}} = \dfrac{x^2-5x+6}{x^2+7x-8} \cdot \dfrac{64-x^2}{9-x^2} = \dfrac{(x-3)(x-2)}{(x+8)(x-1)} \cdot \dfrac{(8-x)(8+x)}{(3-x)(3+x)} = -\dfrac{(x-2)(8-x)}{(x-1)(3+x)}$

ADDITION AND SUBTRACTION OF FRACTIONS

4. a) $\dfrac{1}{3} + \dfrac{1}{6} = \dfrac{2}{6} + \dfrac{1}{6} = \dfrac{3}{6} = \dfrac{1}{2}$

  d) $\dfrac{3t^2}{5} - \dfrac{4t^2}{15} = \dfrac{3t^2(3) - 4t^2(1)}{15} = \dfrac{5t^2}{15} = \dfrac{t^2}{3}$

  b) $\dfrac{5}{18} + \dfrac{7}{24} = \dfrac{5(4)}{72} + \dfrac{7(3)}{72} = \dfrac{41}{72}$

  e) $\dfrac{1}{x} + \dfrac{1}{y} = \dfrac{y + x}{xy}$

  c) $\dfrac{x}{6} + \dfrac{5x}{21} = \dfrac{x(7) + 5x(2)}{42} = \dfrac{17x}{42}$

  f) $\dfrac{3}{x} + \dfrac{4}{3y} = \dfrac{3(3y) + 4(x)}{3xy} = \dfrac{9y + 4x}{3xy}$

  g) $\dfrac{5}{2x} - \dfrac{3}{4x^2} = \dfrac{5(2x) - 3(1)}{4x^2} = \dfrac{10x - 3}{4x^2}$

  h) $\dfrac{3a}{bc} + \dfrac{2b}{ac} = \dfrac{3a(a) + 2b(b)}{abc} = \dfrac{3a^2 + 2b^2}{abc}$

  i) $\dfrac{3t - 1}{10} + \dfrac{5 - 2t}{15} = \dfrac{(3t - 1)3 + (5 - 2t)2}{30} = \dfrac{9t - 3 + 10 - 4t}{30} = \dfrac{5t + 7}{30}$

  j) $\dfrac{3}{x} - \dfrac{2}{x + 1} + \dfrac{2}{x^2} = \dfrac{3x(x + 1) - 2x^2 + 2(x + 1)}{x^2(x + 1)} = \dfrac{x^2 + 5x + 2}{x^2(x + 1)}$

  k) $5 - \dfrac{5}{x + 3} + \dfrac{10}{x^2 - 9} = \dfrac{5(x^2 - 9) - 5(x - 3) + 10}{x^2 - 9} = \dfrac{5(x^2 - x - 4)}{x^2 - 9}$

  l) $\dfrac{3}{y - 2} - \dfrac{2}{y + 2} - \dfrac{y}{y^2 - 4} = \dfrac{3(y + 2) - 2(y - 2) - y}{y^2 - 4} = \dfrac{10}{y^2 - 4}$

  m) $\dfrac{5}{2s + 4} - \dfrac{3}{s^2 + 3s + 2} + \dfrac{s}{s^2 - s - 2} = \dfrac{5}{2(s + 2)} - \dfrac{3}{(s + 1)(s + 2)} + \dfrac{s}{(s - 2)(s + 1)}$

   $= \dfrac{5(s + 1)(s - 2) - 3(2)(s - 2) + s(2)(s + 2)}{2(s + 2)(s + 1)(s - 2)} = \dfrac{7s^2 - 7s + 2}{2(s + 2)(s + 1)(s - 2)}$

  n) $\dfrac{3x - 6}{4x^2 + 12x - 16} - \dfrac{2x - 5}{6x^2 - 6} + \dfrac{3x^2 + 3}{8x^2 + 40x + 32} = \dfrac{3x - 6}{4(x + 4)(x - 1)} - \dfrac{2x - 5}{6(x + 1)(x - 1)} + \dfrac{3x^2 + 3}{8(x + 4)(x + 1)}$

   $= \dfrac{(3x - 6)(6)(x + 1) - (2x - 5)(4)(x + 4) + (3x^2 + 3)(3)(x - 1)}{24(x + 4)(x - 1)(x + 1)} = \dfrac{9x^3 + x^2 - 21x + 35}{24(x + 4)(x - 1)(x + 1)}$

COMPLEX FRACTIONS

5. a) $\dfrac{5/7}{3/4} = \dfrac{5}{7} \cdot \dfrac{4}{3} = \dfrac{20}{21}$

  b) $\dfrac{2/3}{7} = \dfrac{2}{3} \cdot \dfrac{1}{7} = \dfrac{2}{21}$

  c) $\dfrac{10}{5/6} = 10 \cdot \dfrac{6}{5} = \dfrac{60}{5} = 12$

  d) $\dfrac{\frac{2}{3} + \frac{5}{6}}{\frac{3}{8}} = \dfrac{\frac{4}{6} + \frac{5}{6}}{\frac{3}{8}} = \dfrac{\frac{9}{6}}{\frac{3}{8}} = \dfrac{9}{6} \cdot \dfrac{8}{3} = 4$

  f) $\dfrac{\frac{2}{a - b}}{a - b} = \dfrac{2}{a - b} \cdot \dfrac{1}{a - b} = \dfrac{2}{(a - b)^2}$

  g) $\dfrac{\frac{2a}{a}}{x + 1} = 2a \cdot \dfrac{x + 1}{a} = 2(x + 1)$

  e) $\dfrac{\frac{x + y}{3x^2}}{\frac{x - y}{x}} = \dfrac{x + y}{3x^2} \cdot \dfrac{x}{x - y} = \dfrac{x + y}{3x(x - y)}$

h) $\dfrac{\dfrac{a+b}{a-b} - \dfrac{a-b}{a+b}}{1 + \dfrac{a-b}{a+b}} = \dfrac{\dfrac{(a+b)^2 - (a-b)^2}{(a-b)(a+b)}}{\dfrac{(a+b) + (a-b)}{a+b}} = \dfrac{\dfrac{4ab}{(a-b)(a+b)}}{\dfrac{2a}{a+b}} = \dfrac{4ab}{(a-b)(a+b)} \cdot \dfrac{a+b}{2a} = \dfrac{2b}{a-b}$

i) $\dfrac{\dfrac{2}{x+h-3} - \dfrac{2}{x-3}}{h} = \dfrac{\dfrac{2(x-3) - 2(x+h-3)}{(x+h-3)(x-3)}}{h} = \dfrac{\dfrac{-2h}{(x+h-3)(x-3)}}{h} = \dfrac{-2}{(x+h-3)(x-3)}$

j) $3y + \dfrac{1 + \dfrac{2}{y}}{\dfrac{y+2}{y-2}} = 3y + \dfrac{\dfrac{y+2}{y}}{\dfrac{y+2}{y-2}} = 3y + \left(\dfrac{y+2}{y}\right)\left(\dfrac{y-2}{y+2}\right) = 3y + \dfrac{y-2}{y} = \dfrac{3y^2 + y - 2}{y}$

k) $\dfrac{1}{1 - \dfrac{1}{1 + \dfrac{1}{x}}} = \dfrac{1}{1 - \dfrac{1}{\dfrac{x+1}{x}}} = \dfrac{1}{1 - \dfrac{x}{x+1}} = \dfrac{1}{\dfrac{x+1-x}{x+1}} = \dfrac{1}{\dfrac{1}{x+1}} = x+1$

l) $\dfrac{a}{a - b + \dfrac{a+b}{\dfrac{a}{b} - \dfrac{b}{a}}} = \dfrac{a}{a - b + \dfrac{a+b}{\dfrac{a^2-b^2}{ab}}} = \dfrac{a}{a - b + \dfrac{ab(a+b)}{(a+b)(a-b)}} = \dfrac{a}{(a-b) + \dfrac{ab}{a-b}}$

$= \dfrac{a}{\dfrac{(a-b)^2 + ab}{a-b}} = \dfrac{a}{\dfrac{a^2 - ab + b^2}{a-b}} = \dfrac{a(a-b)}{a^2 - ab + b^2}$

m) $1 - \dfrac{1}{2 - \dfrac{1}{3 - \dfrac{2a-1}{2a+1}}} = 1 - \dfrac{1}{2 - \dfrac{1}{\dfrac{3(2a+1) - (2a-1)}{2a+1}}} = 1 - \dfrac{1}{2 - \dfrac{2a+1}{4a+4}}$

$= 1 - \dfrac{1}{\dfrac{2(4a+4) - (2a+1)}{4a+4}} = 1 - \dfrac{4a+4}{6a+7} = \dfrac{6a + 7 - (4a+4)}{6a+7} = \dfrac{2a+3}{6a+7}$

SUPPLEMENTARY PROBLEMS

Show that:

6. a) $\dfrac{24x^3y^2}{18xy^3} = \dfrac{4x^2}{3y}$ d) $\dfrac{4x^2 - 16}{x^2 - 2x} = \dfrac{4(x+2)}{x}$ g) $\dfrac{ax^4 - a^2x^3 - 6a^3x^2}{9a^4x - a^2x^3} = -\dfrac{x(x+2a)}{a(x+3a)}$

b) $\dfrac{36\,x\,y^4 z^2}{-15x^4 y^3 z} = \dfrac{-12yz}{5x^3}$ e) $\dfrac{y^2 - 5y + 6}{4 - y^2} = \dfrac{3-y}{y+2}$ h) $\dfrac{xy - y^2}{x^4 y - xy^4} = \dfrac{1}{x(x^2 + xy + y^2)}$

c) $\dfrac{5a^2 - 10ab}{a - 2b} = 5a$ f) $\dfrac{(x^2 + 4x)^2}{x^2 + 6x + 8} = \dfrac{x^2(x+4)}{x+2}$ i) $\dfrac{3a^2}{4b^3} \cdot \dfrac{2b^4}{9a^3} = \dfrac{b}{6a}$

7. a) $\dfrac{8xyz^2}{3x^3y^2z} \cdot \dfrac{9xy^2z}{4xz^5} = \dfrac{6y}{x^2z^3}$

b) $\dfrac{xy^2}{2x-2y} \cdot \dfrac{x^2-y^2}{x^3y^2} = \dfrac{x+y}{2x^2}$

c) $\dfrac{x^2+3x}{4x^2-4} \cdot \dfrac{2x^2+2x}{x^2-9} \cdot \dfrac{x^2-4x+3}{x^2} = \dfrac{1}{2}$

d) $\dfrac{x^2-4y^2}{3xy+3x} \cdot \dfrac{2y^2-2}{2y^2+xy-x^2} = -\dfrac{2(x+2y)(y-1)}{3x(x+y)}$

e) $\dfrac{y^2-y-6}{y^2-2y+1} \cdot \dfrac{y^2+3y-4}{9y-y^3} = -\dfrac{(y+2)(y+4)}{y(y-1)(y+3)}$

f) $\dfrac{t^3+3t^2+t+3}{4t^2-16t+16} \cdot \dfrac{8-t^3}{t^3+t} = \dfrac{(t+3)(t^2+2t+4)}{4t(2-t)}$

8. a) $\dfrac{3x}{8y} \div \dfrac{9x}{16y} = \dfrac{2}{3}$ b) $\dfrac{24x^3y^2}{5z^2} \div \dfrac{8x^2y^3}{15z^4} = \dfrac{9xz^2}{y}$ c) $\dfrac{x^2-4y^2}{x^2+xy} \div \dfrac{x^2-xy-6y^2}{y^2+xy} = \dfrac{y(x-2y)}{x(x-3y)}$

9. a) $\dfrac{\dfrac{6x^2-x-2}{3x-2}}{2x+1} = (2x+1)^2$ b) $\dfrac{\dfrac{y^2-3y+2}{y^2+4y-21}}{\dfrac{4-4y+y^2}{9-y^2}} = -\dfrac{(y-1)(y+3)}{(y-2)(y+7)}$ c) $\dfrac{\dfrac{x^2y+xy^2}{x-y}}{x+y} = \dfrac{xy}{x-y}$

10. a) $\dfrac{2x}{3} - \dfrac{x}{2} = \dfrac{x}{6}$

e) $\dfrac{1}{x+2} + \dfrac{1}{x-2} - \dfrac{x}{x^2-4} = \dfrac{x}{x^2-4}$

b) $\dfrac{4}{3x} - \dfrac{5}{4x} = \dfrac{1}{12x}$

f) $\dfrac{r-1}{r^2+r-6} - \dfrac{r+2}{r^2+4r+3} + \dfrac{1}{3r-6} = \dfrac{r^2+4r+12}{3(r+3)(r-2)(r+1)}$

c) $\dfrac{3}{2y^2} - \dfrac{8}{y} = \dfrac{3-16y}{2y^2}$

g) $\dfrac{x}{2x^2+3xy+y^2} - \dfrac{x-y}{y^2-4x^2} + \dfrac{y}{2x^2+xy-y^2} = \dfrac{3x^2+xy}{(2x+y)(2x-y)(x+y)}$

d) $\dfrac{x+y^2}{x^2} + \dfrac{x-1}{x} - 1 = \dfrac{y^2}{x^2}$

h) $\dfrac{a}{(c-a)(a-b)} + \dfrac{b}{(a-b)(b-c)} + \dfrac{c}{(b-c)(c-a)} = 0$

11. a) $\dfrac{x+y}{\dfrac{1}{x}+\dfrac{1}{y}} = xy$

d) $\dfrac{\dfrac{x+1}{x-1} - \dfrac{x-1}{x+1}}{\dfrac{1}{x+1} + \dfrac{1}{x-1}} = 2$

b) $\dfrac{2+\dfrac{1}{x}}{2x^2+x} = \dfrac{1}{x^2}$

e) $\dfrac{x}{1 - \dfrac{1}{1+\dfrac{x}{y}}} = x+y$

c) $\dfrac{y+\dfrac{2y}{y-2}}{1+\dfrac{4}{y^2-4}} = y+2$

f) $2 - \dfrac{2}{1 - \dfrac{2}{2-\dfrac{2}{x^2}}} = 2x^2$

CHAPTER 6

Exponents

POSITIVE INTEGRAL EXPONENT. If n is a positive integer, a^n represents the product of n factors each of which is a. Thus $a^4 = a \cdot a \cdot a \cdot a$. In a^n, a is called the base and n the exponent or index. We may read a^n as the "nth power of a" or "a to the nth". If $n = 2$ we read a^2 as "a squared"; a^3 is read "a cubed".

Examples. $x^3 = x \cdot x \cdot x$, $2^5 = 2 \cdot 2 \cdot 2 \cdot 2 \cdot 2 = 32$, $(-3)^3 = (-3)(-3)(-3) = -27$

NEGATIVE INTEGRAL EXPONENT. If n is a positive integer, we define

$$a^{-n} = \frac{1}{a^n} \qquad \text{assuming } a \neq 0.$$

Examples. $2^{-4} = \frac{1}{2^4} = \frac{1}{16}$, $\frac{1}{3^{-3}} = 3^3 = 27$, $-4x^{-2} = \frac{-4}{x^2}$, $(a+b)^{-1} = \frac{1}{(a+b)}$

ROOTS. If n is a positive integer and if a and b are such that $a^n = b$, then a is said to be an nth root of b.

If b is positive, there is only one positive number a such that $a^n = b$. We write this positive number $\sqrt[n]{b}$ and call it the *principal* nth root of b.

Example 1. $\sqrt[4]{16}$ is that positive number which when raised to the 4th power yields 16. Clearly this is $+2$ and so we write $\sqrt[4]{16} = +2$.

Example 2. The number -2 when raised to the 4th power also yields 16. We call -2 a 4th root of 16 but not the principal 4th root of 16.

If b is negative, there is no positive nth root of b, but there is a negative nth root of b if n is odd. We call this negative number the principal nth root of b and we write it $\sqrt[n]{b}$.

Example 3. $\sqrt[3]{-27}$ is that number which raised to the third power (or cubed) yields -27. Clearly this is -3 and so we write $\sqrt[3]{-27} = -3$ as the principal cube root of -27.

Example 4. If n is even, as in $\sqrt[4]{-16}$, there is no principal nth root in terms of real numbers.

Note. In advanced mathematics it can be shown that there are exactly n values of a such that $a^n = b$, $b \neq 0$, provided we allow imaginary (or complex) numbers.

POSITIVE FRACTIONAL EXPONENT. If m and n are positive integers we define

$$a^{m/n} = \sqrt[n]{a^m} \qquad \text{(assume } a \geqq 0 \text{ if } n \text{ is even)}$$

Examples. $4^{3/2} = \sqrt{4^3} = \sqrt{64} = 8$, $\quad (27)^{2/3} = \sqrt[3]{(27)^2} = 9$

NEGATIVE FRACTIONAL EXPONENT. If m and n are positive integers we define

$$a^{-m/n} = \frac{1}{a^{m/n}}$$

Examples. $8^{-2/3} = \dfrac{1}{8^{2/3}} = \dfrac{1}{\sqrt[3]{8^2}} = \dfrac{1}{\sqrt[3]{64}} = \dfrac{1}{4}$, $\qquad x^{-5/2} = \dfrac{1}{x^{5/2}} = \dfrac{1}{\sqrt{x^5}}$

ZERO EXPONENT. We define $a^0 = 1$ if $a \neq 0$.

Examples. $10^0 = 1$, $\quad (-3)^0 = 1$, $\quad (ax)^0 = 1$ (if $ax \neq 0$)

GENERAL LAWS OF EXPONENTS. If p and q are real numbers, the following laws hold.

A) $$a^p \cdot a^q = a^{p+q}$$

Examples. $2^3 \cdot 2^2 = 2^{3+2} = 2^5$, $\quad 5^{-3} \cdot 5^7 = 5^{-3+7} = 5^4$, $\quad 2^{1/2} \cdot 2^{5/2} = 2^3 = 8$

$3^{1/3} \cdot 3^{1/6} = 3^{1/3 + 1/6} = 3^{1/2} = \sqrt{3}$, $\quad 3^9 \cdot 3^{-2} \cdot 3^{-3} = 3^4 = 81$

B) $$(a^p)^q = a^{pq}$$

Examples. $(2^4)^3 = 2^{12}$, $\quad (5^{1/3})^{-3} = 5^{(1/3)(-3)} = 5^{-1} = 1/5$, $\quad (3^2)^0 = 3^0 = 1$

$(x^5)^{-4} = x^{-20}$, $\quad (a^{2/3})^{3/4} = a^{(2/3)(3/4)} = a^{1/2}$

C) $$\frac{a^p}{a^q} = a^{p-q} \qquad a \neq 0$$

Examples. $\dfrac{2^6}{2^4} = 2^{6-4} = 2^2 = 4$, $\quad \dfrac{3^{-2}}{3^4} = 3^{-2-4} = 3^{-6}$, $\quad \dfrac{x^{1/2}}{x^{-1}} = x^{1/2 - (-1)} = x^{3/2}$

$\dfrac{(x+15)^{4/3}}{(x+15)^{5/6}} = (x+15)^{4/3 - 5/6} = (x+15)^{1/2} = \sqrt{x+15}$

D) $$(ab)^p = a^p b^p$$

Examples. $(2 \cdot 3)^4 = 2^4 \cdot 3^4$, $\quad (2x)^3 = 2^3 x^3 = 8x^3$, $\quad (3a)^{-2} = 3^{-2} a^{-2} = \dfrac{1}{9a^2}$

$(4x)^{1/2} = 4^{1/2} x^{1/2} = 2x^{1/2} = 2\sqrt{x}$

$E)$ $\qquad\qquad\qquad (\dfrac{a}{b})^p = \dfrac{a^p}{b^p} \qquad b \neq 0$

Examples. $\quad (\dfrac{2}{3})^5 = \dfrac{2^5}{3^5} = \dfrac{32}{243}, \qquad (\dfrac{x^2}{y^3})^{-3} = \dfrac{(x^2)^{-3}}{(y^3)^{-3}} = \dfrac{x^{-6}}{y^{-9}} = \dfrac{y^9}{x^6}$

$\qquad\qquad (\dfrac{5^3}{2^6})^{-1/3} = \dfrac{(5^3)^{-1/3}}{(2^6)^{-1/3}} = \dfrac{5^{-1}}{2^{-2}} = \dfrac{2^2}{5^1} = \dfrac{4}{5}$

SOLVED PROBLEMS

POSITIVE INTEGRAL EXPONENT

1. a) $2^3 = 2 \cdot 2 \cdot 2 = 8$

   b) $(-3)^4 = (-3)(-3)(-3)(-3) = 81$

   c) $(\dfrac{2}{3})^5 = (\dfrac{2}{3})(\dfrac{2}{3})(\dfrac{2}{3})(\dfrac{2}{3})(\dfrac{2}{3}) = \dfrac{32}{243}$

   d) $(3y)^2 (2y)^3 = (3y)(3y)(2y)(2y)(2y) = 72y^5$

   e) $(-3xy^2)^3 = (-3xy^2)(-3xy^2)(-3xy^2)$
   $\qquad\qquad = -27x^3 y^6$

NEGATIVE INTEGRAL EXPONENT

2. a) $2^{-3} = \dfrac{1}{2^3} = \dfrac{1}{8}$

   b) $3^{-1} = \dfrac{1}{3^1} = \dfrac{1}{3}$

   c) $-4(4)^{-2} = -4(\dfrac{1}{4^2}) = -\dfrac{1}{4}$

   d) $-2b^{-2} = -2(\dfrac{1}{b^2}) = -\dfrac{2}{b^2}$

   e) $(-2b)^{-2} = \dfrac{1}{(-2b)^2} = \dfrac{1}{4b^2}$

   f) $5 \cdot 10^{-3} = 5(\dfrac{1}{10^3}) = \dfrac{5}{1000} = \dfrac{1}{200}$

   g) $\dfrac{8}{10^{-2}} = 8 \cdot 10^2 = 800$

   h) $\dfrac{4}{x^{-2} y^{-2}} = 4x^2 y^2$

   i) $(\dfrac{3}{4})^{-3} = \dfrac{1}{(3/4)^3} = (\dfrac{4}{3})^3 = \dfrac{64}{27}$

   j) $(\dfrac{x}{y})^{-3} = \dfrac{1}{(x/y)^3} = (\dfrac{y}{x})^3 = \dfrac{y^3}{x^3}$

   k) $(.02)^{-1} = (\dfrac{2}{100})^{-1} = \dfrac{100}{2} = 50$

   l) $\dfrac{ab^{-4}}{a^{-2}b} = \dfrac{a \cdot a^2}{b \cdot b^4} = \dfrac{a^3}{b^5}$

   m) $\dfrac{x^{2n+1}}{y^{3n-1}} = x^{2n+1} y^{1-3n}$

   n) $\dfrac{(x-1)^{-2}(x+3)^{-1}}{(2x-4)^{-1}(x+5)^{-3}} = \dfrac{(2x-4)(x+5)^3}{(x-1)^2(x+3)}$

POSITIVE FRACTIONAL EXPONENTS

3. a) $(8)^{2/3} = \sqrt[3]{8^2} = \sqrt[3]{64} = 4$

   b) $(-8)^{2/3} = \sqrt[3]{(-8)^2} = \sqrt[3]{64} = 4$

$c)\ (-x^3)^{1/3} = \sqrt[3]{-x^3} = -x$ $d)\ (\frac{1}{16})^{1/2} = \sqrt{\frac{1}{16}} = \frac{1}{4}$ $e)\ (-\frac{1}{8})^{2/3} = \sqrt[3]{(-\frac{1}{8})^2} = \sqrt[3]{\frac{1}{64}} = \frac{1}{4}$

NEGATIVE FRACTIONAL EXPONENT

4. $a)\ x^{-1/3} = \dfrac{1}{x^{1/3}} = \dfrac{1}{\sqrt[3]{x}}$

$d)\ (-x^3)^{-1/3} = \dfrac{1}{(-x^3)^{1/3}} = \dfrac{1}{\sqrt[3]{-x^3}} = \dfrac{1}{-x} = -\dfrac{1}{x}$

$b)\ (8)^{-2/3} = \dfrac{1}{8^{2/3}} = \dfrac{1}{\sqrt[3]{8^2}} = \dfrac{1}{4}$

$e)\ (-1)^{-2/3} = \dfrac{1}{(-1)^{2/3}} = \dfrac{1}{\sqrt[3]{(-1)^2}} = 1$

$c)\ (-8)^{-2/3} = \dfrac{1}{(-8)^{2/3}} = \dfrac{1}{\sqrt[3]{(-8)^2}} = \dfrac{1}{4}$

$f)\ -(1)^{-2/3} = -\dfrac{1}{1^{2/3}} = -1$

$g)\ -(-1)^{-3/5} = -\dfrac{1}{(-1)^{3/5}} = -\dfrac{1}{\sqrt[5]{(-1)^3}} = -\dfrac{1}{\sqrt[5]{-1}} = -\dfrac{1}{-1} = 1$

ZERO EXPONENT

5. $a)\ 7^0 = 1,\quad (-3)^0 = 1,\quad (-2/3)^0 = 1$

$e)\ 4\cdot10^0 = 4\cdot1 = 4$

$b)\ (x-y)^0 = 1,$ if $x-y \neq 0$

$f)\ (4\cdot10)^0 = (40)^0 = 1$

$c)\ 3x^0 = 3\cdot1 = 3,$ if $x \neq 0$

$g)\ -(1)^0 = -1$

$d)\ (3x)^0 = 1,$ if $3x \neq 0,$ i.e. if $x \neq 0$

$h)\ (-1)^0 = 1$

$i)\ (3x)^0(4y)^0 = 1\cdot1 = 1,$ if $3x \neq 0$ and $4y \neq 0,$ i.e. if $x \neq 0,\ y \neq 0$

$j)\ -2(3x+2y-4)^0 = -2(1) = -2,$ if $3x+2y-4 \neq 0$

$k)\ \dfrac{(5x+3y)}{(5x+3y)^0} = \dfrac{5x+3y}{1} = 5x+3y,$ if $5x+3y \neq 0$

$l)\ 4(x^2+y^2)(x^2+y^2)^0 = 4(x^2+y^2)(1) = 4(x^2+y^2),$ if $x^2+y^2 \neq 0$

GENERAL LAWS OF EXPONENTS

6. $a)\ a^p\cdot a^q = a^{p+q}$

$g)\ 10^7\cdot10^{-3} = 10^{7-3} = 10^4$

$b)\ a^3\cdot a^5 = a^{3+5} = a^8$

$h)\ (4\cdot10^{-6})(2\cdot10^4) = 8\cdot10^{4-6} = 8\cdot10^{-2}$

$c)\ 3^4\cdot3^5 = 3^9$

$i)\ a^x\cdot a^y\cdot a^{-z} = a^{x+y-z}$

$d)\ a^{n+1}\cdot a^{n-2} = a^{2n-1}$

$j)\ (\sqrt{x+y})(x+y) = (x+y)^{1/2}(x+y)^1 = (x+y)^{3/2}$

$e)\ x^{1/2}\cdot x^{1/3} = x^{1/2+1/3} = x^{5/6}$

$k)\ 10^{1.7}\cdot10^{2.6} = 10^{4.3}$

$f)\ x^{1/2}\cdot x^{-1/3} = x^{1/2-1/3} = x^{1/6}$

$l)\ 10^{-4.1}\cdot10^{3.5}\cdot10^{-.1} = 10^{-4.1+3.5-.1} = 10^{-.7}$

m) $(\frac{b}{a})^{3/2} \cdot (\frac{b}{a})^{-2/3} = (\frac{b}{a})^{3/2-2/3} = (\frac{b}{a})^{5/6}$

n) $(\frac{x}{x+y})^{-1} (\frac{x}{x+y})^{1/2} = (\frac{x}{x+y})^{-1/2} = (\frac{x+y}{x})^{1/2} = \sqrt{\frac{x+y}{x}}$

o) $(x^2+1)^{-5/2} (x^2+1)^0 (x^2+1)^2 = (x^2+1)^{-5/2+0+2} = (x^2+1)^{-1/2} = \frac{1}{(x^2+1)^{1/2}} = \frac{1}{\sqrt{x^2+1}}$

7. a) $(a^p)^q = a^{pq}$

     f) $(49)^{3/2} = (7^2)^{3/2} = 7^{2 \cdot 3/2} = 7^3 = 343$

b) $(x^3)^4 = x^{3 \cdot 4} = x^{12}$

     g) $(3^{-1/2})^{-2} = 3^1 = 3$

c) $(a^{m+2})^n = a^{(m+2)n} = a^{mn+2n}$

     h) $(u^{-2})^{-3} = u^{(-2)(-3)} = u^6$

d) $(10^3)^2 = 10^{3 \cdot 2} = 10^6$

     i) $(81)^{3/4} = (3^4)^{3/4} = 3^3 = 27$

e) $(10^{-3})^2 = 10^{-3 \cdot 2} = 10^{-6}$

     j) $(\sqrt{x+y})^5 = [(x+y)^{1/2}]^5 = (x+y)^{5/2}$

k) $(\sqrt[3]{x^3+y^3})^6 = [(x^3+y^3)^{1/3}]^6 = (x^3+y^3)^{1/3 \cdot 6} = (x^3+y^3)^2$

l) $\sqrt{\sqrt[6]{\sqrt[3]{a^2}}} = \sqrt[6]{a^{2/3}} = (a^{2/3})^{1/6} = a^{1/9} = \sqrt[9]{a}$

8. a) $\frac{a^p}{a^q} = a^{p-q}$

     g) $\frac{y^{2/3}}{y^{1/3}} = y^{2/3-1/3} = y^{1/3}$

b) $\frac{a^5}{a^3} = a^{5-3} = a^2$

     h) $\frac{z^{1/2}}{z^{3/4}} = z^{1/2-3/4} = z^{-1/4}$

c) $\frac{7^4}{7^3} = 7^{4-3} = 7^1 = 7$

     i) $\frac{(x+y)^{3a+1}}{(x+y)^{2a+5}} = (x+y)^{a-4}$

d) $\frac{p^{2n+3}}{p^{n+1}} = p^{(2n+3)-(n+1)} = p^{n+2}$

     j) $\frac{8 \cdot 10^2}{2 \cdot 10^{-6}} = \frac{8}{2} \cdot 10^{2+6} = 4 \cdot 10^8$

e) $\frac{10^2}{10^5} = 10^{2-5} = 10^{-3}$

     k) $\frac{9 \cdot 10^{-2}}{3 \cdot 10^4} = \frac{9}{3} \cdot 10^{-2-4} = 3 \cdot 10^{-6}$

f) $\frac{x^{m+3}}{x^{m-1}} = x^4$

     l) $\frac{a^3 b^{-1/2}}{a \, b^{-3/2}} = a^2 b^1 = a^2 b$

m) $\frac{4x^3 y^{-2} z^{-3/2}}{2x^{-1/2} y^{-4} z} = (\frac{4}{2}) x^{3+1/2} y^{-2+4} z^{-3/2-1} = 2x^{7/2} y^2 z^{-5/2}$

n) $\frac{8 \sqrt[3]{x^2} \sqrt[4]{y} \sqrt{1/z}}{-2 \sqrt[3]{x} \sqrt{y^5} \sqrt{z}} = \frac{8x^{2/3} y^{1/4} z^{-1/2}}{-2x^{1/3} y^{5/2} z^{1/2}} = -4x^{1/3} y^{-9/4} z^{-1}$

9. a) $(ab)^p = a^p b^p$

     c) $(3 \cdot 10^2)^4 = 3^4 \cdot 10^8 = 81 \cdot 10^8$

b) $(2a)^4 = 2^4 a^4 = 16a^4$

     d) $(4x^8 y^4)^{1/2} = 4^{1/2} (x^8)^{1/2} (y^4)^{1/2} = 2x^4 y^2$

e) $\sqrt[3]{64a^{12}b^6} = (64a^{12}b^6)^{1/3} = (64)^{1/3}(a^{12})^{1/3}(b^6)^{1/3} = 4a^4b^2$

f) $(x^{2n}y^{-1/2}z^{n-1})^2 = x^{4n}y^{-1}z^{2n-2}$

g) $(27x^{3p}y^{6q}z^{12r})^{1/3} = (27)^{1/3}(x^{3p})^{1/3}(y^{6q})^{1/3}(z^{12r})^{1/3} = 3x^py^{2q}z^{4r}$

10. *a)* $\left(\dfrac{a}{b}\right)^p = \dfrac{a^p}{b^p}$

e) $\left(\dfrac{a^{m+1}}{b}\right)^m = \dfrac{a^{m^2+m}}{b^m}$

b) $\left(\dfrac{2}{3}\right)^4 = \dfrac{2^4}{3^4} = \dfrac{16}{81}$

f) $\left(\dfrac{a^2}{b^4}\right)^{3/2} = \dfrac{(a^2)^{3/2}}{(b^4)^{3/2}} = \dfrac{a^3}{b^6}$ (where $a \geqq 0$, $b \neq 0$)

c) $\left(\dfrac{3a}{4b}\right)^3 = \dfrac{(3a)^3}{(4b)^3} = \dfrac{27a^3}{64b^3}$

g) $\left(\dfrac{2}{5}\right)^{-3} = \left(\dfrac{5}{2}\right)^3 = \dfrac{125}{8}$

d) $\left(\dfrac{x^2}{y^3}\right)^n = \dfrac{x^{2n}}{y^{3n}}$

h) $\left(\dfrac{5^3}{2^6}\right)^{-1/3} = \left(\dfrac{2^6}{5^3}\right)^{1/3} = \dfrac{2^2}{5} = \dfrac{4}{5}$

i) $\sqrt[3]{\dfrac{8x^{3n}}{27y^6}} = \left(\dfrac{8x^{3n}}{27y^6}\right)^{1/3} = \dfrac{(8x^{3n})^{1/3}}{(27y^6)^{1/3}} = \dfrac{8^{1/3}x^n}{27^{1/3}y^2} = \dfrac{2x^n}{3y^2}$

j) $\left(\dfrac{a^{1/3}}{x^{1/3}}\right)^{3/2} = \dfrac{(a^{1/3})^{3/2}}{(x^{1/3})^{3/2}} = \dfrac{a^{1/2}}{x^{1/2}}$

k) $\left(\dfrac{x^{-1/3}y^{-2}}{z^{-4}}\right)^{-3/2} = \dfrac{(x^{-1/3}y^{-2})^{-3/2}}{(z^{-4})^{-3/2}} = \dfrac{(x^{-1/3})^{-3/2}(y^{-2})^{-3/2}}{z^6} = \dfrac{x^{1/2}y^3}{z^6}$

l) $\sqrt{\dfrac{\sqrt[5]{x}\sqrt[4]{y^3}}{\sqrt[3]{z^2}}} = \sqrt{\dfrac{x^{1/5}y^{3/4}}{z^{2/3}}} = \left(\dfrac{x^{1/5}y^{3/4}}{z^{2/3}}\right)^{1/2} = \dfrac{x^{1/10}y^{3/8}}{z^{1/3}}$

MISCELLANEOUS EXAMPLES

11. *a)* $2^3 + 2^2 + 2^1 + 2^0 + 2^{-1} + 2^{-2} + 2^{-3} = 8 + 4 + 2 + 1 + \dfrac{1}{2} + \dfrac{1}{4} + \dfrac{1}{8} = 15\dfrac{7}{8}$

b) $4^{3/2} + 4^{1/2} + 4^{-1/2} + 4^{-3/2} = 8 + 2 + \dfrac{1}{2} + \dfrac{1}{8} = 10\dfrac{5}{8}$

c) $\dfrac{4x^0}{2^{-4}} = 4(1)(2^4) = 4 \cdot 16 = 64$

d) $10^4 + 10^3 + 10^2 + 10^1 + 10^0 + 10^{-1} + 10^{-2} = 10,000 + 1,000 + 100 + 10 + 1 + .1 + .01$
$$= 11,111.11$$

e) $3 \cdot 10^3 + 5 \cdot 10^2 + 2 \cdot 10^1 + 4 \cdot 10^0 = 3524$

f) $\dfrac{4^{3n}}{2^n} = \dfrac{(2^2)^{3n}}{2^n} = \dfrac{2^{6n}}{2^n} = 2^{6n-n} = 2^{5n}$

g) $(0.125)^{1/3} (0.25)^{-1/2} = \dfrac{\sqrt[3]{0.125}}{\sqrt{0.25}} = \dfrac{.5}{.5} = 1$

12. a) Evaluate $4x^{-2/3} + 3x^{1/3} + 2x^0$ when $x = 8$.

$4 \cdot 8^{-2/3} + 3 \cdot 8^{1/3} + 2 \cdot 8^0 = \dfrac{4}{8^{2/3}} + 3 \cdot 8^{1/3} + 2 \cdot 8^0 = \dfrac{4}{4} + 3 \cdot 2 + 2 \cdot 1 = 9$

b) Evaluate $\dfrac{(-3)^2 (-2x)^{-3}}{(x+1)^{-2}}$ when $x = 2$. $\dfrac{(-3)^2 (-4)^{-3}}{3^{-2}} = \dfrac{9(\frac{1}{-4})^3}{\frac{1}{3^2}} = 9(-\frac{1}{4})^3(9) = -\dfrac{81}{64}$

13. a) $\dfrac{2^0 - 2^{-2}}{2 - 2(2)^{-2}} = \dfrac{1 - 1/2^2}{2 - 2/2^2} = \dfrac{1 - 1/4}{2 - 2/4} = \dfrac{3/4}{6/4} = \dfrac{1}{2}$

b) $\dfrac{2a^{-1} + a^0}{a^{-2}} = \dfrac{\frac{2}{a} + 1}{\frac{1}{a^2}} = \dfrac{\frac{2+a}{a}}{\frac{1}{a^2}} = \dfrac{2+a}{a} \cdot a^2 = (2+a)a = 2a + a^2$

c) $(\dfrac{2^0}{8^{1/3}})^{-1} = (\dfrac{1}{2})^{-1} = \dfrac{1}{1/2} = 2$ or $(\dfrac{2^0}{8^{1/3}})^{-1} = (\dfrac{8^{1/3}}{2^0})^1 = \dfrac{2}{1} = 2$

d) $(\dfrac{1}{3})^{-2} - (-3)^{-2} = (3)^2 - (-\dfrac{1}{3})^2 = 9 - \dfrac{1}{9} = \dfrac{80}{9}$

e) $(-\dfrac{1}{27})^{-2/3} + (-\dfrac{1}{32})^{2/5} = (-27)^{2/3} + (-\dfrac{1}{2^5})^{2/5} = [(-3)^3]^{2/3} + [(-\dfrac{1}{2})^5]^{2/5}$

$= (-3)^2 + (-\dfrac{1}{2})^2 = \dfrac{37}{4}$

14. a) $\dfrac{(-3a)^3 \cdot 3a^{-2/3}}{(2a)^{-2} \cdot a^{1/3}} = \dfrac{(-3a)^3 \cdot 3 \cdot (2a)^2}{a^{2/3} \cdot a^{1/3}} = \dfrac{-27a^3 \cdot 3 \cdot 4a^2}{a^{2/3 + 1/3}} = \dfrac{-324a^5}{a} = -324a^4$

b) $\dfrac{(x^{-2})^{-3} \cdot (x^{-1/3})^9}{(x^{1/2})^{-3} \cdot (x^{-3/2})^5} = \dfrac{x^6 \cdot x^{-3}}{x^{-3/2} \cdot x^{-15/2}} = \dfrac{x^{6-3}}{x^{-3/2 - 15/2}} = \dfrac{x^3}{x^{-9}} = x^{12}$

c) $\dfrac{(x + \frac{1}{y})^m \cdot (x - \frac{1}{y})^n}{(y + \frac{1}{x})^m \cdot (y - \frac{1}{x})^n} = \dfrac{(\frac{xy+1}{y})^m \cdot (\frac{xy-1}{y})^n}{(\frac{xy+1}{x})^m \cdot (\frac{xy-1}{x})^n} = \dfrac{\frac{(xy+1)^m}{y^m} \cdot \frac{(xy-1)^n}{y^n}}{\frac{(xy+1)^m}{x^m} \cdot \frac{(xy-1)^n}{x^n}}$

$= \dfrac{\frac{(xy+1)^m (xy-1)^n}{y^{m+n}}}{\frac{(xy+1)^m (xy-1)^n}{x^{m+n}}} = \dfrac{x^{m+n}}{y^{m+n}} = (\dfrac{x}{y})^{m+n}$

d) $\dfrac{3^{pq+q}}{3^{pq+p}} \cdot \dfrac{3^{2p}}{3^{2q}} = \dfrac{3^{pq+q+2p}}{3^{pq+p+2q}} = 3^{(pq+q+2p)-(pq+p+2q)} = 3^{p-q}$

15. a) $\dfrac{(x^{3/4} \cdot x^{1/2})^{1/3}}{(y^{2/3} \cdot y^{4/3})^{1/2}} = \dfrac{(x^{5/4})^{1/3}}{(y^2)^{1/2}} = \dfrac{x^{5/12}}{y}$

b) $\dfrac{(x^{3/4})^{2/3} - (y^{5/4})^{2/5}}{(x^{3/4})^{1/3} + (y^{2/3})^{3/8}} = \dfrac{x^{1/2} - y^{1/2}}{x^{1/4} + y^{1/4}} = \dfrac{(x^{1/4})^2 - (y^{1/4})^2}{x^{1/4} + y^{1/4}} = \dfrac{(x^{1/4} + y^{1/4})(x^{1/4} - y^{1/4})}{x^{1/4} + y^{1/4}}$

$$= x^{1/4} - y^{1/4}$$

c) $\dfrac{1}{1 + x^{p-q}} + \dfrac{1}{1 + x^{q-p}} = \dfrac{1}{1 + \dfrac{x^p}{x^q}} + \dfrac{1}{1 + \dfrac{x^q}{x^p}} = \dfrac{x^q}{x^q + x^p} + \dfrac{x^p}{x^p + x^q} = \dfrac{x^q + x^p}{x^q + x^p} = 1$

d) $\dfrac{x^{3n} - y^{3n}}{x^n - y^n} = \dfrac{(x^n)^3 - (y^n)^3}{x^n - y^n} = \dfrac{(x^n - y^n)(x^{2n} + x^n y^n + y^{2n})}{x^n - y^n} = x^{2n} + x^n y^n + y^{2n}$

e) $\sqrt[a]{a^{2} - a} = [a^{a(a-1)}]^{1/a} = a^{a-1}$ f) $2^{2^{3^2}} = 2^{2^9} = 2^{512}$

16. a) $(.004)(30,000)^2 = (4 \times 10^{-3})(3 \times 10^4)^2 = (4 \times 10^{-3})(3^2 \times 10^8) = 4 \cdot 3^2 \times 10^{-3+8}$

$$= 36 \times 10^5 \quad \text{or} \quad 3,600,000$$

b) $\dfrac{48,000,000}{1200} = \dfrac{48 \times 10^6}{12 \times 10^2} = 4 \times 10^{6-2} = 4 \times 10^4 \quad \text{or} \quad 40,000$

c) $\dfrac{.078}{.00012} = \dfrac{78 \times 10^{-3}}{12 \times 10^{-5}} = 6.5 \times 10^{-3+5} = 6.5 \times 10^2 \quad \text{or} \quad 650$

d) $\dfrac{(80,000,000)^2(.000003)}{(600,000)(.0002)^4} = \dfrac{(8 \times 10^7)^2(3 \times 10^{-6})}{(6 \times 10^5)(2 \times 10^{-4})^4} = \dfrac{8^2 \cdot 3}{6 \cdot 2^4} \cdot \dfrac{10^{14} \cdot 10^{-6}}{10^5 \cdot 10^{-16}} = 2 \times 10^{19}$

e) $\sqrt[3]{\dfrac{(.004)^4(.0036)}{(120,000)^2}} = \sqrt[3]{\dfrac{(4 \times 10^{-3})^4(36 \times 10^{-4})}{(12 \times 10^4)^2}} = \sqrt[3]{\dfrac{256(36)}{144} \cdot \dfrac{10^{-12} \cdot 10^{-4}}{10^8}}$

$$= \sqrt[3]{64 \times 10^{-24}} = 4 \times 10^{-8}$$

17. For what real values of the literals involved will each of the following operations be valid and yield real numbers?

a) $\sqrt{x^2} = (x^2)^{1/2} = x^1 = x$ b) $\sqrt{a^2 + 2a + .1} = \sqrt{(a+1)^2} = a + 1$

c) $\dfrac{a^{-2} - b^{-2}}{a^{-1} - b^{-1}} = \dfrac{(a^{-1})^2 - (b^{-1})^2}{a^{-1} - b^{-1}} = \dfrac{(a^{-1} + b^{-1})(a^{-1} - b^{-1})}{(a^{-1} - b^{-1})} = a^{-1} + b^{-1}$

d) $\sqrt{x^4 + 2x^2 + 1} = \sqrt{(x^2 + 1)^2} = x^2 + 1$

e) $\dfrac{x - 1}{\sqrt{x - 1}} = \dfrac{(x-1)^1}{(x-1)^{1/2}} = (x-1)^{1-1/2} = (x-1)^{1/2} = \sqrt{x - 1}$

a) When x is a real number, $\sqrt{x^2}$ must be positive or zero. Assuming $\sqrt{x^2} = x$ were true for all x, then if $x = -1$ we would have $\sqrt{(-1)^2} = -1$ or $\sqrt{1} = -1$, i.e. $1 = -1$, a contradiction. Thus $\sqrt{x^2} = x$ cannot be true for all values of x. We will have $\sqrt{x^2} = x$ if $x \geq 0$. If $x \leq 0$ we have $\sqrt{x^2} = -x$. A result valid both for $x \geq 0$ and $x \leq 0$ may be written $\sqrt{x^2} = |x|$ (absolute value of x).

b) $\sqrt{a^2 + 2a + 1}$ must be positive or zero and thus will equal $a+1$ if $a+1 \geq 0$, i.e. if $a \geq -1$. A result valid for all real values of a is given by $\sqrt{a^2 + 2a + 1} = |a+1|$.

c) $\dfrac{a^{-2} - b^{-2}}{a^{-1} - b^{-1}}$ is not defined if a or b or both are equal to zero. Similarly it is not defined if the denominator $a^{-1} - b^{-1} = 0$, i.e. if $a^{-1} = b^{-1}$ or $a = b$. Hence the result $\dfrac{a^{-2} - b^{-2}}{a^{-1} - b^{-1}} = a^{-1} + b^{-1}$ is valid if and only if $a \neq 0$, $b \neq 0$ and $a \neq b$.

d) $\sqrt{x^4 + 2x^2 + 1}$ must be positive or zero and will equal x^2+1 if $x^2+1 \geq 0$. Since x^2+1 is greater than zero for all real numbers x, the result is valid for all real values of x.

e) $\sqrt{x-1}$ will not be a real number if $x-1 < 0$, i.e. if $x < 1$. Also $\dfrac{x-1}{\sqrt{x-1}}$ will not be defined if the denominator is zero, i.e. if $x = 1$. Hence $\dfrac{x-1}{\sqrt{x-1}} = \sqrt{x-1}$ if and only if $x > 1$.

18. A student was asked to evaluate the expression $x + 2y + \sqrt{(x-2y)^2}$ for $x = 2$, $y = 4$. He wrote

$$x + 2y + \sqrt{(x-2y)^2} = x + 2y + x - 2y = 2x$$

and thus obtained the value $2x = 2(2) = 4$ for his answer. Was he correct?

Putting $x = 2$, $y = 4$ in the given expression, we obtain

$$x + 2y + \sqrt{(x-2y)^2} = 2 + 2(4) + \sqrt{(2-8)^2} = 2 + 8 + \sqrt{36} = 2 + 8 + 6 = 16.$$

The student made the mistake of writing $\sqrt{(x-2y)^2} = x - 2y$ which is true only if $x \geq 2y$. If $x \leq 2y$, $\sqrt{(x-2y)^2} = 2y - x$. In all cases, $\sqrt{(x-2y)^2} = |x - 2y|$. The required simplification should have been $x + 2y + 2y - x = 4y$ which does give 16 when $y = 4$.

SUPPLEMENTARY PROBLEMS

Evaluate each of the following.

19. a) 3^4

  b) $(-2x)^3$

  c) $(\frac{3y}{4})^3$

  d) 4^{-3}

  e) $(-4x)^{-2}$

  f) $(2y^{-1})^{-1}$

  g) $\frac{3^{-1}x^2y^{-4}}{2^{-2}x^{-3}y^3}$

  h) $(16)^{1/4}$

  i) $\frac{8^{-2/3}(-8)^{2/3}}{8^{1/3}}$

  j) $(-a^3b^3)^{-2/3}$

  k) $-3(-1)^{-1/5}(4)^{-1/2}$

  l) $(10^3)^0$

  m) $(x-y)^0\left[(x-y)^4\right]^{-1/2}$

  n) $x^y \cdot x^{4y}$

  o) $3y^{2/3} \cdot y^{4/3}$

  p) $(4 \cdot 10^3)(3 \cdot 10^{-5})(6 \cdot 10^4)$

20. a) $\frac{2^3 \cdot 2^{-2} \cdot 2^4}{2^{-1} \cdot 2^0 \cdot 2^{-3}}$

  b) $\frac{10^{x+y} \cdot 10^{y-x} \cdot 10^{y+1}}{10^{y+1} \cdot 10^{2y+1}}$

  c) $\frac{3^{1/2} \cdot 3^{-2/3}}{3^{-1/2} \cdot 3^{1/3}}$

  d) $\frac{(x+y)^{2/3}(x+y)^{-1/6}}{\left[(x+y)^2\right]^{1/4}}$

  e) $\frac{(10^2)^{-3}(10^3)^{1/6}}{\sqrt{10} \cdot (10^4)^{-1/2}}$

  f) $\left[(x^{-1})^{-2}\right]^{-3}$

  g) $\frac{4^{-1/2}a^{2/3}b^{-1/6}c^{-3/2}}{8^{2/3}a^{-1/3}b^{-2/3}c^{5/2}}$

  h) $(\frac{2^{-8} \cdot 3^4}{5^{-4}})^{-1/4}$

  i) $\sqrt{\frac{\sqrt[4]{a^2}\sqrt[3]{b^5}}{c^{-2}d^2}}$

21. a) $\sqrt{27^{-2/3}} + 5^{2/3} \cdot 5^{1/3}$

  b) $4(\frac{1}{2})^0 + 2^{-1} - 16^{-1/2} \cdot 4 \cdot 3^0$

  c) $8^{2/3} + 3^{-2} - \frac{1}{9}(10)^0$

  d) $27^{2/3} - 3(3x)^0 + 25^{1/2}$

  e) $8^{2/3} \cdot 16^{-3/4} \cdot 2^0 - 8^{-2/3}$

  f) $\sqrt[3]{(x-2)^{-2}}$ when $x = -6$

  g) $x^{3/2} + 4x^{-1} - 5x^0$ when $x = 4$

  h) $y^{2/3} + 3y^{-1} - 2y^0$ when $y = 1/8$

  i) $64^{-2/3} \cdot 16^{5/4} \cdot 2^0 \cdot (\sqrt{3})^4$

  j) $\frac{\sqrt{a} \cdot a^{-2/3}}{\sqrt[6]{a^5}} + \frac{a^{-5/6}}{\sqrt[3]{a^2} \cdot a^{-1/2}}$

  k) $(\frac{\sqrt{72y^{2n}}}{3} \cdot 9^0)(2y^{n+2})^{-1}$

22. a) $25^0 + 0.25^{1/2} - 8^{1/3} \cdot 4^{-1/2} + 0.027^{1/3}$

  b) $\frac{1}{8^{-2/3}} - 3a^0 + (3a)^0 + (27)^{-1/3} - 1^{3/2}$

  c) $\frac{3^{-2} + 5(2)^0}{3 - 4(3)^{-1}}$

  d) $\frac{3^0 x + 4x^{-1}}{x^{-2/3}}$ if $x = 8$

  e) $\frac{2 + 2^{-1}}{5} + (-8)^0 - 4^{3/2}$

  f) $(64)^{-2/3} - 3(150)^0 + 12(2)^{-2}$

  g) $(0.125)^{-2/3} + \frac{3}{2 + 2^{-1}}$

  h) $\sqrt[n]{\frac{32}{2^{5+n}}}$

  i) $\frac{(60,000)^3(.00002)^4}{100^2(72,000,000)(.0002)^5}$

23. a) $\dfrac{(x^2 + 3x + 4)^{1/3}\left[-\frac{1}{2}(5-x)^{-1/2}\right] - (5-x)^{1/2}\left[(x^2 + 3x + 4)^{-2/3}(2x+3)/3\right]}{(x^2 + 3x + 4)^{2/3}}$ if $x = 1$

    b) $\dfrac{(9x^2 - 5y)^{1/4}(2x) - x^2\left[\frac{1}{4}(9x^2 - 5y)^{-3/4}(18x)\right]}{(9x^2 - 5y)^{1/2}}$ if $x = 2$, $y = 4$

    c) $\dfrac{(x+1)^{2/3}\left[\frac{1}{2}(x-1)^{-1/2}\right] - (x-1)^{1/2}\left[\frac{2}{3}(x+1)^{-1/3}\right]}{(x+1)^{4/3}}$

    d) $x - 1 + \sqrt{x^2 + 2x + 1}$

    e) $3x - 2y - \sqrt{4x^2 - 4xy + y^2}$

ANSWERS TO SUPPLEMENTARY PROBLEMS.

19. a) 81   d) 1/64   g) $\dfrac{4x^5}{3y^7}$ j) $\dfrac{1}{a^2 b^2}$ m) $\dfrac{1}{(x-y)^2}$ o) $3y^2$

    b) $-8x^3$ e) $\dfrac{1}{16x^2}$ h) 2   k) 3/2   n) x^{5y} p) 7200

    c) $\dfrac{27y^3}{64}$ f) $y/2$ i) 1/2   l) 1

20. a) 2^9 b) 1/10   c) 1   d) 1   e) 10^{-4} f) x^{-6} g) $\dfrac{a\sqrt{b}}{8c^4}$ h) $\dfrac{4}{15}$ i) $\dfrac{a^{1/4}b^{5/6}c}{d}$

21. a) $\dfrac{16}{3}$ b) $\dfrac{7}{2}$ c) 4   d) 11   e) $\dfrac{1}{4}$ f) $\dfrac{1}{4}$ g) 4   h) $\dfrac{89}{4}$ i) 18   j) $\dfrac{2}{a}$ k) $\dfrac{\sqrt{2}}{y^2}$

22. a) 0.8   b) $\dfrac{4}{3}$ c) $\dfrac{46}{15}$ d) 34   e) $-\dfrac{13}{2}$ f) $\dfrac{1}{16}$ g) $\dfrac{26}{5}$ h) $\dfrac{1}{2}$ i) 150

23. a) $-\dfrac{1}{3}$

    b) $\dfrac{7}{8}$

    c) $\dfrac{7 - x}{6(x-1)^{1/2}(x+1)^{5/3}}$

    d) $2x$ if $x \geqq -1$, -2 if $x \leqq -1$

    e) $x - y$ if $2x \geqq y$, $5x - 3y$ if $2x \leqq y$

CHAPTER 7

Radicals

A RADICAL is an expression of the form $\sqrt[n]{a}$ which denotes the principal nth root of a. The positive integer n is the index, or order, of the radical and the number a is the radicand. The index is omitted if $n = 2$.

Thus $\sqrt[3]{5}$, $\sqrt[4]{7x^3 - 2y^2}$, $\sqrt{x + 10}$ are radicals which have respectively indices 3, 4, 2 and radicands 5, $7x^3 - 2y^2$, $x + 10$.

LAWS FOR RADICALS are the same as laws for exponents, since $\sqrt[n]{a} = a^{1/n}$. The following are laws most frequently used. Note: If n is even, assume $a, b \geqq 0$.

A) $$(\sqrt[n]{a})^n = a$$

Examples. $(\sqrt[3]{6})^3 = 6$, $(\sqrt[4]{x^2 + y^2})^4 = x^2 + y^2$

B) $$\sqrt[n]{ab} = \sqrt[n]{a} \, \sqrt[n]{b}$$

Examples. $\sqrt[3]{54} = \sqrt[3]{27 \cdot 2} = \sqrt[3]{27} \cdot \sqrt[3]{2} = 3\sqrt[3]{2}$, $\sqrt[7]{x^2 y^5} = \sqrt[7]{x^2} \, \sqrt[7]{y^5}$

C) $$\sqrt[n]{\frac{a}{b}} = \frac{\sqrt[n]{a}}{\sqrt[n]{b}} \qquad b \neq 0$$

Examples. $\sqrt[5]{\dfrac{5}{32}} = \dfrac{\sqrt[5]{5}}{\sqrt[5]{32}} = \dfrac{\sqrt[5]{5}}{2}$, $\sqrt[3]{\dfrac{(x+1)^3}{(y-2)^6}} = \dfrac{\sqrt[3]{(x+1)^3}}{\sqrt[3]{(y-2)^6}} = \dfrac{x+1}{(y-2)^2}$

D) $$\sqrt[n]{a^m} = (\sqrt[n]{a})^m$$

Example. $\sqrt[3]{(27)^4} = (\sqrt[3]{27})^4 = 3^4 = 81$

E) $$\sqrt[m]{\sqrt[n]{a}} = \sqrt[mn]{a}$$

Examples. $\sqrt{\sqrt[3]{5}} = \sqrt[6]{5}$, $\sqrt[4]{\sqrt[3]{2}} = \sqrt[12]{2}$, $\sqrt[5]{\sqrt[3]{x^2}} = \sqrt[15]{x^2}$

A RATIONAL NUMBER is a real number that can be written in the form p/q where p and q are integers.

53

Thus 4, 2/3, 3/8, .36, −2.4, $\sqrt{16}$, $\sqrt{36/49}$, $\sqrt[3]{-27}$ are rational numbers because they can be expressed as the quotient of two integers as follows:
$$\frac{4}{1}, \frac{2}{3}, \frac{3}{8}, \frac{36}{100} \text{ or } \frac{9}{25}, \frac{-24}{10} \text{ or } \frac{-12}{5}, \frac{4}{1}, \frac{6}{7}, \frac{-3}{1}.$$

AN IRRATIONAL NUMBER, OR SURD, is a real number which cannot be written in the form p/q where p and q are integers.

Thus $\sqrt{2}$, $\sqrt{3}$, $\sqrt[3]{2}$, $\sqrt[4]{15}$, $\sqrt{2/3}$, $\sqrt[3]{-4/5}$ are irrational numbers.

An irrational square root of a rational number such as $\sqrt{5}$ or $\sqrt{1/6}$ is sometimes called a quadratic surd.

THE FORM OF A RADICAL may be changed in the following ways.

a) Removal of Perfect nth Powers from the Radicand.

Examples. $\sqrt[3]{32} = \sqrt[3]{2^3(4)} = \sqrt[3]{2^3} \cdot \sqrt[3]{4} = 2\sqrt[3]{4}$

$\sqrt{8x^5 y^7} = \sqrt{(4x^4 y^6)(2xy)} = \sqrt{4x^4 y^6}\sqrt{2xy} = 2x^2 y^3\sqrt{2xy}$

b) Reduction of the Index of the Radical.

Examples. $\sqrt[4]{64} = \sqrt[4]{2^6} = 2^{6/4} = 2^{3/2} = \sqrt{2^3} = \sqrt{8}$, where the index is reduced from 4 to 2.

$\sqrt[6]{25x^6} = \sqrt[6]{(5x^3)^2} = (5x^3)^{2/6} = (5x^3)^{1/3} = \sqrt[3]{5x^3} = x\sqrt[3]{5}$, where the index is reduced from 6 to 3.

Note. $\sqrt[4]{(-4)^2} = \sqrt[4]{16} = 2$.

It is <u>incorrect</u> to write $\sqrt[4]{(-4)^2} = (-4)^{2/4} = (-4)^{1/2} = \sqrt{-4}$.

c) Rationalization of the Denominator in the Radicand.

Example 1. Rationalize the denominator of $\sqrt[3]{9/2}$.

Multiply the numerator and denominator of the radicand (9/2) by such a number as will make the denominator a perfect nth power (here $n=3$) and then remove the denominator from under the radical sign. The number in this case is 2^2. Then
$$\sqrt[3]{\frac{9}{2}} = \sqrt[3]{\frac{9}{2}\left(\frac{2^2}{2^2}\right)} = \sqrt[3]{\frac{9(2^2)}{2^3}} = \frac{\sqrt[3]{36}}{2}.$$

2. Rationalize the denominator of $\sqrt[4]{\dfrac{7a^3 y^2}{8b^6 x^3}}$.

To make $8b^6 x^3$ a perfect 4th power, multiply numerator and denominator by $2b^2 x$. Then
$$\sqrt[4]{\frac{7a^3 y^2}{8b^6 x^3}} = \sqrt[4]{\frac{7a^3 y^2}{8b^6 x^3} \cdot \frac{2b^2 x}{2b^2 x}} = \sqrt[4]{\frac{14a^3 y^2 b^2 x}{16b^8 x^4}} = \frac{\sqrt[4]{14a^3 y^2 b^2 x}}{2b^2 x}$$

A RADICAL is said to be in simplest form if

    a) all perfect *n*th powers have been removed from the radical,
    b) the index of the radical is as small as possible,
    c) no fractions are present in the radicand, i.e. the denominator has been rationalized.

SIMILAR RADICALS. Two or more radicals are said to be similar if after being reduced to simplest form they have the same index and radicand.

Thus $\sqrt{32}$, $\sqrt{1/2}$, $\sqrt{8}$ are similar since $\sqrt{32} = \sqrt{16 \cdot 2} = 4\sqrt{2}$, $\sqrt{\dfrac{1}{2}} = \sqrt{\dfrac{1}{2} \cdot \dfrac{2}{2}} = \dfrac{\sqrt{2}}{2}$, $\sqrt{8} = \sqrt{4 \cdot 2} = 2\sqrt{2}$. Here each radicand is 2 and each index is 2.

However, $\sqrt[3]{32}$ and $\sqrt[3]{2}$ are dissimilar since $\sqrt[3]{32} = \sqrt[3]{8 \cdot 4} = 2\sqrt[3]{4}$.

TO ADD ALGEBRAICALLY two or more radicals, reduce each radical to simplest form and combine terms with similar radicals.

Thus $\sqrt{32} - \sqrt{1/2} - \sqrt{8} = 4\sqrt{2} - \dfrac{\sqrt{2}}{2} - 2\sqrt{2} = (4 - \dfrac{1}{2} - 2)\sqrt{2} = \dfrac{3}{2}\sqrt{2}$.

MULTIPLICATION OF RADICALS.

    a) To multiply two or more radicals having the *same index*, use Law *B*:

$$\sqrt[n]{a}\ \sqrt[n]{b} = \sqrt[n]{ab}.$$

Examples. $(2\sqrt[3]{4})(3\sqrt[3]{16}) = 2 \cdot 3\sqrt[3]{4}\ \sqrt[3]{16} = 6\sqrt[3]{64} = 6 \cdot 4 = 24$

$\qquad\qquad (3\sqrt[4]{x^2 y})(\sqrt[4]{x^3 y^2}) = 3\sqrt[4]{(x^2 y)(x^3 y^2)} = 3\sqrt[4]{x^5 y^3} = 3x\sqrt[4]{xy^3}.$

    b) To multiply radicals with *different indices* it is convenient to use fractional exponents and the laws of exponents.

Examples. $\sqrt[3]{5}\ \sqrt{2} = 5^{1/3} \cdot 2^{1/2} = 5^{2/6} \cdot 2^{3/6} = (5^2 \cdot 2^3)^{1/6} = (25 \cdot 8)^{1/6} = \sqrt[6]{200}.$

$\qquad\qquad \sqrt[3]{4}\ \sqrt{2} = \sqrt[3]{2^2}\sqrt{2} = 2^{2/3} \cdot 2^{1/2} = 2^{4/6} \cdot 2^{3/6} = 2^{7/6} = \sqrt[6]{2^7} = 2\sqrt[6]{2}$

DIVISION OF RADICALS.

    a) To divide two radicals having the *same index*, use Law *C*, $\dfrac{\sqrt[n]{a}}{\sqrt[n]{b}} = \sqrt[n]{\dfrac{a}{b}}$, and simplify.

Example. $\dfrac{\sqrt[3]{5}}{\sqrt[3]{3}} = \sqrt[3]{\dfrac{5}{3}} = \sqrt[3]{\dfrac{5}{3} \cdot \dfrac{3^2}{3^2}} = \sqrt[3]{\dfrac{45}{3^3}} = \dfrac{\sqrt[3]{45}}{3}$

We may also rationalize the denominator directly, as follows.

$\qquad \dfrac{\sqrt[3]{5}}{\sqrt[3]{3}} = \dfrac{\sqrt[3]{5}}{\sqrt[3]{3}} \cdot \dfrac{\sqrt[3]{3^2}}{\sqrt[3]{3^2}} = \dfrac{\sqrt[3]{5 \cdot 3^2}}{\sqrt[3]{3^3}} = \dfrac{\sqrt[3]{45}}{3}$

    b) To divide two radicals with *different indices* it is convenient to use fractional exponents and the laws of exponents.

Examples. $\dfrac{\sqrt{6}}{\sqrt[4]{2}} = \dfrac{6^{1/2}}{2^{1/4}} = \dfrac{6^{2/4}}{2^{1/4}} = \sqrt[4]{\dfrac{6^2}{2}} = \sqrt[4]{\dfrac{36}{2}} = \sqrt[4]{18}$

$\dfrac{\sqrt[3]{4}}{\sqrt{2}} = \dfrac{\sqrt[3]{2^2}}{\sqrt{2}} = \dfrac{2^{2/3}}{2^{1/2}} = \dfrac{2^{4/6}}{2^{3/6}} = 2^{1/6} = \sqrt[6]{2}$

CONJUGATES. The binomial quadratic surds $\sqrt{a} + \sqrt{b}$ and $\sqrt{a} - \sqrt{b}$ are called conjugates of each other.   Thus $2\sqrt{3} + \sqrt{2}$ and $2\sqrt{3} - \sqrt{2}$ are conjugates.

To rationalize a fraction whose denominator is a binomial quadratic surd, multiply numerator and denominator by the conjugate.

Example. $\dfrac{5}{2\sqrt{3} + \sqrt{2}} = \dfrac{5}{2\sqrt{3} + \sqrt{2}} \cdot \dfrac{2\sqrt{3} - \sqrt{2}}{2\sqrt{3} - \sqrt{2}} = \dfrac{5(2\sqrt{3} - \sqrt{2})}{12 - 2} = \dfrac{2\sqrt{3} - \sqrt{2}}{2}$

SOLVED PROBLEMS

REDUCTION OF A RADICAL EXPRESSION TO SIMPLEST FORM

1. a) $\sqrt{18} = \sqrt{9 \cdot 2} = \sqrt{3^2 \cdot 2} = 3\sqrt{2}$ d) $\sqrt[3]{648} = \sqrt[3]{8 \cdot 27 \cdot 3} = \sqrt[3]{2^3 \cdot 3^3 \cdot 3} = 6\sqrt[3]{3}$

b) $\sqrt[3]{80} = \sqrt[3]{8 \cdot 10} = \sqrt[3]{2^3 \cdot 10} = 2\sqrt[3]{10}$ e) $a\sqrt{9b^4 c^3} = a\sqrt{3^2 b^4 c^2 \cdot c} = 3ab^2 c\sqrt{c}$

c) $5\sqrt[3]{243} = 5\sqrt[3]{27 \cdot 9} = 5\sqrt[3]{3^3 \cdot 9} = 15\sqrt[3]{9}$ f) $\sqrt[6]{343} = \sqrt[6]{7^3} = 7^{3/6} = 7^{1/2} = \sqrt{7}$

g) $\sqrt[6]{81a^2} = \sqrt[6]{3^4 a^2} = 3^{4/6} a^{2/6} = 3^{2/3} a^{1/3} = \sqrt[3]{9a}$ Note that $a \geqq 0$. See k) below.

h) $\sqrt[3]{64x^7 y^{-6}} = \sqrt[3]{4^3 x^6 \cdot xy^{-6}} = 4x^2 y^{-2}\sqrt[3]{x} = \dfrac{4x^2}{y^2}\sqrt[3]{x}$

i) $\sqrt[5]{(72)^4} = (72)^{4/5} = (8 \cdot 9)^{4/5} = (2^3 \cdot 3^2)^{4/5} = 2^{12/5} \cdot 3^{8/5}$

$= (2^2 \cdot 2^{2/5})(3 \cdot 3^{3/5}) = 2^2 \cdot 3 \sqrt[5]{2^2 \cdot 3^3} = 12\sqrt[5]{108}$

j) $(7\sqrt[3]{4ab})^2 = 49(4ab)^{2/3} = 49\sqrt[3]{16a^2 b^2} = 98\sqrt[3]{2a^2 b^2}$

k) $2a\sqrt{a^2 + 6a + 9} = 2a\sqrt{(a+3)^2} = 2a(a+3)$.   The student is reminded that $\sqrt{(a+3)^2}$ is a positive number or zero; hence $\sqrt{(a+3)^2} = a+3$ only if $a+3 \geqq 0$. If values of a such that $a + 3 < 0$ are included, we must write $\sqrt{(a+3)^2} = |a+3|$.

l) $\dfrac{x - 25}{\sqrt{x} + 5} = \dfrac{(\sqrt{x} + 5)(\sqrt{x} - 5)}{\sqrt{x} + 5} = \sqrt{x} - 5$

m) $\sqrt{12x^4 - 36x^2 y^2 + 27y^4} = \sqrt{3(4x^4 - 12x^2 y^2 + 9y^4)} = \sqrt{3(2x^2 - 3y^2)^2} = (2x^2 - 3y^2)\sqrt{3}$

Note that this is valid only if $2x^2 \geqq 3y^2$. See k) above.

n) $\sqrt[n]{a^n b^{2n} c^{3n+1} d^{n+2}} = (a^n b^{2n} c^{3n+1} d^{n+2})^{1/n} = ab^2 c^3 c^{1/n} d\, d^{2/n} = ab^2 c^3 d\sqrt[n]{cd^2}$

o) $\sqrt[3]{\sqrt{256}} = \sqrt[3]{16} = \sqrt[3]{8 \cdot 2} = 2\sqrt[3]{2}$

p) $\sqrt[4]{\sqrt[3]{6ab^2}} = [(6ab^2)^{1/3}]^{1/4} = (6ab^2)^{1/12} = \sqrt[12]{6ab^2}$

q) $\sqrt[5]{729\sqrt{a^3}} = \sqrt[5]{729\,a^{3/2}} = (3^6 a^{3/2})^{1/5} = 3^{12/10}a^{3/10} = 3\sqrt[10]{9a^3}$

CHANGE IN FORM OF A RADICAL

2. Express as radicals of the 12th order.

a) $\sqrt[3]{5} = 5^{1/3} = 5^{4/12} = \sqrt[12]{5^4} = \sqrt[12]{625}$

b) $\sqrt{ab} = (ab)^{1/2} = (ab)^{6/12} = \sqrt[12]{(ab)^6} = \sqrt[12]{a^6 b^6}$

c) $\sqrt[6]{x^n} = x^{n/6} = x^{2n/12} = \sqrt[12]{x^{2n}}$

3. Express in terms of radicals of least order.

a) $\sqrt[4]{9} = 9^{1/4} = (3^2)^{1/4} = 3^{1/2} = \sqrt{3}$

b) $\sqrt[12]{8x^3y^6} = \sqrt[12]{(2xy^2)^3} = (2xy^2)^{3/12} = (2xy^2)^{1/4} = \sqrt[4]{2xy^2}$

c) $\sqrt[8]{a^2 + 2ab + b^2} = \sqrt[8]{(a+b)^2} = (a+b)^{2/8} = (a+b)^{1/4} = \sqrt[4]{a+b}$

4. Convert into entire radicals, i.e. radicals having coefficient 1.

a) $6\sqrt{3} = \sqrt{36 \cdot 3} = \sqrt{108}$

b) $4x^2\sqrt[3]{y^2} = \sqrt[3]{(4x^2)^3 y^2} = \sqrt[3]{64x^6 y^2}$

c) $\dfrac{2x}{y}\sqrt[4]{\dfrac{2y}{x}} = \sqrt[4]{(\dfrac{2x}{y})^4 \cdot \dfrac{2y}{x}} = \sqrt[4]{\dfrac{16x^4}{y^4} \cdot \dfrac{2y}{x}} = \sqrt[4]{\dfrac{32x^3}{y^3}}$

d) $\dfrac{a-b}{a+b}\sqrt{\dfrac{a+b}{a-b}} = \sqrt{\dfrac{(a-b)^2}{(a+b)^2} \cdot \dfrac{a+b}{a-b}} = \sqrt{\dfrac{a-b}{a+b}}$

5. Determine which of the following irrational numbers is the greater.

a) $\sqrt[3]{2}, \sqrt[4]{3}$ *b)* $\sqrt{5}, \sqrt[3]{11}$ *c)* $2\sqrt{5}, 3\sqrt{2}$

a) $\sqrt[3]{2} = 2^{1/3} = 2^{4/12} = (2^4)^{1/12} = (16)^{1/12}$; $\sqrt[4]{3} = 3^{1/4} = 3^{3/12} = (3^3)^{1/12} = (27)^{1/12}$.
   Since $(27)^{1/12} > (16)^{1/12}$, $\sqrt[4]{3} > \sqrt[3]{2}$.

b) $\sqrt{5} = 5^{1/2} = 5^{3/6} = (5^3)^{1/6} = (125)^{1/6}$; $\sqrt[3]{11} = (11)^{1/3} = (11)^{2/6} = (11^2)^{1/6} = (121)^{1/6}$
   Since $125 > 121$, $\sqrt{5} > \sqrt[3]{11}$.

c) $2\sqrt{5} = \sqrt{2^2 \cdot 5} = \sqrt{20}$; $3\sqrt{2} = \sqrt{3^2 \cdot 2} = \sqrt{18}$. Hence $2\sqrt{5} > 3\sqrt{2}$.

6. Rationalize the denominator.

a) $\sqrt{\dfrac{2}{3}} = \sqrt{\dfrac{2}{3} \cdot \dfrac{3}{3}} = \sqrt{\dfrac{6}{3^2}} = \dfrac{1}{3}\sqrt{6}$

$b)$ $\dfrac{3}{\sqrt[3]{6}} \;=\; \dfrac{3}{\sqrt[3]{6}} \cdot \dfrac{\sqrt[3]{6^2}}{\sqrt[3]{6^2}} \;=\; \dfrac{3\sqrt[3]{6^2}}{\sqrt[3]{6 \cdot 6^2}} \;=\; \dfrac{3\sqrt[3]{36}}{6} \;=\; \dfrac{1}{2}\sqrt[3]{36}$

Another method: $\dfrac{3}{\sqrt[3]{6}} \;=\; \dfrac{3}{6^{1/3}} \cdot \dfrac{6^{2/3}}{6^{2/3}} \;=\; \dfrac{3 \cdot 6^{2/3}}{6^1} \;=\; \dfrac{3\sqrt[3]{6^2}}{6} \;=\; \dfrac{1}{2}\sqrt[3]{36}$

$c)$ $3x\sqrt[4]{\dfrac{y}{2x}} \;=\; 3x\sqrt[4]{\dfrac{y(2x)^3}{2x(2x)^3}} \;=\; 3x\sqrt[4]{\dfrac{y(8x^3)}{(2x)^4}} \;=\; \dfrac{3x}{2x}\sqrt[4]{8x^3 y} \;=\; \dfrac{3}{2}\sqrt[4]{8x^3 y}$

$d)$ $\sqrt{\dfrac{a-b}{a+b}} \;=\; \sqrt{\dfrac{a-b}{a+b} \cdot \dfrac{a+b}{a+b}} \;=\; \sqrt{\dfrac{a^2 - b^2}{(a+b)^2}} \;=\; \dfrac{1}{a+b}\sqrt{a^2 - b^2}$

$e)$ $\dfrac{4xy^2}{\sqrt[3]{2xy^2}} \;=\; \dfrac{4xy^2}{\sqrt[3]{2xy^2}} \cdot \dfrac{\sqrt[3]{(2xy^2)^2}}{\sqrt[3]{(2xy^2)^2}} \;=\; \dfrac{4xy^2\sqrt[3]{(2xy^2)^2}}{2xy^2} \;=\; 2\sqrt[3]{4x^2 y^4} \;=\; 2y\sqrt[3]{4x^2 y}$

ADDITION AND SUBTRACTION OF SIMILAR RADICALS

$7.$ $a)$ $\sqrt{18} + \sqrt{50} - \sqrt{72} \;=\; \sqrt{9 \cdot 2} + \sqrt{25 \cdot 2} - \sqrt{36 \cdot 2} \;=\; 3\sqrt{2} + 5\sqrt{2} - 6\sqrt{2} \;=\; (3 + 5 - 6)\sqrt{2} \;=\; 2\sqrt{2}$

$b)$ $2\sqrt{27} - 4\sqrt{12} \;=\; 2\sqrt{9 \cdot 3} - 4\sqrt{4 \cdot 3} \;=\; 2 \cdot 3\sqrt{3} - 4 \cdot 2\sqrt{3} \;=\; 6\sqrt{3} - 8\sqrt{3} \;=\; -2\sqrt{3}$

$c)$ $4\sqrt{75} + 3\sqrt{4/3} - 2\sqrt{48} \;=\; 4 \cdot 5\sqrt{3} + 3\sqrt{\dfrac{4}{3} \cdot \dfrac{3}{3}} - 2 \cdot 4\sqrt{3} \;=\; (20 + 3 \cdot \dfrac{2}{3} - 8)\sqrt{3} \;=\; 14\sqrt{3}$

$d)$ $\sqrt[3]{432} - \sqrt[3]{250} + \sqrt[3]{1/32} \;=\; \sqrt[3]{6^3 \cdot 2} - \sqrt[3]{5^3 \cdot 2} + \sqrt[3]{\dfrac{1}{2^5} \cdot \dfrac{2}{2}} \;=\; (6 - 5 + \dfrac{1}{4})\sqrt[3]{2} \;=\; \dfrac{5}{4}\sqrt[3]{2}$

$e)$ $\sqrt{3} + \sqrt[3]{81} - \sqrt{27} + 5\sqrt[3]{3} \;=\; \sqrt{3} + \sqrt[3]{27 \cdot 3} - \sqrt{9 \cdot 3} + 5\sqrt[3]{3}$

            $=\; \sqrt{3} + 3\sqrt[3]{3} - 3\sqrt{3} + 5\sqrt[3]{3} \;=\; -2\sqrt{3} + 8\sqrt[3]{3}$

$f)$ $2a\sqrt[3]{27x^3 y} + 3b\sqrt[3]{8x^3 y} - 6c\sqrt[3]{-x^3 y} \;=\; 6ax\sqrt[3]{y} + 6bx\sqrt[3]{y} + 6cx\sqrt[3]{y} \;=\; 6x(a + b + c)\sqrt[3]{y}$

$g)$ $2\sqrt{\dfrac{2}{3}} + 4\sqrt{\dfrac{3}{8}} - 5\sqrt{\dfrac{1}{24}} \;=\; 2\sqrt{\dfrac{2}{3} \cdot \dfrac{3}{3}} + 4\sqrt{\dfrac{3}{8} \cdot \dfrac{2}{2}} - 5\sqrt{\dfrac{1}{24} \cdot \dfrac{6}{6}}$

            $=\; (\dfrac{2}{3} + 4 \cdot \dfrac{1}{4} - \dfrac{5}{12})\sqrt{6} \;=\; \dfrac{5}{4}\sqrt{6}$

$h)$ $\dfrac{\sqrt{5}}{\sqrt{2}} + \dfrac{3}{\sqrt{.1}} - \sqrt{1.6} \;=\; \sqrt{\dfrac{5}{2} \cdot \dfrac{2}{2}} + \dfrac{3}{\sqrt{1/10}} - \sqrt{(.16)(10)} \;=\; \dfrac{1}{2}\sqrt{10} + 3\sqrt{10} - .4\sqrt{10} \;=\; 3.1\sqrt{10}$

$i)$ $2\sqrt{\dfrac{a}{b}} - 3\sqrt{\dfrac{b}{a}} + \dfrac{4}{\sqrt{ab}} \;=\; 2\sqrt{\dfrac{a}{b} \cdot \dfrac{b}{b}} - 3\sqrt{\dfrac{b}{a} \cdot \dfrac{a}{a}} + \dfrac{4}{\sqrt{ab}} \cdot \dfrac{\sqrt{ab}}{\sqrt{ab}}$

            $=\; \dfrac{2}{b}\sqrt{ab} - \dfrac{3}{a}\sqrt{ab} + \dfrac{4}{ab}\sqrt{ab} \;=\; (\dfrac{2}{b} - \dfrac{3}{a} + \dfrac{4}{ab})\sqrt{ab} \;=\; (\dfrac{2a - 3b + 4}{ab})\sqrt{ab}$

MULTIPLICATION OF RADICALS

8. a) $(2\sqrt{7})(3\sqrt{5}) = (2\cdot3)\sqrt{7\cdot5} = 6\sqrt{35}$

b) $(3\sqrt[3]{2})(5\sqrt[3]{6})(8\sqrt[3]{4}) = (3\cdot5\cdot8)\sqrt[3]{2\cdot6\cdot4} = 120\sqrt[3]{48} = 240\sqrt[3]{6}$

c) $(\sqrt[3]{18x^2})(\sqrt[3]{2x}) = \sqrt[3]{36x^3} = x\sqrt[3]{36}$

d) $\sqrt[4]{ab^{-1}c^5}\cdot\sqrt[4]{a^3b^3c^{-1}} = \sqrt[4]{a^4b^2c^4} = \sqrt{a^2bc^2} = ac\sqrt{b}$

e) $\sqrt{3}\cdot\sqrt[3]{2} = 3^{1/2}\cdot2^{1/3} = 3^{3/6}\cdot2^{2/6} = \sqrt[6]{3^3\cdot2^2} = \sqrt[6]{108}$

f) $(\sqrt[3]{14})(\sqrt[4]{686}) = (\sqrt[3]{7\cdot2})(\sqrt[4]{7^3\cdot2}) = (7^{1/3}\cdot2^{1/3})(7^{3/4}\cdot2^{1/4}) = (7^{4/12}\cdot2^{4/12})(7^{9/12}\cdot2^{3/12})$

$= 7(7^{1/12}\cdot2^{7/12}) = 7\sqrt[12]{7\cdot2^7} = 7\sqrt[12]{896}$

g) $(-\sqrt{5}\,\sqrt[3]{x})^6 = 5^{6/2}x^{6/3} = 5^3x^2 = 125x^2$

h) $(\sqrt{4\times10^{-6}})(\sqrt{8.1\times10^3})(\sqrt{.0016}) = (\sqrt{4\times10^{-6}})(\sqrt{81\times10^2})(\sqrt{16\times10^{-4}})$

$= (2\times10^{-3})(9\times10)(4\times10^{-2}) = 72\times10^{-4} = .0072$

i) $(\sqrt{6}+\sqrt{3})(\sqrt{6}-2\sqrt{3}) = \sqrt{6}\sqrt{6} + \sqrt{3}\sqrt{6} + (\sqrt{6})(-2\sqrt{3}) + (\sqrt{3})(-2\sqrt{3})$

$= 6 + \sqrt{18} - 2\sqrt{18} - 2\cdot3 = -\sqrt{18} = -3\sqrt{2}$

j) $(\sqrt{5}+\sqrt{2})^2 = (\sqrt{5})^2 + 2(\sqrt{5})(\sqrt{2}) + (\sqrt{2})^2 = 5 + 2\sqrt{10} + 2 = 7 + 2\sqrt{10}$

k) $(7\sqrt{5}-4\sqrt{3})^2 = (7\sqrt{5})^2 - 2(7\sqrt{5})(4\sqrt{3}) + (4\sqrt{3})^2$

$= 7^2\cdot5 - 2\cdot7\cdot4\sqrt{15} + 4^2\cdot3 = 245 - 56\sqrt{15} + 48 = 293 - 56\sqrt{15}$

l) $(\sqrt{3}+1)(\sqrt{3}-1) = (\sqrt{3})^2 - (1)^2 = 3 - 1 = 2$

m) $(2\sqrt{3}-\sqrt{5})(2\sqrt{3}+\sqrt{5}) = (2\sqrt{3})^2 - (\sqrt{5})^2 = 4\cdot3 - 5 = 12 - 5 = 7$

n) $(2\sqrt{5}-3\sqrt{2})(2\sqrt{5}+3\sqrt{2}) = (2\sqrt{5})^2 - (3\sqrt{2})^2 = 4\cdot5 - 9\cdot2 = 20 - 18 = 2$

o) $(2+\sqrt[3]{3})(2-\sqrt[3]{3}) = 4 - \sqrt[3]{9}$

p) $(3\sqrt{2}+2\sqrt[3]{4})(3\sqrt{2}-2\sqrt[3]{4}) = (3\sqrt{2})^2 - (2\sqrt[3]{4})^2 = 18 - 4\sqrt[3]{16} = 18 - 8\sqrt[3]{2}$

q) $(3\sqrt{2}-4\sqrt{5})(2\sqrt{3}+3\sqrt{6}) = (3\sqrt{2})(2\sqrt{3}) - (4\sqrt{5})(2\sqrt{3}) + (3\sqrt{2})(3\sqrt{6}) - (4\sqrt{5})(3\sqrt{6})$

$= 6\sqrt{6} - 8\sqrt{15} + 9\sqrt{12} - 12\sqrt{30} = 6\sqrt{6} - 8\sqrt{15} + 18\sqrt{3} - 12\sqrt{30}$

r) $(\sqrt{x+y}-z)(\sqrt{x+y}+z) = x + y - z^2$

s) $(2\sqrt{x-1}-x\sqrt{2})(3\sqrt{x-1}+2x\sqrt{2}) = 6(x-1) - 3x\sqrt{2(x-1)} + 4x\sqrt{2(x-1)} - 4x^2$

$= 6(x-1) + x\sqrt{2(x-1)} - 4x^2$

9. a) $(\sqrt{2}+\sqrt{3}+\sqrt{5})(\sqrt{2}+\sqrt{3}-\sqrt{5}) = [(\sqrt{2}+\sqrt{3})+\sqrt{5}][(\sqrt{2}+\sqrt{3})-\sqrt{5}]$

$= (\sqrt{2}+\sqrt{3})^2 - (\sqrt{5})^2 = 2 + 2\sqrt{6} + 3 - 5 = 2\sqrt{6}$

$b)$ $(2\sqrt{3} + 3\sqrt{2} + 1)(2\sqrt{3} - 3\sqrt{2} - 1) = [2\sqrt{3} + (3\sqrt{2} + 1)][2\sqrt{3} - (3\sqrt{2} + 1)]$

$$= (2\sqrt{3})^2 - (3\sqrt{2} + 1)^2 = 12 - (9 \cdot 2 + 6\sqrt{2} + 1) = -7 - 6\sqrt{2}$$

$c)$ $(\sqrt{2} + \sqrt{3} + \sqrt{5})^2 = (\sqrt{2})^2 + (\sqrt{3})^2 + (\sqrt{5})^2 + 2(\sqrt{2})(\sqrt{3}) + 2(\sqrt{3})(\sqrt{5}) + 2(\sqrt{2})(\sqrt{5})$

$$= 2 + 3 + 5 + 2\sqrt{6} + 2\sqrt{15} + 2\sqrt{10} = 10 + 2\sqrt{6} + 2\sqrt{15} + 2\sqrt{10}$$

$d)$ $(\sqrt{6 + 3\sqrt{3}})(\sqrt{6 - 3\sqrt{3}}) = \sqrt{(6 + 3\sqrt{3})(6 - 3\sqrt{3})} = \sqrt{36 - 9 \cdot 3} = \sqrt{9} = 3$

$e)$ $(\sqrt{a+b} - \sqrt{a-b})^2 = a + b - 2\sqrt{(a+b)(a-b)} + a - b = 2a - 2\sqrt{a^2 - b^2}$

DIVISION OF RADICALS. RATIONALIZATION OF DENOMINATORS

10. $a)$ $\dfrac{10\sqrt{6}}{5\sqrt{2}} = \dfrac{10}{5}\sqrt{\dfrac{6}{2}} = 2\sqrt{3}$ $d)$ $\dfrac{\sqrt{2}}{\sqrt{3}} = \sqrt{\dfrac{2}{3}} = \sqrt{\dfrac{2}{3} \cdot \dfrac{3}{3}} = \sqrt{\dfrac{6}{9}} = \dfrac{1}{3}\sqrt{6}$

$b)$ $\dfrac{2\sqrt[4]{30}}{3\sqrt[4]{5}} = \dfrac{2}{3}\sqrt[4]{\dfrac{30}{5}} = \dfrac{2}{3}\sqrt[4]{6}$ $e)$ $\dfrac{\sqrt[3]{2}}{\sqrt[3]{3}} = \sqrt[3]{\dfrac{2}{3}} = \sqrt[3]{\dfrac{2}{3} \cdot \dfrac{9}{9}} = \sqrt[3]{\dfrac{18}{27}} = \dfrac{1}{3}\sqrt[3]{18}$

$c)$ $\dfrac{4x}{y} \cdot \dfrac{\sqrt[3]{x^2 y^2}}{\sqrt[3]{xy}} = \dfrac{4x}{y}\sqrt[3]{\dfrac{x^2 y^2}{xy}} = \dfrac{4x}{y}\sqrt[3]{xy}$ $f)$ $\sqrt[5]{\dfrac{1}{2}} = \sqrt[5]{\dfrac{1}{2} \cdot \dfrac{16}{16}} = \sqrt[5]{\dfrac{16}{32}} = \dfrac{1}{2}\sqrt[5]{16}$

$g)$ $\dfrac{\sqrt{3} + 4\sqrt{2} - 5\sqrt{8}}{\sqrt{2}} = \dfrac{\sqrt{3} + 4\sqrt{2} - 5\sqrt{8}}{\sqrt{2}} \cdot \dfrac{\sqrt{2}}{\sqrt{2}} = \dfrac{\sqrt{6} + 4 \cdot 2 - 5\sqrt{16}}{2} = \dfrac{\sqrt{6} - 12}{2}$

$h)$ $\dfrac{3}{\sqrt{5} + \sqrt{2}} = \dfrac{3}{\sqrt{5} + \sqrt{2}} \cdot \dfrac{\sqrt{5} - \sqrt{2}}{\sqrt{5} - \sqrt{2}} = \dfrac{3(\sqrt{5} - \sqrt{2})}{5 - 2} = \sqrt{5} - \sqrt{2}$

$i)$ $\dfrac{1 + \sqrt{2}}{1 - \sqrt{2}} = \dfrac{1 + \sqrt{2}}{1 - \sqrt{2}} \cdot \dfrac{1 + \sqrt{2}}{1 + \sqrt{2}} = \dfrac{1 + 2\sqrt{2} + 2}{1 - 2} = -(3 + 2\sqrt{2})$

$j)$ $\dfrac{1}{x - \sqrt{x^2 - y^2}} - \dfrac{1}{x + \sqrt{x^2 - y^2}} = \dfrac{(x + \sqrt{x^2 - y^2}) - (x - \sqrt{x^2 - y^2})}{(x - \sqrt{x^2 - y^2})(x + \sqrt{x^2 - y^2})} = \dfrac{2\sqrt{x^2 - y^2}}{y^2}$

$k)$ $\dfrac{3\sqrt{3}}{4\sqrt[3]{2}} = \dfrac{3\sqrt{3}}{4\sqrt[3]{2}} \cdot \dfrac{\sqrt[3]{4}}{\sqrt[3]{4}} = \dfrac{3(3^{3/6})(4^{2/6})}{4\sqrt[3]{8}} = \dfrac{3\sqrt[6]{3^3 \cdot 4^2}}{8} = \dfrac{3}{8}\sqrt[6]{432}$

$l)$ $\dfrac{\sqrt{x-1} - \sqrt{x+1}}{\sqrt{x-1} + \sqrt{x+1}} \cdot \dfrac{\sqrt{x-1} - \sqrt{x+1}}{\sqrt{x-1} - \sqrt{x+1}} = \dfrac{(x-1) - 2\sqrt{(x-1)(x+1)} + (x+1)}{(x-1) - (x+1)} = \sqrt{x^2 - 1} - x$

$m)$ $\dfrac{x + \sqrt{x}}{1 + \sqrt{x} + x} = \dfrac{x + \sqrt{x}}{1 + x + \sqrt{x}} \cdot \dfrac{1 + x - \sqrt{x}}{1 + x - \sqrt{x}} = \dfrac{x^2 + \sqrt{x}}{(1 + x)^2 - x} = \dfrac{x^2 + \sqrt{x}}{1 + x + x^2}$

$n)$ $\dfrac{1}{\sqrt[3]{3} + \sqrt[3]{4}} = \dfrac{1}{3^{1/3} + 4^{1/3}}$ Let $x = 3^{1/3}$, $y = 4^{1/3}$.   Then

$$\dfrac{1}{x + y} \cdot \dfrac{x^2 - xy + y^2}{x^2 - xy + y^2} = \dfrac{x^2 - xy + y^2}{x^3 + y^3} = \dfrac{3^{2/3} - 3^{1/3} 4^{1/3} + 4^{2/3}}{(3^{1/3})^3 + (4^{1/3})^3} = \dfrac{\sqrt[3]{9} - \sqrt[3]{12} + 2\sqrt[3]{2}}{7}$$

SUPPLEMENTARY PROBLEMS

Show that:

11. a) $\sqrt{72} = 6\sqrt{2}$

b) $\sqrt{27} = 3\sqrt{3}$

c) $3\sqrt{20} = 6\sqrt{5}$

d) $\frac{2}{5}\sqrt{50a^2} = 2a\sqrt{2}$ (assuming $a \geqq 0$)

e) $\frac{a}{b}\sqrt{75a^3b^2} = 5a^2\sqrt{3a}$

f) $\frac{4}{ab}\sqrt{98a^2b^3} = 28\sqrt{2b}$

g) $\sqrt[3]{640} = 4\sqrt[3]{10}$

h) $\sqrt[3]{88x^3y^6z^5} = 2xy^2z\sqrt[3]{11z^2}$

i) $\sqrt{a/b} = \frac{\sqrt{ab}}{b}$

j) $14\sqrt{2/7} = 2\sqrt{14}$

k) $3\sqrt[3]{2/3} = \sqrt[3]{18}$

l) $\frac{3a}{4}\sqrt[3]{\frac{3}{2a}} = \frac{3}{8}\sqrt[3]{12a^2}$

m) $xyz\sqrt{\dfrac{5}{2x^2yz}} = \frac{1}{2}\sqrt{10yz}$

n) $60\sqrt{4/45} = 8\sqrt{5}$

o) $3\sqrt[4]{4/9} = \sqrt{6}$

12. a) $\sqrt{27} + \sqrt{48} - \sqrt{12} = 5\sqrt{3}$

b) $5\sqrt{8} - 3\sqrt{18} = \sqrt{2}$

c) $2\sqrt{150} - 4\sqrt{54} + 6\sqrt{48} = 24\sqrt{3} - 2\sqrt{6}$

d) $5\sqrt{2} - 3\sqrt{50} + 7\sqrt{288} = 74\sqrt{2}$

e) $\sqrt{16a^3 - 48a^2b} = 4a\sqrt{a - 3b}$ $(a \geq 0)$

f) $3\sqrt[3]{16a^3} + 8\sqrt[3]{a^3/4} = 10a\sqrt[3]{2}$

g) $\sqrt{\sqrt[3]{128}} = 2\sqrt[6]{2}$

h) $\sqrt[n]{x^{n+1}y^{2n-1}z^{3n}} = xy^2z^3\sqrt[n]{x/y}$

i) $3\sqrt[4]{9} - 2\sqrt[6]{27} = \sqrt{3}$

j) $6\sqrt{8a^3/3} - 2\sqrt{24ab^2} + a\sqrt{54a} = (7a-4b)\sqrt{6a}$ $a,b\geqq0$

k) $(x+1)\sqrt[3]{16x^2} - 4x\sqrt[3]{x^2/4} = 2\sqrt[3]{2x^2}$

l) $2\sqrt{54} - 6\sqrt{2/3} - \sqrt{96} = 0$

m) $4\sqrt{x/y} + \dfrac{3}{\sqrt[4]{x^2y^2}} - 5\sqrt[6]{y^3/x^3} = \dfrac{4x - 5y + 3}{xy}\sqrt{xy}$

13. a) $(3\sqrt{8})(6\sqrt{5}) = 36\sqrt{10}$

b) $\sqrt{48x^5}\sqrt{3x^3} = 12x^4$

c) $\sqrt[3]{2}\sqrt[3]{32} = 4$

d) $\sqrt{2}(\sqrt{2} + \sqrt{18}) = 8$

e) $(5 + \sqrt{2})(5 - \sqrt{2}) = 23$

f) $(x - \sqrt{y})(x + \sqrt{y}) = x^2 - y$

g) $(2\sqrt{3} - \sqrt{6})(3\sqrt{3} + 3\sqrt{6}) = 9\sqrt{2}$

h) $(3\sqrt{2} - 2\sqrt{3})(4\sqrt{2} + 3\sqrt{3}) = 6 + \sqrt{6}$

i) $(\sqrt{2} - \sqrt{3})^2 + (\sqrt{2} + \sqrt{3})^2 = 10$

j) $(2\sqrt{a} + 5\sqrt{a-b})(\sqrt{a} + \sqrt{a-b})$
     $= 7a - 5b + 7\sqrt{a^2 - ab}$

k) $(\sqrt{3} + \sqrt{5} + \sqrt{7})(\sqrt{3} + \sqrt{5} - \sqrt{7}) = 1 + 2\sqrt{15}$

l) $\sqrt{8 - 2\sqrt{7}}\sqrt{8 + 2\sqrt{7}} = 6$

m) $\dfrac{4 + \sqrt{8}}{2} = 2 + \sqrt{2}$

n) $\dfrac{6 - \sqrt{18}}{3} = 2 - \sqrt{2}$

o) $\dfrac{3}{\sqrt{2}} - \sqrt{\dfrac{1}{2}} = \sqrt{2}$

p) $\dfrac{8 + 4\sqrt{48}}{8} = 1 + 2\sqrt{3}$

q) $\dfrac{36 - 2\sqrt[3]{81}}{6} = 6 - \sqrt[3]{3}$

14.

a) $\dfrac{2\sqrt{24x^3}}{\sqrt{3x}} = 4x\sqrt{2} \quad (x > 0)$

b) $\dfrac{a\sqrt{b}}{b\sqrt{a}} = \dfrac{\sqrt{ab}}{b}$

c) $\dfrac{\sqrt{3}+\sqrt{2}}{\sqrt{2}} = 1 + \dfrac{1}{2}\sqrt{6}$

d) $\dfrac{\sqrt{6}-\sqrt{10}-\sqrt{12}}{\sqrt{18}} = \dfrac{\sqrt{3}-\sqrt{5}-\sqrt{6}}{3}$

e) $\sqrt[3]{\dfrac{9V^2}{16\pi^2}} = \dfrac{1}{4\pi}\sqrt[3]{36\pi\,V^2}$

f) $\dfrac{6a}{\sqrt[3]{12}} = a\sqrt[3]{18}$

g) $\dfrac{\sqrt[5]{3a^7b^6c^5}}{\sqrt[5]{24a^2bc}} = \dfrac{ab}{2}\sqrt[5]{4c^4}$

h) $\sqrt[3]{\dfrac{x^{-2}y^{-3}z^{-1}}{4xyz^2}} = \dfrac{1}{2xy^2z}\sqrt[3]{2y^2}$

i) $\dfrac{\sqrt{2}}{\sqrt[3]{3}} = \dfrac{\sqrt{2}\,\sqrt[3]{9}}{3} = \dfrac{\sqrt[6]{648}}{3}$

j) $\dfrac{\sqrt[3]{20}-\sqrt[3]{18}}{\sqrt[3]{12}} = \dfrac{1}{3}\sqrt[3]{45} - \dfrac{1}{2}\sqrt[3]{12}$

k) $\dfrac{1}{\sqrt{7}-2} = \dfrac{\sqrt{7}+2}{3}$

l) $\dfrac{5}{3+\sqrt{2}} = \dfrac{5}{7}(3-\sqrt{2})$

m) $\dfrac{-2}{2-\sqrt{3}} = -4 - 2\sqrt{3}$

n) $\dfrac{s\sqrt{3}}{\sqrt{3}-1} = \dfrac{3s}{2} + \dfrac{s\sqrt{3}}{2}$

o) $\dfrac{2\sqrt{3}-1}{\sqrt{3}+2} = 5\sqrt{3} - 8$

p) $\dfrac{1-\sqrt{x+1}}{1+\sqrt{x+1}} = \dfrac{2\sqrt{x+1}-x-2}{x}$

15.

a) $\dfrac{4}{2+\sqrt{5}} + \dfrac{3}{5+2\sqrt{5}} = \dfrac{14}{5}\sqrt{5} - 5$

b) $\dfrac{3\sqrt{3}+2\sqrt{5}}{3\sqrt{3}-2\sqrt{5}} = \dfrac{47+12\sqrt{15}}{7}$

c) $\dfrac{\sqrt{2}+\sqrt{3}+\sqrt{6}}{\sqrt{2}+\sqrt{3}} = 1 + 3\sqrt{2} - 2\sqrt{3}$

d) $\dfrac{a\sqrt{b}-b\sqrt{a}}{a\sqrt{b}+b\sqrt{a}} = \dfrac{a+b-2\sqrt{ab}}{a-b}$

e) $\dfrac{x+\sqrt{y}}{x-\sqrt{y}} + \dfrac{x-\sqrt{y}}{x+\sqrt{y}} = \dfrac{2x^2+2y}{x^2-y}$

f) $\sqrt{\dfrac{x-y}{x^3y-2x^2y^2+xy^3}} = \dfrac{\sqrt{xy(x-y)}}{xy(x-y)}$

g) $\dfrac{2+\sqrt{3}+\sqrt{5}}{2+\sqrt{3}-\sqrt{5}} = \dfrac{6+10\sqrt{3}+4\sqrt{5}+3\sqrt{15}}{11}$

h) $\dfrac{1}{2+\sqrt[3]{2}} = \dfrac{4-2\sqrt[3]{2}+\sqrt[3]{4}}{10}$

i) $\dfrac{3}{x+y-\sqrt{x^2-2xy+y^2}} = \dfrac{3}{2y} \quad \text{if } x \geqq y$

$\qquad\qquad\qquad\qquad\qquad = \dfrac{3}{2x} \quad \text{if } x \leqq y$

j) $\dfrac{2}{x^2-\sqrt{x^4+2x^2+1}} = -2$

CHAPTER 8

Simple Operations with Complex Numbers

THE UNIT OF IMAGINARY NUMBERS is $\sqrt{-1}$ and is generally designated by the letter i. Many laws which hold for real numbers hold for imaginary numbers as well.

Thus $\sqrt{-4} = \sqrt{(4)(-1)} = 2\sqrt{-1} = 2i$, $\sqrt{-18} = \sqrt{(18)(-1)} = \sqrt{18}\sqrt{-1} = 3\sqrt{2}\,i$. Also, since $i = \sqrt{-1}$, we have $i^2 = -1$, $i^3 = i^2 \cdot i = (-1)i = -i$, $i^4 = (i^2)^2 = (-1)^2 = 1$, $i^5 = i^4 \cdot i = 1 \cdot i = i$, and similarly for any integral power of i.

Note. One must be very careful in applying some of the laws which hold for real numbers. For example, one might be tempted to write

$$\sqrt{-4}\,\sqrt{-4} = \sqrt{(-4)(-4)} = \sqrt{16} = 4, \quad \text{which is incorrect.}$$

To avoid such difficulties, always express $\sqrt{-m}$, where m is a positive number, as $\sqrt{m}\,i$; and use $i^2 = -1$ whenever it arises. Thus

$$\sqrt{-4}\,\sqrt{-4} = (2i)(2i) = 4i^2 = -4, \quad \text{which is correct.}$$

A COMPLEX NUMBER is an expression of the form $a + bi$, where a and b are real numbers and $i = \sqrt{-1}$. In the complex number $a + bi$, a is called the *real part* and bi the *imaginary part*. When $a = 0$, the complex number is called a *pure imaginary*. If $b = 0$, the complex number reduces to the real number a. Thus complex numbers include all real numbers and all pure imaginary numbers.

Two complex numbers $a + bi$ and $c + di$ are *equal* if and only if $a = c$ and $b = d$. Thus $a + bi = 0$ if and only if $a = 0$, $b = 0$. If $c + di = 3$, then $c = 3$, $d = 0$.

THE CONJUGATE OF A COMPLEX NUMBER $a + bi$ is $a - bi$, and conversely. Thus $5 - 3i$ and $5 + 3i$ are conjugates.

ALGEBRAIC OPERATIONS WITH COMPLEX NUMBERS.

1) *To add* two complex numbers, add the real parts and the imaginary parts separately. Thus

$$(a + bi) + (c + di) = (a + c) + (b + d)i$$
$$(5 + 4i) + (3 + 2i) = (5 + 3) + (4 + 2)i = 8 + 6i$$
$$(-6 + 2i) + (4 - 5i) = (-6 + 4) + (2 - 5)i = -2 - 3i.$$

2) *To subtract* two complex numbers, subtract the real parts and the imaginary parts separately. Thus

$$(a + bi) - (c + di) = (a - c) + (b - d)i$$

$$(3+2i) - (5-3i) = (3-5) + (2+3)i = -2 + 5i$$
$$(-1+i) - (-3+2i) = (-1+3) + (1-2)i = 2 - i.$$

3) *To multiply* two complex numbers, treat the numbers as ordinary binomials and replace i^2 by -1. Thus

$$(a+bi)(c+di) = ac + adi + bci + bdi^2 = (ac-bd) + (ad+bc)i$$
$$(5+3i)(2-2i) = 10 - 10i + 6i - 6i^2 = 10 - 4i - 6(-1) = 16 - 4i.$$

4) *To divide* two complex numbers, multiply the numerator and denominator of the fraction by the conjugate of the denominator, replacing i^2 by -1. Thus

$$\frac{2+i}{3-4i} = \frac{2+i}{3-4i} \cdot \frac{3+4i}{3+4i} = \frac{6+8i+3i+4i^2}{9-16i^2} = \frac{2+11i}{25} = \frac{2}{25} + \frac{11}{25}i.$$

More advanced topics in complex numbers are treated in Chapter 20.

SOLVED PROBLEMS

1. Express each of the following in terms of i.

a) $\sqrt{-25} = \sqrt{(25)(-1)} = \sqrt{25}\sqrt{-1} = 5i$

b) $3\sqrt{-36} = 3\sqrt{36}\sqrt{-1} = 3 \cdot 6 \cdot i = 18i$

c) $-4\sqrt{-81} = -4\sqrt{81}\sqrt{-1} = -4 \cdot 9 \cdot i = -36i$

d) $\sqrt{-\frac{1}{2}} = \sqrt{\frac{1}{2}}\sqrt{-1} = \sqrt{\frac{2}{4}}\,i = \frac{\sqrt{2}}{2}\,i$

e) $2\sqrt{\frac{-16}{25}} - 3\sqrt{\frac{-49}{100}} = 2 \cdot \frac{4}{5}i - 3 \cdot \frac{7}{10}i = \frac{8}{5}i - \frac{21}{10}i = \frac{16}{10}i - \frac{21}{10}i = -\frac{1}{2}i$

f) $\sqrt{-12} - \sqrt{-3} = \sqrt{12}\,i - \sqrt{3}\,i = 2\sqrt{3}\,i - \sqrt{3}\,i = \sqrt{3}\,i$

g) $3\sqrt{-50} + 5\sqrt{-18} - 6\sqrt{-200} = 15\sqrt{2}\,i + 15\sqrt{2}\,i - 60\sqrt{2}\,i = -30\sqrt{2}\,i$

h) $-2 + \sqrt{-4} = -2 + \sqrt{4}\,i = -2 + 2i$ \qquad j) $\sqrt{8} + \sqrt{-8} = \sqrt{8} + \sqrt{8}\,i = 2\sqrt{2} + 2\sqrt{2}\,i$

i) $6 - \sqrt{-50} = 6 - \sqrt{50}\,i = 6 - 5\sqrt{2}\,i$ \qquad k) $\frac{1}{5}(-10 + \sqrt{-125}) = \frac{1}{5}(-10 + 5\sqrt{5}\,i) = -2 + \sqrt{5}\,i$

l) $\frac{1}{4}(\sqrt{32} + \sqrt{-128}) = \frac{1}{4}(4\sqrt{2} + 8\sqrt{2}\,i) = \sqrt{2} + 2\sqrt{2}\,i$

m) $\frac{\sqrt[3]{-8} + \sqrt{-8}}{2} = \frac{-2 + 2\sqrt{2}\,i}{2} = -1 + \sqrt{2}\,i$

2. Perform the indicated operations and simplify.

a) $(5-2i) + (6+3i) = 11 + i$ \qquad\qquad b) $(6+3i) - (4-2i) = 6 + 3i - 4 + 2i = 2 + 5i$

c) $(5 - 3i) - (-2 + 5i) = 5 - 3i + 2 - 5i = 7 - 8i$

d) $(\frac{3}{2} + \frac{5}{8}i) + (-\frac{1}{4} + \frac{1}{4}i) = \frac{3}{2} - \frac{1}{4} + (\frac{5}{8} + \frac{1}{4})i = \frac{5}{4} + \frac{7}{8}i$

e) $(a + bi) + (a - bi) = 2a$

f) $(a + bi) - (a - bi) = a + bi - a + bi = 2bi$

g) $(5 - \sqrt{-125}) - (4 - \sqrt{-20}) = (5 - 5\sqrt{5}\,i) - (4 - 2\sqrt{5}\,i) = 1 - 3\sqrt{5}\,i$

3. a) $\sqrt{-2}\,\sqrt{-32} = (\sqrt{2}\,i)(\sqrt{32}\,i) = \sqrt{2}\,\sqrt{32}\,i^2 = \sqrt{64}\,(-1) = -8$

b) $-3\sqrt{-5}\,\sqrt{-20} = -3(\sqrt{5}\,i)(\sqrt{20}\,i) = -3(\sqrt{5}\,\sqrt{20})i^2 = -3\sqrt{100}\,(-1) = 30$

c) $(4i)(-3i) = -12i^2 = 12$ e) $(2\sqrt{-1})^3 = (2i)^3 = 8i^3 = 8i(i^2) = -8i$

d) $(6i)^2 = 36i^2 = -36$ f) $3i(i + 2) = 3i^2 + 6i = -3 + 6i$

g) $(3 - 2i)(4 + i) = 3 \cdot 4 + 3 \cdot i - (2i)4 - (2i)i = 12 + 3i - 8i + 2 = 14 - 5i$

h) $(5 - 3i)(i + 2) = 5i + 10 - 3i^2 - 6i = 5i + 10 + 3 - 6i = 13 - i$

i) $(5 + 3i)^2 = 5^2 + 2(5)3i + (3i)^2 = 25 + 30i + 9i^2 = 16 + 30i$

j) $(2 - i)(3 + 2i)(1 - 4i) = (6 + 4i - 3i - 2i^2)(1 - 4i) = (8 + i)(1 - 4i)$
$$= 8 - 32i + i - 4i^2 = 12 - 31i$$

k) $(\frac{\sqrt{2}}{2} + \frac{\sqrt{2}}{2}i)^2 = (\frac{\sqrt{2}}{2})^2 + 2(\frac{\sqrt{2}}{2})(\frac{\sqrt{2}}{2}i) + (\frac{\sqrt{2}}{2}i)^2 = \frac{1}{2} + i - \frac{1}{2} = i$

l) $(1 + i)^3 = 1 + 3i + 3i^2 + i^3 = 1 + 3i - 3 - i = -2 + 2i$

m) $(3 - 2i)^3 = 3^3 + 3(3^2)(-2i) + 3(3)(-2i)^2 + (-2i)^3$
$$= 27 + 3(9)(-2i) + 3(3)(4i^2) - 8i^3 = 27 - 54i - 36 + 8i = -9 - 46i$$

n) $(3 + 2i)^3 = 3^3 + 3(3^2)(2i) + 3(3)(2i)^2 + (2i)^3$
$$= 27 + 54i + 36i^2 + 8i^3 = 27 + 54i - 36 - 8i = -9 + 46i$$

o) $(1 + 2i)^4 = [(1 + 2i)^2]^2 = (1 + 4i + 4i^2)^2 = (-3 + 4i)^2 = 9 - 24i + 16i^2 = -7 - 24i$

p) $(-1 + i)^8 = [(-1 + i)^2]^4 = (1 - 2i + i^2)^4 = (-2i)^4 = 16i^4 = 16$

4. a) $\frac{1 + i}{3 - i} = \frac{1 + i}{3 - i} \cdot \frac{3 + i}{3 + i} = \frac{3 + 3i + i + i^2}{3^2 - i^2} = \frac{2 + 4i}{10} = \frac{1}{5} + \frac{2}{5}i$

b) $\frac{1}{i} = \frac{1}{i}(\frac{-i}{-i}) = \frac{-i}{-i^2} = \frac{-i}{1} = -i$

c) $\frac{2\sqrt{3} + \sqrt{2}\,i}{3\sqrt{2} - 4\sqrt{3}\,i} = \frac{2\sqrt{3} + \sqrt{2}\,i}{3\sqrt{2} - 4\sqrt{3}\,i} \cdot \frac{3\sqrt{2} + 4\sqrt{3}\,i}{3\sqrt{2} + 4\sqrt{3}\,i} = \frac{(2\sqrt{3} + \sqrt{2}\,i)(3\sqrt{2} + 4\sqrt{3}\,i)}{(3\sqrt{2})^2 - (4\sqrt{3})^2 i^2}$

$$= \frac{6\sqrt{6} + 8\sqrt{9}\,i + 3\sqrt{4}\,i + 4\sqrt{6}\,i^2}{18 + 48} = \frac{2\sqrt{6} + 30i}{66} = \frac{\sqrt{6}}{33} + \frac{5}{11}i$$

SUPPLEMENTARY PROBLEMS

5. Express each of the following in terms of i.

a) $2\sqrt{-49}$ d) $4\sqrt{-1/8}$

b) $-4\sqrt{-64}$ e) $3\sqrt{-25} - 5\sqrt{-100}$

c) $6\sqrt{-1/9}$ f) $2\sqrt{-72} + 3\sqrt{-32}$

g) $\dfrac{-4 + \sqrt{-4}}{2}$

h) $\dfrac{1}{6}(-12 - \sqrt{-288})$

i) $4\sqrt{-81} - 3\sqrt{-36} + 4\sqrt{25}$

j) $3\sqrt{12} - 3\sqrt{-12}$

Perform each of the indicated operations and simplify.

6. a) $(3 + 4i) + (-1 - 6i)$

b) $(-2 + 5i) - (3 - 2i)$

c) $(\frac{2}{3} - \frac{1}{2}i) - (-\frac{1}{3} + \frac{1}{2}i)$

d) $(3 + \sqrt{-8}) - (2 - \sqrt{-32})$

e) $\sqrt{-3}\,\sqrt{-12}$

f) $(-i\sqrt{2})(i\sqrt{2})$

g) $(2i)^4$

h) $(\frac{1}{2}\sqrt{-3})^6$

i) $5i(2 - i)$

j) $(2 + i)(2 - i)$

k) $(-3 + 4i)(-3 - 4i)$

l) $(2 - 5i)(3 + 2i)$

m) $(3 - 4i)^2$

n) $(1 + i)(2 + 2i)(3 - i)$

o) $(i - 1)^3$

p) $(2 + 3i)^3$

q) $(1 - i)^4$

r) $(i + 2)^5$

7. a) $\dfrac{2 - 5i}{4 + 3i}$

b) $\dfrac{-1}{2 - 2i}$

c) $\dfrac{3\sqrt{2} + 2\sqrt{3}\,i}{3\sqrt{2} - 2\sqrt{3}\,i}$

d) $\dfrac{3 - \sqrt{2}\,i}{\sqrt{2}\,i}$

e) $\dfrac{1}{1 - 2i} + \dfrac{3}{1 + 4i}$

f) $\dfrac{5}{3 - 4i} + \dfrac{10}{4 + 3i}$

g) $\dfrac{i + i^2 + i^3 + i^4}{1 + i}$

h) $\dfrac{i^{26} - i}{i - 1}$

i) $\left(\dfrac{4i^{11} - i}{1 + 2i}\right)^2$

ANSWERS TO SUPPLEMENTARY PROBLEMS.

5. a) $14i$ c) $2i$ e) $-35i$ g) $-2 + i$ i) $18i + 20$

b) $-32i$ d) $\sqrt{2}\,i$ f) $24\sqrt{2}\,i$ h) $-2 - 2\sqrt{2}\,i$ j) $6\sqrt{3} - 6\sqrt{3}\,i$

6. a) $2 - 2i$ d) $1 + 6\sqrt{2}\,i$ g) 16 j) 5 m) $-7 - 24i$ p) $-46 + 9i$

b) $-5 + 7i$ e) -6 h) $-27/64$ k) 25 n) $4 + 12i$ q) -4

c) $1 - i$ f) 2 i) $5 + 10i$ l) $16 - 11i$ o) $2 + 2i$ r) $-38 + 41i$

7. a) $-\dfrac{7}{25} - \dfrac{26}{25}i$ c) $\dfrac{1}{5} + \dfrac{2}{5}\sqrt{6}\,i$ e) $\dfrac{32}{85} - \dfrac{26}{85}i$ g) 0

h) i

b) $-\dfrac{1}{4} - \dfrac{1}{4}i$ d) $-1 - \dfrac{3}{2}\sqrt{2}\,i$ f) $\dfrac{11}{5} - \dfrac{2}{5}i$ i) $3 + 4i$

CHAPTER 9

Equations in General

AN EQUATION is a statement of equality between two expressions called members.

An equation which is true for only certain values of the literals (sometimes called unknowns) involved is called a *conditional equation* or simply an equation.

An equation which is true for all permissible values of the literals (or unknowns) involved is called an *identity*. By permissible values are meant the values for which the members are defined.

For example:

1) $x + 5 = 8$ is true only for $x = 3$; it is a conditional equation.

2) $x^2 - y^2 = (x - y)(x + y)$ is true for all values of x and y; it is an identity.

3) $\dfrac{1}{x - 2} + \dfrac{1}{x - 3} = \dfrac{2x - 5}{(x - 2)(x - 3)}$ is true for all values except for the non-

permissible values $x = 2$, $x = 3$; these excluded values lead to division by zero which is not allowed. Since the equation is true for all permissible values of x, it is an identity.

The symbol $\equiv$ is often used for identities instead of $=$.

THE SOLUTIONS of a conditional equation are those values of the unknowns which make both members equal. These solutions are said to satisfy the equation. If only one unknown is involved the solutions are also called *roots*. To solve an equation means to find all of the solutions.

Thus $x = 2$ is a solution or root of $2x + 3 = 7$, since if we substitute $x = 2$ into the equation we obtain $2(2) + 3 = 7$ and both members are equal, i.e. the equation is satisfied. Similarly, three (of the many) solutions of $2x + y = 4$ are: $x = 0$, $y = 4$; $x = 1$, $y = 2$; $x = 5$, $y = -6$.

OPERATIONS USED IN TRANSFORMING EQUATIONS.

a) If equals are added to equals, the results are equal.

Thus if $x - y = z$, we may add y to both members and obtain $x = y + z$.

b) If equals are subtracted from equals, the results are equal.

Thus if $x + 2 = 5$, we may subtract 2 from both members to obtain $x = 3$.

Note. Because of *a*) and *b*) we may transpose a term from one member of an equation to the other member merely by changing the sign of the term. Thus if $3x + 2y - 5 = x - 3y + 2$,   then $3x - x + 2y + 3y = 5 + 2$ or $2x + 5y = 7$.

c) If equals are multiplied by equals, the results are equal.

Thus if both members of $\frac{1}{4}y = 2x^2$ are multiplied by 4 the result is $y = 8x^2$.

Similarly, if both members of $\frac{9}{5}C = F - 32$ are multiplied by $\frac{5}{9}$ the result is $C = \frac{5}{9}(F - 32)$.

d) If equals are divided by equals, the results are equal provided there is no division by zero.

Thus if $-4x = -12$, we may divide both members by -4 to obtain $x = 3$.

Similarly, if $E = IR$ we may divide both sides by $R \neq 0$ to obtain $I = E/R$.

e) The same powers of equals are equal.

Thus if $T = 2\pi\sqrt{l/g}$, then $T^2 = (2\pi\sqrt{l/g})^2 = 4\pi^2 l/g$.

f) The same roots of equals are equal. Thus if $r^3 = \dfrac{3V}{4\pi}$, then $r = \sqrt[3]{\dfrac{3V}{4\pi}}$.

g) Reciprocals of equals are equal provided the reciprocal of zero does not occur.

Thus if $\dfrac{1}{x} = \dfrac{1}{3}$, then $x = 3$.   Similarly, if $\dfrac{1}{R} = \dfrac{R_1 + R_2}{R_1 R_2}$ then $R = \dfrac{R_1 R_2}{R_1 + R_2}$.

Operations a)-f) are sometimes called axioms of equality.

EQUIVALENT EQUATIONS are equations having the same solutions.

Thus $x - 2 = 0$ and $2x = 4$ have the same solution $x = 2$ and so are equivalent.   However, $x - 2 = 0$ and $x^2 - 2x = 0$ are not equivalent since $x^2 - 2x = 0$ has the additional solution $x = 0$.

The above operations used in transforming equations may not all yield equations equivalent to the original equations. The use of such operations may yield derived equations with either more or fewer solutions than the original equation.

If the operations yield an equation with more solutions than the original, the extra solutions are called *extraneous* and the derived equation is said to be *redundant* with respect to the original equation. If the operations yield an equation with fewer solutions than the original, the derived equation is said to be *defective* with respect to the original equation.

Operations a) and b) always yield equivalent equations. Operations c) and e) may give rise to redundant equations and extraneous solutions. Operations d) and f) may give rise to defective equations.

A FORMULA is an equation which expresses a general fact, rule, or principle.

For example, in geometry the formula $A = \pi r^2$ gives the area A of a circle in terms of its radius r.

In physics the formula $s = \frac{1}{2}gt^2$, where g is approximately 32.2, gives the relation between the distance s, in feet, which an object will fall freely from rest during a time t, in seconds.

To solve a formula for one of the literals involved is to perform the same operations on both members of the formula until the desired literal appears on one side of the equation but not on the other side.

Thus if $F = ma$, we may divide by m to obtain $a = F/m$ and the formula is solved for a in terms of the other literals F and m. To check the work, substitute $a = F/m$ into the original equation to obtain $F = m(F/m)$, an identity.

A RATIONAL INTEGRAL TERM in a number of unknowns $x, y, z, \ldots$ has the form $ax^p y^q z^r \ldots$ where the exponents $p, q, r, \ldots$ are either positive integers or zero and the coefficient a is independent of the unknowns. The sum of the exponents $p + q + r + \ldots$ is called the *degree* of the term in the unknowns $x, y, z, \ldots$

Examples. $3x^2 z^3$, $\frac{1}{2}x^4$, 6 are rational integral terms.

$3x^2 z^3$ is of degree 2 in x, 3 in z, and 5 in x and z.

$\frac{1}{2}x^4$ is of fourth degree. 6 is of degree zero.

$\frac{4y}{x} = 4yx^{-1}$ is not integral in x; $3x\sqrt{y}\,z^3$ is not rational in y.

When reference is made to degree without specifying the unknowns considered, the degree in all unknowns is implied.

A RATIONAL INTEGRAL EXPRESSION or polynomial in various unknowns consists of terms each of which is rational and integral. The degree of such a rational integral expression is defined as the degree of the terms of highest degree.

Example. $3x^3 y^4 z + xy^2 z^5 - 8x + 3$ is a rational integral expression of degree 3 in x, 4 in y, 5 in z, 7 in x and y, 7 in y and z, 6 in x and z, and 8 in x, y and z.

A RATIONAL INTEGRAL EQUATION is a statement of equality between two rational integral expressions. The degree of such an equation is the degree of the term of highest degree present in the equation.

Example. $xyz^2 + 3xz = 2x^3 y + 3z^2$ is of degree 3 in x, 1 in y, 2 in z, 4 in x and y, 3 in y and z, 3 in x and z, and 4 in x, y and z.

It should be understood that like terms in the equation have been combined. Thus $4x^3 y + x^2 z - xy^2 = 4x^3 y + z$ should be written $x^2 z - xy^2 = z$.

An equation is called *linear* if it is of degree 1 and *quadratic* if it is of degree 2. Similarly the words *cubic*, *quartic* and *quintic* refer to equations of degree 3, 4 and 5 respectively.

Examples. $2x + 3y = 7z$ is a linear equation in x, y and z.

$x^2 - 4xy + 5y^2 = 10$ is a quadratic equation in x and y.

$x^3 + 3x^2 - 4x - 6 = 0$ is a cubic equation in x.

A RATIONAL INTEGRAL EQUATION OF DEGREE n in the unknown x may be written

$$a_0 x^n + a_1 x^{n-1} + a_2 x^{n-2} + \ldots + a_{n-1} x + a_n = 0 \qquad a_0 \neq 0$$

where a_0, a_1, $\ldots$, a_n are given constants and n is a positive integer.

As special cases we see that

$a_0 x + a_1 = 0$ or $ax + b = 0$ is of degree 1 (linear equation),

$a_0 x^2 + a_1 x + a_2 = 0$ or $ax^2 + bx + c = 0$ is of degree 2 (quadratic equation),

$a_0 x^3 + a_1 x^2 + a_2 x + a_3 = 0$ is of degree 3 (cubic equation),

$a_0 x^4 + a_1 x^3 + a_2 x^2 + a_3 x + a_4 = 0$ is of degree 4 (quartic equation).

SOLVED PROBLEMS

1. Which of the following are conditional equations and which are identities?

a) $3x - (x + 4) = 2(x - 2)$, $2x - 4 = 2x - 4$; identity.

b) $(x - 1)(x + 1) = (x - 1)^2$, $x^2 - 1 = x^2 - 2x + 1$; conditional equation.

c) $(y - 3)^2 + 3(2y - 3) = y(y + 1) - y$, $y^2 - 6y + 9 + 6y - 9 = y^2 + y - y$, $y^2 = y^2$; identity.

d) $x + 3y - 5 = 2(x + 2y) + 3$, $x + 3y - 5 = 2x + 4y + 3$; conditional equation.

2. Check each of the following equations for the indicated solution or solutions.

a) $\frac{x}{2} + \frac{x}{3} = 10$; $x = 12$. $\frac{12}{2} + \frac{12}{3} = 10$, $6 + 4 = 10$, and $x = 12$ is a solution.

b) $\frac{x^2 + 6x}{x + 2} = 3x - 2$; $x = 2$, $x = -1$. $\frac{2^2 + 6(2)}{2 + 2} = 3(2) - 2$, $\frac{16}{4} = 4$, and $x = 2$ is a solution.

$\frac{(-1)^2 + 6(-1)}{-1 + 2} = 3(-1) - 2$, $\frac{-5}{1} = -5$, and $x = -1$ is a solution.

c) $x^2 - xy + y^2 = 19$; $x = -2$, $y = 3$; $x = 4$, $y = 2 + \sqrt{7}$; $x = 2$, $y = -1$.

$x = -2$, $y = 3$: $(-2)^2 - (-2)3 + 3^2 = 19$, $19 = 19$,   and $x = -2$, $y = 3$ is a solution.

$x = 4$, $y = 2 + \sqrt{7}$: $4^2 - 4(2 + \sqrt{7}) + (2 + \sqrt{7})^2 = 19$, $16 - 8 - 4\sqrt{7} + (4 + 4\sqrt{7} + 7) = 19$, $19 = 19$, and $x = 4$, $y = 2 + \sqrt{7}$ is a solution.

$x = 2$, $y = -1$: $2^2 - 2(-1) + (-1)^2 = 19$, $7 = 19$,   and $x = 2$, $y = -1$ is not a solution.

3. Use the axioms of equality to solve each equation.

a) $2(x + 3) = 3(x - 1)$, $2x + 6 = 3x - 3$.

Transposing terms: $2x - 3x = -6 - 3$, $-x = -9$.   Multiplying by -1: $x = 9$.

Check: $2(9 + 3) = 3(9 - 1)$, $24 = 24$.

b) $\frac{x}{3} + \frac{x}{6} = 1$.   Multiplying by 6: $2x + x = 6$, $3x = 6$.   Dividing by 3: $x = 2$.

Check: $2/3 + 2/6 = 1$, $1 = 1$.

c) $3y - 2(y - 1) = 4(y + 2)$, $3y - 2y + 2 = 4y + 8$, $y + 2 = 4y + 8$.

Transposing: $y - 4y = 8 - 2$, $-3y = 6$.   Dividing by -3: $y = \frac{6}{-3} = -2$.

Check: $3(-2) - 2(-2 - 1) = 4(-2 + 2)$, $0 = 0$.

d) $\dfrac{2x-3}{x-1} = \dfrac{4x-5}{x-1}$.   Multiplying by $x-1$, $2x-3 = 4x-5$ or $x=1$.

Check: Substituting $x=1$ into the given equation, we find $-1/0 = -1/0$. This is meaningless since division by zero is an excluded operation, and the given equation has no solution.

Note that 1) $\dfrac{2x-3}{x-1} = \dfrac{4x-5}{x-1}$ and 2) $2x-3 = 4x-5$ are not equivalent equations. When 1) is multiplied by $x-1$ an *extraneous solution* $x=1$ is introduced, and equation 2) is *redundant* with respect to equation 1).

e) $x(x-3) = 2(x-3)$.   Dividing each member by $x-3$ gives a solution $x=2$.

Now $x-3=0$ or $x=3$ is also a solution of the given equation which was lost in the division. The required roots are $x=2$ and $x=3$.

The equation $x=2$ is *defective* with respect to the given equation.

f) $\sqrt{x+2} = -1$.   Squaring both sides, $x+2=1$ or $x=-1$.

Check: Substituting $x=-1$ into the given equation, $\sqrt{1} = -1$ or $1=-1$ which is false.

Thus $x=-1$ is an *extraneous solution*. The given equation has no solution.

g) $\sqrt{2x-4} = 6$.   Squaring both sides, $2x-4 = 36$ or $x=20$.

Check: If $x=20$, $\sqrt{2(20)-4} = 6$ or $\sqrt{36} = 6$ which is true.

Hence $x=20$ is a solution. In this case no extraneous root was introduced.

4. In each of the following formulas, solve for the indicated letter.

a) $E = IR$, for R.   Dividing both sides by $I \neq 0$, we have $R = E/I$.

b) $s = v_0 t + \tfrac{1}{2}at^2$, for a.

Transposing, $\tfrac{1}{2}at^2 = s - v_0 t$. Multiplying by 2, $at^2 = 2(s - v_0 t)$.

Dividing by $t^2 \neq 0$, $a = \dfrac{2(s - v_0 t)}{t^2}$.

c) $\dfrac{1}{f} = \dfrac{1}{p} + \dfrac{1}{q}$, for p.   Transposing, $\dfrac{1}{p} = \dfrac{1}{f} - \dfrac{1}{q} = \dfrac{q-f}{fq}$.

Taking reciprocals, $p = \dfrac{fq}{q-f}$ (assuming $q \neq f$).

d) $T = 2\pi\sqrt{l/g}$, for g. Squaring both sides, $T^2 = \dfrac{4\pi^2 l}{g}$.

Multiplying by g, $gT^2 = 4\pi^2 l$. Dividing by T^2, $g = 4\pi^2 l/T^2$.

5. In each of the following formulas, find the value of the indicated letter, given the values of the other letters.

a) $F = \dfrac{9}{5}C + 32$, $F = 68$; find C. $68 = \dfrac{9}{5}C + 32$, $36 = \dfrac{9}{5}C$, $C = \dfrac{5}{9}(36) = 20$.

Another method. $\dfrac{9}{5}C = F - 32$, $C = \dfrac{5}{9}(F - 32) = \dfrac{5}{9}(68 - 32) = \dfrac{5}{9}(36) = 20$.

b) $\dfrac{1}{R} = \dfrac{1}{R_1} + \dfrac{1}{R_2}$, $R = 6, R_1 = 15$; find R_2. $\dfrac{1}{6} = \dfrac{1}{15} + \dfrac{1}{R_2}$, $\dfrac{1}{R_2} = \dfrac{1}{6} - \dfrac{1}{15} = \dfrac{5-2}{30} = \dfrac{1}{10}$, $R_2 = 10$.

Another method. $\dfrac{1}{R_2} = \dfrac{1}{R} - \dfrac{1}{R_1} = \dfrac{R_1 - R}{RR_1}$, $R_2 = \dfrac{RR_1}{R_1 - R} = \dfrac{6(15)}{15 - 6} = 10$.

c) $V = \dfrac{4}{3}\pi r^3$, $V = 288\pi$; find r. $288\pi = \dfrac{4}{3}\pi r^3$, $r^3 = \dfrac{288\pi}{4\pi/3} = 216$, $r = 6$.

Another method. $3V = 4\pi r^3$, $r^3 = \dfrac{3V}{4\pi}$, $r = \sqrt[3]{\dfrac{3V}{4\pi}} = \sqrt[3]{\dfrac{3(288\pi)}{4\pi}} = \sqrt[3]{216} = 6$.

6. Determine the degree of each of the following equations in each of the indicated unknowns.

a) $2x^2 + xy - 3 = 0$: x; y; x and y.

Degree 2 in x, 1 in y, 2 in x and y.

b) $3xy^2 - 4y^2z + 5x - 3y = x^4 + 2$: x; z; y and z; x, y and z.

Degree 4 in x, 1 in z, 3 in y and z, 4 in x, y and z.

c) $x^2 = \dfrac{3}{y + z}$: x; x and z; x, y and z.

As it stands it is not a rational integral equation. However, it can be transformed into one by multiplying by $y + z$ to obtain $x^2(y + z) = 3$ or $x^2y + x^2z = 3$. The derived equation is a rational integral equation of degree 2 in x, 3 in x and z, and 3 in x, y and z.

d) $\sqrt{x + 3} = x + y$: y; x and y.

As given it is not a rational integral equation, but it can be transformed into one by squaring both sides. Thus we obtain $x + 3 = x^2 + 2xy + y^2$ which is of degree 2 in y and 2 in x and y.

It should be mentioned, however, that the equations are not equivalent since $x^2 + 2xy + y^2 = x + 3$ includes both $\sqrt{x + 3} = x + y$ and $-\sqrt{x + 3} = x + y$.

7. Find all values of x for which *a*) $x^2 = 81$, *b*) $(x - 1)^2 = 4$.

a) There is nothing to indicate whether x is a positive or negative number, so we must assume either is possible. Taking the square root of both sides of the given equation, we obtain $\sqrt{x^2} = \sqrt{81} = 9$. Now $\sqrt{x^2}$ represents a positive number (or zero) if x is real. Hence we have $\sqrt{x^2} = x$ if x is positive while $\sqrt{x^2} = -x$ if x is negative. Thus when writing $\sqrt{x^2}$ we must consider that it is either x (if $x > 0$) or $-x$ (if $x < 0$). Therefore the equation $\sqrt{x^2} = 9$ may be written as either $x = 9$ or $-x = 9$ (i.e. $x = -9$). The two solutions may be written $x = \pm 9$.

b) $(x - 1)^2 = 4$, $\pm(x - 1) = 2$ or $(x - 1) = \pm 2$,   and the two roots are $x = 3$ and $x = -1$.

8. Explain the fallacy.
 a) Let $x = y$: $x = y$
 b) Multiply both sides by x: $x^2 = xy$
 c) Subtract y^2 from both sides: $x^2 - y^2 = xy - y^2$
 d) Write as: $(x - y)(x + y) = y(x - y)$
 e) Divide by $x - y$: $x + y = y$
 f) Replace x by its equal, y: $y + y = y$
 g) Hence: $2y = y$
 h) Divide by y: $2 = 1$

There is nothing wrong in steps *a*), *b*), *c*), *d*).

However, in *e*) we divide by $x - y$ which from the original assumption is zero. Since division by zero is not defined, everything we do from *e*) on is to be looked upon with disfavor.

9. Show that $\sqrt{2}$ is an irrational number, i.e. it cannot be the quotient of two integers.

Assume that $\sqrt{2} = p/q$ where p and q are integers having no common factor except ± 1, i.e. p/q is in lowest terms. Squaring, we have $p^2/q^2 = 2$ or $p^2 = 2q^2$. Since $2q^2$ is an even number, p^2 is even and hence p is even (if p were odd, p^2 would be odd); then $p = 2k$, where k is an integer. Thus $p^2 = 2q^2$ becomes $(2k)^2 = 2q^2$ or $q^2 = 2k^2$; hence q^2 is even and q is even. But if p and q are both even they would have a common factor 2, thus contradicting the assumption that they have no common factor except ± 1. Hence $\sqrt{2}$ is irrational.

SUPPLEMENTARY PROBLEMS

10. State which of the following are conditional equations and which are identities.

a) $2x + 3 - (2 - x) = 4x - 1$

b) $(2y - 1)^2 + (2y + 1)^2 = (2y)^2 + 6$

c) $2\{x + 4 - 3(2x - 1)\} = 3(4 - 3x) + 2 - x$

d) $(x + 2y)(x - 2y) - (x - 2y)^2 + 4y(2y - x) = 0$

e) $\dfrac{9x^2 - 4y^2}{3x - 2y} = 2x + 3y$

f) $(x - 3)(x^2 + 3x + 9) = x^3 - 27$

g) $\dfrac{x^2}{4} + \dfrac{x^2}{12} = x^2$

h) $(x^2 - y^2)^2 + (2xy)^2 = (x^2 + y^2)^2$

11. Check each of the following equations for the indicated solution or solutions.

a) $\dfrac{y^2 - 4}{y - 2} = 2y - 1$; $y = 3$

b) $x^2 - 3x = 4$; $-1, -4$

c) $\sqrt{3x - 2} - \sqrt{x + 2} = 4$; $34, 2$

d) $x^3 - 6x^2 + 11x - 6 = 0$; $1, 2, 3$

e) $\dfrac{1}{x} + \dfrac{1}{2x} = \dfrac{1}{x - 1}$; $x = 3$

f) $y^3 + y^2 - 5y - 5 = 0$; $\pm\sqrt{5}, -1$

g) $x^2 - 2y = 3y^2$; $x = 4, y = 2$; $x = 1, y = -1$

h) $(x + y)^2 + (x - y)^2 = 2(x^2 + y^2)$; any values of x, y

12. Use the axioms of equality to solve each equation. Check the solutions obtained.

a) $5(x - 4) = 2(x + 1) - 7$

b) $\dfrac{2y}{3} - \dfrac{y}{6} = 2$

c) $\dfrac{1}{y} = 8 - \dfrac{3}{y}$

d) $\dfrac{x + 1}{x - 1} = \dfrac{x - 1}{x - 2}$

e) $\dfrac{3x - 2}{x - 2} = \dfrac{x + 2}{x - 2}$

f) $\sqrt{3x - 2} = 4$

g) $\sqrt{2x + 1} + 5 = 0$

h) $\sqrt[3]{2x - 3} + 1 = 0$

i) $(y + 1)^2 = 16$

j) $(2x + 1)^2 + (2x - 1)^2 = 34$

13. In each of the following formulas, solve for the indicated letter.

a) $\dfrac{P_1 V_1}{T_1} = \dfrac{P_2 V_2}{T_2}$; T_2

b) $t = \sqrt{\dfrac{2s}{g}}$; s

c) $m = \dfrac{1}{2}\sqrt{2a^2 + 2b^2 - c^2}$; c

d) $v^2 = v_0^2 + 2as$; a

e) $T = 2\pi\sqrt{\dfrac{m}{k}}$; k

f) $S = \dfrac{n}{2}[2a + (n - 1)d]$; d

14. In each formula find the value of the indicated letter given the values of the other letters.

    a) $v = v_0 + at$; find a if $v = 20$, $v_0 = 30$, $t = 5$.

    b) $S = \dfrac{n}{2}(a+d)$; find d if $S = 570$, $n = 20$, $a = 40$.

    c) $\dfrac{1}{f} = \dfrac{1}{p} + \dfrac{1}{q}$; find q if $f = 30$, $p = 10$.

    d) $Fs = \tfrac{1}{2}mv^2$; find v if $F = 100$, $s = 5$, $m = 2.5$.

    e) $f = \dfrac{1}{2\pi\sqrt{LC}}$; find C to four decimal places if $f = 1000$, $L = 4 \cdot 10^{-6}$.

15. Determine the degree of each equation in each of the indicated unknowns.

    a) $x^3 - 3x + 2 = 0$: x

    b) $x^2 + xy + 3y^4 = 6$: x; y; x and y

    c) $2xy^3 - 3x^2y^2 + 4xy = 2x^3$: x; y; x and y

    d) $xy + yz + xz + z^2x = y^4$: x; y; z; x and z; y and z; x, y and z

16. Classify each equation according as it is (or can be transformed into) an equation which is linear, quadratic, cubic, quartic or quintic in all of the unknowns present.

    a) $2x^4 + 3x^3 - x - 5 = 0$ e) $\sqrt{x^2 + y^2 - 1} = x + y$

    b) $x - 2y = 4$ f) $\dfrac{2x + y}{x - 3y} = 4$

    c) $2x^2 + 3xy + y^2 = 10$ g) $3y^2 - 4y + 2 = 2(y - 3)^2$

    d) $x^2y^3 - 2xyz = 4 + y^5$ h) $(z + 1)^2 (z - 2) = 0$

17. Is the equation $\sqrt{(x+4)^2} = x + 4$ an identity? Explain.

18. Prove that $\sqrt{3}$ is irrational.

ANSWERS TO SUPPLEMENTARY PROBLEMS.

10. a) Conditional equation   c) Identity   e) Conditional equation   g) Conditional equation
    b) Conditional equation   d) Identity   f) Identity   h) Identity

11. a) $y = 3$ is a solution.   e) $x = 3$ is a solution.
    b) $x = -1$ is a solution, $x = -4$ is not.   f) $y = \pm\sqrt{5}$, -1 are all solutions.
    c) $x = 34$ is a solution, $x = 2$ is not.   g) $x = 4$, $y = 2$; $x = 1$, $y = -1$ are solutions.
    d) $x = 1, 2, 3$ are all solutions.   h) The equation is an identity; hence any values of x and y are solutions.

12. a) $x = 5$ c) $y = 1/2$ e) no solution   g) no solution   i) $y = 3, -5$
    b) $y = 4$ d) $x = 3$ f) $x = 6$ h) $x = 1$ j) $x = \pm 2$

13. a) $T_2 = \dfrac{P_2V_2T_1}{P_1V_1}$ c) $c = \pm\sqrt{2a^2 + 2b^2 - 4m^2}$ e) $k = \dfrac{4\pi^2 m}{T^2}$

    b) $s = \tfrac{1}{2}gt^2$ d) $a = \dfrac{v^2 - v_0^2}{2s}$ f) $d = \dfrac{2S - 2an}{n(n-1)}$

14. a) $a = -2$ b) $d = 17$ c) $q = -15$ d) $v = \pm 20$ e) $C = .0063$

15. a) 3   b) 2, 4, 4   c) 3, 3, 4   d) 1, 4, 2, 3, 4, 4

16. a) quartic   c) quadratic   e) quadratic   g) quadratic
    b) linear   d) quintic   f) linear   h) cubic

17. $\sqrt{(x+4)^2} = x + 4$ only if $x + 4 \geqq 0$; $\sqrt{(x+4)^2} = -(x+4)$ if $x + 4 \leqq 0$.
    The given equation is not an identity.

CHAPTER 10

Functions and Graphs

A VARIABLE is a symbol which may assume any one of a set of values during a discussion. A *constant* is a symbol which stands for only one particular value during the discussion.

Letters at the end of the alphabet, as x, y, z, u, v, w, are usually employed to represent variables, and letters at the beginning of the alphabet, as a, b, c, are used as constants.

FUNCTION OF A VARIABLE. A variable y is said to be a *function* of a variable x if there exists a relation between y and x such that to each value which x may assume there corresponds one or more values of y. In this chapter we shall consider only real numbers and real functions.

Example 1. $y = x^2 - 5x + 2$ defines a relation between the variables x and y. Thus if $x = 0, 1, 2, -1$ then $y = 2, -2, -4, 8$ respectively.

Example 2. The circumference C of a circle is a function of the radius r of the circle, the relation being given by $C = 2\pi r$. Thus circles of radii 1, 3, 5 ft have circumferences $2\pi, 6\pi, 10\pi$ ft respectively.

Example 3. The population y of the United States is a function of the year x. The following table shows this relationship for the years 1880-1950 in ten-year intervals.

Year x	1880	1890	1900	1910	1920	1930	1940	1950
Population y in millions	50.2	62.9	76.0	92.0	105.7	122.8	131.7	150.7

If there is one and only one value of y corresponding to each value which x may assume, then y is a *single-valued* function of x, otherwise it is a *many-valued* function of x.

Thus in examples 1 and 3 above, y is a single-valued function of x because for a given value of x there is one and only one value for y. Similarly in example 2, C is a single-valued function of r.

However, $y = \pm\sqrt{x}$ defines y as a two-valued function of x because given x there correspond two values of y (an exception being $x = 0$). Thus if $x = 4$, $y = \pm 2$; if $x = 5$, $y = \pm\sqrt{5}$; etc.

INDEPENDENT AND DEPENDENT VARIABLES. The variable to which values are assigned (in the above examples, x) is called the *independent variable*; the other variable which is thereby determined (in the above examples, y) is called the *dependent variable*. Saying that y is a function of x is equivalent to saying that y depends on x.

The set of values which the independent variable may assume is called the *range* of the variable.

In example 1, x may be any real number.

In example 2, the independent variable is the radius r of the circle. The range of r is the set of all positive numbers and zero.

In example 3, the years 1880, ..., 1950 constitute the range of x.

In $y = \pm\sqrt{x}$, x may assume only non-negative values if we require real values for y. The range in this case is the set of all numbers greater than or equal to zero, written $x \geqq 0$.

In $y = \dfrac{x}{(x-1)(x+2)}$, x may assume any real value except $x = 1$ and $x = -2$ for which y is not defined. Thus the range consists of all real numbers except 1 and −2.

THE FUNCTIONAL NOTATION $y = f(x)$, read "y equals f of x", is often used to designate that y is a function of x. With this notation $f(a)$ represents the value of the dependent variable y when $x = a$ (provided that there is a value).

Thus $y = x^2 - 5x + 2$ may be written $f(x) = x^2 - 5x + 2$. Then $f(2)$, i.e. the value of $f(x)$ or y when $x = 2$, is $f(2) = 2^2 - 5(2) + 2 = -4$. Similarly, $f(-1) = (-1)^2 - 5(-1) + 2 = 8$.

Any letter may be used in the functional notation; thus $g(x)$, $h(x)$, $F(x)$, etc., may represent functions of x.

A RECTANGULAR COORDINATE SYSTEM is used to give a picture of the functional relationship between two variables.

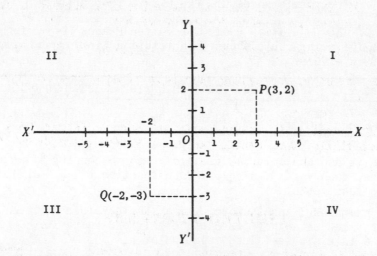

Consider two mutually perpendicular lines $X'X$ and $Y'Y$ intersecting in the point O, as shown in the above figure.

The line $X'X$, called the x axis, is usually horizontal.

The line $Y'Y$, called the y axis, is usually vertical.

The point O is called the origin.

Using a convenient unit of length, lay off points on the x axis at successive units to the right and left of the origin O, labeling those points to the right $1, 2, 3, 4, \ldots$ and those to the left $-1, -2, -3, -4, \ldots$ Here we have arbitrarily chosen OX to be the positive direction; this is customary but not necessary.

Do the same on the y axis, choosing OY as the positive direction. It is customary (but not necessary) to use the same unit of length on both axes.

The x and y axes divide the plane into 4 parts known as *quadrants*, which are labeled I, II, III, IV as in the figure.

Given a point P in this xy plane, drop perpendiculars from P to the x and y axes. The values of x and y at the points where these perpendiculars meet the x and y axes determine respectively the x *coordinate* (or abscissa) of the point and the y *coordinate* (or ordinate) of the point P. These coordinates are indicated by the symbol (x, y).

Conversely, given the coordinates of a point, we may locate or plot the point in the xy plane.

For example, the point P in the figure has coordinates $(3, 2)$; the point having coordinates $(-2, -3)$ is Q.

THE GRAPH OF A FUNCTION $y = f(x)$ is the locus of all points (x, y) satisfied by the equation $y = f(x)$.

FUNCTION OF TWO VARIABLES. The variable z is said to be a function of the variables x and y if there exists a relation such that to each pair of values of x and y there corresponds one or more values of z. Here x and y are independent variables and z is the dependent variable.

The functional notation used in this case is $z = f(x, y)$, read "z equals f of x and y". Then $f(a, b)$ denotes the value of z when $x = a$ and $y = b$, provided the function is defined for these values.

Thus if $f(x, y) = x^3 + xy^2 - 2x$, then $f(2, 3) = 2^3 + 2 \cdot 3^2 - 2 \cdot 3 = 20$.

In like manner we may define functions of more than two independent variables.

STATISTICAL GRAPHS depict the relationship of data obtained from observations such as scientific experiments, census, business conditions, etc.

SOLVED PROBLEMS

1. Express the area A of a square as a function of its a) side x, b) perimeter P, c) diagonal D.

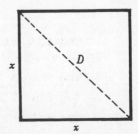

a) $A = x^2$

b) $P = 4x$ or $x = \dfrac{P}{4}$.   Then $A = x^2 = (\dfrac{P}{4})^2$ or $A = \dfrac{P^2}{16}$.

c) $D = \sqrt{x^2 + x^2} = x\sqrt{2}$ or $x = \dfrac{D}{\sqrt{2}}$.   Then $A = x^2 = (\dfrac{D}{\sqrt{2}})^2$ or $A = \dfrac{D^2}{2}$.

2. Express the a) area A, b) perimeter P and c) diagonal D of a rectangle as a function of its sides x and y. Refer to Fig.(a) below.

a) $A = xy$, b) $P = 2x + 2y$, c) $D = \sqrt{x^2 + y^2}$

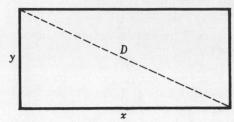

Fig.(a) Prob. 2

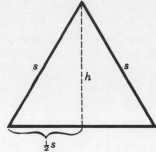

Fig.(b) Prob. 3

3. Express the a) altitude h and b) area A of an equilateral triangle as a function of its side s. Refer to Fig.(b) above.

a) $h = \sqrt{s^2 - (\frac{1}{2}s)^2} = \sqrt{\frac{3}{4}s^2} = \frac{s\sqrt{3}}{2}$ b) $A = \frac{1}{2}hs = \frac{1}{2}(\frac{s\sqrt{3}}{2})s = \frac{s^2\sqrt{3}}{4}$

4. The surface area S and volume V of a sphere of radius r are given by $S = 4\pi r^2$ and $V = \frac{4}{3}\pi r^3$.
Express a) r as a function of S and also as a function of V, b) V as a function of S and (c) S as a function of V.

a) From $S = 4\pi r^2$ obtain $r = \sqrt{\frac{S}{4\pi}} = \frac{1}{2}\sqrt{\frac{S}{\pi}}$.   From $V = \frac{4}{3}\pi r^3$ obtain $r = \sqrt[3]{\frac{3V}{4\pi}}$.

b) Set $r = \frac{1}{2}\sqrt{\frac{S}{\pi}}$ in $V = \frac{4}{3}\pi r^3$ and obtain $V = \frac{4}{3}\pi(\frac{1}{2}\sqrt{\frac{S}{\pi}})^3 = \frac{S}{6}\sqrt{\frac{S}{\pi}}$.

c) Set $r = \sqrt[3]{\frac{3V}{4\pi}}$ in $S = 4\pi r^2$ and obtain $S = 4\pi\sqrt[3]{(\frac{3V}{4\pi})^2} = 4\pi\sqrt[3]{\frac{9V^2}{16\pi^2}\cdot\frac{4\pi}{4\pi}} = \sqrt[3]{36\pi V^2}$.

5. Given $y = 3x^2 - 4x + 1$, find the values of y corresponding to $x = -2, -1, 0, 1, 2$.

For $x = -2$, $y = 3(-2)^2 - 4(-2) + 1 = 21$;   for $x = -1$, $y = 3(-1)^2 - 4(-1) + 1 = 8$;   for $x = 0$, $y = 3(0)^2 - 4(0) + 1 = 1$;   for $x = 1$, $y = 3(1)^2 - 4(1) + 1 = 0$;   for $x = 2$, $y = 3(2)^2 - 4(2) + 1 = 5$. These values of x and y are conveniently listed in the following table.

x	−2	−1	0	1	2
y	21	8	1	0	5

6. Extend the table of values in Problem 5 by finding the values of y corresponding to $x = -3/2$, $-1/2, 1/2, 3/2$.

For $x = -3/2$, $y = 3(-3/2)^2 - 4(-3/2) + 1 = 13\frac{3}{4}$; etc. The following table of values summarizes the results.

x	−2	$-\frac{3}{2}$	−1	$-\frac{1}{2}$	0	$\frac{1}{2}$	1	$\frac{3}{2}$	2
y	21	$13\frac{3}{4}$	8	$3\frac{3}{4}$	1	$-\frac{1}{4}$	0	$1\frac{3}{4}$	5

FUNCTIONAL NOTATION.

7. If $f(x) = x^3 - 5x - 2$, find $f(-2)$, $f(-3/2)$, $f(-1)$, $f(0)$, $f(1)$, $f(2)$.

$$f(-2) = (-2)^3 - 5(-2) - 2 = 0 \qquad\qquad f(0) = 0^3 - 5(0) - 2 = -2$$

$$f(-3/2) = (-3/2)^3 - 5(-3/2) - 2 = 17/8 \qquad f(1) = 1^3 - 5(1) - 2 = -6$$

$$f(-1) = (-1)^3 - 5(-1) - 2 = 2 \qquad\qquad f(2) = 2^3 - 5(2) - 2 = -4$$

We may arrange these values in a table.

x	-2	-3/2	-1	0	1	2
$f(x)$	0	17/8	2	-2	-6	-4

8. If $F(t) = \dfrac{t^3 + 2t}{t - 1}$, find $F(-2)$, $F(x)$, $F(-x)$.

$$F(-2) = \frac{(-2)^3 + 2(-2)}{-2 - 1} = \frac{-8 - 4}{-3} = 4$$

$$F(x) = \frac{x^3 + 2x}{x - 1}$$

$$F(-x) = \frac{(-x)^3 + 2(-x)}{-x - 1} = \frac{-x^3 - 2x}{-x - 1} = \frac{x^3 + 2x}{x + 1}$$

9. Given $R(x) = \dfrac{3x - 1}{4x + 2}$, find $a)$ $R(\dfrac{x - 1}{x + 2})$, $b)$ $\dfrac{R(x + h) - R(x)}{h}$, $c)$ $R[R(x)]$.

$a)$ $R(\dfrac{x - 1}{x + 2}) = \dfrac{3(\frac{x - 1}{x + 2}) - 1}{4(\frac{x - 1}{x + 2}) + 2} = \dfrac{\frac{2x - 5}{x + 2}}{\frac{6x}{x + 2}} = \dfrac{2x - 5}{6x}$

$b)$ $\dfrac{R(x + h) - R(x)}{h} = \dfrac{1}{h}\{R(x + h) - R(x)\} = \dfrac{1}{h}\{\dfrac{3(x + h) - 1}{4(x + h) + 2} - \dfrac{3x - 1}{4x + 2}\}$

$$= \dfrac{1}{h}\{\dfrac{[3(x + h) - 1][4x + 2] - [3x - 1][4(x + h) + 2]}{[4(x + h) + 2][4x + 2]}\} = \dfrac{5}{2(2x + 2h + 1)(2x + 1)}$$

$c)$ $R[R(x)] = R(\dfrac{3x - 1}{4x + 2}) = \dfrac{3(\frac{3x - 1}{4x + 2}) - 1}{4(\frac{3x - 1}{4x + 2}) + 2} = \dfrac{5x - 5}{20x} = \dfrac{x - 1}{4x}$

10. If $F(x,y) = x^3 - 3xy + y^2$, find $a)$ $F(2,3)$, $b)$ $F(-3,0)$, $c)$ $\dfrac{F(x, y+k) - F(x,y)}{k}$.

$a)$ $F(2,3) = 2^3 - 3(2)(3) + 3^2 = -1$

$b)$ $F(-3,0) = (-3)^3 - 3(-3)(0) + 0^2 = -27$

$c)$ $\dfrac{F(x, y+k) - F(x,y)}{k} = \dfrac{x^3 - 3x(y + k) + (y + k)^2 - [x^3 - 3xy + y^2]}{k} = -3x + 2y + k$

RECTANGULAR COORDINATES AND GRAPHS.

11. Plot the following points on a rectangular coordinate system: $(2,1)$, $(4,3)$, $(-2,4)$, $(-4,2)$, $(-4,-2)$, $(-5/2,-9/2)$, $(4,-3)$, $(2,-\sqrt{2})$. See Figure (c) below.

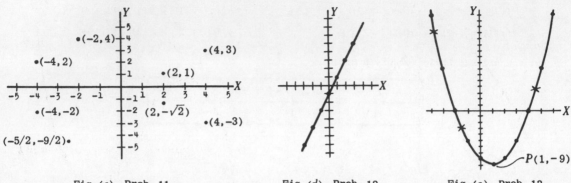

Fig. (c) Prob. 11   Fig. (d) Prob. 12   Fig. (e) Prob. 13

12. Given $y = 2x - 1$, obtain the values of y corresponding to $x = -3,-2,-1,0,1,2,3$ and plot the points (x,y) thus obtained.

The following table lists the values of y corresponding to the given values of x.

x	-3	-2	-1	0	1	2	3
y	-7	-5	-3	-1	1	3	5

The points $(-3,-7)$, $(-2,-5)$, $(-1,-3)$, $(0,-1)$, $(1,1)$, $(2,3)$, $(3,5)$ are plotted, as shown in Figure (d) above.

Note that all points satisfying $y = 2x - 1$ lie on a straight line. In general the graph of $y = ax + b$, where a and b are constants, is a straight line; hence $y = ax + b$ or $f(x) = ax + b$ is called a *linear function*. Since two points determine a straight line, only two points need be plotted and the line drawn connecting them.

13. Obtain the graph of the function defined by $y = x^2 - 2x - 8$ or $f(x) = x^2 - 2x - 8$.

The following table gives the values of y or $f(x)$ for various values of x.

x	-4	-3	-2	-1	0	1	2	3	4	5	6
y or $f(x)$	16	7	0	-5	-8	-9	-8	-5	0	7	16

Thus the following points lie on the graph: $(-4,16)$, $(-3,7)$, $(-2,0)$, $(-1,-5)$, etc.

In plotting these points it is convenient to use different scales on the x and y axes, as shown in Figure (e) above. The points marked × were added to those already obtained in order to get a more accurate picture.

The curve thus obtained is called a *parabola*. The lowest point P, called a minimum point, is the *vertex* of the parabola.

14. Graph the function defined by $y = 3 - 2x - x^2$.

x	-5	-4	-3	-2	-1	0	1	2	3	4
y	-12	-5	0	3	4	3	0	-5	-12	-21

The curve obtained is a parabola, as shown in Figure (f) below. The point $Q(-1,4)$, the vertex of the parabola, is a maximum point. In general, $y = ax^2 + bx + c$ represents a parabola whose vertex is either a maximum or minimum point depending on whether a is − or +, respectively. The function $f(x) = ax^2 + bx + c$ is sometimes called a quadratic function.

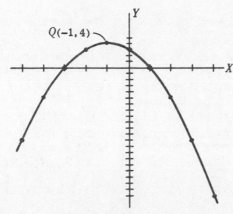

Fig.(f) Prob. 14

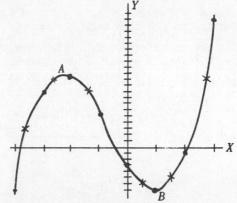

Fig.(g) Prob. 15

15. Obtain the graph of $y = x^3 + 2x^2 - 7x - 3$.

x	−4	−3	−2	−1	0	1	2	3
y	−7	9	11	5	−3	−7	−1	21

The graph is shown in Figure (g) above. Points marked × are not listed in the table; they were added in order to improve the accuracy of the graph.

Point A is called a *relative maximum point*; it is not the highest point on the entire curve, but points on either side are lower. Point B is called a *relative minimum point*. The calculus enables us to determine such relative maximum and minimum points.

16. Obtain the graph of $x^2 + y^2 = 36$.

We may write $y^2 = 36 - x^2$ or $y = \pm\sqrt{36 - x^2}$. Note that x must have a value between −6 and +6 if y is to be a real number.

x	−6	−5	−4	−3	−2	−1	0	1	2	3	4	5	6
y	0	$\pm\sqrt{11}$	$\pm\sqrt{20}$	$\pm\sqrt{27}$	$\pm\sqrt{32}$	$\pm\sqrt{35}$	±6	$\pm\sqrt{35}$	$\pm\sqrt{32}$	$\pm\sqrt{27}$	$\pm\sqrt{20}$	$\pm\sqrt{11}$	0

The points to be plotted are $(-6,0)$, $(-5,\sqrt{11})$, $(-5,-\sqrt{11})$, $(-4,\sqrt{20})$, $(-4,-\sqrt{20})$, etc.

The adjoining figure shows the graph, a circle of radius 6.

In general, the graph of $x^2 + y^2 = a^2$ is a circle with center at the origin and radius a.

It should be noted that if the units had not been taken the same on the x and y axes, the graph would not have looked like a circle.

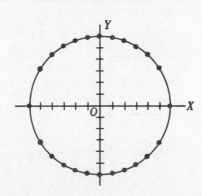

SOME APPLICATIONS OF GRAPHS.

17. A man has 40 ft of wire fencing with which to form a rectangular garden. The fencing is to be used only on three sides of the garden, his house providing the fourth side. Determine the maximum area which can be enclosed?

Let x = length of each of two of the fenced sides of the rectangle; then $40 - 2x$ = length of the third fenced side.

The area A of the garden is $A = x(40 - 2x) = 40x - 2x^2$. We wish to find the maximum value of A. A table of values and the graph of A plotted against x are shown. It is clear that x must have a value between 0 and 20 ft if A is to be positive.

x	0	5	8	10	12	15	20
A	0	150	192	200	192	150	0

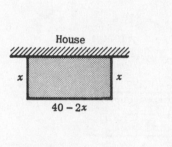

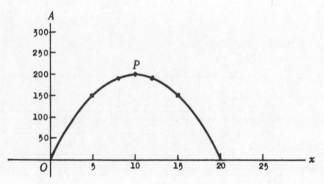

From the graph, the maximum point P has coordinates (10, 200) so that the dimensions of the garden are 10 ft by 20 ft and the area is 200 ft^2.

18. A rectangular piece of tin has dimensions 12 in. by 18 in. It is desired to make an open box from this material by cutting out equal squares from the corners and then bending up the sides. What are the dimensions of the squares cut out if the volume of the box is to be as large as possible?

Let x be the length of the side of the square cut out from each corner. The volume V of the box thus obtained is seen from the figure to be $V = x(12 - 2x)(18 - 2x)$. It is clear that x must be between 0 and 6 in. if there is to be a box.

x	0	1	2	2½	3	3½	4	5	6
V	0	160	224	227½	216	192½	160	80	0

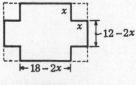

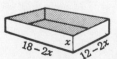

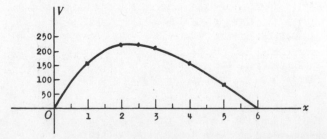

From the graph, the value of x corresponding to the maximum value of V lies between 2 and 2.5 in. By plotting more points it is seen that $x = 2.4$ in. approximately.

Problems such as this and Problem 17 may often be solved easily and exactly by methods of the calculus.

19. A cylindrical can is to have a volume of 200 cubic inches. Find the dimensions of the can which is made of the least amount of material.

Let x be the radius and y the height of the cylinder.

The area of the top or bottom of the can is πx^2 and the lateral area is $2\pi xy$; then the total area $S = 2\pi x^2 + 2\pi xy$.

The volume of the cylinder is $\pi x^2 y$, so that $\pi x^2 y = 200$ and $y = \dfrac{200}{\pi x^2}$. Then

$$S = 2\pi x^2 + 2\pi x \left(\frac{200}{\pi x^2}\right) \quad \text{or} \quad S = 2\pi x^2 + \frac{400}{x}.$$

A table of values and the graph of S plotted against x are shown below. We take $\pi = 3.14$ approximately.

x	1	2	3	3.2	3.5	4	4.5	5	6	7	8
S	406	225	190	189	191	200	216	237	293	365	452

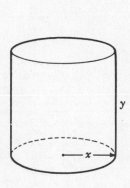
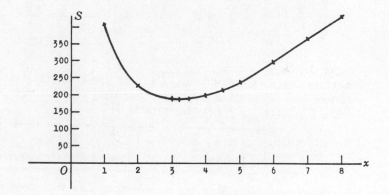

From the graph, minimum $S = 189$ in^2 occurs when $x = 3.2$ in. approximately; and from $y = \dfrac{200}{\pi x^2}$ we have $y = 6.2$ in. approximately.

20. Find approximately the values of x for which $x^3 + 2x^2 - 7x - 3 = 0$.

Consider $y = x^3 + 2x^2 - 7x - 3$. We must find values of x for which $y = 0$.

From the graph of $y = x^3 + 2x^2 - 7x - 3$, which is shown in Problem 15, it is clear that there are three real values of x for which $y = 0$ (the values of x where the curve intersects the x axis). These values are $x = -3.7$, $x = -.4$, $x = 2.1$ approximately. More exact values may be obtained by advanced techniques.

STATISTICAL GRAPHS.

21. The following table shows the population of the United States (in millions) for the years
1840, 1850, ..., 1950. Graph these data.

Year	1840	1850	1860	1870	1880	1890	1900	1910	1920	1930	1940	1950
Population (millions)	17.1	23.2	31.4	39.8	50.2	62.9	76.0	92.0	105.7	122.8	131.7	150.7

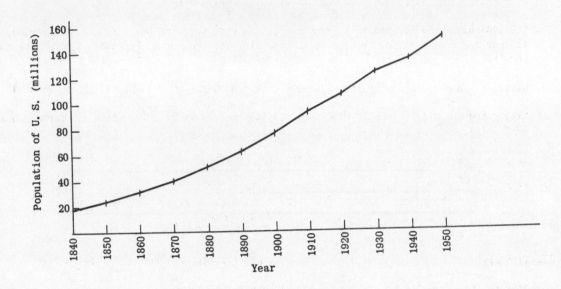

22. The time T (in seconds) required for one complete vibration of a simple pendulum of length
l (in centimeters) is given by the following observations obtained in a physics laboratory.
Exhibit graphically T as a function of l.

l	16.2	22.2	33.8	42.0	53.4	66.7	74.5	86.6	100.0
T	.81	.95	1.17	1.30	1.47	1.65	1.74	1.87	2.01

The observation points are connected by a smooth curve, as is usually done in science and
engineering.

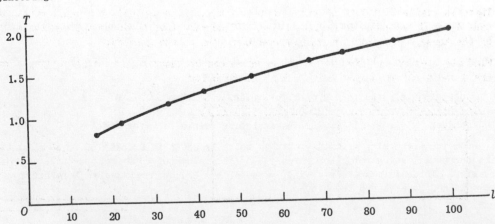

SUPPLEMENTARY PROBLEMS

23. A rectangle has sides of lengths x and $2x$. Express the area A of the rectangle as a function of its a) side x, b) perimeter P, c) diagonal D.

24. Express the area S of a circle in terms of its a) radius r, b) diameter d, c) perimeter P.

25. Express the area A of an isosceles triangle as a function of x and y, where x is the length of the two equal sides and y is the length of the third side.

26. The side of a cube has length x. Express a) x as a function of the volume V of the cube, b) the surface area S of the cube as a function of x, c) the volume V as a function of the surface S.

27. Given $y = 5 + 3x - 2x^2$, find the values of y corresponding to $x = -3, -2, -1, 0, 1, 2, 3$.

28. Extend the table of values in Problem 27 by finding the values of y which correspond to $x = -5/2, -3/2, -1/2, 1/2, 3/2, 5/2$.

29. If $f(x) = 2x^2 + 6x - 1$, find $f(-3)$, $f(-2)$, $f(0)$, $f(1/2)$, $f(3)$.

30. If $F(u) = \dfrac{u^2 - 2u}{1 + u}$, find a) $F(1)$, b) $F(2)$, c) $F(x)$, d) $F(-x)$.

31. If $G(x) = \dfrac{x - 1}{x + 1}$, find a) $G(\dfrac{x}{x + 1})$, b) $\dfrac{G(x + h) - G(x)}{h}$, c) $G(x^2 + 1)$.

32. If $F(x, y) = 2x^2 + 4xy - y^2$, find a) $F(1, 2)$, b) $F(-2, -3)$, c) $F(x + 1, y - 1)$.

33. Plot the following points on a rectangular coordinate system:
    a) $(1, 3)$, b) $(-2, 1)$, c) $(-1/2, -2)$, d) $(-3, 2/3)$, e) $(-\sqrt{3}, 3)$.

34. If $y = 3x + 2$, a) obtain the values of y corresponding to $x = -2, -1, 0, 1, 2$ and b) plot the points (x, y) thus obtained.

35. Graph the functions a) $f(x) = 1 - 2x$, b) $f(x) = x^2 - 4x + 3$, c) $f(x) = 4 - 3x - x^2$.

36. Graph $y = x^3 - 6x^2 + 11x - 6$.

37. Graph a) $x^2 + y^2 = 16$, b) $x^2 + 4y^2 = 16$.

38. There is available 120 ft of wire fencing with which to enclose two equal rectangular gardens A and B, as shown in Fig. (a) below. If no wire fencing is used along the sides formed by the house, determine the maximum combined area of the gardens.

39. Find the area of the largest rectangle which can be inscribed in a right triangle whose legs are 6 and 8 inches respectively. See Fig. (b) below.

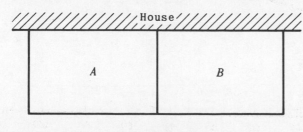

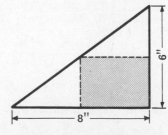

Fig. (a) Prob. 38   Fig. (b) Prob. 39

40. Obtain the relative maximum and minimum values of the function $f(x) = 2x^3 - 15x^2 + 36x - 23$.

41. From the graph of $y = x^3 - 7x + 6$ obtain the roots of the equation $x^3 - 7x + 6 = 0$.

42. Show that the equation $x^3 - x^2 + 2x - 3 = 0$ has only one real root.

43. Show that $x^4 - x^2 + 1 = 0$ cannot have any real roots.

44. The percentage of workers in the United States employed in agriculture during the years 1860, 1870, ..., 1950 is given in the following table. Graph the data.

Year	1860	1870	1880	1890	1900	1910	1920	1930	1940	1950
% of all workers in agriculture	58.9	53.0	49.4	42.6	37.5	31.0	27.0	21.4	18.0	12.8

45. The total time required to bring an automobile to a stop after perceiving danger is composed of the *reaction time* (time between recognition of danger and application of brakes) plus *braking time* (time for stopping after application of brakes). The following table gives the stopping distances d (feet) of an automobile traveling at speeds v (miles per hour) at the instant danger is sighted. Graph d against v.

Speed v (mi/hr)	20	30	40	50	60	70
Stopping distance d (ft)	54	90	138	206	292	396

46. The time t taken for an object to fall freely from rest through various heights h is given in the following table.

Time t (sec)	1	2	3	4	5	6
Height h (ft)	16	64	144	256	400	576

a) Graph h against t.
b) How long would it take an object to fall freely from rest through 48 ft? 300 ft?
c) Through what distance can an object fall freely from rest in 3.6 seconds?

ANSWERS TO SUPPLEMENTARY PROBLEMS.

23. $A = 2x^2$, $A = \dfrac{P^2}{18}$, $A = \dfrac{2D^2}{5}$ 25. $A = \dfrac{y}{2}\sqrt{x^2 - y^2/4} = \dfrac{y}{4}\sqrt{4x^2 - y^2}$

24. $S = \pi r^2$, $S = \dfrac{\pi d^2}{4}$, $S = \dfrac{P^2}{4\pi}$ 26. $x = \sqrt[3]{V}$, $S = 6x^2$, $V = \sqrt{\dfrac{S^3}{216}} = \dfrac{S}{36}\sqrt{6S}$

27.

x	-3	-2	-1	0	1	2	3
y	-22	-9	0	5	6	3	-4

28.

x	-5/2	-3/2	-1/2	1/2	3/2	5/2
y	-15	-4	3	6	5	0

29. $f(-3) = -1$, $f(-2) = -5$, $f(0) = -1$, $f(1/2) = 5/2$, $f(3) = 35$

30. *a*) $-1/2$,   *b*) 0,   *c*) $\dfrac{x^2 - 2x}{1 + x}$,   *d*) $\dfrac{x^2 + 2x}{1 - x}$

31. *a*) $\dfrac{-1}{2x + 1}$,   *b*) $\dfrac{2}{(x + 1)(x + h + 1)}$,   *c*) $\dfrac{x^2}{x^2 + 2}$

32. *a*) 6,   *b*) 23,   *c*) $2x^2 + 4xy - y^2 + 6y - 3$ 38. 1200 ft^2 39. 12 in^2

40. Maximum value of $f(x)$ is 5 (at $x = 2$); minimum value of $f(x)$ is 4 (at $x = 3$).

41. Roots are $x = -3$, $x = 1$, $x = 2$.   46. *b*) 1.7 sec, 4.3 sec;   *c*) 207 ft

CHAPTER 11

Linear Equations in One Unknown

A LINEAR EQUATION IN ONE UNKNOWN has the form $ax + b = 0$, where $a \neq 0$ and b are constants. The solution of this equation is given by $x = -\dfrac{b}{a}$.

SOLVED PROBLEMS

1. Solve each of the following equations.

a) $x + 1 = 5$, $\quad x = 5 - 1$, $\quad x = 4$.
Check: Put $x = 4$ in the original equation and obtain $4 + 1 = 5$, $\quad 5 = 5$.

b) $3x - 7 = 14$, $\quad 3x = 14 + 7$, $\quad 3x = 21$, $\quad x = 7$. $\qquad$ Check: $3(7) - 7 = 14$, $\quad 14 = 14$.

c) $3x + 2 = 6x - 4$, $\quad 3x - 6x = -4 - 2$, $\quad -3x = -6$, $\quad x = 2$.

d) $x + 3(x - 2) = 2x - 4$, $\quad x + 3x - 6 = 2x - 4$, $\quad 4x - 2x = 6 - 4$, $\quad 2x = 2$, $\quad x = 1$.

e) $3x - 2 = 7 - 2x$, $\quad 3x + 2x = 7 + 2$, $\quad 5x = 9$, $\quad x = 9/5$.

f) $2(t + 3) = 5(t - 1) - 7(t - 3)$, $\quad 2t + 6 = 5t - 5 - 7t + 21$, $\quad 4t = 10$, $\quad t = 10/4 = 5/2$.

g) $3x + 4(x - 2) = x - 5 + 3(2x - 1)$, $\quad 3x + 4x - 8 = x - 5 + 6x - 3$, $\quad 7x - 8 = 7x - 8$.
This is an identity and is true for all values of x.

h) $\dfrac{x - 3}{2} = \dfrac{2x + 4}{5}$, $\quad 5(x - 3) = 2(2x + 4)$, $\quad 5x - 15 = 4x + 8$, $\quad x = 23$.

i) $3 + 2[y - (2y + 2)] = 2[y + (3y - 1)]$, $\quad 3 + 2[y - 2y - 2] = 2[y + 3y - 1]$,
$\qquad 3 + 2y - 4y - 4 = 2y + 6y - 2$, $\quad -2y - 1 = 8y - 2$, $\quad -10y = -1$, $\quad y = 1/10$.

j) $(s + 3)^2 = (s - 2)^2 - 5$, $\quad s^2 + 6s + 9 = s^2 - 4s + 4 - 5$, $\quad 6s + 4s = -9 - 1$, $\quad s = -1$.

k) $\dfrac{x - 2}{x + 2} = \dfrac{x - 4}{x + 4}$, $\quad (x - 2)(x + 4) = (x - 4)(x + 2)$, $\quad x^2 + 2x - 8 = x^2 - 2x - 8$, $\quad 4x = 0$, $\quad x = 0$.

Check: $\dfrac{0 - 2}{0 + 2} = \dfrac{0 - 4}{0 + 4}$, $\quad -1 = -1$.

l) $\dfrac{3x + 1}{x + 2} = \dfrac{3x - 2}{x + 1}$, $\quad (x + 1)(3x + 1) = (x + 2)(3x - 2)$, $\quad 3x^2 + 4x + 1 = 3x^2 + 4x - 4$ or $1 = -4$.

There is no value of x which satisfies this equation.

$n)$ $\dfrac{5}{x} + \dfrac{5}{2x} = 6$.   Multiplying by $2x$, $5(2) + 5 = 12x$, $12x = 15$, $x = 5/4$.

$o)$ $\dfrac{x+3}{2x} + \dfrac{5}{x-1} = \dfrac{1}{2}$.   Multiplying by $2x(x-1)$, the L.C.D. of the fractions,

$\quad$ $(x+3)(x-1) + 5(2x) = x(x-1)$, $x^2 + 2x - 3 + 10x = x^2 - x$, $13x = 3$, $x = 3/13$.

$p)$ $\dfrac{2}{x-3} - \dfrac{4}{x+3} = \dfrac{16}{x^2-9}$.   Multiplying by $(x-3)(x+3)$ or $x^2 - 9$,

$\quad$ $2(x+3) - 4(x-3) = 16$, $2x + 6 - 4x + 12 = 16$, $-2x = -2$, $x = 1$.

$q)$ $\dfrac{1}{y} - \dfrac{1}{y+3} = \dfrac{1}{y+2} - \dfrac{1}{y+5}$, $\dfrac{(y+3)-y}{y(y+3)} = \dfrac{(y+5)-(y+2)}{(y+2)(y+5)}$, $\dfrac{3}{y(y+3)} = \dfrac{3}{(y+2)(y+5)}$,

$\quad$ $(y+2)(y+5) = y(y+3)$, $y^2 + 7y + 10 = y^2 + 3y$, $4y = -10$, $y = -5/2$.

$r)$ $\dfrac{3}{x^2-4x} - \dfrac{2}{2x^2-5x-12} = \dfrac{9}{2x^2+3x}$ or $\dfrac{3}{x(x-4)} - \dfrac{2}{(2x+3)(x-4)} = \dfrac{9}{x(2x+3)}$.

$\quad$ Multiplying by $x(x-4)(2x+3)$, the L.C.D. of the fractions,

$\quad$ $3(2x+3) - 2x = 9(x-4)$, $6x + 9 - 2x = 9x - 36$, $45 = 5x$, $x = 9$.

LITERAL EQUATIONS.

2. Solve for x.

$a)$ $2x - 4p = 3x + 2p$, $2x - 3x = 2p + 4p$, $-x = 6p$, $x = -6p$.

$b)$ $ax + a = bx + b$, $ax - bx = b - a$, $x(a-b) = b - a$, $x = \dfrac{b-a}{a-b} = -1$ provided $a \neq b$.

$\quad$ If $a = b$ the equation is an identity and is true for all values of x.

$c)$ $2cx + 4d = 3ax - 4b$, $2cx - 3ax = -4b - 4d$, $x = \dfrac{-4b-4d}{2c-3a} = \dfrac{4b+4d}{3a-2c}$ provided $3a \neq 2c$.

$\quad$ If $3a = 2c$ there is no solution unless $d = -b$, in which case the original equation is an identity.

$d)$ $\dfrac{3x+a}{b} = \dfrac{4x+b}{a}$, $3ax + a^2 = 4bx + b^2$, $3ax - 4bx = b^2 - a^2$, $x = \dfrac{b^2-a^2}{3a-4b}$ (provided $3a \neq 4b$).

TRANSLATION OF WORDS INTO SYMBOLS.

3. Express each statement in terms of algebraic symbols.

$a)$ One more than twice a certain number.

$\quad$ Let x = the number. Then $2x$ = twice the number, and one more than twice the number = $2x + 1$.

$b)$ Three less than five times a certain number.

$\quad$ Let x = the number. Then three less than five times the number = $5x - 3$.

$c)$ Each of two numbers whose sum is 100.

$\quad$ If x = one of the numbers, then $100 - x$ = the other number.

d) Three consecutive integers (for example, 5, 6, 7).

If x is the smallest integer, then $(x+1)$ and $(x+2)$ are the other two integers.

e) Each of two numbers whose difference is 10.

Let x = the smaller number; then $(x+10)$ = the larger number.

f) The amount by which 100 exceeds three times a given number.

Let x = given number. Then the excess of 100 over $3x$ is $(100-3x)$.

g) Any odd integer.

Let x = any integer. Then $2x$ is always an even integer, and $(2x+1)$ is an odd integer.

h) Four consecutive odd integers (for example, 1, 3, 5, 7; 17, 19, 21, 23).

The difference between two consecutive odd integers is 2.
Let $2x+1$ = smallest odd integer. Then the required numbers are $2x+1$, $2x+3$, $2x+5$, $2x+7$.

i) The number of cents in x dollars.

Since 1 dollar = 100 cents, x dollars = $100x$ cents.

j) John is twice as old as Mary, and Mary is three times as old as Bill. Express each of their ages in terms of a single unknown.

Let x = Bill's age. Then Mary's age is $3x$ and John's age is $2(3x) = 6x$.

Another method. Let y = John's age. Then Mary's age = $\frac{1}{2}y$ and Bill's age = $\frac{1}{3}(\frac{1}{2}y) = \frac{1}{6}y$.

k) The three angles A, B, C of a triangle if angle A has $10°$ more than twice the number of degrees in angle C.

Let $C = x°$; then $A = (2x+10)°$. Since $A+B+C = 180°$, $B = 180° - (A+C) = (170-3x)°$.

l) The time it takes a boat traveling at a speed of 20 mi/hr to cover a distance of x miles.
Distance = speed × time. Then time = $\dfrac{\text{distance}}{\text{speed}} = \dfrac{x \text{ mi}}{20 \text{ mi/hr}} = \dfrac{x}{20}$ hr.

m) The perimeter and area of a rectangle if one side is 4 ft longer than twice the other side.

Let x ft = length of shorter side; then $(2x+4)$ft = length of longer side.
The perimeter = $2(x) + 2(2x+4) = (6x+8)$ ft, and the area = $x(2x+4)$ ft^2.

n) The fraction whose numerator is 3 less than 4 times its denominator.

Let x = denominator; then numerator = $4x-3$. The fraction is $\dfrac{4x-3}{x}$.

o) The number of quarts of alcohol contained in a tank holding x gallons of a mixture which is 40% alcohol by volume.

In x gallons of mixture are $0.40x$ gallons of alcohol or $4(0.40x) = 1.6x$ quarts of alcohol.

ABSTRACT NUMBER PROBLEMS.

4. The sum of two numbers is 21, and one number is twice the other. Find the numbers.

Let x and $2x$ be the two numbers.
Then $x + 2x = 21$ or $x = 7$, and the required numbers are $x = 7$ and $2x = 14$.
Check. $7 + 14 = 21$ and $14 = 2(7)$, as required.

5. Ten less than four times a certain number is 14. Determine the number.

Let x = required number. Then $4x - 10 = 14$, $4x = 24$, and $x = 6$.

Check. Ten less than four times 6 is $4(6) - 10 = 14$, as required.

6. The sum of three consecutive integers is 24. Find the integers.

Let the three consecutive integers be x, $x + 1$, $x + 2$.

Then $x + (x + 1) + (x + 2) = 24$ or $x = 7$, and the required integers are 7, 8, 9.

7. The sum of two numbers is 37. If the larger is divided by the smaller, the quotient is 3 and the remainder is 5. Find the numbers.

Let x = smaller number, $37 - x$ = larger number.

Then $\dfrac{\text{larger number}}{\text{smaller number}} = 3 + \dfrac{5}{\text{smaller number}}$ or $\dfrac{37 - x}{x} = 3 + \dfrac{5}{x}$.

Solving, $37 - x = 3x + 5$, $4x = 32$, $x = 8$. The required numbers are 8, 29.

AGE PROBLEMS.

8. A man is 41 years old and his son is 9. In how many years will the father be three times as old as the son?

Let x = required number of years.

$$\text{Father's age in } x \text{ years} = 3(\text{son's age in } x \text{ years})$$
$$41 + x = 3(9 + x) \qquad \text{and } x = 7 \text{ yr.}$$

9. Ten years ago John was four times as old as Bill. Now he is only twice as old as Bill. Find their present ages.

Let x = Bill's present age; then $2x$ = John's present age.

$$\text{John's age ten years ago} = 4(\text{Bill's age ten years ago})$$
$$2x - 10 = 4(x - 10) \qquad \text{and } x = 15 \text{ yr.}$$

Hence Bill's present age is $x = 15$ yr and John's present age is $2x = 30$ yr.

Check. Ten years ago Bill was 5 and John 20, i.e. John was four times as old as Bill.

COIN PROBLEMS.

10. Robert has 50 coins, all in nickels and dimes, amounting to \$3.50. How many nickels does he have? Let x = number of nickels; then $50 - x$ = number of dimes.

$$\text{Amount in nickels} + \text{amount in dimes} = 350\cent$$
$$5x\cent \quad + \quad 10(50 - x)\cent = 350\cent \quad \text{from which } x = 30 \text{ nickels.}$$

11. In a purse are nickels, dimes and quarters amounting to \$1.85. There are twice as many dimes as quarters, and the number of nickels is two less than twice the number of dimes. Determine the number of coins of each kind.

Let x = number of quarters; then $2x$ = no. of dimes, and $2(2x) - 2 = 4x - 2$ = no. of nickels.

$$\text{Amount in quarters} + \text{amount in dimes} + \text{amount in nickels} = 185\cent$$
$$25(x)\cent \quad + \quad 10(2x)\cent \quad + \quad 5(4x - 2)\cent = 185\cent \quad \text{from which } x = 3.$$

Hence there are $x = 3$ quarters, $2x = 6$ dimes, and $4x - 2 = 10$ nickels.

Check. 3 quarters = 75¢, 6 dimes = 60¢, 10 nickels = 50¢, and their sum = \$1.85.

DIGIT PROBLEMS.

12. The tens digit of a certain two digit number exceeds the units digit by 4 and is 1 less than twice the units digit. Find the two digit number.

Let x = units digit; then $x + 4$ = tens digit.

Since the tens digit = 2(units digit) – 1, we have $x + 4 = 2(x) - 1$ or $x = 5$.

Thus $x = 5$, $x + 4 = 9$, and the required number is 95.

13. The sum of the digits of a two digit number is 12. If the digits are reversed, the new number is 4/7 times the original number. Determine the original number.

Let x = units digit; $12 - x$ = tens digit.

Original number = $10(12 - x) + x$; reversing digits, the new number = $10x + (12 - x)$. Then

$$\text{new number} = \frac{4}{7}(\text{original number}) \quad \text{or} \quad 10x + (12 - x) = \frac{4}{7}\left[10(12 - x) + x\right].$$

Solving, $x = 4$, $12 - x = 8$, and the original number is 84.

BUSINESS PROBLEMS.

14. A man has $4000 invested, part at 5% and the remainder at 3% simple interest. The total income per year from these investments is $168. How much does he have invested at each rate?

Let x = amount invested at 5%; $4000 - x$ = amount at 3%.

$$\text{Interest from 5\% investment} + \text{interest from 3\% investment} = \$168$$
$$.05x \qquad + \qquad .03(4000 - x) \qquad = \qquad 168.$$

Solving, $x = \$2400$ at 5%, $\$4000 - x = \$1600$ at 3%.

15. What amount should an employee receive as bonus so that he would net $500 after deducting 30% for taxes?

Let x = required amount. Then required amount – taxes = $500

or $x - 0.30x = \$500$ and $x = \$714.29$.

16. At what price should a merchant mark a sofa that costs him $120 in order that he may offer a discount of 20% on the marked price and still make a profit of 25% on the selling price?

Let x = marked price; then sale price = $x - 0.20x = 0.80x$.

Since profit = 25% of sale price, cost = 75% of sale price. Then

$$\text{cost} = 0.75(\text{sale price})$$
$$\$120 = 0.75(0.8x), \qquad \$120 = 0.6x \quad \text{and} \quad x = \$200.$$

MENSURATION PROBLEMS.

17. When each side of a given square is increased by 4 feet the area is increased by 64 square feet. Determine the dimensions of the original square.

Let x = side of given square; $x + 4$ = side of new square.

$$\text{New area} = \text{old area} + 64$$
$$(x + 4)^2 = x^2 + 64 \qquad \text{from which } x = 6 \text{ ft.}$$

18. One leg of a right triangle is 20 inches and the hypotenuse is 10 inches longer than the other leg. Find the lengths of the unknown sides.

   Let x = length of unknown leg; $x + 10$ = length of hypotenuse.

$$\text{Square of hypotenuse} = \text{sum of squares of legs}$$
$$(x + 10)^2 = x^2 + (20)^2 \qquad \text{from which } x = 15 \text{ in.}$$

   The required sides are $x = 15$ in. and $x + 10 = 25$ in.

19. Temperature fahrenheit = $\frac{9}{5}$(temperature centigrade) + 32. At what temperature have the fahrenheit and centigrade readings the same value?

   Let x = required temperature = temperature fahrenheit = temperature centigrade.

   Then $x = \frac{9}{5}x + 32$ or $x = -40°$. Thus $-40°F = -40°C$.

MIXTURE PROBLEMS.

20. A mixture of 40 lb of candy worth 60¢ a pound is to be made up by taking some worth 45¢/lb and some worth 85¢/lb. How many pounds of each should be taken?

   Let x = weight of 45¢ candy; $40 - x$ = weight of 85¢ candy.

   Value of 45¢/lb candy + value of 85¢/lb candy = value of mixture
   or $\qquad x(45¢) \qquad + \qquad (40-x)(85¢) \qquad = \qquad 40(60¢)$.

   Solving, $x = 25$ lb of 45¢/lb candy; $40 - x = 15$ lb of 85¢/lb candy.

21. A tank contains 20 gallons of a mixture of alcohol and water which is 40% alcohol by volume. How much of the mixture should be removed and replaced by an equal volume of water so that the resulting solution will be 25% alcohol by volume?

   Let x = volume of 40% solution to be removed.

   Volume of alcohol in final solution = volume of alcohol in 20 gal of 25% solution
   or $\qquad\qquad 0.40(20 - x) = 0.25(20) \qquad$ from which $x = 7.5$ gallons.

22. What weight of water must be evaporated from 40 lb of a 20% salt solution to produce a 50% solution? All percents are by weight.

   Let x = weight of water to be evaporated.

   Weight of salt in 20% solution = weight of salt in 50% solution
   or $\qquad\qquad 0.20(40 \text{ lb}) = 0.50(40 \text{ lb} - x) \qquad$ from which $x = 24$ lb.

23. How many quarts of a 60% alcohol solution must be added to 40 quarts of a 20% alcohol solution to obtain a mixture which is 30% alcohol? All percents are by volume.

   Let x = number of quarts of 60% alcohol to be added.

   Alcohol in 60% solution + alcohol in 20% solution = alcohol in 30% solution
   or $\qquad 0.60x \qquad + \qquad 0.20(40) \qquad = 0.30(x + 40) \qquad$ and $x = 13\frac{1}{3}$ qt.

24. Two unblended manganese (Mn) ores contain 40% and 25% of manganese respectively. How many tons of each must be mixed to give 100 tons of blended ore containing 35% of manganese? All percents are by weight.

Let x = weight of 40% ore required; $100-x$ = weight of 25% ore required.

$$\text{Mn from 40\% ore} + \text{Mn from 25\% ore} = \text{total Mn in 100 tons of mixture}$$
$$0.40x + 0.25(100 - x) = 0.35(100)$$

from which x = 67 tons of 40% ore and $100-x$ = 33 tons of 25% ore.

MOTION PROBLEMS.

25. Two cars A and B having average speeds of 30 and 40 mi/hr respectively are 280 miles apart. They start moving toward each other at 3:00 P.M. At what time and where will they meet?

Let t = time in hours each car travels before they meet. Distance = speed × time.

$$\text{Distance traveled by } A + \text{distance traveled by } B = 280 \text{ miles}$$
$$30t + 40t = 280 \quad \text{from which } t = 4 \text{ hr.}$$

They meet at 7:00 P.M. at a distance $30t$ = 120 mi from initial position of A or at a distance $40t$ = 160 mi from initial position of B.

26. A and B start from a given point and travel on a straight road at average speeds of 30 and 50 mi/hr respectively. If B starts 3 hr after A, find a) the time and b) the distance they travel before meeting.

Let t and $(t-3)$ be the number of hours A and B respectively travel before meeting.

a) Distance in miles = average speed in mi/hr × time in hours. When they meet,

$$\text{distance covered by } A = \text{distance covered by } B$$
or
$$30t = 50(t-3) \quad \text{from which } t = 7\tfrac{1}{2} \text{ hr.}$$

Hence A travels t = 7½ hr and B travels $(t-3)$ = 4½ hr.

b) Distance = $30t$ = 30(7½) = 225 mi, or distance = $50(t-3)$ = 50(4½) = 225 mi.

27. A and B can run around a circular mile track in 6 and 10 minutes respectively. If they start at the same instant from the same place, in how many minutes will they pass each other if they run around the track a) in the same direction, b) in opposite directions?

Let t = required time in minutes.

a) They will pass each other when A covers 1 mile more than B. The speeds of A and B are 1/6 and 1/10 mi/min respectively. Then, since distance = speed × time:

$$\text{Distance by } A - \text{distance by } B = 1 \text{ mile}$$
$$\frac{1}{6}t - \frac{1}{10}t = 1 \quad \text{and } t = 15 \text{ minutes.}$$

b)
$$\text{Distance by } A + \text{distance by } B = 1 \text{ mile}$$
$$\frac{1}{6}t + \frac{1}{10}t = 1 \quad \text{and } t = 15/4 \text{ minutes.}$$

28. A boat, propelled to move at 25 mi/hr in still water, travels 4.2 mi against the river current in the same time that it can travel 5.8 mi with the current. Find the speed of the current.

Let v = speed of current. Then, since time = distance/speed,

$$\text{time against the current} = \text{time in direction of the current}$$
or
$$\frac{4.2 \text{ mi}}{(25-v) \text{ mi/hr}} = \frac{5.8 \text{ mi}}{(25+v) \text{ mi/hr}} \quad \text{and } v = 4 \text{ mi/hr.}$$

WORK PROBLEMS.

29. *A* can do a job in 3 days, and *B* can do the same job in 6 days. How long will it take them if they work together?

Let *n* = number of days it will take them working together.

In 1 day *A* does 1/3 of the job and *B* does 1/6 of the job, thus together completing 1/*n* of the job (in 1 day). Then

$$\frac{1}{3} + \frac{1}{6} = \frac{1}{n} \qquad \text{from which } n = 2 \text{ days.}$$

Another method. In *n* days *A* and *B* together complete

$$n\left(\frac{1}{3} + \frac{1}{6}\right) = 1 \text{ complete job.} \qquad \text{Solving, } n = 2 \text{ days.}$$

30. A tank can be filled by three pipes separately in 20, 30, and 60 minutes respectively. In how many minutes can it be filled by the three pipes acting together?

Let *t* = time required, in minutes.

In 1 minute the three pipes together fill $\left(\frac{1}{20} + \frac{1}{30} + \frac{1}{60}\right)$ of the tank. Then in *t* minutes they together fill

$$t\left(\frac{1}{20} + \frac{1}{30} + \frac{1}{60}\right) = 1 \text{ complete tank.} \qquad \text{Solving, } t = 10 \text{ minutes.}$$

31. *A* and *B* working together can complete a job in 6 days. *A* works twice as fast as *B*. How many days would it take each of them, working alone, to complete the job?

Let *n*, 2*n* = number of days required by *A* and *B* respectively, working alone, to do the job.

In 1 day *A* can do 1/*n* of job and *B* can do 1/2*n* of job. Then in 6 days they can do

$$6\left(\frac{1}{n} + \frac{1}{2n}\right) = 1 \text{ complete job.} \quad \text{Solving, } n = 9 \text{ days, } 2n = 18 \text{ days.}$$

32. *A*'s rate of doing work is three times that of *B*. On a given day *A* and *B* work together for 4 hours; then *B* is called away and *A* finishes the rest of the job in 2 hours. How long would it take *B* to do the complete job alone?

Let *t*, 3*t* = time in hours required by *A* and *B* respectively, working alone, to do the job.

In 1 hour *A* does 1/*t* of job and *B* does 1/3*t* of job. Then

$$4\left(\frac{1}{t} + \frac{1}{3t}\right) + 2\left(\frac{1}{t}\right) = 1 \text{ complete job.} \qquad \text{Solving, } 3t = 22 \text{ hours.}$$

33. A man is paid \$18 for each day he works and forfeits \$3 for each day he is idle. If at the end of 40 days he nets \$531, how many days was he idle?

Let *x* = number of days idle; 40 − *x* = number of days worked.

Amount earned − amount forfeited = \$531

or   \$18(40 − *x*) −   3*x*   = \$531   and *x* = 9 days idle.

SUPPLEMENTARY PROBLEMS

34. Solve each of the following equations.

a) $3x - 2 = 7$

b) $y + 3(y - 4) = 4$

c) $4x - 3 = 5 - 2x$

d) $x - 3 - 2(6 - 2x) = 2(2x - 5)$

e) $\dfrac{2t - 9}{3} = \dfrac{3t + 4}{2}$

f) $\dfrac{2x + 3}{2x - 4} = \dfrac{x - 1}{x + 1}$

g) $(x - 3)^2 + (x + 1)^2 = (x - 2)^2 + (x + 3)^2$

h) $(2x + 1)^2 = (x - 1)^2 + 3x(x + 2)$

i) $\dfrac{3}{z} - \dfrac{4}{5z} = \dfrac{1}{10}$

j) $\dfrac{2x + 1}{x} + \dfrac{x - 4}{x + 1} = 3$

k) $\dfrac{5}{y - 1} - \dfrac{5}{y + 1} = \dfrac{2}{y - 2} - \dfrac{2}{y + 3}$

l) $\dfrac{7}{x^2 - 4} + \dfrac{2}{x^2 - 3x + 2} = \dfrac{4}{x^2 + x - 2}$

35. Solve for the indicated letter.

a) $2(x - p) = 3(6p - x)$: x

b) $2by - 2a = ay - 4b$: y

c) $\dfrac{2x - a}{b} = \dfrac{2x - b}{a}$: x

d) $\dfrac{x - a}{x - b} = \dfrac{x - c}{x - d}$: x

e) $\dfrac{1}{ay} + \dfrac{1}{by} = \dfrac{1}{c}$: y

36. Express each of the following statements in terms of algebraic symbols.

a) Two more than five times a certain number.

b) Six less than twice a certain number.

c) Each of two numbers whose difference is 25.

d) The squares of three consecutive integers.

e) The amount by which five times a certain number exceeds 40.

f) The square of any odd integer.

g) The excess of the square of a number over twice the number.

h) The number of pints in x gallons.

i) The difference between the squares of two consecutive even integers.

j) Bob is six years older than Jane who is half as old as Jack. Express each of their ages in terms of a single unknown.

k) The three angles A, B, C of a triangle ABC if angle A exceeds twice angle B by $20°$.

l) The perimeter and area of a rectangle if one side is 3 ft shorter than three times the other side.

m) The fraction whose denominator is 4 more than twice the square of the numerator.

n) The amount of salt in a tank holding x quarts of water if the concentration is 2 lb of salt per gallon.

37. Abstract Number Problems.

a) One half of a certain number is 10 more than one sixth of the number. Find the number.

b) The difference between two numbers is 20 and their sum is 48. Find the numbers.

c) Find two consecutive even integers such that twice the smaller exceeds the larger by 18.

d) The sum of two numbers is 36. If the larger is divided by the smaller, the quotient is 2 and the remainder is 3. Find the numbers.

e) Find two consecutive positive odd integers such that the difference of their squares is 64.

f) The first of three numbers exceeds twice the second number by 4, while the third number is twice the first. If the sum of the three numbers is 54, find the numbers.

38. Age Problems.

a) A father is 24 years older than his son. In 8 years he will be twice as old as his son. Determine their present ages.

b) Mary is fifteen years older than her sister Jane. Six years ago Mary was six times as old as Jane. Find their present ages.

c) Larry is now twice as old as Bill. Five years ago Larry was three times as old as Bill. Find their present ages.

39. Coin Problems.

a) In a purse is $3.05 in nickels and dimes, 19 more nickels than dimes. How many coins are there of each kind?

b) Richard has twice as many dimes as quarters, amounting to $6.75 in all. How many coins does he have?

c) Admission tickets to a theater were 60¢ for adults and 25¢ for children. Receipts for the day showed that 280 persons attended and $140 was collected. How many children attended that day?

40. Digit Problems.

a) The tens digit of a certain two digit number exceeds the units digit by 3. The sum of the digits is 1/7 of the number. Find the number.

b) The sum of the digits of a certain two digit number is 10. If the digits are reversed, a new number is formed which is one less than twice the original number. Find the original number.

c) The tens digit of a certain two digit number is 1/3 of the units digit. When the digits are reversed, the new number exceeds twice the original number by 2 more than the sum of the digits. Find the original number.

41. Business and Investment Problems.

a) Goods cost a merchant $72. At what price should he mark them so that he may sell them at a discount of 10% from his marked price and still make a profit of 20% on the selling price?

b) A man is paid $20 for each day he works and forfeits $5 for each day he is idle. At the end of 25 days he nets $450. How many days did he work?

c) A labor report states that in a certain factory a total of 400 men and women are employed. The average daily wage is $16 for a man and $12 for a woman. If the labor cost is $5720 per day, how many women are employed?

d) A man has $450 invested, part at 2% and the remainder at 3% simple interest. How much is invested at each rate if the total annual income from these investments is $11.

e) A man has $2000 invested at 7% and $5000 at 4% simple interest. What additional sum must he invest at 6% to give him an overall return of 5%?

42. Mensuration Problems.

   a) The perimeter of a rectangle is 110 ft. Find the dimensions if the length is 5 ft less than twice the width.

   b) The length of a rectangular floor is 8 ft greater than its width. If each dimension is increased by 2 ft, the area is increased by 60 ft^2. Find the dimensions of the floor.

   c) The area of a square exceeds the area of a rectangle by 3 in.^{2} The width of the rectangle is 3 in. shorter and the length 4 in. longer than the side of the square. Find the side of the square.

   d) A piece of wire 40 in. long is bent into the form of a right triangle one of whose legs is 15 in. long. Determine the lengths of the other two sides.

   e) The length of a rectangular swimming pool is twice its width. The pool is surrounded by a cement walk 4 ft wide. If the area of the walk is 784 ft^2, determine the dimensions of the pool.

43. Mixture Problems.

   a) An excellent solution for cleaning grease stains from cloth or leather consists of the following: carbon tetrachloride 80% (by volume), ligroin 16%, amyl alcohol 4%. How many pints of each should be taken to make up 75 pints of solution?

   b) Lubricating oil worth 28 cents/quart is to be mixed with oil worth 33 cents/quart to make up 45 quarts of a mixture to sell at 30 ¢/qt. What volume of each grade should be taken?

   c) What weight of water must be added to 50 lb of a 36% sulfuric acid solution to yield a 20% solution? All percents are by weight.

   d) How many quarts of pure alcohol must be added to 10 quarts of a 15% alcohol solution to obtain a mixture which is 25% alcohol? All percents are by volume.

   e) There is available 60 gallons of a 50% solution of glycerin and water. What volume of water must be added to the solution to reduce the glycerin concentration to 12%? All percents are by volume.

   f) The radiator of a jeep has a capacity of 4 gallons. It is filled with an anti-freeze solution of water and glycol which analyzes 10% glycol. What volume of the mixture must be drawn off and replaced with glycol to obtain a 25% glycol solution? All percents are by volume.

   g) One thousand quarts of milk testing 4% butterfat are to be reduced to 3%. How many quarts of cream testing 23% butterfat must be separated from the milk to produce the required result? All percents are by volume.

   h) There are available 10 tons of coal containing 2.5% sulfur, and also supplies of coal containing 0.80% and 1.10% sulfur respectively. How many tons of each of the latter should be mixed with the original 10 tons to give 20 tons containing 1.7% sulfur?

   i) A clay contains 45% silica and 10% water. Determine the percentage of silica in the clay on a dry (water-free) basis. All percents are by weight.

   j) A nugget of gold and quartz weighs 100 grams. Gold weighs 19.3 g/cm^3 (grams per cubic centimeter), quartz weighs 2.6 g/cm^3, and the nugget weighs 6.4 g/cm^3. Find the weight of gold in the nugget.
   Hint: Let x = weight of gold in nugget; $(100 - x)$ = weight of quartz in nugget.
      Volume of nugget = volume of gold in nugget + volume of quartz in nugget.

   k) A cold cream sample weighing 8.41 grams lost 5.83 grams of moisture on heating to 110°C. The residue on extracting with water and drying lost 1.27 grams of water-soluble glycerin. The balance was oil. Calculate the percentage composition of this cream.

   l) A coal contains 2.4% water. After drying, the moisture-free residue contains 71.0% carbon. Determine the percentage of carbon on the "wet basis". All percents are by weight.

44. Motion Problems.

a) Two motorists start toward each other at 4:30 P.M. from towns 255 miles apart. If their respective average speeds are 40 and 45 mi/hr, at what time will they meet?

b) Two planes start from Chicago at the same time and fly in opposite directions, one averaging a speed of 40 mi/hr greater than the other. If they are 2000 miles apart after 5 hours, find their average speeds.

c) At what rate must motorist A travel to overtake motorist B who is traveling at a rate 20 mi/hr slower if A starts two hours after B and wishes to overtake B in 4 hours?

d) A motorist starts from city A at 2:00 P.M. and travels to city B at an average speed of 30 mi/hr. After resting at B for one hour, he returns over the same route at an average speed of 40 mi/hr and arrives at A that evening at 6:30 P.M. Determine the distance between A and B.

e) Tom traveled a distance of 265 miles. He drove at 40 mi/hr during the first part of the trip and at 35 mi/hr during the remaining part. If he made the trip in 7 hours, how long did he travel at 40 mi/hr?

f) A boat can move at 8 mi/hr in still water. If it can travel 20 miles downstream in the same time it can travel 12 miles upstream, determine the rate of the stream.

g) The speed of a plane is 120 mi/hr in a calm. With the wind it can cover a certain distance in 4 hours, but against the wind it can cover only 3/5 of that distance in the same time. Find the velocity of the wind.

h) An army of soldiers is marching down a road at 5 mi/hr. A messenger on horseback rides from the front to the rear and returns immediately, the total time taken being 10 minutes. Assuming that the messenger rides at the rate of 10 mi/hr, determine the distance from the front to the rear.

i) A train moving at r mi/hr can cover a given distance in h hours. By how many mi/hr must its speed be increased in order to cover the same distance in one hour less time?

45. Work Problems.

a) A farmer can plow a certain field three times as fast as his son. Working together, it would take them 6 hours to plow the field. How long would it take each to do it alone?

b) A painter can do a given job in 6 hours. His helper can do the same job in 10 hours. The painter begins the work and after two hours is joined by the helper. In how many hours will they complete the job?

c) One group of workers can do a job in 8 days. After this group has worked 3 days, another group joins it and together they complete the job in 3 more days. In what time could the second group have done the job alone?

d) A tank can be filled by two pipes separately in 10 and 15 minutes respectively. When a third pipe is used simultaneously with the first two pipes, the tank can be filled in 4 minutes. How long would it take the third pipe alone to fill the tank?

ANSWERS TO SUPPLEMENTARY PROBLEMS.

34. a) $x = 3$ d) $x = 5$ g) $x = -1/2$ j) $x = 1/4$
    b) $y = 4$ e) $t = -6$ h) all values of x (identity)   k) $y = 5$
    c) $x = 4/3$ f) $x = 1/11$ i) $z = 22$ l) $x = -1$

35. a) $x = 4p$

    b) $y = -2$ if $a \neq 2b$ c) $x = \dfrac{a+b}{2}$ if $a \neq b$ d) $x = \dfrac{bc - ad}{b + c - a - d}$ e) $y = \dfrac{ac + bc}{ab}$

36. *a*) $5x + 2$
    b) $2x - 6$
    c) $x + 25$, x
    d) x^2, $(x+1)^2$, $(x+2)^2$
    e) $5x - 40$
    f) $(2x+1)^2$ where x = integer
    g) $x^2 - 2x$
    h) $8x$
    i) $(2x+2)^2 - (2x)^2$, x = integer

    j) Jane's age x, Bob's age $x+6$, Jack's age $2x$
    k) $B = x°$, $A = (2x+20)°$, $C = (160-3x)°$
    l) One side is x, adjacent side is $3x-3$.
    Perimeter = $8x-6$, Area = $3x^2 - 3x$
    m) $\dfrac{x}{2x^2+4}$
    n) $\dfrac{x}{2}$ lb salt

37. *a*) 30   *b*) 34, 14   *c*) 20, 22   *d*) 25, 11   *e*) 15, 17   *f*) 16, 6, 32

38. *a*) Father 40, son 16   *b*) Mary 24, Jane 9   *c*) Larry 20, Bill 10

39. *a*) 14 dimes, 33 nickels   *b*) 15 quarters, 30 dimes   *c*) 200 adults, 80 children

40. *a*) 63   *b*) 37   *c*) 26

41. *a*) $100   *b*) 23 days   *c*) 170 women   *d*) $200 at 3%, $250 at 2%   *e*) $1000

42. *a*) width 20 ft, length 35 ft
    b) width 10 ft, length 18 ft
    c) 9 in.
    d) other leg 8 ft, hypotenuse 17 ft
    e) 30 ft by 60 ft

43. *a*) 60, 12, 3 pints
    b) 18 qt of 33¢, 27 qt of 28¢
    c) 40 lb
    d) 4/3 qt
    e) 190 gal
    f) 2/3 gal
    g) 50 qt
    h) 6.7 tons of 0.80%, 3.3 tons of 1.10%
    i) 50% silica
    j) 69 grams gold
    k) 69.3% moisture, 15.1% glycerin, 15.6% oil
    l) 69.3% carbon

44. *a*) 7:30 P.M.   *c*) 60 mi/hr   *e*) 4 hr   *g*) 30 mi/hr   *i*) $\dfrac{r}{h-1}$
    b) 180, 220 mi/hr   *d*) 60 mi   *f*) 2 mi/hr   *h*) 5/8 mi

45. *a*) Father 8 hr, son 24 hr   *b*) $2\frac{1}{2}$ hr   *c*) 12 days   *d*) 12 minutes

CHAPTER 12

Simultaneous Linear Equations

A LINEAR EQUATION IN TWO UNKNOWNS (or variables) x and y is of the form $ax + by = c$ where a, b, c are constants and a, b are not both zero. If we consider two such equations

$$a_1 x + b_1 y = c_1$$
$$a_2 x + b_2 y = c_2$$

we say that we have two simultaneous linear equations in two unknowns or a system of two linear equations in two unknowns. A pair of values for x and y which satisfies both equations is called a *simultaneous solution* of the given equations.

Thus the simultaneous solution of $x + y = 7$ and $x - y = 3$ is $x = 5$, $y = 2$.

SYSTEMS OF TWO LINEAR EQUATIONS IN TWO UNKNOWNS. Three methods of solving such systems of linear equations are illustrated here.

A) SOLUTION BY ADDITION OR SUBTRACTION. If necessary, multiply the given equations by such numbers as will make the coefficients of one unknown in the resulting equations numerically equal. If the signs of the equal coefficients are unlike, add the resulting equations; if like, subtract them. Consider

(1) $2x - y = 4$
(2) $x + 2y = -3$.

To eliminate y, multiply (1) by 2 and add with (2) to obtain

$$\begin{aligned} 2 \times (1): \quad 4x - 2y &= 8 \\ (2): \quad \underline{x + 2y} &= \underline{-3} \\ \text{Addition: } 5x &= 5 \qquad \text{or } x = 1. \end{aligned}$$

Substitute $x = 1$ in (1) and obtain $2 - y = 4$ or $y = -2$.
Thus the simultaneous solution of (1) and (2) is $x = 1$, $y = -2$.

Check: Put $x = 1$, $y = -2$ in (2) and obtain $1 + 2(-2) = -3$, $-3 = -3$.

B) SOLUTION BY SUBSTITUTION. Find the value of one unknown in either of the given equations and substitute this value in the other equation.

For example, consider the system (1), (2) above. From (1) obtain $y = 2x - 4$ and substitute this value into (2) to get $x + 2(2x - 4) = -3$ which reduces to $x = 1$. Then put $x = 1$ into either (1) or (2) and obtain $y = -2$.

C) GRAPHICAL SOLUTION. Graph both equations, obtaining two straight lines. The simultaneous solution is given by the coordinates (x, y) of the point of

100

intersection of these lines. Fig. (a) below shows that the simultaneous solution of (1) $2x - y = 4$ and (2) $x + 2y = -3$ is $x = 1$, $y = -2$, also written $(1,-2)$.

If the lines are parallel, the equations are *inconsistent* and have no simultaneous solution. For example, (3) $x + y = 2$ and (4) $2x + 2y = 8$ are inconsistent, as indicated in Fig. (b) below. Note that if equation (3) is multiplied by 2 we obtain $2x + 2y = 4$ which is obviously inconsistent with (4).

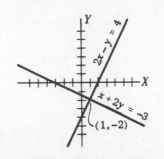 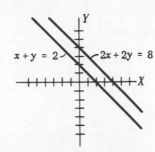 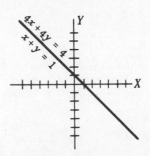

a) Consistent equations	b) Inconsistent equations	c) Dependent equations
(1) $2x - y = 4$	(3) $x + y = 2$	(5) $x + y = 1$
(2) $x + 2y = -3$	(4) $2x + 2y = 8$	(6) $4x + 4y = 4$

Dependent equations are represented by the same line. Thus every point on the line represents a solution and, since there are an infinite number of points, there are an infinite number of simultaneous solutions. For example, (5) $x + y = 1$ and (6) $4x + 4y = 4$ are dependent equations; note that if (5) is multiplied by 4 the result is (6).

A SYSTEM OF THREE LINEAR EQUATIONS IN THREE UNKNOWNS is solved by eliminating one unknown from any two of the equations and then eliminating the same unknown from any other pair of equations.

SOLVED PROBLEMS

TWO LINEAR EQUATIONS IN TWO UNKNOWNS.

Solve the following systems.

1.   (1) $2x - y = 4$
      (2) $x + y = 5$

Add (1) and (2) and obtain $3x = 9$, $x = 3$.

Now put $x = 3$ in (1) or (2) and get $y = 2$. The solution is $x = 3$, $y = 2$ or $(3,2)$.

Another method. From (1) obtain $y = 2x - 4$ and substitute this value into equation (2) to get $x + 2x - 4 = 5$, $3x = 9$, $x = 3$. Now put $x = 3$ into (1) or (2) and obtain $y = 2$.

Check: $2x - y = 2(3) - 2 = 4$ and $x + y = 3 + 2 = 5$.

Graphical solution. The graph of a linear equation is a straight line. Since a straight

line is determined by two points, we need plot only two
points for each equation. However, to insure accuracy we
shall plot three points for each line.

For $2x - y = 4$:

x	-1	0	1
y	-6	-4	-2

For $x + y = 5$:

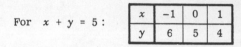

x	-1	0	1
y	6	5	4

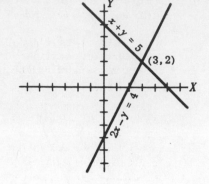

The simultaneous solution is the point of intersection $(3, 2)$ of the lines.

2.
　　　　(1) $5x + 2y = 3$
　　　　(2) $2x + 3y = -1$

To eliminate y, multiply (1) by 3 and (2) by 2 and subtract the results.

$$3 \times (1): \quad 15x + 6y = 9$$
$$2 \times (2): \quad 4x + 6y = -2$$
$$\text{Subtraction:} \quad 11x \qquad = 11 \quad \text{or} \quad x = 1.$$

Now put $x = 1$ in (1) or (2) and obtain $y = -1$. The simultaneous solution is $(1, -1)$.

3.
　　　　(1) $2x + 3y = 3$　　　　Rearranging (2),　　　　(1) $2x + 3y = 3$
　　　　(2) $6y - 6x = 1$　　　　　　　　　　　　　　　　(2) $-6x + 6y = 1$

To eliminate x, multiply (1) by 3 and add the result with (2) to get

$$3 \times (1): \quad 6x + 9y = 9$$
$$(2): -6x + 6y = 1$$
$$15y = 10 \qquad \text{or} \quad y = 2/3.$$

Now put $y = 2/3$ into (1) or (2) and obtain $x = 1/2$. The solution is $(1/2, 2/3)$.

4.
　　　　(1) $5y = 3 - 2x$　　　　Solve by substitution.
　　　　(2) $3x = 2y + 1$

From (1), $y = \dfrac{3 - 2x}{5}$. Put this value into (2) and obtain $3x = 2(\dfrac{3 - 2x}{5}) + 1$ or $x = \dfrac{11}{19}$.

Then $y = \dfrac{3 - 2x}{5} = \dfrac{3 - 2(11/19)}{5} = \dfrac{7}{19}$ and the solution is $(\dfrac{11}{19}, \dfrac{7}{19})$.

5.
　　　　(1) $\dfrac{x - 2}{3} + \dfrac{y + 1}{6} = 2$

　　　　(2) $\dfrac{x + 3}{4} - \dfrac{2y - 1}{2} = 1$

To eliminate fractions, multiply (1) by 6 and (2) by 4 and simplify to obtain

$$(1_1) \ 2x + y = 15 \qquad \text{Solving, we find} \quad x = \dfrac{59}{9}, \ y = \dfrac{17}{9}.$$
$$(2_1) \ x - 4y = -1$$

6.   (1) $x - 3y = 2a$
      (2) $2x + y = 5a$

To eliminate x, multiply (1) by 2 and subtract (2); then $y = a/7$.
To eliminate y, multiply (2) by 3 and add with (1); then $x = 17a/7$.
The solution is $x = 17a/7$, $y = a/7$.

7.   (1) $3u + 2v = 7r + s$
      (2) $2u - v = 3s$ Solve for u and v in terms of r and s.

To eliminate v, multiply (2) by 2 and add with (1); then $7u = 7r + 7s$ or $u = r + s$.
To eliminate u, multiply (1) by 2, (2) by -3, and add the results; then $v = 2r - s$.
The solution is $u = r + s$, $v = 2r - s$.

8.   (1) $ax + by = 2a^2 - 3b^2$
      (2) $x + 2y = 2a - 6b$

Multiply (2) by a and subtract from (1); then $by - 2ay = 6ab - 3b^2$, $y(b - 2a) = 3b(2a - b)$,

and $y = \dfrac{3b(2a - b)}{b - 2a} = \dfrac{-3b(b - 2a)}{b - 2a} = -3b$ provided $b - 2a \neq 0$.

Similarly, we obtain $x = 2a$ provided $b - 2a \neq 0$.

Check: (1) $a(2a) + b(-3b) = 2a^2 - 3b^2$,   (2) $2a + 2(-3b) = 2a - 6b$.

Note. If $b - 2a = 0$, or $b = 2a$, the given equations become

$$(1_1) \ ax + 2ay = -10a^2$$
$$(2_1) \ x + 2y = -10a$$

which are dependent since (1_1) may be obtained by multiplying (2_1) by a. Thus if $b = 2a$ the system possesses an infinite number of solutions, i.e. any values of x and y which satisfy $x + 2y = -10a$.

9. The sum of two numbers is 28 and their difference is 12. Find the numbers.

Let x and y be the two numbers. Then (1) $x + y = 28$ and (2) $x - y = 12$.

Add (1) and (2) to obtain $2x = 40$, $x = 20$. Subtract (2) from (1) to obtain $2y = 16$, $y = 8$.

Note. Of course this problem may also be solved easily by using one unknown. Let the numbers be n and $28 - n$. Then $n - (28 - n) = 12$ or $n = 20$, and $28 - n = 8$.

10. If the numerator of a certain fraction is increased by 2 and the denominator is increased by 1, the resulting fraction equals 1/2. If, however, the numerator is increased by 1 and the denominator decreased by 2, the resulting fraction equals 3/5. Find the fraction.

Let x = numerator, y = denominator, and x/y = required fraction. Then

(1) $\dfrac{x + 2}{y + 1} = \dfrac{1}{2}$ or $2x - y = -3$,   and (2) $\dfrac{x + 1}{y - 2} = \dfrac{3}{5}$ or $5x - 3y = -11$.

Solve (1) and (2) simultaneously and obtain $x = 2$, $y = 7$. The required fraction is 2/7.

11. Two years ago a man was six times as old as his son. In 18 years he will be twice as old as his son. Determine their present ages.

Let x = father's present age, y = son's present age.

Equation for condition 2 years ago: (1) $(x - 2) = 6(y - 2)$.
Equation for condition 18 years hence: (2) $(x + 18) = 2(y + 18)$.

Solve (1) and (2) simultaneously and obtain $x = 32$, $y = 7$.

12. Find the two digit number satisfying the following two conditions. (1) Four times the units digit is six less than twice the tens digit. (2) The number is nine less than three times the number obtained by reversing the digits.

Let t = tens digit, u = units digit.

The required number = $10t + u$; reversing digits, the new number = $10u + t$. Then

$$(1) \ 4u = 2t - 6 \qquad \text{and} \qquad (2) \ 10t + u = 3(10u + t) - 9.$$

Solving (1) and (2) simultaneously, $t = 7$, $u = 2$, and the required number is 72.

13. Five tables and eight chairs cost $115; three tables and five chairs cost $70. Determine the cost of each table and of each chair.

Let x = cost of a table, y = cost of a chair. Then

$$(1) \ 5x + 8y = \$115 \qquad \text{and} \qquad (2) \ 3x + 5y = \$70.$$

Solve (1) and (2) simultaneously and obtain $x = \$15$, $y = \$5$.

14. A merchant sold his entire stock of shirts and ties for $1000, the shirts being priced at 3 for $10 and the ties at $2 each. If he had sold only 1/2 of the shirts and 2/3 of the ties he would have collected $600. How many of each kind did he sell?

Let s = number of shirts sold, t = number of ties sold. Then

$$(1) \ \frac{10}{3}s + 2t = 1000 \qquad \text{and} \qquad (2) \ \frac{10}{3}(\tfrac{1}{2}s) + 2(\tfrac{2}{3}t) = 600.$$

Solving (1) and (2) simultaneously, $s = 120$, $t = 300$.

15. An investor has $1100 income from bonds bearing 4% and 5%. If the amounts at 4% and 5% were interchanged he would earn $50 more per year. Find the total sum invested.

Let x = amount invested at 4%, y = amount at 5%. Then

$$(1) \ .04x + .05y = 1100 \qquad \text{and} \qquad (2) \ .05x + .04y = 1150.$$

Solving (1) and (2) simultaneously, $x = \$15,000$, $y = \$10,000$, and their sum is $25,000.

16. Tank A contains a mixture of 10 gallons water and 5 gallons pure alcohol. Tank B has 12 gallons water and 3 gallons alcohol. How many gallons should be taken from each tank and combined in order to obtain an 8 gallon solution containing 25% alcohol by volume?

In 8 gal of required mixture are $0.25(8) = 2$ gal alcohol.
Let x, y = volumes taken from tanks A, B respectively; then (1) $x + y = 8$.

Fraction of alcohol in tank $A = \dfrac{5}{10 + 5} = \dfrac{1}{3}$, in tank $B = \dfrac{3}{12 + 5} = \dfrac{1}{5}$. Thus in x gal of A are $x/3$ gal alcohol, and in y gal of B are $y/5$ gal alcohol; then (2) $x/3 + y/5 = 2$.
Solving (1) and (2) simultaneously, $x = 3$ gal, $y = 5$ gal.

Another method, using only one unknown.

Let x = volume taken from tank A, $8 - x$ = volume taken from tank B.

Then $\dfrac{1}{3}x + \dfrac{1}{5}(8 - x) = 2$ from which $x = 3$ gal, $8 - x = 5$ gal.

17. A given alloy contains 20% copper and 5% tin. How many pounds of copper and of tin must be melted with 100 lb of the given alloy to produce another alloy analyzing 30% copper and 10% tin? All percents are by weight.

Let x, y = number of pounds of copper and tin to be added, respectively.

In 100 lb of given alloy are 20 lb copper and 5 lb tin. Then, in the new alloy,

$$\text{fraction of copper} \ = \ \frac{\text{pounds copper}}{\text{pounds alloy}} \quad \text{or} \quad (1) \ 0.30 = \frac{20 + x}{100 + x + y}$$

$$\text{fraction of tin} \ = \ \frac{\text{pounds tin}}{\text{pounds alloy}} \quad \text{or} \quad (2) \ 0.10 = \frac{5 + y}{100 + x + y} \ .$$

The simultaneous solution of (1) and (2) is $x = 17.5$ lb copper, $y = 7.5$ lb tin.

18. Determine the rate of a man's rowing in still water and the rate of the river current, if it takes him 2 hours to row 9 miles with the current and 6 hours to return against the current.

Let x = rate of rowing in still water, y = rate of current.

With current: 2 hr $\times (x + y)$ mi/hr = 9 mi   or   (1) $2x + 2y = 9.$
Against current: 6 hr $\times (x - y)$ mi/hr = 9 mi   or   (2) $6x - 6y = 9.$

Solving (1) and (2) simultaneously, $x = 3$ mi/hr, $y = 3/2$ mi/hr.

19. Two particles move at different but constant speeds along a circle of circumference 276 ft. Starting at the same instant and from the same place, when they move in opposite directions they pass each other every 6 sec and when they move in the same direction they pass each other every 23 sec. Determine their rates.

Let x, y = their respective rates in ft/sec.

Opposite directions: 6 sec $\times (x + y)$ ft/sec = 276   or   (1) $6x + 6y = 276.$
Same direction: 23 sec $\times (x - y)$ ft/sec = 276   or   (2) $23x - 23y = 276.$

Solving (1) and (2) simultaneously, $x = 29$ ft/sec, $y = 17$ ft/sec.

20. Fahrenheit temperature = m(centigrade temperature) + n, or $F = mC + n$, where m and n are constants. At one atmosphere pressure, the boiling point of water is 212°F or 100°C and the freezing point of water is 32°F or 0°C. *a*) Find m and n. *b*) What fahrenheit temperature corresponds to –273°C, the lowest temperature obtainable?

a) (1) $212 = m(100) + n$ and   (2) $32 = m(0) + n$.   Solving, $m = 9/5$, $n = 32$.

b) $F = \dfrac{9}{5}C + 32 = \dfrac{9}{5}(-273) + 32 = -491.4 + 32 = -459°F$, to the nearest degree.

THREE LINEAR EQUATIONS IN THREE UNKNOWNS.

Solve the following systems.

21.   (1) $2x - y + z = 3$
           (2) $x + 3y - 2z = 11$
           (3) $3x - 2y + 4z = 1$

To eliminate y between (1) and (2) multiply (1) by 3 and add with (2) to obtain

$$(1_1) \ 7x + z = 20.$$

To eliminate y between (2) and (3) multiply (2) by 2, (3) by 3, and add the results to get

$$(2_1) \ 11x + 8z = 25.$$

Solving (1_1) and (2_1) simultaneously, we find $x = 3$, $z = -1$. Substituting these values into any of the given equations, we find $y = 2$.

Thus the solution is $x = 3$, $y = 2$, $z = -1$.

22.   (1) $\dfrac{x}{3} + \dfrac{y}{2} - \dfrac{z}{4} = 2$,   (2) $\dfrac{x}{4} + \dfrac{y}{3} - \dfrac{z}{2} = \dfrac{1}{6}$,   (3) $\dfrac{x}{2} - \dfrac{y}{4} + \dfrac{z}{3} = \dfrac{23}{6}$.

To remove fractions, multiply the equations by 12 and obtain the system

$$(1_1)\ 4x + 6y - 3z = 24$$
$$(2_1)\ 3x + 4y - 6z = 2$$
$$(3_1)\ 6x - 3y + 4z = 46.$$

To eliminate x between (1_1) and (2_1) multiply (1_1) by 3, (2_1) by -4, and add the results to obtain

$$(1_2)\ 2y + 15z = 64.$$

To eliminate x between (2_1) and (3_1) multiply (2_1) by 2 and subtract (3_1) to obtain

$$(2_2)\ 11y - 16z = -42.$$

The simultaneous solution of (1_2) and (2_2) is $y = 2$, $z = 4$. Substituting these values of y and z into any of the given equations, we find $x = 6$.

Thus the simultaneous solution of the three given equations is $x = 6$, $y = 2$, $z = 4$.

23.   (1) $\dfrac{1}{x} - \dfrac{2}{y} - \dfrac{2}{z} = 0$,   (2) $\dfrac{2}{x} + \dfrac{3}{y} + \dfrac{1}{z} = 1$,   (3) $\dfrac{3}{x} - \dfrac{1}{y} - \dfrac{3}{z} = 3$.

Let $\dfrac{1}{x} = u$, $\dfrac{1}{y} = v$, $\dfrac{1}{z} = w$ so that the given equations may be written

$$(1_1)\ \ u - 2v - 2w = 0$$
$$(2_1)\ \ 2u + 3v + w = 1$$
$$(3_1)\ \ 3u - v - 3w = 3 \quad \text{from which we find } u = -2,\ v = 3,\ w = -4.$$

Thus $\dfrac{1}{x} = -2$ or $x = -1/2$, $\dfrac{1}{y} = 3$ or $y = 1/3$, $\dfrac{1}{z} = -4$ or $z = -1/4$.

Check: (1) $\dfrac{1}{-1/2} - \dfrac{2}{1/3} - \dfrac{2}{-1/4} = 0$,   (2) $\dfrac{2}{-1/2} + \dfrac{3}{1/3} + \dfrac{1}{-1/4} = 1$,   (3) $\dfrac{3}{-1/2} - \dfrac{1}{1/3} - \dfrac{3}{-1/4} = 3$.

24.   (1) $3x + y - z = 4$,   (2) $x + y + 4z = 3$,   (3) $9x + 5y + 10z = 8$.

Subtracting (2) from (1), we obtain $(1_1)\ 2x - 5z = 1$.
Multiplying (2) by 5 and subtracting (3), we obtain $(2_1)\ -4x + 10z = 7$.

Now (1_1) and (2_1) are inconsistent since (1_1) multiplied by -2 gives $-4x + 10z = -2$, thus contradicting (2_1). This indicates that the original system is inconsistent and hence has no simultaneous solution.

25. A and B working together can do a given job in 4 days, B and C together can do the job in 3 days, and A and C together can do it in 2.4 days. In how many days can each do the job working alone?

Let a, b, c = number of days required by each working alone to do the job, respectively.

Then $\dfrac{1}{a}, \dfrac{1}{b}, \dfrac{1}{c}$ = fraction of complete job done by each in 1 day, respectively. Thus

$$(1)\ \dfrac{1}{a} + \dfrac{1}{b} = \dfrac{1}{4},\quad (2)\ \dfrac{1}{b} + \dfrac{1}{c} = \dfrac{1}{3},\quad (3)\ \dfrac{1}{a} + \dfrac{1}{c} = \dfrac{1}{2.4}.$$

Solving (1),(2),(3) simultaneously, we find $a = 6$, $b = 12$, $c = 4$ days.

SUPPLEMENTARY PROBLEMS

SIMULTANEOUS LINEAR EQUATIONS IN TWO UNKNOWNS.

26. Solve each of the following pairs of simultaneous equations by the methods indicated.

a) $\begin{cases} 2x - 3y = 7 \\ 3x + y = 5 \end{cases}$ Solve (1) by addition or subtraction, (2) by substitution.

b) $\begin{cases} 3x - y = -6 \\ 2x + 3y = 7 \end{cases}$ Solve (1) graphically, (2) by addition or subtraction.

c) $\begin{cases} 4x + 2y = 5 \\ 5x - 3y = -2 \end{cases}$ Solve (1) graphically, (2) by addition or subtraction, (3) by substitution.

27. Solve each of the following pairs of simultaneous equations by any method.

a) $\begin{cases} 2x - 5y = 10 \\ 4x + 3y = 7 \end{cases}$ e) $\begin{cases} 2x - 3y = 9t \\ 4x - y = 8t \end{cases}$

b) $\begin{cases} 2y - x = 1 \\ 2x + y = 8 \end{cases}$ f) $\begin{cases} 2x + y + 1 = 0 \\ 3x - 2y + 5 = 0 \end{cases}$

c) $\begin{cases} \dfrac{2x}{3} + \dfrac{y}{5} = 6 \\[2mm] \dfrac{x}{6} - \dfrac{y}{2} = -4 \end{cases}$ g) $\begin{cases} 2u - v = -5s \\ 3u + 2v = 7r - 4s \end{cases}$ Find u and v in terms of r and s.

h) $\begin{cases} 5/x - 3/y = 1 \\ 2/x + 1/y = 7 \end{cases}$

d) $\begin{cases} \dfrac{2x-1}{3} + \dfrac{y+2}{4} = 4 \\[2mm] \dfrac{x+3}{2} - \dfrac{x-y}{3} = 3 \end{cases}$ i) $\begin{cases} ax - by = a^2 + b^2 \\ 2bx - ay = 2b^2 + 3ab - a^2 \end{cases}$ Find x and y in terms of a and b.

28. Indicate which of the following systems are (1) consistent, (2) dependent, (3) inconsistent.

a) $\begin{cases} x + 3y = 4 \\ 2x - y = 1 \end{cases}$ c) $\begin{cases} 3x = 2y + 3 \\ x - 2y/3 = 1 \end{cases}$ e) $\begin{cases} 2x - y = 1 \\ 2y - x = 1 \end{cases}$

b) $\begin{cases} 2x - y = 5 \\ 2y = 7 + 4x \end{cases}$ d) $\begin{cases} (x+3)/4 = (2y-1)/6 \\ 3x - 4y = 2 \end{cases}$ f) $\begin{cases} (x+2)/4 - (y-2)/12 = 5/4 \\ y = 3x - 7 \end{cases}$

29. Abstract Number Problems.

a) When the first of two numbers is added to twice the second the result is 21, but when the second number is added to twice the first the result is 18. Find the two numbers.

b) If the numerator and denominator of a certain fraction are both increased by 3, the resulting fraction equals 2/3. If, however, the numerator and denominator are both decreased by 2, the resulting fraction equals 1/2. Determine the fraction.

c) Twice the sum of two numbers exceeds three times their difference by 8, while half the sum is one more than the difference. What are the numbers?

d) If three times the larger of two numbers is divided by the smaller, the quotient is 6 and the remainder is 6. If five times the smaller is divided by the larger, the quotient is 2 and the remainder is 3. Find the numbers.

30. Age Problems.

a) Six years ago Bob was four times as old as Mary. In four years he will be twice as old as Mary. How old are they now?

b) *A* is eleven times as old as *B*. In a certain number of years *A* will be five times as old as *B*, and five years after that he will be three times as old as *B*. How old are they now?

31. Digit Problems.

a) Three times the tens digit of a certain two digit number is two more than four times the units digit. The difference between the given number and the number obtained by reversing the digits is two less than twice the sum of the digits. Find the number.

b) When a certain two digit number is divided by the number obtained by reversing the digits, the quotient is 2 and the remainder is 7. If the number is divided by the sum of its digits, the quotient is 7 and the remainder 6. Find the number.

32. Business Problems.

a) Two pounds of coffee and 3 lb of butter cost $4.20. A month later the price of coffee advanced 10% and that of butter 20%, making the total cost of a similar order $4.86. Determine the original cost of a pound of each.

b) If 3 gallons of Grade A oil are mixed with 7 gal of Grade B oil the resulting mixture is worth 43 ¢/gal. However, if 3 gal of Grade A oil are mixed with 2 gal of Grade B oil the resulting mixture is worth 46 ¢/gal. Find the price per gallon of each grade.

c) An investor has $116 annual income from bonds bearing 3% and 5% interest. Then he buys 25% more of the 3% bonds and 40% more of the 5% bonds, thereby increasing his annual income by $41. Find his initial investment in each type of bond.

33. Mixture Problems.

a) Tank A contains 32 gallons of solution which is 25% alcohol by volume. Tank B has 50 gal of solution which is 40% alcohol by volume. What volume should be taken from each tank and combined in order to make up 40 gal of solution containing 30% alcohol by volume?

b) Tank A holds 40 gal of a salt solution containing 80 lb of dissolved salt. Tank B has 120 gal of solution containing 60 lb of dissolved salt. What volume should be taken from each tank and combined in order to make up 30 gal of solution having a salt concentration of 1.5 lb/gal?

c) A given alloy contains 10% zinc and 20% copper. How many pounds of zinc and of copper must be melted with 1000 lb of the given alloy to produce another alloy analyzing 20% zinc and 24% copper? All percents are by weight.

d) An alloy weighing 600 lb is composed of 100 lb copper and 50 lb tin. Another alloy weighing 1000 lb is composed of 300 lb copper and 150 lb tin. What weights of copper and tin must be melted with the two given alloys to produce a third alloy analyzing 32% copper and 28% tin. All percents are by weight.

34. Motion Problems.

a) Determine the speed of a motor boat in still water and the speed of the river current, if it takes 3 hr to travel a distance of 45 mi upstream and 2 hr to travel 50 mi downstream.

b) When two cars race around a circular mile track starting from the same place and at the same instant, they pass each other every 18 seconds when traveling in opposite directions and every 90 seconds when traveling in the same direction. Find their speeds in mi/hr.

c) A passenger on the front of train A observes that he passes the complete length of train B in 33 seconds when traveling in the same direction as B and in 3 seconds when traveling in the opposite direction. If B is 330 ft long, find the speeds of the two trains.

SIMULTANEOUS LINEAR EQUATIONS IN THREE UNKNOWNS.

35. Solve each of the following systems of equations.

a) $\begin{cases} 2x - y + 2z = -8 \\ x + 2y - 3z = 9 \\ 3x - y - 4z = 3 \end{cases}$

b) $\begin{cases} x = y - 2z \\ 2y = x + 3z + 1 \\ z = 2y - 2x - 3 \end{cases}$

c) $\begin{cases} \dfrac{x}{3} + \dfrac{y}{2} - z = 7 \\ \dfrac{x}{4} - \dfrac{3y}{2} + \dfrac{z}{2} = -6 \\ \dfrac{x}{6} - \dfrac{y}{4} - \dfrac{z}{3} = 1 \end{cases}$

d) $\begin{cases} \dfrac{1}{x} + \dfrac{1}{y} + \dfrac{1}{z} = 5 \\ \dfrac{2}{x} - \dfrac{3}{y} - \dfrac{4}{z} = -11 \\ \dfrac{3}{x} + \dfrac{2}{y} - \dfrac{1}{z} = -6 \end{cases}$

36. Indicate which of the following systems are (1) consistent, (2) dependent, (3) inconsistent.

a) $\begin{cases} x + y - z = 2 \\ x - 3y + 2z = 1 \\ 3x - 5y + 3z = 4 \end{cases}$

b) $\begin{cases} 2x - y + z = 1 \\ x + 2y - 3z = -2 \\ 3x - 4y + 5z = 1 \end{cases}$

c) $\begin{cases} x + y + 2z = 3 \\ 3x - y + z = 1 \\ 2x + 3y - 4z = 8 \end{cases}$

37. The first of three numbers exceeds the third by one-half of the second. The sum of the second and third numbers is one more than the first. If the second is subtracted from the sum of the first and third numbers the result is 5. Determine the numbers.

38. When a certain three digit number is divided by the number with digits reversed, the quotient is 2 and the remainder 25. The tens digit is one less than twice the sum of the hundreds digit and units digit. If the units digit is subtracted from the tens digit, the result is twice the hundreds digit. Find the number.

ANSWERS TO SUPPLEMENTARY PROBLEMS.

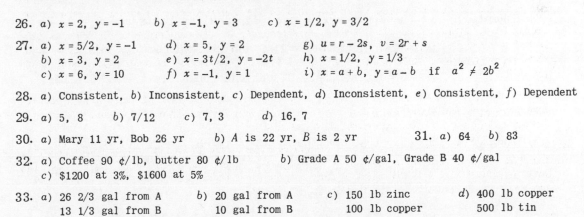

26. a) $x = 2$, $y = -1$ b) $x = -1$, $y = 3$ c) $x = 1/2$, $y = 3/2$

27. a) $x = 5/2$, $y = -1$ d) $x = 5$, $y = 2$ g) $u = r - 2s$, $v = 2r + s$
    b) $x = 3$, $y = 2$ e) $x = 3t/2$, $y = -2t$ h) $x = 1/2$, $y = 1/3$
    c) $x = 6$, $y = 10$ f) $x = -1$, $y = 1$ i) $x = a + b$, $y = a - b$ if $a^2 \neq 2b^2$

28. a) Consistent, b) Inconsistent, c) Dependent, d) Inconsistent, e) Consistent, f) Dependent

29. a) 5, 8   b) 7/12   c) 7, 3   d) 16, 7

30. a) Mary 11 yr, Bob 26 yr   b) A is 22 yr, B is 2 yr   31. a) 64   b) 83

32. a) Coffee 90 ¢/lb, butter 80 ¢/lb   b) Grade A 50 ¢/gal, Grade B 40 ¢/gal
    c) \$1200 at 3%, \$1600 at 5%

33. a) 26 2/3 gal from A   b) 20 gal from A   c) 150 lb zinc   d) 400 lb copper
       13 1/3 gal from B   10 gal from B   100 lb copper   500 lb tin

34. a) boat 20 mi/hr, river 5 mi/hr   b) 120 mi/hr, 80 mi/hr   c) 60 ft/sec, 50 ft/sec

35. a) $x = -1$, $y = 2$, $z = -2$ c) $x = 6$, $y = 4$, $z = -3$
    b) $x = 0$, $y = 2$, $z = 1$ d) $x = 1/2$, $y = -1/3$, $z = 1/6$

36. a) Dependent   b) Inconsistent   c) Consistent

37. 4, 2, 3

38. 371

CHAPTER 13

Quadratic Equations in One Unknown

A QUADRATIC EQUATION IN THE UNKNOWN x has the form $ax^2 + bx + c = 0$ where a, b, and c are constants and $a \neq 0$.

Thus $x^2 - 6x + 5 = 0$, $2x^2 + x - 6 = 0$, and $3x^2 - 5 = 0$ are quadratic equations in one unknown. The last two equations may be divided by 2 and 3 respectively to obtain $x^2 + \frac{1}{2}x - 3 = 0$ and $x^2 - \frac{5}{3} = 0$ where the coefficient of x^2 in each case is 1.

A *pure quadratic equation* is one which has no x term; e.g., $4x^2 - 5 = 0$.

TO SOLVE A QUADRATIC EQUATION $ax^2 + bx + c = 0$ is to find values of x which satisfy the equation. These values of x are called *solutions* or *roots* of the equation.

For example, $x^2 - 5x + 6 = 0$ is satisfied by $x = 2$ and $x = 3$. Then $x = 2$ and $x = 3$ are solutions or roots of the equation.

METHODS OF SOLVING QUADRATIC EQUATIONS.

A) PURE QUADRATIC EQUATIONS.

Example 1. $x^2 - 4 = 0$. Then $x^2 = 4$, $x = \pm 2$, and the roots are $x = 2, -2$.

2. $2x^2 - 21 = 0$. Then $x^2 = 21/2$ and the roots are $x = \pm\sqrt{21/2} = \pm\frac{1}{2}\sqrt{42}$.

3. $x^2 + 9 = 0$. Then $x^2 = -9$ and the roots are $x = \pm\sqrt{-9} = \pm 3i$.

B) SOLUTION BY FACTORING.

Example 4. $x^2 - 5x + 6 = 0$ may be written $(x-3)(x-2) = 0$. Since the product of the two factors is zero, either factor or both factors are zero. If $x - 3 = 0$, $x = 3$; if $x - 2 = 0$, $x = 2$. Thus the solutions are $x = 3$, $x = 2$.

5. $3x^2 + 2x - 5 = 0$ may be written $(3x+5)(x-1) = 0$. Then $3x + 5 = 0$, $x - 1 = 0$ give the solutions $x = -5/3$, $x = 1$.

6. $x^2 - 4x + 4 = 0$ may be written $(x-2)(x-2) = 0$. Then the equation has the *double root* $x = 2$.

C) SOLUTION BY COMPLETING THE SQUARE.

Example 7.   Solve $x^2 - 6x - 2 = 0$.

Write the unknowns on one side and the constant term on the other; then

$$x^2 - 6x = 2.$$

Add 9 to both sides, thus making the left hand side a perfect square; then

$$x^2 - 6x + 9 = 2 + 9 \quad \text{or} \quad (x - 3)^2 = 11.$$

Hence $x - 3 = \pm\sqrt{11}$ and the required roots are $x = 3 \pm \sqrt{11}$.

Note. In the method of completing the square (1) the coefficient of the x^2 term must be 1 and (2) the number added to both sides is the square of half the coefficient of x.

Example 8. Solve $3x^2 - 5x + 1 = 0$.

Dividing by 3, $x^2 - \dfrac{5x}{3} = -\dfrac{1}{3}$.

Adding $[\dfrac{1}{2}(-\dfrac{5}{3})]^2 = \dfrac{25}{36}$ to both sides,

$$x^2 - \frac{5}{3}x + \frac{25}{36} = -\frac{1}{3} + \frac{25}{36} = \frac{13}{36}, \quad (x - \frac{5}{6})^2 = \frac{13}{36},$$

$$x - \frac{5}{6} = \pm\frac{\sqrt{13}}{6} \quad \text{and} \quad x = \frac{5}{6} \pm \frac{\sqrt{13}}{6}.$$

D) SOLUTION BY QUADRATIC FORMULA.

The solutions of the quadratic equation $ax^2 + bx + c = 0$ are given by the formula

$$x = \frac{-b \pm \sqrt{b^2 - 4ac}}{2a}$$

where $b^2 - 4ac$ is called the *discriminant* of the quadratic equation.

For a derivation of the quadratic formula see Problem 5.

Example 9. Solve $3x^2 - 5x + 1 = 0$. Here $a = 3$, $b = -5$, $c = 1$ so that

$$x = \frac{-(-5) \pm \sqrt{(-5)^2 - 4(3)(1)}}{2(3)} = \frac{5 \pm \sqrt{13}}{6} \quad \text{as in Example 8.}$$

E) GRAPHICAL SOLUTION.

The real roots or solutions of $ax^2 + bx + c = 0$ are the values of x corresponding to $y = 0$ on the graph of the parabola $y = ax^2 + bx + c$. Thus the solutions are the abscissas of the points where the parabola intersects the x-axis. If the graph does not intersect the x-axis the roots are imaginary.

THE SUM S AND THE PRODUCT P OF THE ROOTS of the quadratic equation $ax^2 + bx + c = 0$ are given by $S = -\dfrac{b}{a}$ and $P = \dfrac{c}{a}$.

Thus in $2x^2 + 7x - 6 = 0$ we have $a = 2$, $b = 7$, $c = -6$ so that $S = -7/2$ and $P = -6/2 = -3$.

It follows that a quadratic equation whose roots are r_1, r_2 is given by $x^2 - Sx + P = 0$ where $S = r_1 + r_2$ and $P = r_1 r_2$. Thus the quadratic equation whose roots are $x = 2$ and $x = -5$ is $x^2 - (2 - 5)x + 2(-5) = 0$ or $x^2 + 3x - 10 = 0$.

THE CHARACTER OF THE ROOTS of the quadratic equation $ax^2 + bx + c = 0$ is determined by the discriminant $b^2 - 4ac$.

Assuming a, b, c are *real numbers* then

1) if $b^2 - 4ac > 0$, the roots are *real* and *unequal*,
2) if $b^2 - 4ac = 0$, the roots are *real* and *equal*,
3) if $b^2 - 4ac < 0$, the roots are *imaginary*.

Assuming a, b, c are *rational numbers* then

1) if $b^2 - 4ac$ is a perfect square $\neq 0$, the roots are real, rational and unequal,
2) if $b^2 - 4ac = 0$, the roots are real, rational and equal,
3) if $b^2 - 4ac > 0$ but not a perfect square, the roots are real, irrational and unequal,
4) if $b^2 - 4ac < 0$, the roots are imaginary.

Thus $2x^2 + 7x - 6 = 0$, with discriminant $b^2 - 4ac = 7^2 - 4(2)(-6) = 97$, has roots which are real, irrational and unequal.

A RADICAL EQUATION is an equation having one or more unknowns under a radical.

Thus $\sqrt{x + 3} - \sqrt{x} = 1$ and $\sqrt[3]{y} = \sqrt{y - 4}$ are radical equations.

To solve a radical equation, isolate one of the radical terms on one side of the equation and transpose all other terms to the other side. If both members of the equation are then raised to a power equal to the index of the isolated radical, the radical will be removed. This process is continued until radicals are no longer present.

Example 10.   Solve $\sqrt{x + 3} - \sqrt{x} = 1$.

Transposing, $\sqrt{x + 3} = \sqrt{x} + 1$.

Squaring, $x + 3 = x + 2\sqrt{x} + 1$ or $\sqrt{x} = 1$.

Finally, squaring both sides of $\sqrt{x} = 1$ gives $x = 1$.

Check. $\sqrt{1 + 3} - \sqrt{1} = 1$, $2 - 1 = 1$.

It is very important to check values obtained, as this method often introduces extraneous roots which are to be rejected.

AN EQUATION OF QUADRATIC TYPE has the form $az^{2n} + bz^n + c = 0$ where $a \neq 0$, b, c, and $n \neq 0$ are constants and where z depends on x. Upon letting $z^n = u$ this equation becomes $au^2 + bu + c = 0$ which may be solved for u. These values of u may be used to obtain z from which it may be possible to obtain x.

SOLVED PROBLEMS

PURE QUADRATIC EQUATIONS.

1. Solve.

a) $x^2 - 16 = 0$. Then $x^2 = 16$, $x = \pm 4$.

b) $4t^2 - 9 = 0$. Then $4t^2 = 9$, $t^2 = 9/4$, $t = \pm 3/2$.

c) $3 - x^2 = 2x^2 + 1$. Then $3x^2 = 2$, $x^2 = 2/3$, $x = \pm\sqrt{2/3} = \pm\frac{1}{3}\sqrt{6}$.

d) $4x^2 + 9 = 0$. Then $x^2 = -9/4$, $x = \pm\sqrt{-9/4} = \pm\frac{3}{2}i$.

e) $\dfrac{2x^2 - 1}{x - 3} = x + 3 + \dfrac{17}{x - 3}$. Then $2x^2 - 1 = (x + 3)(x - 3) + 17$, $2x^2 - 1 = x^2 - 9 + 17$, $x^2 = 9$,

and $x = \pm 3$.

Check. If $x = 3$ is substituted into the original equation, we have division by zero which is not allowed. Hence $x = 3$ is not a solution.

If $x = -3$, $\dfrac{2(-3)^2 - 1}{-3 - 3} = -3 + 3 + \dfrac{17}{-3 - 3}$ or $\dfrac{17}{6} = \dfrac{17}{6}$ and $x = -3$ is a solution.

SOLUTION BY FACTORING.

2. Solve by factoring.

a) $x^2 + 5x - 6 = 0$, $(x + 6)(x - 1) = 0$, $x = -6, 1$.

b) $t^2 = 4t$, $t^2 - 4t = 0$, $t(t - 4) = 0$, $t = 0, 4$.

c) $x^2 + 3x = 28$, $x^2 + 3x - 28 = 0$, $(x + 7)(x - 4) = 0$, $x = -7, 4$.

d) $5x - 2x^2 = 2$, $2x^2 - 5x + 2 = 0$, $(2x - 1)(x - 2) = 0$, $x = 1/2, 2$.

e) $\dfrac{1}{t - 1} + \dfrac{1}{t - 4} = \dfrac{5}{4}$. Multiplying by $4(t - 1)(t - 4)$,

$4(t - 4) + 4(t - 1) = 5(t - 1)(t - 4)$, $5t^2 - 33t + 40 = 0$, $(t - 5)(5t - 8) = 0$, $t = 5, 8/5$.

f) $\dfrac{y}{2p} = \dfrac{3p}{6y - 5p}$, $6y^2 - 5py - 6p^2 = 0$, $(3y + 2p)(2y - 3p) = 0$, $y = -2p/3, 3p/2$.

SOLUTION BY COMPLETING THE SQUARE.

3. What term must be added to each of the following expressions in order to make it a perfect square trinomial?

a) $x^2 - 2x$. Add $\left[\frac{1}{2}(\text{coefficient of } x)\right]^2 = \left[\frac{1}{2}(-2)\right]^2 = 1$. Check: $x^2 - 2x + 1 = (x - 1)^2$.

b) $x^2 + 4x$. Add $\left[\frac{1}{2}(\text{coefficient of } x)\right]^2 = \left[\frac{1}{2}(4)\right]^2 = 4$. Check: $x^2 + 4x + 4 = (x + 2)^2$.

c) $u^2 + \frac{5}{4}u$. Add $\left[\frac{1}{2}(\frac{5}{4})\right]^2 = \dfrac{25}{64}$. Check: $u^2 + \frac{5}{4}u + \dfrac{25}{64} = (u + \frac{5}{8})^2$.

d) $x^4 + px^2$. Add $\left[\frac{1}{2}(p)\right]^2 = p^2/4$. Check: $x^4 + px^2 + p^2/4 = (x^2 + p/2)^2$.

4. Solve by completing the square.

a) $x^2 - 6x + 8 = 0$.   Then $x^2 - 6x = -8$, $x^2 - 6x + 9 = -8 + 9$, $(x - 3)^2 = 1$.

Hence $x - 3 = \pm 1$, $x = 3 \pm 1$, and the roots are $x = 4$ and $x = 2$.

Check: For $x = 4$, $4^2 - 6(4) + 8 = 0$, $0 = 0$.   For $x = 2$, $2^2 - 6(2) + 8 = 0$, $0 = 0$.

b) $t^2 = 4 - 3t$.   Then $t^2 + 3t = 4$, $t^2 + 3t + (\frac{3}{2})^2 = 4 + (\frac{3}{2})^2$, $(t + \frac{3}{2})^2 = \frac{25}{4}$.

Hence $t + \frac{3}{2} = \pm \frac{5}{2}$, $t = -\frac{3}{2} \pm \frac{5}{2}$, and the roots are $t = 1, -4$.

c) $3x^2 + 8x + 5 = 0$.   Then $x^2 + \frac{8}{3}x = -\frac{5}{3}$, $x^2 + \frac{8}{3}x + (\frac{4}{3})^2 = -\frac{5}{3} + (\frac{4}{3})^2$, $(x + \frac{4}{3})^2 = \frac{1}{9}$.

Hence $x + \frac{4}{3} = \pm \frac{1}{3}$, $x = -\frac{4}{3} \pm \frac{1}{3}$, and the roots are $x = -1, -5/3$.

d) $x^2 + 4x + 1 = 0$.   Then $x^2 + 4x = -1$, $x^2 + 4x + 4 = 3$, $(x + 2)^2 = 3$.

Hence $x + 2 = \pm \sqrt{3}$, and the roots are $x = -2 \pm \sqrt{3}$.

Check: For $x = -2 + \sqrt{3}$, $(-2 + \sqrt{3})^2 + 4(-2 + \sqrt{3}) + 1 = (4 - 4\sqrt{3} + 3) - 8 + 4\sqrt{3} + 1 = 0$.

For $x = -2 - \sqrt{3}$, $(-2 - \sqrt{3})^2 + 4(-2 - \sqrt{3}) + 1 = (4 + 4\sqrt{3} + 3) - 8 - 4\sqrt{3} + 1 = 0$.

e) $5x^2 - 6x + 5 = 0$.   Then $5x^2 - 6x = -5$, $x^2 - \frac{6x}{5} + (\frac{3}{5})^2 = -1 + (\frac{3}{5})^2$, $(x - \frac{3}{5})^2 = -\frac{16}{25}$.

Hence $x - 3/5 = \pm \sqrt{-16/25}$, and the roots are $x = \frac{3}{5} \pm \frac{4}{5}i$.

5. Solve the equation $ax^2 + bx + c = 0$, $a \neq 0$, by the method of completing the square.

Dividing both sides by a, $x^2 + \frac{b}{a}x + \frac{c}{a} = 0$ or $x^2 + \frac{b}{a}x = -\frac{c}{a}$.

Adding $[\frac{1}{2}(\frac{b}{a})]^2 = \frac{b^2}{4a^2}$ to both sides, $x^2 + \frac{b}{a}x + \frac{b^2}{4a^2} = -\frac{c}{a} + \frac{b^2}{4a^2} = \frac{b^2 - 4ac}{4a^2}$.

Then $(x + \frac{b}{2a})^2 = \frac{b^2 - 4ac}{4a^2}$, $x + \frac{b}{2a} = \pm \frac{\sqrt{b^2 - 4ac}}{2a}$,   and $x = \frac{-b \pm \sqrt{b^2 - 4ac}}{2a}$.

SOLUTION BY QUADRATIC FORMULA.

6. Solve by the quadratic formula.

a) $x^2 - 3x + 2 = 0$.   Here $a = 1$, $b = -3$, $c = 2$.   Then

$$x = \frac{-b \pm \sqrt{b^2 - 4ac}}{2a} = \frac{-(-3) \pm \sqrt{(-3)^2 - 4(1)(2)}}{2(1)} = \frac{3 \pm 1}{2} \quad \text{or} \quad x = 1, 2.$$

b) $4t^2 + 12t + 9 = 0$.   Here $a = 4$, $b = 12$, $c = 9$.   Then

$$t = \frac{-12 \pm \sqrt{(12)^2 - 4(4)(9)}}{2(4)} = \frac{-12 \pm 0}{8} = -\frac{3}{2} \quad \text{and } t = -\frac{3}{2} \text{ is a double root.}$$

c) $9x^2 + 18x - 17 = 0$.   Here $a = 9$, $b = 18$, $c = -17$.   Then

$$x = \frac{-18 \pm \sqrt{(18)^2 - 4(9)(-17)}}{2(9)} = \frac{-18 \pm \sqrt{936}}{18} = \frac{-18 \pm 6\sqrt{26}}{18} = \frac{-3 \pm \sqrt{26}}{3}.$$

d) $6u(2-u) = 7$.   Then $6u^2 - 12u + 7 = 0$ and

$$u = \frac{-(-12) \pm \sqrt{(-12)^2 - 4(6)(7)}}{2(6)} = \frac{12 \pm \sqrt{-24}}{12} = \frac{12 \pm 2\sqrt{6}i}{12} = 1 \pm \frac{\sqrt{6}}{6}i.$$

GRAPHICAL SOLUTION.

7. Solve graphically: *a)* $2x^2 + 3x - 5 = 0$,   *b)* $4x^2 - 12x + 9 = 0$,   *c)* $4x^2 - 4x + 5 = 0$.

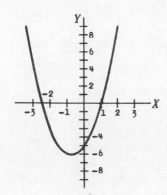 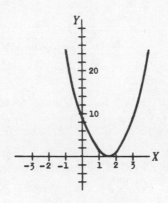 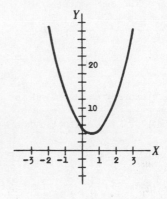

Real distinct roots	Real equal roots	Imaginary roots
(a)	*(b)*	*(c)*

a) $y = 2x^2 + 3x - 5$

x	-3	-2	-1	0	1	2
y	4	-3	-6	-5	0	9

The graph of $y = 2x^2 + 3x - 5$ indicates that when $y = 0$, $x = 1$ and -2.5.

Thus the roots of $2x^2 + 3x - 5 = 0$ are $x = 1, -2.5$.

b) $y = 4x^2 - 12x + 9$

x	-1	0	1	2	3	4
y	25	9	1	1	9	25

The graph of $y = 4x^2 - 12x + 9$ is tangent to the x-axis at $x = 1.5$, i.e. when $y = 0$, $x = 1.5$.

Thus $4x^2 - 12x + 9 = 0$ has the equal roots $x = 1.5$.

c) $y = 4x^2 - 4x + 5$

x	-2	-1	0	1	2	3
y	29	13	5	5	13	29

The graph of $y = 4x^2 - 4x + 5$ does not intersect the x-axis, i.e. there is no real value of x for which $y = 0$.

Hence the roots of $4x^2 - 4x + 5 = 0$ are imaginary.

(By the quadratic formula the roots are $x = \frac{1}{2} \pm i$.)

SUM AND PRODUCT OF ROOTS.

8. Prove that the sum S and product P of the roots of the quadratic equation $ax^2 + bx + c = 0$ are $S = -\dfrac{b}{a}$ and $P = \dfrac{c}{a}$.

By the quadratic formula the roots are $\dfrac{-b + \sqrt{b^2 - 4ac}}{2a}$ and $\dfrac{-b - \sqrt{b^2 - 4ac}}{2a}$.

The sum of the roots is $S = \dfrac{-2b}{2a} = -\dfrac{b}{a}$. The product of the roots is

$$P = \left(\dfrac{-b + \sqrt{b^2 - 4ac}}{2a}\right)\left(\dfrac{-b - \sqrt{b^2 - 4ac}}{2a}\right) = \dfrac{(-b)^2 - (b^2 - 4ac)}{4a^2} = \dfrac{c}{a}.$$

9. Without solving, find the sum S and product P of the roots.

a) $x^2 - 7x + 6 = 0$.   Here $a = 1$, $b = -7$, $c = 6$; then $S = -\dfrac{b}{a} = 7$, $P = \dfrac{c}{a} = 6$.

b) $2x^2 + 6x - 3 = 0$.   Here $a = 2$, $b = 6$, $c = -3$; then $S = -\dfrac{6}{2} = -3$, $P = \dfrac{-3}{2}$.

c) $x + 3x^2 + 5 = 0$.   Write as $3x^2 + x + 5 = 0$. Then $S = -\dfrac{1}{3}$, $P = \dfrac{5}{3}$.

d) $3x^2 - 5x = 0$.   Here $a = 3$, $b = -5$, $c = 0$; then $S = \dfrac{5}{3}$, $P = 0$.

e) $2x^2 + 3 = 0$.   Here $a = 2$, $b = 0$, $c = 3$; then $S = 0$, $P = \dfrac{3}{2}$.

f) $mnx^2 + (m^2 + n^2)x + mn = 0$.   Then $S = -\dfrac{m^2 + n^2}{mn}$, $P = \dfrac{mn}{mn} = 1$.

g) $0.3x^2 - 0.01x + 4 = 0$.   Then $S = -\dfrac{-0.01}{0.3} = \dfrac{1}{30}$, $P = \dfrac{4}{0.3} = \dfrac{40}{3}$.

10. Find the discriminant $b^2 - 4ac$ of each of the following equations and thus determine the character of their roots.

a) $x^2 - 8x + 12 = 0$. $b^2 - 4ac = (-8)^2 - 4(1)(12) = 16$; the roots are real, rational, unequal.

b) $3y^2 + 2y - 4 = 0$. $b^2 - 4ac = 52$; the roots are real, irrational, unequal.

c) $2x^2 - x + 4 = 0$. $b^2 - 4ac = -31$; the roots are conjugate imaginaries.

d) $4z^2 - 12z + 9 = 0$. $b^2 - 4ac = 0$; the roots are real, rational, equal.

e) $2x - 4x^2 = 1$ or $4x^2 - 2x + 1 = 0$. $b^2 - 4ac = -12$; the roots are conjugate imaginaries.

f) $\sqrt{2}\,x^2 - 4\sqrt{3}\,x + 4\sqrt{2} = 0$.   Here the coefficients are real but not rational numbers.

$b^2 - 4ac = 16$; the roots are real and unequal.

11. Find a quadratic equation with integer coefficients having the given pair of roots.
(S = sum of roots, P = product of roots.)

a) 1, 2

Method 1. $S = 1 + 2 = 3$, $P = 2$; hence $x^2 - 3x + 2 = 0$.

Method 2. $(x - 1)$ and $(x - 2)$ must be factors of the quadratic expression.

Then $(x - 1)(x - 2) = 0$ or $x^2 - 3x + 2 = 0$.

b) $-3, 2$

    Method 1. $S = -1$, $P = -6$; hence $x^2 + x - 6 = 0$.

    Method 2. $[x - (-3)]$ and $(x - 2)$ are factors of the quadratic expression.

        Then $(x + 3)(x - 2) = 0$ or $x^2 + x - 6 = 0$.

c) $\frac{4}{3}$, $-\frac{3}{5}$. $S = \frac{11}{15}$, $P = -\frac{4}{5}$; hence $x^2 - \frac{11}{15}x - \frac{4}{5} = 0$ or $15x^2 - 11x - 12 = 0$.

d) $2 + \sqrt{2}$, $2 - \sqrt{2}$

    Method 1. $S = 4$, $P = (2 + \sqrt{2})(2 - \sqrt{2}) = 2$; hence $x^2 - 4x + 2 = 0$.

    Method 2. $[x - (2 + \sqrt{2})]$ and $[x - (2 - \sqrt{2})]$ are factors of the quadratic expression.

        Then $[x - (2 + \sqrt{2})][x - (2 - \sqrt{2})] = [(x - 2) - \sqrt{2}][(x - 2) + \sqrt{2}] = 0$,

        $(x - 2)^2 - 2 = 0$ or $x^2 - 4x + 2 = 0$.

    Method 3. Since $x = 2 \pm \sqrt{2}$, $x - 2 = \pm \sqrt{2}$. Squaring, $(x - 2)^2 = 2$ or $x^2 - 4x + 2 = 0$.

e) $-3 + 2i$, $-3 - 2i$

    Method 1. $S = -6$, $P = (-3 + 2i)(-3 - 2i) = 13$; hence $x^2 + 6x + 13 = 0$.

    Method 2. $[x - (-3 + 2i)]$ and $[x - (-3 - 2i)]$ are factors of the quadratic expression.

        Then $[(x + 3) - 2i][(x + 3) + 2i] = 0$, $(x + 3)^2 + 4 = 0$ or $x^2 + 6x + 13 = 0$.

12. In each quadratic equation find the value of the constant p subject to the given condition.

    a) $2x^2 - px + 4 = 0$ has one root equal to -3.

    Since $x = -3$ is a root, it must satisfy the given equation.

    Then $2(-3)^2 - p(-3) + 4 = 0$ and $p = -22/3$.

    b) $(p + 2)x^2 + 5x + 2p = 0$ has the product of its roots equal to $2/3$.

    The product of the roots is $\frac{2p}{p + 2}$; then $\frac{2p}{p + 2} = \frac{2}{3}$ and $p = 1$.

    c) $2px^2 + px + 2x = x^2 + 7p + 1$ has the sum of its roots equal to $-4/3$.

    Write the equation in the form $(2p - 1)x^2 + (p + 2)x - (7p + 1) = 0$.

    Then the sum of the roots is $-\frac{p + 2}{2p - 1} = -\frac{4}{3}$ and $p = 2$.

    d) $3x^2 + (p + 1)x + 24 = 0$ has one root equal to twice the other. Let the roots be r, $2r$.

    Product of the roots is $r(2r) = 8$; then $r^2 = 4$ and $r = \pm 2$.

    Sum of the roots is $3r = -\frac{p + 1}{3}$. Substitute $r = 2$ and $r = -2$ into this equation and obtain $p = -19$ and $p = 17$ respectively.

    e) $2x^2 - 12x + p + 2 = 0$ has the difference between its roots equal to 2.

    Let the roots be r, s; then (1) $r - s = 2$. The sum of the roots is 6; then (2) $r + s = 6$. The simultaneous solution of (1) and (2) is $r = 4$, $s = 2$.

    Now put $x = 2$ or $x = 4$ into the given equation and obtain $p = 14$.

13. Find the roots of each quadratic equation subject to the given conditions.

    a) $(2k + 2)x^2 + (4 - 4k)x + k - 2 = 0$ has roots which are reciprocals of each other.

Let r and $1/r$ be the roots, their product being 1.

Product of roots is $\dfrac{k-2}{2k+2} = 1$, from which $k = -4$.

Put $k = -4$ into the given equation; then $3x^2 - 10x + 3 = 0$ and the roots are 1/3, 3.

b) $kx^2 - (1+k)x + 3k + 2 = 0$ has the sum of its roots equal to twice the product of its roots.

Sum of roots = 2(product of roots); then $\dfrac{1+k}{k} = 2\left(\dfrac{3k+2}{k}\right)$ and $k = -\dfrac{3}{5}$.

Put $k = -3/5$ into the given equation; then $3x^2 + 2x - 1 = 0$ and the roots are -1, 1/3.

c) $(x+k)^2 = 2 - 3k$ has equal roots.

Write the equation as $x^2 + 2kx + (k^2 + 3k - 2) = 0$ where $a = 1$, $b = 2k$, $c = k^2 + 3k - 2$. The roots are equal if the discriminant $(b^2 - 4ac) = 0$.

Then from $b^2 - 4ac = (2k)^2 - 4(1)(k^2 + 3k - 2) = 0$ we get $k = 2/3$.

Put $k = 2/3$ into the given equation and solve to obtain the double root $-2/3$.

RADICAL EQUATIONS.

14. Solve.

a) $\sqrt{2x+1} = 3$.   Squaring both sides, $2x + 1 = 9$ and $x = 4$.

Check. $\sqrt{2(4)+1} = 3$, $3 = 3$.

b) $\sqrt{5+2x} = x + 1$. Squaring both sides, $5 + 2x = x^2 + 2x + 1$, $x^2 = 4$ and $x = \pm 2$.

Check. For $x = 2$, $\sqrt{5+2(2)} = 2 + 1$ or $3 = 3$.

For $x = -2$, $\sqrt{5+2(-2)} = -2 + 1$ or $\sqrt{1} = -1$ which is not true since $\sqrt{1} = 1$.

Thus $x = 2$ is the only solution; $x = -2$ is an extraneous root.

c) $\sqrt{3x-5} = x - 1$. Squaring, $3x - 5 = x^2 - 2x + 1$, $x^2 - 5x + 6 = 0$ and $x = 3, 2$.

Check. For $x = 3$, $\sqrt{3(3)-5} = 3 - 1$ or $2 = 2$. For $x = 2$, $\sqrt{3(2)-5} = 2 - 1$ or $1 = 1$.

Thus both $x = 3$ and $x = 2$ are solutions of the given equation.

d) $\sqrt[3]{x^2 - x + 6} - 2 = 0$. Then $\sqrt[3]{x^2 - x + 6} = 2$, $x^2 - x + 6 = 8$, $x^2 - x - 2 = 0$ and $x = 2, -1$.

Check. For $x = 2$, $\sqrt[3]{2^2 - 2 + 6} - 2 = 0$ or $2 - 2 = 0$.

For $x = -1$, $\sqrt[3]{(-1)^2 - (-1) + 6} - 2 = 0$ or $2 - 2 = 0$.

15. Solve.

a) $\sqrt{2x+1} - \sqrt{x} = 1$.   Rearranging, (1) $\sqrt{2x+1} = \sqrt{x} + 1$.

Squaring both sides of (1), $2x + 1 = x + 2\sqrt{x} + 1$ or (2) $x = 2\sqrt{x}$.

Squaring (2), $x^2 = 4x$; then $x(x-4) = 0$ and $x = 0, 4$.

Check. For $x = 0$, $\sqrt{2(0)+1} - \sqrt{0} = 1$, $1 = 1$. For $x = 4$, $\sqrt{2(4)+1} - \sqrt{4} = 1$, $1 = 1$.

b) $\sqrt{4x-1} + \sqrt{2x+3} = 1$. Rearranging, (1) $\sqrt{4x-1} = 1 - \sqrt{2x+3}$.

Squaring (1), $4x - 1 = 1 - 2\sqrt{2x+3} + 2x + 3$ or (2) $2\sqrt{2x+3} = 5 - 2x$.

Squaring (2), $4(2x+3) = 25 - 20x + 4x^2$, $4x^2 - 28x + 13 = 0$ and $x = 1/2, 13/2$.

Check. For $x = 1/2$, $\sqrt{4(1/2) - 1} + \sqrt{2(1/2) + 3} = 1$ or $3 = 1$ which is not true.

For $x = 13/2$, $\sqrt{4(13/2) - 1} + \sqrt{2(13/2) + 3} = 1$ or $9 = 1$ which is not true.

Hence $x = 1/2$ and $x = 13/2$ are extraneous roots; the equation has no solution.

c) $\sqrt{\sqrt{x + 16} - \sqrt{x}} = 2$.   Squaring, $\sqrt{x + 16} - \sqrt{x} = 4$ or (1) $\sqrt{x + 16} = \sqrt{x} + 4$.

Squaring (1), $x + 16 = x + 8\sqrt{x} + 16$, $8\sqrt{x} = 0$, and $x = 0$ is a solution.

16. Solve.

a) $\sqrt{x^2 + 6x} = x + \sqrt{2x}$.

Squaring, $x^2 + 6x = x^2 + 2x\sqrt{2x} + 2x$, $2x\sqrt{2x} = 4x$, $x(\sqrt{2x} - 2) = 0$.

Then $x = 0$; and from $\sqrt{2x} - 2 = 0$, $\sqrt{2x} = 2$, $2x = 4$, $x = 2$.

Both $x = 0$ and $x = 2$ satisfy the given equation.

b) $\sqrt{x} - \dfrac{2}{\sqrt{x}} = 1$.   Multiply by $\sqrt{x}$ and obtain (1) $x - 2 = \sqrt{x}$.

Squaring (1), $x^2 - 4x + 4 = x$, $x^2 - 5x + 4 = 0$, $(x - 1)(x - 4) = 0$, and $x = 1, 4$.

Only $x = 4$ satisfies the given equation; $x = 1$ is extraneous.

17. Solve the equation $x^2 - 6x - \sqrt{x^2 - 6x - 3} = 5$.

Let $x^2 - 6x = u$; then $u - \sqrt{u - 3} = 5$ or (1) $\sqrt{u - 3} = u - 5$.

Squaring (1), $u - 3 = u^2 - 10u + 25$, $u^2 - 11u + 28 = 0$, and $u = 7, 4$.

Since only $u = 7$ satisfies (1), substitute $u = 7$ in $x^2 - 6x = u$ and obtain

$$x^2 - 6x - 7 = 0, \quad (x - 7)(x + 1) = 0, \quad \text{and} \quad x = 7, -1.$$

Both $x = 7$ and $x = -1$ satisfy the original equation and are thus solutions.

Note. If we write the given equation as $\sqrt{x^2 - 6x - 3} = x^2 - 6x - 5$ and square both sides, the resulting fourth degree equation would be difficult to solve.

18. Solve the equation $\dfrac{4 - x}{\sqrt{x^2 - 8x + 32}} = \dfrac{3}{5}$.

Squaring, $\dfrac{16 - 8x + x^2}{x^2 - 8x + 32} = \dfrac{9}{25}$; then $25(16 - 8x + x^2) = 9(x^2 - 8x + 32)$, $x^2 - 8x + 7 = 0$, and

$x = 7, 1$. The only solution is $x = 1$; reject $x = 7$, an extraneous solution.

EQUATIONS OF QUADRATIC TYPE.

19. Solve.

a) $x^4 - 10x^2 + 9 = 0$.   Let $x^2 = u$; then $u^2 - 10u + 9 = 0$ and $u = 1, 9$.

For $u = 1$, $x^2 = 1$ and $x = \pm 1$; for $u = 9$, $x^2 = 9$ and $x = \pm 3$.

The four solutions are $x = \pm 1, \pm 3$; each satisfies the given equation.

b) $2x^4 + x^2 - 1 = 0$.   Let $x^2 = u$; then $2u^2 + u - 1 = 0$ and $u = \frac{1}{2}, -1$.

If $u = \frac{1}{2}$, $x^2 = \frac{1}{2}$ and $x = \pm\frac{1}{2}\sqrt{2}$; if $u = -1$, $x^2 = -1$ and $x = \pm i$.

The four solutions are $x = \pm\frac{1}{2}\sqrt{2}, \pm i$.

c) $\sqrt{x} - \sqrt[4]{x} - 2 = 0$.   Let $\sqrt[4]{x} = u$; then $u^2 - u - 2 = 0$ and $u = 2, -1$.

If $u = 2$, $\sqrt[4]{x} = 2$ and $x = 2^4 = 16$. Since $\sqrt[4]{x}$ is positive, it cannot equal -1.

Hence $x = 16$ is the only solution of the given equation.

d) $2(x + \frac{1}{x})^2 - 7(x + \frac{1}{x}) + 5 = 0$.   Let $x + \frac{1}{x} = u$; then $2u^2 - 7u + 5 = 0$ and $u = 5/2, 1$.

For $u = \frac{5}{2}$, $x + \frac{1}{x} = \frac{5}{2}$, $2x^2 - 5x + 2 = 0$ and $x = 2, \frac{1}{2}$.

For $u = 1$, $x + \frac{1}{x} = 1$, $x^2 - x + 1 = 0$ and $x = \frac{1}{2} \pm \frac{1}{2}\sqrt{3}\,i$.

The four solutions are $x = 2, \frac{1}{2}, \frac{1}{2} \pm \frac{1}{2}\sqrt{3}\,i$.

e) $9(x + 2)^{-4} + 17(x + 2)^{-2} - 2 = 0$.   Let $(x + 2)^{-2} = u$; then $9u^2 + 17u - 2 = 0$ and $u = 1/9, -2$.

If $(x + 2)^{-2} = 1/9$, $(x + 2)^2 = 9$, $(x + 2) = \pm 3$ and $x = 1, -5$.

If $(x + 2)^{-2} = -2$, $(x + 2)^2 = -\frac{1}{2}$, $(x + 2) = \pm\frac{1}{2}\sqrt{2}\,i$ and $x = -2 \pm \frac{1}{2}\sqrt{2}\,i$.

The four solutions are $x = 1, -5, -2 \pm \frac{1}{2}\sqrt{2}\,i$.

20. Find the values of x which satisfy each of the following equations.

a) $16(\frac{x}{x+1})^4 - 25(\frac{x}{x+1})^2 + 9 = 0$.   Let $(\frac{x}{x+1})^2 = u$; then $16u^2 - 25u + 9 = 0$ and $u = 1, 9/16$.

If $u = 1$, $(\frac{x}{x+1})^2 = 1$ or $\frac{x}{x+1} = \pm 1$. The equation $\frac{x}{x+1} = 1$ has no solution; the equation $\frac{x}{x+1} = -1$ has solution $x = -1/2$.

If $u = 9/16$, $(\frac{x}{x+1})^2 = \frac{9}{16}$ or $\frac{x}{x+1} = \pm\frac{3}{4}$ so that $x = 3, -3/7$.

The required solutions are $x = -1/2, -3/7, 3$.

b) $(x^2 + 3x + 2)^2 - 8(x^2 + 3x) = 4$.   Let $x^2 + 3x = u$; then $(u + 2)^2 - 8u = 4$ and $u = 0, 4$.

If $u = 0$, $x^2 + 3x = 0$ and $x = 0, -3$; if $u = 4$, $x^2 + 3x = 4$ and $x = -4, 1$.

The solutions are $x = -4, -3, 0, 1$.

WORD PROBLEMS.

21. One positive number exceeds three times another positive number by 5. The product of the two numbers is 68. Find the numbers.

Let x = smaller number; then $3x + 5$ = larger number.

Then $x(3x + 5) = 68$, $3x^2 + 5x - 68 = 0$, $(3x + 17)(x - 4) = 0$, and $x = 4, -17/3$.

We exclude $-17/3$ since the problem states that the numbers are positive.

The required numbers are $x = 4$ and $3x + 5 = 17$.

22. When three times a certain number is added to twice its reciprocal, the result is 5. Find the number.

Let x = the required number and $1/x$ = its reciprocal.

Then $3x + 2(1/x) = 5$, $3x^2 - 5x + 2 = 0$, $(3x - 2)(x - 1) = 0$, and $x = 1, 2/3$.

Check. For $x = 1$, $3(1) + 2(1/1) = 5$; for $x = 2/3$, $3(2/3) + 2(3/2) = 5$.

23. Determine the dimensions of a rectangle having perimeter 50 feet and area 150 square feet.

Sum of four sides = 50 ft; hence, sum of two adjacent sides = 25 ft.
Let x and $25 - x$ be the lengths of two adjacent sides.

The area is $x(25 - x) = 150$; then $x^2 - 25x + 150 = 0$, $(x - 10)(x - 15) = 0$, and $x = 10, 15$.

Then $25 - x = 15, 10$; and the rectangle has dimensions 10 ft by 15 ft.

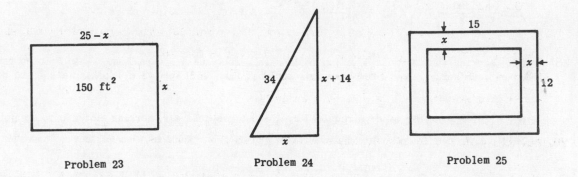

Problem 23   Problem 24   Problem 25

24. The hypotenuse of a right triangle is 34 inches. Find the lengths of the two legs if one leg is 14 inches longer than the other.

Let x and $x + 14$ be the lengths of the legs.

Then $x^2 + (x + 14)^2 = (34)^2$, $x^2 + 14x - 480 = 0$, $(x + 30)(x - 16) = 0$, and $x = -30, 16$.

Since $x = -30$ has no physical significance, we have $x = 16$ in. and $x + 14 = 30$ in.

25. A picture frame of uniform width has outer dimensions 12 in. by 15 in. Find the width of the frame a) if 88 square inches of picture shows, b) if 100 square inches of picture shows.

Let x = width of frame; then the dimensions of the picture are $(15 - 2x)$, $(12 - 2x)$.

a) Area of picture = $(15 - 2x)(12 - 2x) = 88$; then $2x^2 - 27x + 46 = 0$, $(x - 2)(2x - 23) = 0$, and $x = 2, 11\frac{1}{2}$. Clearly, the width cannot be $11\frac{1}{2}$ in. Hence the width of the frame is 2 in.

Check. The area of the picture is $(15 - 4)(12 - 4) = 88$ in.^{2}, as given.

b) Here $(15 - 2x)(12 - 2x) = 100$, $2x^2 - 27x + 40 = 0$ and, by the quadratic formula,

$$x = \frac{-b \pm \sqrt{b^2 - 4ac}}{2a} = \frac{27 \pm \sqrt{409}}{4} \quad \text{or} \quad x = 11.8, 1.7 \text{ (approximately)}.$$

Reject $x = 11.8$ in., which cannot be the width. The required width is 1.7 in.

26. A pilot flies a distance of 600 miles. He could fly the same distance in 30 minutes less time by increasing his average speed by 40 mi/hr. Find his actual average speed.

Let x = actual average speed in mi/hr.   Time in hours = $\dfrac{\text{distance in mi}}{\text{speed in mi/hr}}$.

Time to fly 600 mi at x mi/hr − time to fly 600 mi at $(x+40)$mi/hr = $\frac{1}{2}$ hr.

Then $\dfrac{600}{x} - \dfrac{600}{x+40} = \dfrac{1}{2}$.   Solving, the required speed is $x = 200$ mi/hr.

27. A retailer bought a number of shirts for $180 and sold all but 6 at a profit of $2 per shirt. With the total amount received he could buy 30 more shirts than before. Find the cost per shirt.

Let x = cost per shirt in dollars; $180/x$ = number of shirts bought.

Then $\left(\dfrac{180}{x} - 6\right)(x + 2) = x\left(\dfrac{180}{x} + 30\right)$.   Solving, $x = \$3$ per shirt.

28. A and B working together can do a job in 10 days. It takes A 5 days longer than B to do the job when each works alone. How many days would it take each of them, working alone, to do the job?

Let n, $n-5$ = number of days required by A and B respectively, working alone, to do the job.

In 1 day A does $1/n$ of job and B does $1/(n-5)$ of job. Thus in 10 days they do together

$$10\left(\frac{1}{n} + \frac{1}{n-5}\right) = 1 \text{ complete job.}$$

Then $10(2n-5) = n(n-5)$, $n^2 - 25n + 50 = 0$,   and $n = \dfrac{25 \pm \sqrt{625 - 200}}{2} = 22.8,\ 2.2$.

Rejecting $n = 2.2$, the required solution is $n = 22.8$ days, $n - 5 = 17.8$ days.

29. A ball projected vertically upward with initial speed v_0 ft/sec is at time t sec at a distance s ft from the point of projection as given by the formula $s = v_0 t - 16t^2$. If the ball is given an initial upward speed of 128 ft/sec, at what times would it be 100 ft above the point of projection?

$s = v_0 t - 16t^2$, $100 = 128t - 16t^2$, $4t^2 - 32t + 25 = 0$,   and $t = \dfrac{32 \pm \sqrt{624}}{8} = 7.12, 0.88$.

At $t = 0.88$ sec, $s = 100$ ft and the ball is rising; at $t = 7.12$ sec, $s = 100$ ft and the ball is falling. This is seen from the graph of s plotted against t.

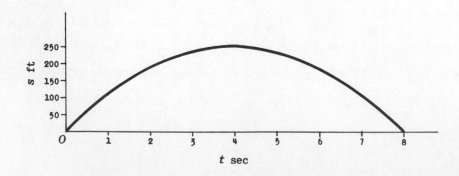

SUPPLEMENTARY PROBLEMS

30. Solve each equation.

a) $x^2 - 40 = 9$

b) $2x^2 - 400 = 0$

c) $x^2 + 36 = 9 - 2x^2$

d) $\dfrac{x}{16} = \dfrac{4}{x}$

e) $\dfrac{y^2}{3} = \dfrac{y^2}{6} + 2$

f) $\dfrac{1 - 2x}{3 - x} = \dfrac{x - 2}{3x - 1}$

g) $\dfrac{1}{2x - 1} - \dfrac{1}{2x + 1} = \dfrac{1}{4}$

h) $x - \dfrac{2x}{x + 1} = \dfrac{5}{x + 1} - 1$

31. Solve each equation by factoring.

a) $x^2 - 7x = -12$

b) $x^2 + x = 6$

c) $x^2 = 5x + 24$

d) $2x^2 + 2 = 5x$

e) $9x^2 = 9x - 2$

f) $4x - 5x^2 = -12$

g) $\dfrac{x}{2a} = \dfrac{4a}{x + 2a}$

h) $\dfrac{1}{4 - x} - \dfrac{1}{2 + x} = \dfrac{1}{4}$

i) $\dfrac{2x - 1}{x + 2} + \dfrac{x + 2}{2x - 1} = \dfrac{10}{3}$

j) $\dfrac{2c - 3y}{y - c} - \dfrac{y}{2y - c} = \dfrac{2}{3}$

32. Solve each equation by completing the square.

a) $x^2 + 4x - 5 = 0$

b) $x(x - 3) = 4$

c) $2x^2 = x + 1$

d) $3x^2 - 2 = 5x$

e) $4x^2 = 12x - 7$

f) $6y^2 = 19y - 15$

g) $2x^2 + 3a^2 = 7ax$

h) $12x - 9x^2 = 5$

33. Solve each equation by the quadratic formula.

a) $x^2 - 5x = 6$

b) $x^2 - 6 = x$

c) $3x^2 - 2x = 8$

d) $16x^2 - 8x + 1 = 0$

e) $x(5x - 4) = 2$

f) $9x^2 + 6x = -4$

g) $\dfrac{5x^2 - 2p^2}{x} = \dfrac{p}{3}$

h) $\dfrac{2x + 3}{4x - 1} = \dfrac{3x - 2}{3x + 2}$

34. Solve each equation graphically.

a) $2x^2 + x - 3 = 0$

b) $4x^2 - 8x + 4 = 0$

c) $x^2 - 2x = 2$

d) $2x^2 + 2 = 3x$

e) $6x^2 - 7x - 5 = 0$

f) $2x^2 + 8x + 3 = 0$

35. Without solving, find the sum S and product P of the roots of each equation.

a) $2x^2 + 3x + 1 = 0$

b) $x - x^2 = 2$

c) $2x(x + 3) = 1$

d) $2x^2 + 6x - 5 = 0$

e) $3x^2 - 4 = 0$

f) $4x^2 + 3x = 0$

g) $2x^2 + 5kx + 3k^2 = 0$

h) $0.2x^2 - 0.1x + 0.03 = 0$

i) $\sqrt{2}x^2 - \sqrt{3}x + 1 = 0$

36. Find the discriminant $b^2 - 4ac$ and thus determine the character of the roots.

a) $2x^2 - 7x + 4 = 0$

b) $3x^2 = 5x - 2$

c) $3x - x^2 = 4$

d) $x(4x + 3) = 5$

e) $2x^2 = 5 + 3x$

f) $4x\sqrt{3} = 4x^2 + 3$

g) $1 + 2x + 2x^2 = 0$

h) $3x + 25/3x = 10$

37. Find a quadratic equation with integer coefficients (if possible) having the given roots.

a) $2, -3$

b) $-3, 0$

c) $8, -4$

d) $-2, -5$

e) $-1/3, 1/2$

f) $2 + \sqrt{3}, \ 2 - \sqrt{3}$

g) $-1 + i, \ -1 - i$

h) $-2 - \sqrt{6}, \ -2 + \sqrt{6}$

i) $2 + \dfrac{3}{2}i, \ 2 - \dfrac{3}{2}i$

j) $\sqrt{3} - \sqrt{2}, \ \sqrt{3} + \sqrt{2}$

k) $a + bi, \ a - bi \quad a, b$ integers

l) $\dfrac{m + \sqrt{n}}{2}, \ \dfrac{m - \sqrt{n}}{2} \quad m, n$ integers

38. In each quadratic equation, evaluate the constant p subject to the given condition.

a) $px^2 - x + 5 - 3p = 0$ has one root equal to 2.

b) $(2p + 1)x^2 + px + p = 4(px + 2)$ has the sum of its roots equal to the product of its roots.

c) $3x^2 + p(x - 2) + 1 = 0$ has roots which are reciprocals.

d) $4x^2 - 8x + 2p - 1 = 0$ has one root equal to three times the other.

e) $4x^2 - 20x + p^2 - 4 = 0$ has one root equal to two more than the other.

f) $x^2 = 5x - 3p + 3$ has the difference between its roots equal to 11.

39. Find the roots of each equation subject to the given condition.

a) $2px^2 - 4px + 5p = 3x^2 + x - 8$ has the product of its roots equal to twice their sum.

b) $x^2 - 3(x - p) - 2 = 0$ has one root equal to 3 less than twice the other root.

c) $p(x^2 + 3x - 9) = x - x^2$ has one root equal to the negative of the other.

d) $(m + 3)x^2 + 2m(x + 1) + 3 = 0$ has one root equal to half the reciprocal of the other.

e) $(2m + 1)x^2 - 4mx = 1 - 3m$ has equal roots.

40. Solve each equation.

a) $\sqrt{x^2 - x + 2} = 2$

b) $\sqrt{2x - 2} = x - 1$

c) $\sqrt{4x + 1} = 3 - 3x$

d) $2 - \sqrt[3]{x^2 + 2x} = 0$

e) $\sqrt{2x + 7} = \sqrt{x} + 2$

f) $\sqrt{2x^2 - 7} - x = 3$

g) $\sqrt{2 + x} - 4 + \sqrt{10 - 3x} = 0$

h) $2\sqrt{x} - \sqrt{4x - 3} = \dfrac{1}{\sqrt{4x - 3}}$

i) $\sqrt{x^2 - \sqrt{2x + 1}} = 2 - x$

j) $\sqrt{2x - 10} + \sqrt{x + 9} = 2$

k) $\sqrt{2x + 8} + \sqrt{2x + 5} = \sqrt{8x + 25}$

l) $\sqrt[3]{2x - 1} = \sqrt[6]{x + 1}$

41. Solve each equation.

a) $x^4 - 13x^2 + 36 = 0$

b) $x^4 - 3x^2 - 10 = 0$

c) $4x^{-4} - 17x^{-2} + 4 = 0$

d) $x^{-4/3} - 5x^{-2/3} + 4 = 0$

e) $(x^2 - 6x)^2 - 2(x^2 - 6x) = 35$

f) $x^2 + x = 7\sqrt{x^2 + x + 2} - 12$

g) $\left(x + \dfrac{1}{x}\right)^2 - \dfrac{7}{2}\left(x + \dfrac{1}{x}\right) = 2$

h) $\sqrt{x + 2} - \sqrt[4]{x + 2} = 6$

i) $x^3 - 7x^{3/2} - 8 = 0$

j) $\dfrac{x^2 + 2}{x} + \dfrac{8x}{x^2 + 2} = 6$

42. Abstract Number Problems.

a) The sum of the squares of two numbers is 34, the first number being one less than twice the second number. Determine the numbers.

b) The sum of the squares of three consecutive integers is 110. Find the numbers.

c) The difference between two positive numbers is 3, and the sum of their reciprocals is 1/2. Determine the numbers.

d) A number exceeds twice its square root by 3. Find the number.

43. Geometrical Problems.

a) The length of a rectangle is three times its width. If the width is diminished by 1 ft and the length increased by 3 ft, the area will be 72 ft^2. Find the dimensions of the original rectangle.

b) A piece of wire 60 in. long is bent into the form of a right triangle having hypotenuse 25 in. Find the other two sides of the triangle.

c) A picture 8 in. by 12 in. is placed in a frame which has uniform width. If the area of the frame equals the area of the picture, find the width of the frame.

d) A rectangular piece of tin 9 in. by 12 in. is to be made into an open box with base area 60 in.^{2} by cutting out equal squares from the four corners and then bending up the edges. Find to the nearest tenth of an inch the length of the side of the square cut from each corner.

44. Digit Problems.

a) The tens digit of a certain two digit number is twice the units digit. If the number is multiplied by the sum of its digits, the product is 63. Find the number.

b) Find a number consisting of two digits such that the tens digit exceeds the units digit by 3 and the number is 4 less than the sum of the squares of its digits.

45. Motion Problems.

a) Two men start at the same time from the same place and travel along roads that are at right angles to each other. One man travels 4 mi/hr faster than the other, and at the end of 2 hours they are 40 miles apart. Determine their rates of travel.

b) By increasing his average speed by 10 mi/hr a motorist could save 36 minutes in traveling a distance of 120 miles. Find his actual average speed.

c) A man travels 36 miles down a river and back in 8 hours. If the rate of his boat in still water is 12 mi/hr, what is the rate of the river current?

46. Cost Problems.

a) A merchant purchased a number of coats, each at the same price, for a total of $720. He sold them at $40 each, thus realizing a profit equal to his cost of 8 coats. How many did he buy?

b) A grocer bought a number of cans of corn for $14.40. Later the price increased 2 cents a can and as a result he received 24 fewer cans for the same amount of money. How many cans were in his first purchase and what was the cost per can?

47. Work Problems.

a) It takes B 6 hours longer than A to assemble a machine. Together they can do it in 4 hr. How long would it take each working alone to do the job?

b) Pipe A can fill a given tank in 4 hr. If pipe B works alone, it takes 3 hr longer to fill the tank than if pipes A and B act together. How long will it take pipe B working alone?

48. A ball projected vertically upward is distant s ft from the point of projection after t seconds, where $s = 64t - 16t^2$.
a) At what times will the ball be 40 ft above the ground?
b) Will the ball ever be 80 ft above the ground?
c) What is the maximum height reached?

ANSWERS TO SUPPLEMENTARY PROBLEMS.

30. *a)* $x = \pm 7$ *c)* $x = \pm 3i$ *e)* $y = \pm 2\sqrt{3}$ *g)* $x = \pm 3/2$
    b) $x = \pm 10\sqrt{2}$ *d)* $x = \pm 8$ *f)* $x = \pm 1$ *h)* $x = \pm 2$

31. *a)* 3, 4   *c)* 8, −3   *e)* 1/3, 2/3   *g)* $2a$, $-4a$ *i)* 1, −7
    b) 2, −3   *d)* 2, 1/2   *f)* 2, −6/5   *h)* 2, −8   *j)* $2c/5$, $4c/5$

32. *a)* 1, −5   *c)* 1, −1/2   *e)* $\dfrac{3 \pm \sqrt{2}}{2}$ *f)* 3/2, 5/3   *h)* $\dfrac{2}{3} \pm \dfrac{i}{3}$
    b) 4, −1   *d)* 2, −1/3   *g)* $3a$, $a/2$

33. *a)* 6, −1   *c)* 2, −4/3   *e)* $\dfrac{2 \pm \sqrt{14}}{5}$ *f)* $\dfrac{-1 \pm i\sqrt{3}}{3}$ *g)* $\dfrac{2p}{3}$, $-\dfrac{3p}{5}$ *h)* $\dfrac{6 \pm \sqrt{42}}{3}$
    b) 3, −2   *d)* 1/4, 1/4

35. *a)* S = −3/2, P = 1/2   *d)* S = −3, P = −5/2   *g)* S = $-5k/2$, P = $3k^2/2$
    b) S = 1, P = 2   *e)* S = 0, P = −4/3   *h)* S = 0.5, P = 0.15
    c) S = −3, P = −1/2   *f)* S = −3/4, P = 0   *i)* S = $\frac{1}{2}\sqrt{6}$, P = $\frac{1}{2}\sqrt{2}$

36. *a)* 17; real, irrational, unequal   *e)* 49; real, rational, unequal
    b) 1; real, rational, unequal   *f)* 0; real, equal
    c) −7; imaginary   *g)* −4; imaginary
    d) 89; real, irrational, unequal   *h)* 0; real, rational, equal

37. *a)* $x^2 + x - 6 = 0$ *e)* $6x^2 - x - 1 = 0$ *i)* $4x^2 - 16x + 25 = 0$

    b) $x^2 + 3x = 0$ *f)* $x^2 - 4x + 1 = 0$ *j)* not possible

    c) $x^2 - 4x - 32 = 0$ *g)* $x^2 + 2x + 2 = 0$ *k)* $x^2 - 2ax + a^2 + b^2 = 0$

    d) $x^2 + 7x + 10 = 0$ *h)* $x^2 + 4x - 2 = 0$ *l)* $4x^2 - 4mx + m^2 - n = 0$

38. *a)* $p = -3$ *b)* $p = -4$ *c)* $p = -1$ *d)* $p = 2$ *e)* $p = \pm 5$ *f)* $p = -7$

39. *a)* 3, 6   *b)* 1, 2   *c)* $\pm 3/2$ *d)* $1/2 \pm i/2$
    e) If $m = -1$, the roots are 2, 2; if $m = 1/2$, the roots are 1/2, 1/2.

40. *a)* 2, −1   *c)* 4/9   *e)* 9, 1   *g)* ± 2 *i)* 3/2   *k)* −2
    b) 1, 3   *d)* −4, 2   *f)* 8, −2   *h)* 1   *j)* no solution   *l)* 5/4

41. *a)* ± 2, ± 3 *d)* ± 1, $\pm 1/8$ *g)* $2 \pm \sqrt{3}$, $-1/4 \pm i\sqrt{15}/4$
    b) $\pm\sqrt{5}$, $\pm i\sqrt{2}$ *e)* 7, 5, ± 1 *h)* 79   *j)* $1 \pm i$, $2 \pm \sqrt{2}$
    c) ± 2, $\pm 1/2$ *f)* 1, −2, $(-1 \pm \sqrt{93})/2$ *i)* 4

42. *a)* 5, 3 or −27/5, −11/5   *b)* 5, 6, 7 or −7, −6, −5   *c)* 3, 6   *d)* 9

43. *a)* 5, 15 ft   *b)* 15, 20 in.   *c)* 2 in.   *d)* 1.3 in.

44. *a)* 21   *b)* 85

45. *a)* 12, 16 mi/hr   *b)* 40 mi/hr   *c)* 6 mi/hr

46. *a)* 24   *b)* 144, 10¢

47. *a)* A, 6 hr; B, 12 hr   *b)* 5.3 hr approx.

48. *a)* 0.78 and 3.22 seconds after projection   *b)* No   *c)* 64 ft

CHAPTER 14

Quadratic Equations in Two Unknowns

THE GENERAL QUADRATIC EQUATION in the two unknowns (or variables) x and y has the form

$$ax^2 + bxy + cy^2 + dx + ey + f = 0 \qquad (1)$$

where a, b, c, d, e, f are given constants and a, b, c are not all zero.

Thus $3x^2 + 5xy = 2$, $x^2 - xy + y^2 + 2x + 3y = 0$, $y^2 = 4x$, $xy = 4$ are quadratic equations in x and y.

THE GRAPH OF EQUATION (1), if a, b, c, d, e, f are real, depends on the value of $b^2 - 4ac$.

1) If $b^2 - 4ac < 0$, the graph is in general an ellipse. However if $b = 0$ and $a = c$ the graph may be a circle, a point, or non-existent.

2) If $b^2 - 4ac = 0$, the graph is a parabola, two parallel or coincident lines, or non-existent.

3) If $b^2 - 4ac > 0$, the graph is a hyperbola or two intersecting lines.

These graphs are the intersections of a plane and a right circular cone and for this reason are called conic sections.

SIMULTANEOUS EQUATIONS INVOLVING QUADRATICS

GRAPHICAL SOLUTION. The real simultaneous solutions of two quadratic equations in x and y are the values of x and y corresponding to the points of intersection of the graphs of the two equations. If the graphs do not intersect, the simultaneous solutions are imaginary.

ALGEBRAIC SOLUTION.

A) ONE LINEAR AND ONE QUADRATIC EQUATION.

Solve the linear equation for one of the unknowns and substitute in the quadratic equation.

Example. Solve the system 1) $x + y = 7$
2) $x^2 + y^2 = 25$.

Solving 1) for y, $y = 7 - x$. Substitute in 2) and obtain

$x^2 + (7 - x)^2 = 25$, $x^2 - 7x + 12 = 0$, $(x - 3)(x - 4) = 0$, and $x = 3, 4$.

When $x = 3$, $y = 7 - x = 4$; when $x = 4$, $y = 7 - x = 3$. Thus the simultaneous solutions are $x = 3$, $y = 4$ and $x = 4$, $y = 3$.

B) TWO EQUATIONS OF THE FORM $ax^2 + by^2 = c$.

Use the method of addition or subtraction.

Example. Solve the system   1) $2x^2 - y^2 = 7$
2) $3x^2 + 2y^2 = 14$.

To eliminate y, multiply 1) by 2 and add with 2); then

$$7x^2 = 28, \quad x^2 = 4 \quad \text{and} \quad x = \pm 2.$$

Now put $x = 2$ or $x = -2$ in 1) and obtain $y = \pm 1$.

The four solutions are:
$x = 2, \; y = 1; \; x = -2, \; y = 1; \; x = 2, \; y = -1; \; x = -2, \; y = -1$.

C) TWO EQUATIONS OF THE FORM $ax^2 + bxy + cy^2 = d$.

Example. Solve the system   1) $x^2 + xy = 6$
2) $x^2 + 5xy - 4y^2 = 10$.

Method 1. Eliminate the constant term between both equations.

Multiply 1) by 5, 2) by 3, and subtract; then

$$x^2 - 5xy + 6y^2 = 0, \quad (x - 2y)(x - 3y) = 0, \quad x = 2y \text{ and } x = 3y.$$

Now put $x = 2y$ in 1) or 2) and obtain $y^2 = 1, \; y = \pm 1$.
When $y = 1, \; x = 2y = 2$; when $y = -1, \; x = 2y = -2$.
Thus two solutions are: $x = 2, \; y = 1; \; x = -2, \; y = -1$.

Then put $x = 3y$ in 1) or 2) and get $\; y^2 = \dfrac{1}{2}, \; y = \pm \dfrac{\sqrt{2}}{2}$.

When $y = \dfrac{\sqrt{2}}{2}, \; x = 3y = \dfrac{3\sqrt{2}}{2}$; when $y = -\dfrac{\sqrt{2}}{2}, \; x = -\dfrac{3\sqrt{2}}{2}$.

Thus the four solutions are: $x = 2, \; y = 1; \; x = -2, \; y = -1; \; x = \dfrac{3\sqrt{2}}{2}, \; y = \dfrac{\sqrt{2}}{2}; \; x = -\dfrac{3\sqrt{2}}{2}, \; y = -\dfrac{\sqrt{2}}{2}$.

Method 2. Let $y = mx$ in both equations.

From 1): $x^2 + mx^2 = 6, \quad x^2 = \dfrac{6}{1 + m}$.

From 2): $x^2 + 5mx^2 - 4m^2x^2 = 10, \quad x^2 = \dfrac{10}{1 + 5m - 4m^2}$.

Then $\dfrac{6}{1 + m} = \dfrac{10}{1 + 5m - 4m^2}$ from which $m = \dfrac{1}{2}, \; \dfrac{1}{3}$; hence

$y = x/2, \; y = x/3$. The solution proceeds as in Method 1.

D) MISCELLANEOUS METHODS.

1) Some systems of equations may be solved by replacing them by equivalent and simpler systems. (See Problems 10-12.)

2) An equation is called symmetric in x and y if interchange of x and y does not change the equation. Thus $x^2 + y^2 - 3xy + 4x + 4y = 8$ is symmetric in x and y. Systems of symmetric equations may often be solved by the substitutions $x = u + v, \; y = u - v$. (See Problems 13-14.)

SOLVED PROBLEMS

1. Draw the graph of each of the following equations:

   a) $4x^2 + 9y^2 = 36$,   b) $4x^2 - 9y^2 = 36$,   c) $4x + 9y^2 = 36$.

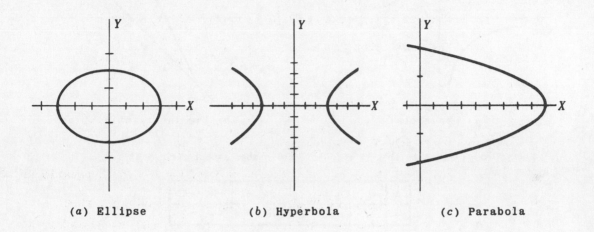

| (a) Ellipse | (b) Hyperbola | (c) Parabola |

a) $4x^2 + 9y^2 = 36$, $y^2 = \frac{4}{9}(9 - x^2)$, $y = \pm\frac{2}{3}\sqrt{9 - x^2}$.

Note that y is real when $9 - x^2 \geq 0$, i.e. when $-3 \leq x \leq 3$. Hence values of x greater than 3 or less than -3 are excluded.

x	-3	-2	-1	0	1	2	3
y	0	±1.49	±1.89	±2	±1.89	±1.49	0

The graph is an ellipse with center at the origin.

b) $4x^2 - 9y^2 = 36$, $y^2 = \frac{4}{9}(x^2 - 9)$, $y = \pm\frac{2}{3}\sqrt{x^2 - 9}$.

Note that x cannot have a value between -3 and 3 if y is to be real.

x	6	5	4	3	-3	-4	-5	-6
y	±3.46	±2.67	±1.76	0	0	±1.76	±2.67	±3.46

The graph consists of two branches and is called a hyperbola.

c) $4x + 9y^2 = 36$, $y^2 = \frac{4}{9}(9 - x)$, $y = \pm\frac{2}{3}\sqrt{9 - x}$.

Note that if x is greater than 9, y is imaginary.

x	-1	0	1	5	8	9
y	±2.11	±2	±1.89	±1.33	±0.67	0

The graph is a parabola.

2. Plot the graph of each of the following equations:

    a) $xy = 8$,   *b*) $2x^2 - 3xy + y^2 + x - 2y - 3 = 0$,   *c*) $x^2 + y^2 - 4x + 8y + 25 = 0$.

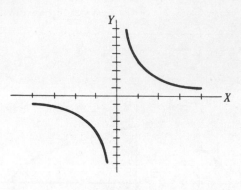

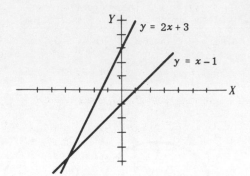

      (*a*) Hyperbola   (*b*) Two Intersecting Lines

a) $xy = 8$, $y = 8/x$. Note that if x is any real number except zero, y is real. The graph is a hyperbola.

x	4	2	1	$\frac{1}{2}$	$-\frac{1}{2}$	-1	-2	-4
y	2	4	8	16	-16	-8	-4	-2

b) $2x^2 - 3xy + y^2 + x - 2y - 3 = 0$.   Write as $y^2 - (3x + 2)y + (2x^2 + x - 3) = 0$ and solve by the

quadratic formula to obtain $y = \dfrac{3x + 2 \pm \sqrt{x^2 + 8x + 16}}{2} = \dfrac{(3x + 2) \pm (x + 4)}{2}$ or $y = 2x + 3$, $y =$

$x - 1$. The given equation is equivalent to two linear equations, as can be seen by writing the the given equation as $(2x - y + 3)(x - y - 1) = 0$. The graph consists of two intersecting lines.

c) Write as $y^2 + 8y + (x^2 - 4x + 25) = 0$; solving, $y = \dfrac{-4 \pm \sqrt{-4(x^2 - 4x + 9)}}{2}$.

    Since $x^2 - 4x + 9 = x^2 - 4x + 4 + 5 = (x - 2)^2 + 5$ is always positive, the quantity under the radical sign is negative. Thus y is imaginary for all real values of x and the graph does not exist.

3. Solve graphically the following systems: *a*) $\begin{array}{l} x^2 + y^2 = 25 \\ x + 2y = 10 \end{array}$, *b*) $\begin{array}{l} x^2 + 4y^2 = 16 \\ xy = 4 \end{array}$, *c*) $\begin{array}{l} x^2 + 2y = 9 \\ 2x^2 - 3y^2 = 1. \end{array}$

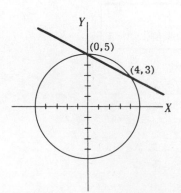

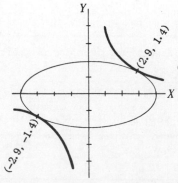

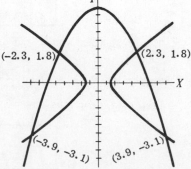

    (*a*) $x^2 + y^2 = 25$ circle   (*b*) $x^2 + 4y^2 = 16$ ellipse   (*c*) $x^2 + 2y = 9$ parabola

        $x + 2y = 10$ line $xy = 4$ hyperbola $2x^2 - 3y^2 = 1$ hyperbola

4. Solve the following systems: $a)$ $\begin{array}{l} x + 2y = 4 \\ y^2 - xy = 7 \end{array}$, $b)$ $\begin{array}{l} 3x - 1 + 2y = 0 \\ 3x^2 - y^2 + 4 = 0. \end{array}$

$a)$ Solving the linear equation for x, $x = 4 - 2y$. Substituting in the quadratic equation,
$$y^2 - y(4 - 2y) = 7, \quad 3y^2 - 4y - 7 = 0, \quad (y + 1)(3y - 7) = 0 \quad \text{and} \quad y = -1, \ 7/3.$$
If $y = -1$, $x = 4 - 2y = 6$; if $y = 7/3$, $x = 4 - 2y = -2/3$.

The solutions are $x = 6$, $y = -1$ and $x = -2/3$, $y = 7/3$.

$b)$ Solving the linear equation for y, $y = \frac{1}{2}(1 - 3x)$. Substituting in the quadratic equation,
$$3x^2 - \left[\frac{1}{2}(1 - 3x)\right]^2 + 4 = 0, \quad x^2 + 2x + 5 = 0 \quad \text{and} \quad x = \frac{-2 \pm \sqrt{2^2 - 4(1)(5)}}{2(1)} = -1 \pm 2i.$$
If $x = -1 + 2i$, $y = \frac{1}{2}(1 - 3x) = \frac{1}{2}[1 - 3(-1 + 2i)] = \frac{1}{2}(4 - 6i) = 2 - 3i$.

If $x = -1 - 2i$, $y = \frac{1}{2}(1 - 3x) = \frac{1}{2}[1 - 3(-1 - 2i)] = \frac{1}{2}(4 + 6i) = 2 + 3i$.

The solutions are $x = -1 + 2i$, $y = 2 - 3i$ and $x = -1 - 2i$, $y = 2 + 3i$.

5. Solve the system: (1) $2x^2 - 3y^2 = 6$, (2) $3x^2 + 2y^2 = 35$.

To eliminate y, multiply (1) by 2, (2) by 3 and add; then $13x^2 = 117$, $x^2 = 9$, $x = \pm 3$.

Now put $x = 3$ or $x = -3$ in (1) and obtain $y = \pm 2$.

The solutions are: $x = 3$, $y = 2$; $x = -3$, $y = 2$; $x = 3$, $y = -2$; $x = -3$, $y = -2$.

6. Solve the system: (1) $\dfrac{8}{x^2} - \dfrac{3}{y^2} = 5$, (2) $\dfrac{5}{x^2} + \dfrac{2}{y^2} = 38$.

The equations are quadratic in $\dfrac{1}{x}$ and $\dfrac{1}{y}$. Substituting $u = \dfrac{1}{x}$ and $v = \dfrac{1}{y}$, we obtain
$$8u^2 - 3v^2 = 5 \quad \text{and} \quad 5u^2 + 2v^2 = 38.$$
Solving simultaneously, $u^2 = 4$, $v^2 = 9$ or $x^2 = 1/4$, $y^2 = 1/9$; then $x = \pm 1/2$, $y = \pm 1/3$.

The solutions are: $x = 1/2$, $y = 1/3$; $x = -1/2$, $y = 1/3$; $x = 1/2$, $y = -1/3$; $x = -1/2$, $y = -1/3$.

7. Solve the system $\begin{array}{l} (1) \ 5x^2 + 4y^2 = 48 \\ (2) \ x^2 + 2xy = 16 \end{array}$ by eliminating the constant terms.

Multiply (2) by 3 and subtract from (1) to obtain
$$2x^2 - 6xy + 4y^2 = 0, \quad x^2 - 3xy + 2y^2 = 0, \quad (x - y)(x - 2y) = 0 \quad \text{and} \quad x = y, \ x = 2y.$$
Substituting $x = y$ in (1) or (2), we have $y^2 = \dfrac{16}{3}$ and $y = \pm \dfrac{4}{3}\sqrt{3}$.

Substituting $x = 2y$ in (1) or (2), we have $y^2 = 2$ and $y = \pm\sqrt{2}$.

The four solutions are:
$$x = \frac{4}{3}\sqrt{3}, \ y = \frac{4}{3}\sqrt{3}; \ x = -\frac{4}{3}\sqrt{3}, \ y = -\frac{4}{3}\sqrt{3}; \ x = 2\sqrt{2}, \ y = \sqrt{2}; \ x = -2\sqrt{2}, \ y = -\sqrt{2}.$$

8. Solve the system $\begin{array}{l} (1) \ 3x^2 - 4xy = 4 \\ (2) \ x^2 - 2y^2 = 2 \end{array}$ by using the substitution $y = mx$.

Put $y = mx$ in (1); then $3x^2 - 4mx^2 = 4$ and $x^2 = \dfrac{4}{3 - 4m}$.

Put $y = mx$ in (2); then $x^2 - 2m^2x^2 = 2$ and $x^2 = \dfrac{2}{1 - 2m^2}$.

Thus $\dfrac{4}{3 - 4m} = \dfrac{2}{1 - 2m^2}$, $4m^2 - 4m + 1 = 0$, $(2m - 1)^2 = 0$ and $m = \frac{1}{2}, \frac{1}{2}$.

Now substitute $y = mx = \frac{1}{2}x$ in (1) or (2) and obtain $x^2 = 4$, $x = \pm 2$.

The solutions are $x = 2$, $y = 1$ and $x = -2$, $y = -1$.

9. Solve the system: (1) $x^2 + y^2 = 40$, (2) $xy = 12$.

From (2), $y = 12/x$; substituting in (1), we have

$$x^2 + \frac{144}{x^2} = 40, \quad x^4 - 40x^2 + 144 = 0, \quad (x^2 - 36)(x^2 - 4) = 0 \quad \text{and} \quad x = \pm 6, \pm 2.$$

For $x = \pm 6$, $y = 12/x = \pm 2$; for $x = \pm 2$, $y = \pm 6$.

The four solutions are: $x = 6$, $y = 2$; $x = -6$, $y = -2$; $x = 2$, $y = 6$; $x = -2$, $y = -6$.

Note. Equation (2) indicates that those solutions in which the product xy is negative (e.g., $x = 2$, $y = -6$) are extraneous.

10. Solve the system: (1) $x^2 + y^2 + 2x - y = 14$, (2) $x^2 + y^2 + x - 2y = 9$.

Subtract (2) from (1): $x + y = 5$ or $y = 5 - x$.

Substitute $y = 5 - x$ in (1) or (2): $2x^2 - 7x + 6 = 0$, $(2x - 3)(x - 2) = 0$ and $x = 3/2, 2$.

The solutions are $x = 3/2$, $y = 7/2$ and $x = 2$, $y = 3$.

11. Solve the system: (1) $x^3 + y^3 = 35$, (2) $x + y = 5$.

Dividing (1) by (2), $\dfrac{x^3 + y^3}{x + y} = \dfrac{35}{5}$ and (3) $x^2 - xy + y^2 = 7$.

From (2), $y = 5 - x$; substituting in (3), we have

$$x^2 - x(5 - x) + (5 - x)^2 = 7, \quad x^2 - 5x + 6 = 0, \quad (x - 3)(x - 2) = 0 \quad \text{and} \quad x = 3, 2.$$

The solutions are $x = 3$, $y = 2$ and $x = 2$, $y = 3$.

12. Solve the system: (1) $x^2 + 3xy + 2y^2 = 3$, (2) $x^2 + 5xy + 6y^2 = 15$.

Dividing (1) by (2), $\dfrac{x^2 + 3xy + 2y^2}{x^2 + 5xy + 6y^2} = \dfrac{(x + y)(x + 2y)}{(x + 3y)(x + 2y)} = \dfrac{x + y}{x + 3y} = \dfrac{1}{5}$.

From $\dfrac{x + y}{x + 3y} = \dfrac{1}{5}$, $y = -2x$. Substituting $y = -2x$ in (1) or (2), $x^2 = 1$ and $x = \pm 1$.

The solutions are $x = 1$, $y = -2$ and $x = -1$, $y = 2$.

13. Solve the system: (1) $x^2 + y^2 + 2x + 2y = 32$, (2) $x + y + 2xy = 22$.

The equations are symmetric in x and y since interchange of x and y yields the same equation. Substituting $x = u + v$, $y = u - v$ in (1) and (2), we obtain

$$(3) \ u^2 + v^2 + 2u = 16 \quad \text{and} \quad (4) \ u^2 - v^2 + u = 11.$$

Adding (3) and (4), we get $2u^2 + 3u - 27 = 0$, $(u-3)(2u+9) = 0$ and $u = 3, -9/2$.

When $u = 3$, $v^2 = 1$ and $v = \pm 1$; when $u = -9/2$, $v^2 = 19/4$ and $v = \pm\sqrt{19}/2$. Thus the solutions of (3) and (4) are: $u = 3$, $v = 1$; $u = 3$, $v = -1$; $u = -9/2$, $v = \sqrt{19}/2$; $u = -9/2$, $v = -\sqrt{19}/2$.

Then, since $x = u+v$, $y = u-v$, the four solutions of (1) and (2) are:
$x = 4$, $y = 2$; $x = 2$, $y = 4$; $x = -9/2 + \sqrt{19}/2$, $y = -9/2 - \sqrt{19}/2$; $x = -9/2 - \sqrt{19}/2$, $y = -9/2 + \sqrt{19}/2$.

14. Solve the system: (1) $x^2 + y^2 = 180$, (2) $\dfrac{1}{x} + \dfrac{1}{y} = \dfrac{1}{4}$.

From (2) obtain (3) $4x + 4y - xy = 0$. Since (1) and (3) are symmetric in x and y, substitute $x = u+v$, $y = u-v$ in (1) and (3) and obtain

$$(4) \ u^2 + v^2 = 90 \quad \text{and} \quad (5) \ 8u - u^2 + v^2 = 0.$$

Subtracting (5) from (4), we have $u^2 - 4u - 45 = 0$, $(u-9)(u+5) = 0$ and $u = 9, -5$.

When $u = 9$, $v = \pm 3$; when $u = -5$, $v = \pm\sqrt{65}$. Thus the solutions of (4) and (5) are:
$u = 9$, $v = 3$; $u = 9$, $v = -3$; $u = -5$, $v = \sqrt{65}$; $u = -5$, $v = -\sqrt{65}$.

Hence the four solutions of (1) and (2) are:
$x = 12$, $y = 6$; $x = 6$, $y = 12$; $x = -5 + \sqrt{65}$, $y = -5 - \sqrt{65}$; $x = -5 - \sqrt{65}$, $y = -5 + \sqrt{65}$.

15. The sum of two numbers is 25 and their product is 144. What are the numbers?

Let the numbers be x, y. Then (1) $x + y = 25$ and (2) $xy = 144$.

The simultaneous solutions of (1) and (2) are $x = 9$, $y = 16$ and $x = 16$, $y = 9$. Hence the required numbers are 9, 16.

16. The difference of two positive numbers is 3 and the sum of their squares is 65. Find the numbers.

Let the numbers be p, q. Then (1) $p - q = 3$ and (2) $p^2 + q^2 = 65$.

The simultaneous solutions of (1) and (2) are $p = 7$, $q = 4$ and $p = -4$, $q = -7$. Hence the required (positive) numbers are 7, 4.

17. A rectangle has perimeter 60 ft and area 216 ft^2. Find its dimensions.

Let the rectangle have sides of lengths x, y. Then (1) $2x + 2y = 60$ and (2) $xy = 216$.

Solving (1) and (2) simultaneously, the required sides are 12 and 18 ft.

18. The hypotenuse of a right triangle is 41 ft long and the area of the triangle is 180 ft^2. Find the lengths of the two legs.

Let the legs have lengths x, y. Then (1) $x^2 + y^2 = (41)^2$ and (2) $\frac{1}{2}(xy) = 180$.

Solving (1) and (2) simultaneously, we find the legs have lengths 9 and 40 ft.

SUPPLEMENTARY PROBLEMS

19. Graph each of the following equations.

   a) $x^2 + y^2 = 9$ e) $y^2 = 4x$

   b) $xy = -4$ f) $x^2 + 3y^2 - 1 = 0$ i) $x^2 + y^2 - 2x + 2y + 2 = 0$

   c) $4x^2 + y^2 = 16$ g) $x^2 + 3xy + y^2 = 16$ j) $2x^2 - xy - y^2 - 7x - 2y + 3 = 0$

   d) $x^2 - 4y^2 = 36$ h) $x^2 + 4y = 4$

20. Solve the following systems graphically.

   a) $x^2 + y^2 = 20$, $3x - y = 2$ c) $y^2 = x$, $x^2 + 2y^2 = 24$

   b) $x^2 + 4y^2 = 25$, $x^2 - y^2 = 5$ d) $x^2 + 1 = 4y$, $3x - 2y = 2$

21. Solve the following systems algebraically.

   a) $2x^2 - y^2 = 14$, $x - y = 1$ h) $x^2 + 3xy = 18$, $x^2 - 5y^2 = 4$

   b) $xy + x^2 = 24$, $y - 3x + 4 = 0$ i) $x^2 + 2xy = 16$, $3x^2 - 4xy + 2y^2 = 6$

   c) $3xy - 10x = y$, $2 - y + x = 0$ j) $x^2 - xy + y^2 = 7$, $x^2 + y^2 = 10$

   d) $4x + 5y = 6$, $xy = -2$ k) $x^2 - 3y^2 + 10y = 19$, $x^2 - 3y^2 + 5x = 9$

   e) $2x^2 - y^2 = 5$, $3x^2 + 4y^2 = 57$ l) $x^3 - y^3 = 9$, $x - y = 3$

   f) $9/x^2 + 16/y^2 = 5$, $18/x^2 - 12/y^2 = -1$ m) $x^3 - y^3 = 19$, $x^2y - xy^2 = 6$

   g) $x^2 - xy = 12$, $xy - y^2 = 3$ n) $1/x^3 + 1/y^3 = 35$, $1/x^2 - 1/xy + 1/y^2 = 7$

22. The square of a certain number exceeds twice the square of another number by 16. Find the numbers if the sum of their squares is 208.

23. The diagonal of a rectangle is 85 ft. If the short side is increased by 11 ft and the long side decreased by 7 ft, the length of the diagonal remains the same. Find the dimensions of the original rectangle.

ANSWERS TO SUPPLEMENTARY PROBLEMS.

19. a) circle   c) ellipse   e) parabola   g) hyperbola   i) a single point $(1,-1)$
        b) hyperbola   d) hyperbola   f) ellipse   h) parabola   j) two intersecting lines

20. a) $(2,4)$, $(-0.8, -4.4)$ c) $(4,2)$, $(4,-2)$
       b) $(3,2)$, $(-3,2)$, $(3,-2)$, $(-3,-2)$ d) $(1, 0.5)$, $(5, 6.5)$

21. a) $(3,2)$, $(-5,-6)$ i) $(2,3)$, $(-2,-3)$
       b) $(3,5)$, $(-2, -10)$ j) $(1,3)$, $(-1,-3)$, $(3,1)$, $(-3,-1)$
       c) $(2,4)$, $(-1/3, 5/3)$ k) $(-12,-5)$, $(4,3)$
       d) $(-1,2)$, $(5/2, -4/5)$ l) $(1,-2)$, $(2,-1)$
       e) $(\sqrt{7}, 3)$, $(\sqrt{7}, -3)$, $(-\sqrt{7}, 3)$, $(-\sqrt{7}, -3)$ m) $(-2,-3)$, $(3,2)$
       f) $(3,2)$, $(3,-2)$, $(-3,2)$, $(-3,-2)$ n) $(1/2, 1/3)$, $(1/3, 1/2)$
       g) $(4,1)$, $(-4,-1)$

       h) $(3,1)$, $(-3,-1)$, $(3i\sqrt{5}, \dfrac{-7i\sqrt{5}}{5})$, $(-3i\sqrt{5}, \dfrac{7i\sqrt{5}}{5})$

22. 12, 8;   -12, -8;   12, -8;   -12, 8

23. 40 ft, 75 ft

CHAPTER 15

Ratio, Proportion, and Variation

THE RATIO of two numbers a and b, written $a:b$, is the fraction $\frac{a}{b}$ provided $b \neq 0$.

Thus $a:b = \frac{a}{b}$, $b \neq 0$. If $a = b \neq 0$, the ratio is $1:1$ or $\frac{1}{1} = 1$.

Examples. 1) The ratio of 4 to 6 $= 4:6 = \frac{4}{6} = \frac{2}{3}$.

2) $\frac{2}{3} : \frac{4}{5} = \frac{2/3}{4/5} = \frac{5}{6}$ 3) $5x : \frac{3y}{4} = \frac{5x}{3y/4} = \frac{20x}{3y}$

A PROPORTION is an equality of two ratios. Thus $a:b = c:d$, or $\frac{a}{b} = \frac{c}{d}$, is a proportion in which a and d are called the *extremes*, b and c the *means*, while d is called the *fourth proportional* to a, b and c.

In the proportion $a:b = b:c$, c is called the *third proportional* to a and b, and b is called a *mean proportional* between a and c.

LAWS OF PROPORTION. If $\frac{a}{b} = \frac{c}{d}$, then

1) $ad = bc$ 3) $\frac{a}{c} = \frac{b}{d}$ 5) $\frac{a-b}{b} = \frac{c-d}{d}$

2) $\frac{b}{a} = \frac{d}{c}$ 4) $\frac{a+b}{b} = \frac{c+d}{d}$ 6) $\frac{a+b}{a-b} = \frac{c+d}{c-d}$.

VARIATION.

1) If x varies *directly* as y or if x is *proportional* to y (written $x \propto y$), then $x = ky$ or $\frac{x}{y} = k$, where k is called the constant of proportionality or the constant of variation.

2) If x varies directly as y^2, then $x \propto y^2$ and $x = ky^2$.

3) If x varies *inversely* as y, then $x = \frac{k}{y}$.

4) If x varies inversely as y^2, then $x = \frac{k}{y^2}$.

5) If x varies *jointly* as y and z, then $x = kyz$.

6) If x varies *directly* as y^2 and *inversely* as z, then $x = \frac{ky^2}{z}$.

The constant k may be determined if one set of values of the variables is known.

SOLVED PROBLEMS

RATIO AND PROPORTION.

1. Express each of the following ratios as a simplified fraction.

a) $96 : 128 = \dfrac{96}{128} = \dfrac{3}{4}$ b) $\dfrac{2}{3} : \dfrac{3}{4} = \dfrac{2/3}{3/4} = \dfrac{8}{9}$ c) $xy^2 : x^2y = \dfrac{xy^2}{x^2y} = \dfrac{y}{x}$

d) $(xy^2 - x^2y) : (x - y)^2 = \dfrac{xy^2 - x^2y}{(x - y)^2} = \dfrac{xy(y - x)}{(x - y)^2} = \dfrac{xy}{y - x}$

2. Find the ratio of each of the following quantities.

a) 6 pounds to 12 ounces.
It is customary to express the quantities in the same units.
Then the ratio of 96 ounces to 12 ounces is $96 : 12 = 8 : 1$.

b) 3 quarts to 2 gallons.
The required ratio is 3 quarts to 8 quarts or $3 : 8$.

c) 3 square yards to 6 square feet.
Since 1 square yard = 9 square feet, the required ratio is $27 \text{ ft}^2 : 6 \text{ ft}^2 = 9 : 2$.

3. In each of the following proportions determine the value of x.

a) $(3 - x) : (x + 1) = 2 : 1$, $\dfrac{3 - x}{x + 1} = \dfrac{2}{1}$ and $x = \dfrac{1}{3}$.

b) $(x + 3) : 10 = (3x - 2) : 8$, $\dfrac{x + 3}{10} = \dfrac{3x - 2}{8}$ and $x = 2$.

c) $(x - 1) : (x + 1) = (2x - 4) : (x + 4)$, $\dfrac{x - 1}{x + 1} = \dfrac{2x - 4}{x + 4}$, $x^2 - 5x = 0$, $x(x - 5) = 0$ and $x = 0, 5$.

4. Find the fourth proportional to each of the following sets of numbers.

In each case let x be the fourth proportional.

a) 2, 3, 6.   Here $2 : 3 = 6 : x$, $\dfrac{2}{3} = \dfrac{6}{x}$ and $x = 9$.

b) 4, −5, 10.   Here $4 : -5 = 10 : x$ and $x = -\dfrac{25}{2}$.

c) a^2, ab, 2.   Here $a^2 : ab = 2 : x$, $a^2x = 2ab$ and $x = \dfrac{2b}{a}$.

5. Find the third proportional to each of the following pairs of numbers.

In each case let x be the third proportional.

a) 2, 3.   Here $2 : 3 = 3 : x$ and $x = 9/2$.

b) $-2, \dfrac{8}{3}$.   Here $-2 : \dfrac{8}{3} = \dfrac{8}{3} : x$ and $x = -\dfrac{32}{9}$.

6. Find the mean proportional between 2 and 8.

   Let x be the required mean proportional. Then $2 : x = x : 8$, $x^2 = 16$ and $x = \pm 4$.

7. A line segment 30 inches long is divided into two parts whose lengths have the ratio $2 : 3$. Find the lengths of the parts.

   Let the required lengths be x and $30 - x$. Then $\dfrac{x}{30 - x} = \dfrac{2}{3}$ and $x = 12$ in., $30 - x = 18$ in.

8. Two brothers are respectively 5 and 8 years old. In how many years (x) will the ratio of their ages be $3 : 4$?

   In x years their respective ages will be $5 + x$ and $8 + x$.

   Then $(5 + x) : (8 + x) = 3 : 4$, $4(5 + x) = 3(8 + x)$ and $x = 4$.

9. Divide 253 into four parts proportional to 2, 5, 7, 9.

   Let the four parts be $2k$, $5k$, $7k$, $9k$.

   Then $2k + 5k + 7k + 9k = 253$ and $k = 11$. Thus the four parts are 22, 55, 77, 99.

10. If $x : y : z = 2 : -5 : 4$ and $x - 3y + z = 63$, find x, y, z.

    Let $x = 2k$, $y = -5k$, $z = 4k$.

    Substitute these values in $x - 3y + z = 63$ and obtain $2k - 3(-5k) + 4k = 63$ or $k = 3$.

    Hence $x = 2k = 6$, $y = -5k = -15$, $z = 4k = 12$.

VARIATION.

11. For each of the following statements write an equation, employing k as the constant of proportionality.

    a) The circumference C of a circle varies as its diameter d. *Ans.* $C = kd$

    b) The period of vibration T of a simple pendulum at a given place is proportional to the square root of its length l. *Ans.* $T = k\sqrt{l}$

    c) The rate of emission of radiant energy E per unit area of a perfect radiator is proportional to the fourth power of its absolute temperature T. *Ans.* $E = kT^4$

    d) The heat H in calories developed in a conductor of resistance R ohms when using a current of I amperes, varies jointly as the square of the current, the resistance of the conductor, and the time t during which the conductor draws the current. *Ans.* $H = kI^2Rt$

    e) The intensity I of a sound wave varies jointly as the square of its frequency n, the square of its amplitude r, the speed of sound v, and the density d of the undisturbed medium.
    Ans. $I = kn^2r^2vd$

    f) The force of attraction F between two masses m_1 and m_2 varies directly as the product of the masses and inversely as the square of the distance r between them. *Ans.* $F = \dfrac{km_1m_2}{r^2}$

    g) At constant temperature, the volume V of a given mass of an ideal gas varies inversely as the pressure p to which it is subjected. *Ans.* $V = k/p$

    h) An unbalanced force F acting on a body produces in it an acceleration a which is directly proportional to the force and inversely proportional to the mass m of the body.
    Ans. $a = k\dfrac{F}{m}$

12. The kinetic energy E of a body is proportional to its weight W and to the square of its velocity v. An 8 lb body moving at 4 ft/sec has 2 ft-lb of kinetic energy. Find the kinetic energy of a 3 ton (6000 lb) truck speeding at 60 mi/hr (88 ft/sec).

To find k: $E = kWv^2$ or $k = \dfrac{E}{Wv^2} = \dfrac{2}{8(4^2)} = \dfrac{1}{64}$.

Thus the kinetic energy of the truck is $E = \dfrac{Wv^2}{64} = \dfrac{6000(88)^2}{64} = 726{,}000$ ft-lb.

13. The pressure p of a given mass of ideal gas varies inversely as the volume V and directly as the absolute temperature T. To what pressure must 100 cubic feet of helium at 1 atmosphere pressure and $253°$ temperature be subjected to be compressed to 50 cubic feet when the temperature is $313°$?

To find k: $p = k\dfrac{T}{V}$ or $k = \dfrac{pV}{T} = \dfrac{1(100)}{253} = \dfrac{100}{253}$.

Thus the required pressure is $p = \dfrac{100}{253}\dfrac{T}{V} = \dfrac{100}{253}\left(\dfrac{313}{50}\right) = 2.47$ atmospheres.

Another Method. Let the subscripts 1 and 2 refer to the initial and final conditions of the gas, respectively.

Then $k = \dfrac{p_1 V_1}{T_1} = \dfrac{p_2 V_2}{T_2}$, $\dfrac{p_1 V_1}{T_1} = \dfrac{p_2 V_2}{T_2}$, $\dfrac{1(100)}{253} = \dfrac{p_2(50)}{313}$ and $p_2 = 2.47$ atm.

14. If 8 men take 12 days to assemble 16 machines, how many days will it take 15 men to assemble 50 machines?

The number of days (x) varies directly as the number of machines (y) and inversely as the number of men (z).

Then $x = \dfrac{ky}{z}$ where $k = \dfrac{xz}{y} = \dfrac{12(8)}{16} = 6$.

Hence the required number of days is $x = \dfrac{6y}{z} = \dfrac{6(50)}{15} = 20$ days.

SUPPLEMENTARY PROBLEMS

15. Express each ratio as a simplified fraction.

  a) $40 : 64$ *b)* $4/5 : 8/3$ *c)* $x^2 y^3 : 3xy^4$ *d)* $(a^2 b + ab^2) : (a^2 b^3 + a^3 b^2)$

16. Find the ratio of the following quantities.
  a) 20 yards to 40 feet   *c)* 2 square feet to 96 square inches
  b) 8 pints to 5 quarts   *d)* 6 gallons to 30 pints

17. In each proportion determine the value of x.
  a) $(x+3) : (x-2) = 3 : 2$ *c)* $(x+1) : 4 = (x+6) : 2x$
  b) $(x+4) : 1 = (2-x) : 2$ *d)* $(2x+1) : (x+1) = 5x : (x+4)$

18. Find the fourth proportional to each set of numbers.
  a) 3, 4, 12   *c)* a, b, c
  b) $-2, 5, 6$ *d)* $m+2, m-2, 3$

19. Find the third proportional to each pair of numbers.

    a) 3, 5   b) −2, 4   c) a, b   d) ab, $\sqrt{ab}$

20. Find the mean proportional between each pair of numbers.

    a) 3, 27   b) −4, −8   c) $3\sqrt{2}$ and $6\sqrt{2}$ d) $m + 2$ and $m + 1$

21. If $(x + y) : (x - y) = 5 : 2$, find $x : y$.

22. Two numbers have the ratio 3:4. If 4 is added to each of the numbers the resulting ratio is 4:5. Find the numbers.

23. A line segment of length 120 inches is divided into three parts whose lengths are proportional to 3,4,5. Find the lengths of the parts.

24. If $x : y : z = 4 : -3 : 2$ and $2x + 4y - 3z = 20$, find x, y, z.

25. a) If x varies directly as y and if $x = 8$ when $y = 5$, find y when $x = 20$.
    b) If x varies directly as y^2 and if $x = 4$ when $y = 3$, find x when $y = 6$.
    c) If x varies inversely as y and if $x = 8$ when $y = 3$, find y when $x = 2$.

26. The distance covered by an object falling freely from rest varies directly as the square of the time of falling. If an object falls 144 ft in 3 sec, how far will it fall in 10 sec?

27. The force of wind on a sail varies jointly as the area of the sail and the square of the wind velocity. On a square foot of sail the force is 1 lb when the wind velocity is 15 mi/hr. Find the force of a 45 mi/hr wind on a sail of area 20 square yards.

28. If 2 men can plow 6 acres of land in 4 hours, how many men are needed to plow 18 acres in 8 hours?

ANSWERS TO SUPPLEMENTARY PROBLEMS.

15. a) 5/8   b) 3/10   c) x/3y   d) 1/ab
16. a) 3 : 2   b) 4 : 5   c) 3 : 1   d) 8 : 5
17. a) 12   b) −2   c) 4, −3   d) 2, −2/3
18. a) 16   b) −15   c) bc/a   d) 3(m − 2)/(m + 2)
19. a) 25/3   b) −8   c) b^2/a d) 1

20. a) ±9   b) $\pm 4\sqrt{2}$ c) ±6   d) $\pm\sqrt{m^2 + 3m + 2}$

21. 7/3
22. 12, 16
23. 30, 40, 50 in.
24. −8, 6, −4
25. a) $12\frac{1}{2}$ b) 16   c) 12
26. 1600 ft
27. 1620 lb
28. 3 men

CHAPTER 16

Progressions

A SEQUENCE of numbers is a set of numbers in a definite order of arrangement and formed according to a definite rule. The numbers of the sequence are called *terms*.

ARITHMETIC PROGRESSIONS.

A) *An arithmetic progression* (A.P.) is a sequence of numbers each of which, after the first, is obtained by adding to the preceding number a constant number called the *common difference*.

Thus $3, 7, 11, 15, 19, \cdots$ is an arithmetic progression because each term is obtained by adding 4 to the preceding number. In the arithmetic progression $50, 45, 40, \cdots$ the common difference is $45 - 50 = 40 - 45 = -5$.

B) *Formulas for Arithmetic Progressions.*

1) The nth term, or last term: $\qquad l = a + (n-1)d$

2) The sum of the first n terms: $\qquad S = \dfrac{n}{2}(a+l) = \dfrac{n}{2}[2a + (n-1)d]$

where a = first term of the progression; d = common difference;
$\qquad\qquad n$ = number of terms; l = nth term, or last term;
$\qquad\qquad S$ = sum of first n terms.

Example. Consider the arithmetic progression $3, 7, 11, \cdots$ where $a = 3$ and $d = 7 - 3 = 11 - 7 = 4$. The sixth term is $l = a + (n-1)d = 3 + (6-1)4 = 23$.

The sum of the first six terms is

$$S = \frac{n}{2}(a+l) = \frac{6}{2}(3+23) = 78 \quad \text{or} \quad S = \frac{n}{2}[2a + (n-1)d] = \frac{6}{2}[2(3) + (6-1)4] = 78.$$

GEOMETRIC PROGRESSIONS.

A) *A geometric progression* (G.P.) is a sequence of numbers each of which, after the first, is obtained by multiplying the preceding number by a constant number called the *common ratio*.

Thus $5, 10, 20, 40, 80, \cdots$ is a geometric progression because each number is obtained by multiplying the preceding number by 2. In the geometric progression $9, -3, 1, -\dfrac{1}{3}, \dfrac{1}{9}, \cdots$ the common ratio is $\dfrac{-3}{9} = \dfrac{1}{-3} = \dfrac{-1/3}{1} = \dfrac{1/9}{-1/3} = -\dfrac{1}{3}$.

B) Formulas for Geometric Progressions.

1) The nth term, or last term: $\quad l = ar^{n-1}$

2) The sum of the first n terms: $\quad S = \dfrac{a(r^n - 1)}{r - 1} = \dfrac{rl - a}{r - 1}, \quad r \ne 1$

where $\quad a$ = first term; r = common ratio; n = number of terms;
$\quad\quad l$ = nth term, or last term; S = sum of first n terms.

Example. Consider the geometric progression $5, 10, 20, \cdots$ where $a = 5$ and $r = \dfrac{10}{5} = \dfrac{20}{10} = 2$. The seventh term is $l = ar^{n-1} = 5(2^{7-1}) = 5(2^6) = 320$.

The sum of the first seven terms is $S = \dfrac{a(r^n - 1)}{r - 1} = \dfrac{5(2^7 - 1)}{2 - 1} = 635$.

INFINITE GEOMETRIC PROGRESSIONS.

The sum to infinity (S_∞) of any geometric progression in which the common ratio r is numerically less than 1 is given by

$$S_\infty = \frac{a}{1 - r}, \quad \text{where} \quad |r| < 1.$$

Example. Consider the infinite geometric series $1 - \dfrac{1}{2} + \dfrac{1}{4} - \dfrac{1}{8} + \cdots$ where $a = 1$ and $r = -\dfrac{1}{2}$. Its sum to infinity is $S_\infty = \dfrac{a}{1 - r} = \dfrac{1}{1 - (-1/2)} = \dfrac{1}{3/2} = \dfrac{2}{3}$.

HARMONIC PROGRESSIONS.

A *harmonic progression* is a sequence of numbers whose reciprocals form an arithmetic progression.

Thus $\dfrac{1}{2}, \dfrac{1}{4}, \dfrac{1}{6}, \dfrac{1}{8}, \dfrac{1}{10}, \ldots$ is a harmonic progression because $2, 4, 6, 8, 10,$
$\ldots$ is an arithmetic progression.

MEANS. The terms between any two given terms of a progression are called the *means* between these *two* terms.

Thus in the arithmetic progression $3, 5, 7, 9, 11, \cdots$ the arithmetic mean between 3 and 7 is 5, and *four* arithmetic means between 3 and 13 are $5, 7, 9,$ 11.

In the geometric progression $2, -4, 8, -16, \cdots$ *two* geometric means between 2 and -16 are $-4, 8$.

In the harmonic progression $\dfrac{1}{2}, \dfrac{1}{3}, \dfrac{1}{4}, \dfrac{1}{5}, \dfrac{1}{6}, \ldots$ the harmonic mean between $\dfrac{1}{2}$ and $\dfrac{1}{4}$ is $\dfrac{1}{3}$, and *three* harmonic means between $\dfrac{1}{2}$ and $\dfrac{1}{6}$ are $\dfrac{1}{3}, \dfrac{1}{4}, \dfrac{1}{5}$.

SOLVED PROBLEMS

ARITHMETIC PROGRESSIONS.

1. Which of the following sequences are arithmetic progressions?

    a) 1, 6, 11, 16, ...   is an A.P. since $6 - 1 = 11 - 6 = 16 - 11 = 5.$ $(d = 5)$

    b) $\frac{1}{3}$, 1, $\frac{5}{3}$, $\frac{7}{3}$, ...   is an A.P. since $1 - \frac{1}{3} = \frac{5}{3} - 1 = \frac{7}{3} - \frac{5}{3} = \frac{2}{3}.$ $(d = \frac{2}{3})$

    c) 4, −1, −6, −11, ...   is an A.P. since $-1 - 4 = -6 - (-1) = -11 - (-6) = -5.$ $(d = -5)$

    d) 9, 12, 16, ...   is not an A.P. since $12 - 9 \neq 16 - 12.$

    e) $\frac{1}{2}$, $\frac{1}{3}$, $\frac{1}{4}$, ...   is not an A.P. since $\frac{1}{3} - \frac{1}{2} \neq \frac{1}{4} - \frac{1}{3}.$

    f) 7, $9 + 3p$, $11 + 6p$, ...   is an A.P. with $d = 2 + 3p.$

2. Prove the formula $S = \frac{n}{2}(a + l)$ for the sum of the first n terms of an arithmetic progression.

    The sum of the first n terms of an arithmetic progression may be written

$$S = a + (a + d) + (a + 2d) + \ldots + l \qquad (n \text{ terms})$$

or
$$S = l + (l - d) + (l - 2d) + \ldots + a \qquad (n \text{ terms})$$

where the sum is written in reversed order.

    Adding, $\qquad 2S = (a + l) + (a + l) + (a + l) + \ldots + (a + l) \qquad$ to n terms.

    Hence $\qquad 2S = n(a + l) \quad$ and $\quad S = \frac{n}{2}(a + l).$

3. Find the 16th term of the A.P.: 4, 7, 10, ...

    Here $a = 4$, $n = 16$, $d = 7 - 4 = 10 - 7 = 3$, and $l = a + (n - 1)d = 4 + (16 - 1)3 = 49.$

4. Determine the sum of the first 12 terms of the A.P.: 3, 8, 13, ...

    Here $a = 3$, $d = 8 - 3 = 13 - 8 = 5$, $n = 12$, and

$$S = \frac{n}{2}[2a + (n - 1)d] = \frac{12}{2}[2(3) + (12 - 1)5] = 366.$$

    Otherwise: $l = a + (n - 1)d = 3 + (12 - 1)5 = 58$ and

$$S = \frac{n}{2}(a + l) = \frac{12}{2}(3 + 58) = 366.$$

5. Find the 40th term and the sum of the first 40 terms of the A.P.: 10, 8, 6, ...

    Here $d = 8 - 10 = 6 - 8 = -2$, $a = 10$, $n = 40$.

    Then $\qquad l = a + (n - 1)d = 10 + (40 - 1)(-2) = -68$ and

$$S = \frac{n}{2}(a + l) = \frac{40}{2}(10 - 68) = -1160.$$

6. Which term of the sequence 5, 14, 23, ... is 239?

$l = a + (n-1)d$, $239 = 5 + (n-1)9$, $9n = 243$ and the required term is $n = 27$.

7. Compute the sum of the first 100 positive integers exactly divisible by 7.

The sequence is 7, 14, 21, ... an A.P. in which $a = 7$, $d = 7$, $n = 100$.

Hence $S = \frac{n}{2}[2a + (n-1)d] = \frac{100}{2}[2(7) + (100-1)7] = 35,350$.

8. How many consecutive integers beginning with 10 must be taken for their sum to equal 2035?

The sequence is 10, 11, 12, ... an A.P. in which $a = 10$, $d = 1$, $S = 2035$.

Using $S = \frac{n}{2}[2a + (n-1)d]$, we obtain $2035 = \frac{n}{2}[20 + (n-1)1]$, $2035 = \frac{n}{2}(n + 19)$,

$n^2 + 19n - 4070 = 0$, $(n-55)(n+74) = 0$, $n = 55, -74$.   Hence 55 integers must be taken.

9. How long will it take to pay off a debt of \$880 if \$25 is paid the first month, \$27 the second month, \$29 the third month, etc.?

From $S = \frac{n}{2}[2a + (n-1)d]$,   we obtain $880 = \frac{n}{2}[2(25) + (n-1)2]$, $880 = 24n + n^2$,

$n^2 + 24n - 880 = 0$, $(n-20)(n+44) = 0$, $n = 20, -44$. The debt will be paid off in 20 months.

10. How many terms of the A.P. 24, 22, 20, ... are needed to give the sum of 150? Write the terms.

$150 = \frac{n}{2}[48 + (n-1)(-2)]$, $n^2 - 25n + 150 = 0$, $(n-10)(n-15) = 0$, $n = 10, 15$.

For $n = 10$:   24, 22, 20, 18, 16, 14, 12, 10, 8, 6.
For $n = 15$:   24, 22, 20, 18, 16, 14, 12, 10, 8, 6, 4, 2, 0, -2, -4.

11. Determine the A.P. whose sum to n terms is $n^2 + 2n$.

The nth term $=$ sum to n terms $-$ sum to $n-1$ terms
$$= n^2 + 2n - [(n-1)^2 + 2(n-1)] = 2n + 1. \quad \text{Thus the A.P. is } 3, 5, 7, 9, \ldots$$

12. Show that the sum of n consecutive odd integers beginning with 1 equals n^2.

We are to find the sum of the A.P. 1, 3, 5, ... to n terms.

Then $a = 1$, $d = 2$, $n = n$ and $S = \frac{n}{2}[2a + (n-1)d] = \frac{n}{2}[2(1) + (n-1)2] = n^2$.

13. Find three numbers in A.P. such that the sum of the first and third is 12 and the product of the first and second is 24.

Let the numbers in A.P. be $(a-d)$, a, $(a+d)$. Then $(a-d) + (a+d) = 12$ or $a = 6$.

Since $(a-d)a = 24$, $(6-d)6 = 24$ or $d = 2$. Hence the numbers are 4, 6, 8.

14. Find three numbers in A.P. whose sum is 21 and whose product is 280.

Let the numbers be $(a-d)$, a, $(a+d)$.   Then $(a-d) + a + (a+d) = 21$ or $a = 7$.

Since $(a-d)(a)(a+d) = 280$, $a(a^2 - d^2) = 7(49 - d^2) = 280$ and $d = \pm 3$.

The required numbers are 4, 7, 10 or 10, 7, 4.

15. Three numbers are in the ratio of $2:5:7$. If 7 is subtracted from the second, the resulting numbers form an A.P. Determine the original numbers.

Let the original numbers be $2x$, $5x$, $7x$. The resulting numbers in A.P. are $2x$, $(5x-7)$, $7x$.

Then $(5x-7)-2x = 7x-(5x-7)$ or $x = 14$. Hence the original numbers are 28, 70, 98.

16. Compute the sum of all integers between 100 and 800 that are divisible by 3.

The A.P. is $102, 105, 108, \ldots, 798$. Then $l = a+(n-1)d$, $798 = 102+(n-1)3$, $n = 233$, and $S = \dfrac{n}{2}(a+l) = \dfrac{233}{2}(102+798) = 104,850$.

17. A slide of uniform grade is to be built on a level surface and is to have 10 supports equidistant from each other. The heights of the longest and shortest supports will be $42\frac{1}{2}$ feet and 2 feet respectively. Determine the required height of each support.

From $l = a+(n-1)d$ we have $42\frac{1}{2} = 2+(10-1)d$ and $d = 4\frac{1}{2}$ ft.

Thus the heights are 2, $6\frac{1}{2}$, 11, $15\frac{1}{2}$, 20, $24\frac{1}{2}$, 29, $33\frac{1}{2}$, 38, $42\frac{1}{2}$ feet respectively.

18. A freely falling body, starting from rest, falls 16 ft during the first second, 48 ft during the second second, 80 ft during the third second, etc. Calculate the distance it falls during the fifteenth second and the total distance it falls in 15 seconds from rest.

Here $d = 48-16 = 80-48 = 32$.

During the 15th second it falls a distance $l = a+(n-1)d = 16+(15-1)32 = 464$ ft.

Total distance covered during 15 sec is $S = \dfrac{n}{2}(a+l) = \dfrac{15}{2}(16+464) = 3600$ ft.

19. In a potato race, 8 potatoes are placed 6 ft apart on a straight line, the first being 6 ft from the basket. A contestant starts from the basket and puts one potato at a time into the basket. Find the total distance he must run in order to finish the race.

Here $a = 2\cdot 6 = 12$ ft and $l = 2(6\cdot 8) = 96$ ft. Then $S = \dfrac{n}{2}(a+l) = \dfrac{8}{2}(12+96) = 432$ ft.

20. Show that if the sides of a right triangle are in A.P., their ratio is $3:4:5$.

Let the sides be $(a-d)$, a, $(a+d)$, where the hypotenuse is $(a+d)$.

Then $(a+d)^2 = a^2+(a-d)^2$ or $a = 4d$. Hence $(a-d):a:(a+d) = 3d:4d:5d = 3:4:5$.

ARITHMETIC MEANS.

21. Derive the formula for the arithmetic mean (x) between two numbers p and q.

Since p,x,q are in A.P., we have $x-p = q-x$ or $x = \frac{1}{2}(p+q)$.

22. Find the arithmetic mean between each of the following pairs of numbers.

$a)$ 4 and 56.　　　　Arithmetic mean $= \dfrac{4+56}{2} = 30$.

$b)$ $3\sqrt{2}$ and $-6\sqrt{2}$.　　Arithmetic mean $= \dfrac{3\sqrt{2}+(-6\sqrt{2})}{2} = -\dfrac{3\sqrt{2}}{2}$.

$c)$ $a+5d$ and $a-3d$.　　Arithmetic mean $= \dfrac{(a+5d)+(a-3d)}{2} = a+d$.

23. Insert 5 arithmetic means between 8 and 26.

We require an A.P. of the form 8, —, —, —, —, —, 26; thus $a = 8$, $l = 26$ and $n = 7$.

Then $l = a + (n-1)d$, $26 = 8 + (7-1)d$, $d = 3$.

The five arithmetic means are 11, 14, 17, 20, 23.

24. Insert between 1 and 36 a number of arithmetic means so that the sum of the resulting arithmetic progression will be 148.

$S = \frac{1}{2}n(a + l)$, $148 = \frac{1}{2}n(1 + 36)$, $37n = 296$ and $n = 8$.

$l = a + (n-1)d$, $36 = 1 + (8-1)d$, $7d = 35$ and $d = 5$.

The complete arithmetic progression is 1, 6, 11, 16, 21, 26, 31, 36.

GEOMETRIC PROGRESSIONS.

25. Which of the following sequences are geometric progressions?

a) 3, 6, 12, ... is a G.P. since $\frac{6}{3} = \frac{12}{6} = 2$. $(r = 2)$

b) 16, 12, 9, ... is a G.P. since $\frac{12}{16} = \frac{9}{12} = \frac{3}{4}$. $(r = \frac{3}{4})$

c) −1, 3, −9, ... is a G.P. since $\frac{3}{-1} = \frac{-9}{3} = -3$. $(r = -3)$

d) 1, 4, 9, ... is not a G.P. since $\frac{4}{1} \neq \frac{9}{4}$.

e) $\frac{1}{2}, \frac{1}{3}, \frac{2}{9}, \ldots$ is a G.P. since $\frac{1/3}{1/2} = \frac{2/9}{1/3} = \frac{2}{3}$. $(r = \frac{2}{3})$

f) $2h, \frac{1}{h}, \frac{1}{2h^3}, \ldots$ is a G.P. since $\frac{1/h}{2h} = \frac{1/2h^3}{1/h} = \frac{1}{2h^2}$. $(r = \frac{1}{2h^2})$

26. Prove the formula $S = \frac{a(r^n - 1)}{r - 1}$ for the sum of the first n terms of a geometric progression.

The sum of the first n terms of a geometric progression may be written

1) $S = a + ar + ar^2 + ar^3 + \ldots + ar^{n-1}$ (n terms). Multiplying 1) by r, we obtain

2) $rS = ar + ar^2 + ar^3 + \ldots + ar^{n-1} + ar^n$ (n terms).

Subtracting 1) from 2), $rS - S = ar^n - a$, $(r-1)S = a(r^n - 1)$ and $S = \frac{a(r^n - 1)}{r - 1}$.

27. Find the 8th term and the sum of the first eight terms of the progression 4, 8, 16, ...

Here $a = 4$, $r = 8/4 = 16/8 = 2$, $n = 8$.

The 8th term is $l = ar^{n-1} = 4(2)^{8-1} = 4(2^7) = 4(128) = 512$.

The sum of the first eight terms is $S = \frac{a(r^n - 1)}{r - 1} = \frac{4(2^8 - 1)}{2 - 1} = \frac{4(256 - 1)}{1} = 1020$.

28. Find the 7th term and the sum of the first seven terms of the progression 9, −6, 4, ...

Here $a = 9$, $r = \frac{-6}{9} = \frac{4}{-6} = -\frac{2}{3}$. Then the 7th term is $l = ar^{n-1} = 9(-\frac{2}{3})^{7-1} = \frac{64}{81}$.

$$S = \frac{a(r^n - 1)}{r - 1} = \frac{a(1 - r^n)}{1 - r} = \frac{9[1 - (-2/3)^7]}{1 - (-2/3)} = \frac{9[1 - (-128/2187)]}{5/3} = \frac{463}{81}$$

29. The second term of a G.P. is 3 and the fifth term is 81/8. Find the eighth term.

5th term = $ar^4 = \frac{81}{8}$, 2nd term = $ar = 3$. Then $\frac{ar^4}{ar} = \frac{81/8}{3}$, $r^3 = \frac{27}{8}$ and $r = \frac{3}{2}$.

Hence the 8th term = $ar^7 = (ar^4)r^3 = \frac{81}{8}(\frac{27}{8}) = \frac{2187}{64}$.

30. Find three numbers in G.P. whose sum is 26 and whose product is 216.

Let the numbers in G.P. be $a/r, a, ar$. Then $(a/r)(a)(ar) = 216$, $a^3 = 216$ and $a = 6$.

Also $a/r + a + ar = 26$, $6/r + 6 + 6r = 26$, $6r^2 - 20r + 6 = 0$ and $r = 1/3, 3$.

For $r = 1/3$, the numbers are 18, 6, 2; for $r = 3$, the numbers are 2, 6, 18.

31. The first term of a G.P. is 375 and the fourth term is 192. Find the common ratio and the sum of the first four terms.

1st term = $a = 375$, 4th term = $ar^3 = 192$. Then $375r^3 = 192$, $r^3 = 64/125$ and $r = 4/5$.

The sum of the first four terms is $S = \frac{a(1 - r^n)}{1 - r} = \frac{375[1 - (4/5)^4]}{1 - 4/5} = 1107$.

32. The first term of a G.P. is 160 and the common ratio is 3/2. How many consecutive terms must be taken to give a sum of 2110?

$$S = \frac{a(r^n - 1)}{r - 1}, \quad 2110 = \frac{160[(3/2)^n - 1]}{3/2 - 1}, \quad (\frac{3}{2})^n - 1 = \frac{211}{32}, \quad (\frac{3}{2})^n = \frac{243}{32} = (\frac{3}{2})^5, \quad n = 5.$$

The five consecutive terms are 160, 240, 360, 540, 810.

33. In a geometric progression consisting of four terms in which the ratio is positive, the sum of the first two terms is 8 and the sum of the last two terms is 72. Find the progression.

The four terms are a, ar, ar^2, ar^3. Then $a + ar = 8$ and $ar^2 + ar^3 = 72$.

Hence $\frac{ar^2 + ar^3}{a + ar} = \frac{ar^2(1 + r)}{a(1 + r)} = r^2 = \frac{72}{8} = 9$, so that $r = 3$.

Since $a + ar = 8$, $a = 2$ and the progression is 2, 6, 18, 54.

34. Prove that $x, x + 3, x + 6$ cannot be a geometric progression.

If $x, x + 3, x + 6$ is a G.P. then $r = \frac{x + 3}{x} = \frac{x + 6}{x + 3}$, $x^2 + 6x + 9 = x^2 + 6x$ or $9 = 0$.

Since this equality can never be true, $x, x + 3, x + 6$ cannot be a G.P.

35. A boy agrees to work at the rate of one cent the first day, two cents the second day, four cents the third day, eight cents the fourth day, etc. How much would he receive at the end of 12 days?

Here $a = 1, r = 2, n = 12$.

$$S = \frac{a(r^n - 1)}{r - 1} = 2^{12} - 1 = 4096 - 1 = 4095¢ = \$40.95.$$

36. It is estimated that the population of a certain town will increase 10% each year for four years. What is the percentage increase in population after four years?

Let p denote the initial population. After one year the population is $1.10p$, after two years $(1.10)^2 p$, after three years $(1.10)^3 p$, after four years $(1.10)^4 p = 1.46p$. Thus the population increases 46%.

37. From a tank filled with 240 gallons of alcohol, 60 gallons are drawn off and the tank is filled up with water. Then 60 gallons of the mixture are removed and replaced with water, etc. How many gallons of alcohol remain in the tank after 5 drawings of 60 gallons each are made?

After the first drawing, $240 - 60 = 180$ gal of alcohol remain in the tank.

After the second drawing, $180(\dfrac{240 - 60}{240}) = 180(\dfrac{3}{4})$ gal of alcohol remain; etc.

The G.P. for the number of gallons of alcohol remaining in the tank after successive drawings is $180, \ 180(\dfrac{3}{4}), \ 180(\dfrac{3}{4})^2, \ \ldots$ where $a = 180, \ r = \dfrac{3}{4}$.

After the fifth drawing $(n = 5)$: $l = ar^{n-1} = 180(\dfrac{3}{4})^4 = 57$ gal of alcohol remain.

38. A sum of \$400 is invested today at 6% per year. To what amount will it accumulate in five years if interest is compounded a) annually, b) semiannually, c) quarterly?

Let P = initial principal, i = interest rate per period,
S = compound amount after n periods.

At end of 1st period: Interest = Pi, new amount = $P + Pi = P(1 + i)$.
At end of 2nd period: Interest = $P(1 + i)i$, new amount = $P(1 + i) + P(1 + i)i = P(1 + i)^2$.

Compound amount at end of n periods is $S = P(1 + i)^n$.

a) Since there is 1 interest period per year, $n = 5$ and $i = .06$.

$$S = P(1 + i)^n = 400(1 + .06)^5 = 400(1.3382) = \$535.28$$

b) Since there are 2 interest periods per year, $n = 2(5) = 10$ and $i = \frac{1}{2}(.06) = .03$.

$$S = P(1 + i)^n = 400(1 + .03)^{10} = 400(1.3439) = \$537.56.$$

c) Since there are 4 interest periods per year, $n = 4(5) = 20$ and $i = \frac{1}{4}(.06) = .015$.

$$S = P(1 + i)^n = 400(1 + .015)^{20} = 400(1.3469) = \$538.76.$$

39. What sum (P) should be invested in a loan association at 4% per annum compounded semiannually, so that the compound amount (S) will be \$500 at the end of $3\frac{1}{2}$ years.

Since there are 2 interest periods per year, $n = 2(3\frac{1}{2}) = 7$ (periods) and the interest rate per period is $i = \frac{1}{2}(.04) = .02$.

Then $S = P(1 + i)^n$ or $P = S(1 + i)^{-n} = 500(1 + .02)^{-7} = 500(.87056) = \$435.28.$

GEOMETRIC MEANS.

40. Derive the formula for the geometric mean, G, between two numbers p and q.

Since p, G, q are in geometric progression, we have $\dfrac{G}{p} = \dfrac{q}{G}$, $G^2 = pq$ and $G = \pm\sqrt{pq}$.

It is customary to take $G = \sqrt{pq}$ if p and q are positive,

and $G = -\sqrt{pq}$ if p and q are negative.

41. Find the geometric mean between each of the following pairs of numbers.

$a)$ 4 and 9. $\qquad G = \sqrt{4(9)} = 6$

$b)$ -2 and -8. $\qquad G = -\sqrt{(-2)(-8)} = -4$

$c)$ $\sqrt{7} + \sqrt{3}$ and $\sqrt{7} - \sqrt{3}$. $\qquad G = \sqrt{(\sqrt{7} + \sqrt{3})(\sqrt{7} - \sqrt{3})} = \sqrt{7 - 3} = 2$

42. Show that the arithmetic mean A of two positive numbers p and q is greater than or equal to their geometric mean G.

Arithmetic mean of p and q is $A = \frac{1}{2}(p+q)$. Geometric mean of p and q is $G = \sqrt{pq}$.

Then $A - G = \frac{1}{2}(p+q) - \sqrt{pq} = \frac{1}{2}(p - 2\sqrt{pq} + q) = \frac{1}{2}(\sqrt{p} - \sqrt{q})^2$.

Now $\frac{1}{2}(\sqrt{p} - \sqrt{q})^2$ is always positive or zero; hence $A \geqq G$. ($A = G$ if and only if $p = q$.)

43. Insert two geometric means between 686 and 2.

We require a G.P. of the form $686, -, -, 2$ where $a = 686$, $l = 2$, $n = 4$.

Then $l = ar^{n-1}$, $2 = 686r^3$, $r^3 = 1/343$ and $r = 1/7$.

Thus the G.P. is $686, 98, 14, 2$ and the means are $98, 14$.

Note. Actually, $r^3 = 1/343$ is satisfied by three different values of r, one of the roots being real and two imaginary. It is customary to exclude geometric progressions with imaginary numbers.

44. Insert five geometric means between 9 and 576.

We require a G.P. of the form $9, -, -, -, -, -, 576$ where $a = 9$, $l = 576$, $n = 7$.

Then $l = ar^{n-1}$, $576 = 9r^6$, $r^6 = 64$, $r^3 = \pm 8$ and $r = \pm 2$.

Thus the progressions are $9, 18, 36, 72, 144, 288, 576$ and $9, -18, 36, -72, 144, -288, 576$; and the corresponding means are $18, 36, 72, 144, 288$ and $-18, 36, -72, 144, -288$.

INFINITE GEOMETRIC PROGRESSIONS.

45. Find the sum of the infinite geometric series.

$a)$ $2 + 1 + \dfrac{1}{2} + \dfrac{1}{4} + \cdots$ $\qquad S_\infty = \dfrac{a}{1 - r} = \dfrac{2}{1 - 1/2} = 4$

$b)$ $\dfrac{1}{3} - \dfrac{2}{9} + \dfrac{4}{27} - \dfrac{8}{81} + \cdots$ $\qquad S_\infty = \dfrac{a}{1 - r} = \dfrac{1/3}{1 - (-2/3)} = \dfrac{1}{5}$

$c)$ $1 + \dfrac{1}{1.04} + \dfrac{1}{(1.04)^2} + \cdots$ $\qquad S_\infty = \dfrac{a}{1 - r} = \dfrac{1}{1 - 1/1.04} = \dfrac{1.04}{1.04 - 1} = \dfrac{104}{4} = 26$

46. Express each of the following repeating decimals as a rational fraction.
$\quad a)$ $.444\ldots$ $\qquad b)$ $.4272727\ldots$ $\qquad c)$ $6.305305\ldots$ $\qquad d)$ $.78367836\ldots$

$a)$ $.444\ldots = .4 + .04 + .004 + \ldots$, where $a = .4$, $r = .1$.

$$S_\infty = \frac{a}{1 - r} = \frac{.4}{1 - .1} = \frac{.4}{.9} = \frac{4}{9}$$

$b)$ $.4272727\ldots = .4 + .0272727\ldots$
$\quad .0272727\ldots = .027 + .00027 + .0000027 + \ldots$, where $a = .027$, $r = .01$.

$$S_\infty = .4 + \frac{a}{1 - r} = .4 + \frac{.027}{1 - .01} = .4 + \frac{27}{990} = \frac{4}{10} + \frac{3}{110} = \frac{47}{110}$$

c) $6.305305... = 6 + .305305...$

$.305305... = .305 + .000305 + ...$, where $a = .305$, $r = .001$.

$$S_\infty = 6 + \frac{a}{1-r} = 6 + \frac{.305}{1-.001} = 6 + \frac{305}{999} = 6\frac{305}{999}$$

d) $.78367836... = .7836 + .00007836 + ...$, where $a = .7836$, $r = .0001$.

$$S_\infty = \frac{a}{1-r} = \frac{.7836}{1-.0001} = \frac{7836}{9999} = \frac{2612}{3333}$$

47. The distances passed over by a certain pendulum bob in succeeding swings form the geometric progression 16, 12, 9, ... inches respectively. Calculate the total distance traversed by the bob before coming to rest.

$$S_\infty = \frac{a}{1-r} = \frac{16}{1-3/4} = \frac{16}{1/4} = 64 \text{ inches}$$

48. Find the least number of terms of the series $\frac{1}{3} + \frac{1}{6} + \frac{1}{12} + ...$ that should be taken so that their sum will differ from their sum to infinity by less than $\frac{1}{1000}$.

Let S_∞ = sum to infinity, S_n = sum to *n* terms. Then

$$S_\infty - S_n = \frac{a}{1-r} - \frac{a(1-r^n)}{1-r} = \frac{ar^n}{1-r}.$$

It is required that $\frac{ar^n}{1-r} < \frac{1}{1000}$, where $a = 1/3$, $r = 1/2$.

Then $\frac{(1/3)(1/2)^n}{1-1/2} < \frac{1}{1000}$, $\frac{1}{3(2^n)} < \frac{1}{2000}$, $3(2^n) > 2000$, $2^n > 666\frac{2}{3}$.

When $n = 9$, $2^n < 666\frac{2}{3}$; when $n = 10$, $2^n > 666\frac{2}{3}$. Thus at least 10 terms should be taken.

HARMONIC PROGRESSIONS.

49. Which of the following sequences are harmonic progressions?

a) $\frac{1}{3}, \frac{1}{5}, \frac{1}{7}, ...$ is a harmonic progression since 3, 5, 7, ... is an A.P.

b) 2, 4, 6, ...   is not a harmonic progression since $\frac{1}{2}, \frac{1}{4}, \frac{1}{6}, ...$ is not an A.P.

c) $\frac{1}{12}, \frac{2}{15}, \frac{1}{3}, ...$ is a harmonic progression since $12, \frac{15}{2}, 3, ...$ is an A.P.

50. Compute the 15th term of the harmonic progression $\frac{1}{4}, \frac{1}{7}, \frac{1}{10}, ...$

The corresponding A.P. is 4, 7, 10, ...; its 15th term is $l = a + (n-1)d = 4 + (15-1)3 = 46$

Hence the 15th term of the harmonic progression is $\frac{1}{46}$.

51. Derive the formula for the harmonic mean, *H*, between two numbers *p* and *q*.

Since *p*, *H*, *q* is a harmonic progression, $\frac{1}{p}, \frac{1}{H}, \frac{1}{q}$ is an arithmetic progression.

Then $\frac{1}{H} - \frac{1}{p} = \frac{1}{q} - \frac{1}{H}$, $\frac{2}{H} = \frac{1}{p} + \frac{1}{q} = \frac{p+q}{pq}$ and $H = \frac{2pq}{p+q}$.

Another Method.

Harmonic mean between p and q = reciprocal of the arithmetic mean between $\frac{1}{p}$ and $\frac{1}{q}$.

Arithmetic mean between $\frac{1}{p}$ and $\frac{1}{q}$ = $\frac{1}{2}(\frac{1}{p} + \frac{1}{q})$ = $\frac{p + q}{2pq}$.

Hence the harmonic mean between p and q = $\frac{2pq}{p + q}$.

52. What is the harmonic mean between 3/8 and 4?

Arithmetic mean between $\frac{8}{3}$ and $\frac{1}{4}$ = $\frac{1}{2}(\frac{8}{3} + \frac{1}{4})$ = $\frac{35}{24}$. Hence the harmonic mean between $\frac{3}{8}$ and 4 = 24/35.

Or, by formula, harmonic mean = $\frac{2pq}{p + q}$ = $\frac{2(3/8)(4)}{3/8 + 4}$ = $\frac{24}{35}$.

53. Insert four harmonic means between 1/4 and 1/64.

To insert four arithmetic means between 4 and 64: $l = a + (n-1)d$, $64 = 4 + (6-1)d$, $d = 12$. Thus the four arithmetic means between 4 and 64 are 16, 28, 40, 52.

Hence the four harmonic means between $\frac{1}{4}$ and $\frac{1}{64}$ are $\frac{1}{16}, \frac{1}{28}, \frac{1}{40}, \frac{1}{52}$.

54. Insert three harmonic means between 10 and 20.

To insert three arithmetic means between $\frac{1}{10}$ and $\frac{1}{20}$: $l = a + (n-1)d$, $\frac{1}{20} = \frac{1}{10} + (5-1)d$, $d = -\frac{1}{80}$. Thus the three arithmetic means between $\frac{1}{10}$ and $\frac{1}{20}$ are $\frac{7}{80}, \frac{6}{80}, \frac{5}{80}$.

Hence the three harmonic means between 10 and 20 are $\frac{80}{7}, \frac{40}{3}, 16$.

55. Determine whether the sequence −1, −4, 2 is in arithmetric, geometric or harmonic progression.

Since $-4 - (-1) \neq 2 - (-4)$, it is not in arithmetic progression.

Since $\frac{-4}{-1} \neq \frac{2}{-4}$, it is not in geometric progression.

Since $\frac{1}{-1}, \frac{1}{-4}, \frac{1}{2}$ is in arithmetic progression, i.e. $\frac{1}{-4} - (-1) = \frac{1}{2} - (\frac{1}{-4})$, the given sequence is in harmonic progression.

SUPPLEMENTARY PROBLEMS

ARITHMETIC PROGRESSIONS.

56. Find the nth term and the sum of the first n terms of each arithmetic progression for the indicated value of n.

     a) 1, 7, 13, ... $n = 100$ c) −26, −24, −22, ... $n = 40$ e) 3, $4\frac{1}{2}$, 6, ... $n = 37$
     b) 2, $5\frac{1}{2}$, 9, ... $n = 23$ d) 2, 6, 10, ... $n = 16$ f) $x-y$, x, $x+y$, ... $n = 30$

57. Find the sum of the first n terms of each arithmetic progression.
     a) 1, 2, 3, ... b) 2, 8, 14, ... c) $1\frac{1}{2}$, 5, $8\frac{1}{2}$, ...

58. An arithmetic progression has first term 4 and last term 34. If the sum of its terms is 247, find the number of terms and their common difference.

59. An arithmetic progression consisting of 49 terms has last term 28. If the common difference of its terms is 1/2, find the first term and the sum of the terms.

60. Find the sum of all even integers between 17 and 99.

61. Find the sum of all integers between 84 and 719 which are exactly divisible by 5.

62. How many terms of the A.P. 3, 7, 11, ... are needed to yield the sum 1275?

63. Find three numbers in A.P. whose sum is 48 and such that the sum of their squares is 800.

64. A ball starting from rest rolls down an inclined plane and passes over 3 in. during the 1st second, 5 in. during the 2nd second, 7 in. during the 3rd second, etc. In what time from rest will it cover 120 inches?

65. If 1¢ is saved the 1st day, 2¢ the 2nd day, 3¢ the third day, etc., find the sum that will accumulate at the end of 365 days.

66. The sum of 40 terms of a certain A.P. is 430, while the sum of 60 terms is 945. Determine the nth term of the A.P.

67. Find an A.P. which has the sum of its first n terms equal to $2n^2 + 3n$.

68. Determine the arithmetic mean between a) 15 and 41, b) −16 and 23, c) $2 - \sqrt{3}$ and $4 + 3\sqrt{3}$, d) $x - 3y$ and $5x + 2y$.

69. a) Insert 4 arithmetic means between 9 and 24.
     b) Insert 2 arithmetic means between −1 and 11.
     c) Insert 3 arithmetic means between $x + 2y$ and $x + 10y$.
     d) Insert between 5 and 26 a number of arithmetic means such that the sum of the resulting A.P. will be 124.

GEOMETRIC PROGRESSIONS.

70. Find the nth term and the sum of the first n terms of each geometric progression for the indicated value of n.

     a) 2, 3, 9/2, ... $n = 5$ d) 1, 3, 9, ... $n = 8$
     b) 6, −12, 24, ... $n = 9$ e) 8, 4, 2, ... $n = 12$
     c) 1, 1/2, 1/4, ... $n = 10$ f) $\sqrt{3}$, 3, $3\sqrt{3}$, ... $n = 8$

71. Find the sum of the first n terms of each geometric progression.
    a) 1, 1/3, 1/9, ...   b) 4/3, 2, 3, ...   c) 1, −2, 4, ...

72. A geometric progression has first term 3 and last term 48. If each term is twice the previous term, find the number of terms and the sum of the G.P.

73. Prove that the sum S of the terms of a G.P. in which the first term is a, the last term is l and the common ratio is r is given by $S = \dfrac{rl - a}{r - 1}$.

74. In a G.P. the second term exceeds the first term by 4, and the sum of the second and third terms is 24. Show that there are two possible G.P.'s satisfying these conditions and find the sum of the first 5 terms of each G.P.

75. In a G.P. consisting of four terms, in which the ratio is positive, the sum of the first two terms is 10 and the sum of the last two terms is 22½. Find the progression.

76. The first two terms of a G.P. are $b/(1+c)$ and $b/(1+c)^2$. Show that the sum of n terms of this progression is given by the formula $S = b\left[\dfrac{1 - (1+c)^{-n}}{c}\right]$

77. Find the sum of the first n terms of the G.P.: $a - 2b$, $ab^2 - 2b^3$, $ab^4 - 2b^5$, ...

78. The third term of a G.P. is 6 and the fifth term is 81 times the first term. Write the first five terms of the progression, assuming the terms are positive.

79. Find three numbers in G.P. whose sum is 42 and whose product is 512.

80. The third term of a G.P. is 144 and the sixth term is 486. Find the sum of the first five terms of the G.P.

81. A tank contains a salt water solution in which is dissolved 972 lb of salt. One third of the solution is drawn off and the tank is filled with pure water. After stirring so that the solution is uniform, one third of the mixture is again drawn off and the tank is again filled with water. If this process is performed four times, what weight of salt remains in the tank?

82. The sum of the first three terms of a G.P. is 26 and the sum of the first six terms is 728. What is the nth term of the G.P.?

83. The sum of three numbers in G.P. is 14. If the first two terms are each increased by 1 and the third term decreased by 1, the resulting numbers are in A.P. Find the G.P.

84. Determine the geometric mean between :
    a) 2 and 18,   b) 4 and 6,   c) −4 and −16,   d) $a + b$ and $4a + 4b$.

85. a) Insert two geometric means between 3 and 192.
    b) Insert four geometric means between $\sqrt{2}$ and 8.
    c) The geometric mean of two numbers is 8. If one of the numbers is 6, find the other.

MISCELLANEOUS PROBLEMS INVOLVING A.P. AND G.P.

86. The first term of an A.P. is 2; and the first, third and eleventh terms are also the first three terms of a G.P. Find the sum of the first eleven terms of the A.P.

87. How many terms of the A.P. 9, 11, 13, ... must be added in order that the sum should equal the sum of nine terms of the G.P. 3, −6, 12, −24, ... ?

88. In a set of four numbers, the first three are in G.P. and the last three are in A.P. with a common difference of 6. If the first number is the same as the fourth, find the four numbers.

89. Find two numbers whose difference is 32 and whose arithmetic mean exceeds the geometric mean by 4.

INFINITE GEOMETRIC PROGRESSIONS.

90. Find the sum of the infinite geometric series.

   a) $3 + 1 + 1/3 + \ldots$ c) $1 + 1/2^2 + 1/2^4 + \ldots$ e) $4 - 8/3 + 16/9 - \ldots$

   b) $4 + 2 + 1 + \ldots$ d) $6 - 2 + 2/3 - \ldots$ f) $1 + .1 + .01 + \ldots$

91. The sum of the first two terms of a decreasing G.P. is 5/4 and the sum to infinity is 9/4. Write the first three terms of the geometric series.

92. The sum of an infinite number of terms of a decreasing G.P. is 3, and the sum of their squares is also 3. Write the first three terms of the series.

93. The successive distances traversed by a swinging pendulum bob are respectively 36, 24, 16, ... inches. Find the distance which the bob will travel before coming to rest.

94. Express each repeating decimal as a rational fraction.
   a) .121212...   c) .270270...   e) .1363636...
   b) .090909...   d) 1.424242...   f) .428571428571428...

HARMONIC PROGRESSIONS.

95. a) Find the 8th term of the harmonic progression 2/3, 1/2, 2/5, ...
   b) Find the 10th term of the harmonic progression 5, 30/7, 15/4, ...
   c) What is the nth term of the harmonic progression 10/3, 2, 10/7, ...?

96. Find the harmonic mean between each pair of numbers.
   a) 3 and 6   b) 1/2 and 1/3   c) $\sqrt{3}$ and $\sqrt{2}$ d) $a+b$ and $a-b$

97. a) Insert two harmonic means between 5 and 10.
   b) Insert four harmonic means between 3/2 and 3/7.

98. An object moves at uniform speed a from A to B and then travels at uniform speed b from B to A. Show that the average speed in making the round trip is $\dfrac{2ab}{a+b}$, the harmonic mean between a and b. Calculate the average speed if $a = 30$ ft/sec and $b = 60$ ft/sec.

ANSWERS TO SUPPLEMENTARY PROBLEMS.

56. a) $l = 595$, $S = 29800$ c) $l = 52$, $S = 520$ e) $l = 57$, $S = 1110$
   b) $l = 79$, $S = 931\frac{1}{2}$ d) $l = 62$, $S = 512$ f) $l = x + 28y$, $S = 30x + 405y$

57. a) $\dfrac{n(n+1)}{2}$ b) $n(3n-1)$ c) $\dfrac{n(7n-1)}{4}$

58. $n = 13$, $d = 5/2$ 59. $a = 4$, $S = 784$ 60. 2378   61. 50800   62. 25

63. 12, 16, 20   64. 10 sec   65. $667.95   66. $\dfrac{n+1}{2}$

67. 5, 9, 13, 17, ... nth term $= 4n + 1$ 68. a) 28,   b) 7/2,   c) $3 + \sqrt{3}$,   d) $3x - y/2$

69. *a*) 12, 15, 18, 21   *c*) $x+4y$, $x+6y$, $x+8y$
    b) 3, 7   *d*) The A.P. is   5, 8, 11, 14, 17, 20, 23, 26.

70. *a*) $l = 81/8$, $S = 211/8$ *d*) $l = 2187$, $S = 3280$
    b) $l = 1536$, $S = 1026$ *e*) $l = 1/256$, $S = 4095/256$
    c) $l = 1/512$, $S = 1023/512$ *f*) $l = 81$, $S = 120 + 40\sqrt{3}$

71. *a*) $\frac{3}{2}\left[1 - (\frac{1}{3})^n\right]$ *b*) $\frac{8}{3}\left[(\frac{3}{2})^n - 1\right]$ *c*) $\dfrac{1 - (-2)^n}{3}$

72. $n = 5$, $S = 93$ 74. $2, 6, 18, \ldots$ and $S = 242$; $\ 4, 8, 16, \ldots$ and $S = 124$

75. $4, 6, 9, 27/2$ 77. $\dfrac{(a - 2b)(b^{2n} - 1)}{b^2 - 1}$ 78. $2/3$, 2, 6, 18, 54   79. 2, 8, 32

80. 844   81. 192 lb   82. $2 \cdot 3^{n-1}$ 83. 2, 4, 8

84. *a*) 6   *b*) $2\sqrt{6}$ *c*) -8 *d*) $2a + 2b$

85. *a*) 12, 48   *b*) 2, $2\sqrt{2}$, 4, $4\sqrt{2}$ *c*) $32/3$

86. 187 or 22   87. 19   88. 8, -4, 2, 8   89. 18, 50

90. *a*) $9/2$ *b*) 8   *c*) $4/3$ *d*) $9/2$ *e*) $12/5$ *f*) $10/9$

91. $3/4$, $1/2$, $1/3$ 92. $3/2$, $3/4$, $3/8$ 93. 108 in.

94. *a*) $4/33$ *b*) $1/11$ *c*) $10/37$ *d*) $47/33$ *e*) $3/22$ *f*) $3/7$

95. *a*) $1/5$ *b*) 2   *c*) $\dfrac{10}{2n + 1}$

96. *a*) 4   *b*) $2/5$ *c*) $6\sqrt{2} - 4\sqrt{3}$ *d*) $\dfrac{a^2 - b^2}{a}$

97. *a*) 6, $15/2$ *b*) 1, $3/4$, $3/5$, $1/2$ 98. 40 ft/sec

CHAPTER 17

The Binomial Theorem

FACTORIAL NOTATION. The following identities indicate the meaning of "factorial n", or $n!$.

$$2! = 1 \cdot 2 = 2, \qquad 3! = 1 \cdot 2 \cdot 3 = 6, \qquad 4! = 1 \cdot 2 \cdot 3 \cdot 4 = 24$$

$$5! = 1 \cdot 2 \cdot 3 \cdot 4 \cdot 5 = 120, \qquad n! = 1 \cdot 2 \cdot 3 \dots n, \qquad (r-1)! = 1 \cdot 2 \cdot 3 \dots (r-1)$$

EXPANSION OF $(a+x)^n$.

A) For Positive Integral Values of n.

$$(a+x)^n = a^n + na^{n-1}x + \frac{n(n-1)}{2!}a^{n-2}x^2 + \frac{n(n-1)(n-2)}{3!}a^{n-3}x^3$$

$$+ \dots + \frac{n(n-1)(n-2)\dots(n-r+2)}{(r-1)!}a^{n-r+1}x^{r-1} + \dots + x^n$$

This equation is called the binomial theorem, or binomial formula.

The rth term of the expansion of $(a+x)^n$ is

$$r\text{th term} = \frac{n(n-1)(n-2)\dots(n-r+2)}{(r-1)!}a^{n-r+1}x^{r-1}.$$

B) For Negative and Fractional Values of n.

The binomial theorem holds also for negative and fractional values of n provided the binomial is expanded in the form $(a+x)^n$, where x is numerically less than a. However, when n is a negative or fractional number the expansion does not terminate.

SOLVED PROBLEMS

1. Evaluate.

  a) $\dfrac{7!}{4!} = \dfrac{1 \cdot 2 \cdot 3 \cdot 4 \cdot 5 \cdot 6 \cdot 7}{1 \cdot 2 \cdot 3 \cdot 4} = 5 \cdot 6 \cdot 7 = 210$ *b)* $\dfrac{6!\,3!}{5!} = \dfrac{(1 \cdot 2 \cdot 3 \cdot 4 \cdot 5 \cdot 6)(1 \cdot 2 \cdot 3)}{1 \cdot 2 \cdot 3 \cdot 4 \cdot 5} = 36$

  c) $\dfrac{m!}{(m-1)!} = \dfrac{1 \cdot 2 \cdot 3 \dots (m-2)(m-1)m}{1 \cdot 2 \cdot 3 \dots (m-2)(m-1)} = m.$ Thus $\dfrac{15!}{14!} = 15.$

d) $\dfrac{n!}{(n-2)!} = \dfrac{1\cdot2\cdot3\ldots(n-3)(n-2)(n-1)n}{1\cdot2\cdot3\ldots(n-3)(n-2)} = n(n-1).$ Thus $\dfrac{18!}{16!} = 18\cdot17 = 306.$

e) $\dfrac{(p-2)!}{(p-4)!} = \dfrac{1\cdot2\cdot3\ldots(p-4)(p-3)(p-2)}{1\cdot2\cdot3\ldots(p-4)} = (p-3)(p-2)$

f) $\dfrac{a!\,(a+b-2)!}{(a-2)!\,(a+b)!} = \dfrac{[a(a-1)(a-2)!]\,(a+b-2)!}{(a-2)!\,[(a+b)(a+b-1)(a+b-2)!]} = \dfrac{a(a-1)}{(a+b)(a+b-1)}$

BINOMIAL EXPANSION FOR POSITIVE INTEGRAL EXPONENTS.

Expand by the binomial formula.

2. $(a+x)^3 = a^3 + 3a^2x + \dfrac{3\cdot2}{1\cdot2}ax^2 + \dfrac{3\cdot2\cdot1}{1\cdot2\cdot3}x^3 = a^3 + 3a^2x + 3ax^2 + x^3$

3. $(a+x)^4 = a^4 + 4a^3x + \dfrac{4\cdot3}{1\cdot2}a^2x^2 + \dfrac{4\cdot3\cdot2}{1\cdot2\cdot3}ax^3 + \dfrac{4\cdot3\cdot2\cdot1}{1\cdot2\cdot3\cdot4}x^4$

$\qquad = a^4 + 4a^3x + 6a^2x^2 + 4ax^3 + x^4$

4. $(a+x)^5 = a^5 + 5a^4x + \dfrac{5\cdot4}{1\cdot2}a^3x^2 + \dfrac{5\cdot4\cdot3}{1\cdot2\cdot3}a^2x^3 + \dfrac{5\cdot4\cdot3\cdot2}{1\cdot2\cdot3\cdot4}ax^4 + x^5$

$\qquad = a^5 + 5a^4x + 10a^3x^2 + 10a^2x^3 + 5ax^4 + x^5$

Note that in the expansion of $(a+x)^n$:

1) The exponent of a + the exponent of x = n (i.e. the degree of each term is n).

2) The number of terms is $n+1$, when n is a positive integer.

3) There are *two* middle terms when n is an odd positive integer.

4) There is only *one* middle term when n is an even positive integer.

5) The coefficients of the terms which are equidistant from the ends are the same. It is interesting to note that these coefficients may be arranged as follows.

$$
\begin{array}{llccccccccc}
(a+x)^0 & & & & & & 1 & & & & \\
(a+x)^1 & & & & & 1 & & 1 & & & \\
(a+x)^2 & & & & 1 & & 2 & & 1 & & \\
(a+x)^3 & & & 1 & & 3 & & 3 & & 1 & \\
(a+x)^4 & & 1 & & 4 & & 6 & & 4 & & 1 \\
(a+x)^5 & 1 & & 5 & & 10 & & 10 & & 5 & & 1 \\
(a+x)^6 & 1 & 6 & & 15 & & 20 & & 15 & & 6 & 1
\end{array}
$$

etc.

This array of numbers is known as *Pascal's Triangle*. The first and last numbers in each row are 1, while any other number in the array can be obtained by adding the two numbers to the right and left of it in the preceding row.

5. $(x-y^2)^6 = x^6 + 6x^5(-y^2) + \dfrac{6\cdot5}{1\cdot2}x^4(-y^2)^2 + \dfrac{6\cdot5\cdot4}{1\cdot2\cdot3}x^3(-y^2)^3 + \dfrac{6\cdot5\cdot4\cdot3}{1\cdot2\cdot3\cdot4}x^2(-y^2)^4$

$\qquad + \dfrac{6\cdot5\cdot4\cdot3\cdot2}{1\cdot2\cdot3\cdot4\cdot5}x(-y^2)^5 + (-y^2)^6$

$\qquad = x^6 - 6x^5y^2 + 15x^4y^4 - 20x^3y^6 + 15x^2y^8 - 6xy^{10} + y^{12}$

In the expansion of a binomial of the form $(a - b)^n$, where n is a positive integer, the terms are alternately + and −.

6. $(3a^3 - 2b)^4 = (3a^3)^4 + 4(3a^3)^3(-2b) + \dfrac{4 \cdot 3}{1 \cdot 2}(3a^3)^2(-2b)^2 + \dfrac{4 \cdot 3 \cdot 2}{1 \cdot 2 \cdot 3}(3a^3)(-2b)^3 + (-2b)^4$

$\qquad = 81a^{12} - 216a^9 b + 216a^6 b^2 - 96a^3 b^3 + 16b^4$

7. $(x - 1)^7 = x^7 + 7x^6(-1) + \dfrac{7 \cdot 6}{1 \cdot 2}x^5(-1)^2 + \dfrac{7 \cdot 6 \cdot 5}{1 \cdot 2 \cdot 3}x^4(-1)^3 + \dfrac{7 \cdot 6 \cdot 5 \cdot 4}{1 \cdot 2 \cdot 3 \cdot 4}x^3(-1)^4$

$\qquad + \dfrac{7 \cdot 6 \cdot 5 \cdot 4 \cdot 3}{1 \cdot 2 \cdot 3 \cdot 4 \cdot 5}x^2(-1)^5 + \dfrac{7 \cdot 6 \cdot 5 \cdot 4 \cdot 3 \cdot 2}{1 \cdot 2 \cdot 3 \cdot 4 \cdot 5 \cdot 6}x(-1)^6 + (-1)^7$

$\qquad = x^7 - 7x^6 + 21x^5 - 35x^4 + 35x^3 - 21x^2 + 7x - 1$

8. $(1 - 2x)^5 = 1 + 5(-2x) + \dfrac{5 \cdot 4}{1 \cdot 2}(-2x)^2 + \dfrac{5 \cdot 4 \cdot 3}{1 \cdot 2 \cdot 3}(-2x)^3 + \dfrac{5 \cdot 4 \cdot 3 \cdot 2}{1 \cdot 2 \cdot 3 \cdot 4}(-2x)^4 + (-2x)^5$

$\qquad = 1 - 10x + 40x^2 - 80x^3 + 80x^4 - 32x^5$

9. $\left(\dfrac{x}{3} + \dfrac{2}{y}\right)^4 = \left(\dfrac{x}{3}\right)^4 + 4\left(\dfrac{x}{3}\right)^3\left(\dfrac{2}{y}\right) + \dfrac{4 \cdot 3}{1 \cdot 2}\left(\dfrac{x}{3}\right)^2\left(\dfrac{2}{y}\right)^2 + \dfrac{4 \cdot 3 \cdot 2}{1 \cdot 2 \cdot 3}\left(\dfrac{x}{3}\right)\left(\dfrac{2}{y}\right)^3 + \left(\dfrac{2}{y}\right)^4$

$\qquad = \dfrac{x^4}{81} + \dfrac{8x^3}{27y} + \dfrac{8x^2}{3y^2} + \dfrac{32x}{3y^3} + \dfrac{16}{y^4}$

10. $(\sqrt{x} + \sqrt{y})^6 = (x^{1/2})^6 + 6(x^{1/2})^5(y^{1/2}) + \dfrac{6 \cdot 5}{1 \cdot 2}(x^{1/2})^4(y^{1/2})^2 + \dfrac{6 \cdot 5 \cdot 4}{1 \cdot 2 \cdot 3}(x^{1/2})^3(y^{1/2})^3$

$\qquad + \dfrac{6 \cdot 5 \cdot 4 \cdot 3}{1 \cdot 2 \cdot 3 \cdot 4}(x^{1/2})^2(y^{1/2})^4 + \dfrac{6 \cdot 5 \cdot 4 \cdot 3 \cdot 2}{1 \cdot 2 \cdot 3 \cdot 4 \cdot 5}(x^{1/2})(y^{1/2})^5 + (y^{1/2})^6$

$\qquad = x^3 + 6x^{5/2}y^{1/2} + 15x^2 y + 20x^{3/2}y^{3/2} + 15xy^2 + 6x^{1/2}y^{5/2} + y^3$

11. $\left(\sqrt{x} - \dfrac{1}{\sqrt{x}}\right)^4 = (x^{1/2} - x^{-1/2})^4$

$\qquad = (x^{1/2})^4 + 4(x^{1/2})^3(-x^{-1/2}) + \dfrac{4 \cdot 3}{1 \cdot 2}(x^{1/2})^2(-x^{-1/2})^2$

$\qquad + \dfrac{4 \cdot 3 \cdot 2}{1 \cdot 2 \cdot 3}(x^{1/2})(-x^{-1/2})^3 + (-x^{-1/2})^4 = x^2 - 4x + 6 - 4x^{-1} + x^{-2}$

12. $(a^{-2} + b^{3/2})^4 = (a^{-2})^4 + 4(a^{-2})^3(b^{3/2}) + \dfrac{4 \cdot 3}{1 \cdot 2}(a^{-2})^2(b^{3/2})^2 + \dfrac{4 \cdot 3 \cdot 2}{1 \cdot 2 \cdot 3}(a^{-2})(b^{3/2})^3 + (b^{3/2})^4$

$\qquad = a^{-8} + 4a^{-6}b^{3/2} + 6a^{-4}b^3 + 4a^{-2}b^{9/2} + b^6$

13. $(e^x - e^{-x})^7 = (e^x)^7 + 7(e^x)^6(-e^{-x}) + \dfrac{7 \cdot 6}{1 \cdot 2}(e^x)^5(-e^{-x})^2 + \dfrac{7 \cdot 6 \cdot 5}{1 \cdot 2 \cdot 3}(e^x)^4(-e^{-x})^3$

$\qquad + \dfrac{7 \cdot 6 \cdot 5 \cdot 4}{1 \cdot 2 \cdot 3 \cdot 4}(e^x)^3(-e^{-x})^4 + \dfrac{7 \cdot 6 \cdot 5 \cdot 4 \cdot 3}{1 \cdot 2 \cdot 3 \cdot 4 \cdot 5}(e^x)^2(-e^{-x})^5$

$\qquad + \dfrac{7 \cdot 6 \cdot 5 \cdot 4 \cdot 3 \cdot 2}{1 \cdot 2 \cdot 3 \cdot 4 \cdot 5 \cdot 6}(e^x)(-e^{-x})^6 + (-e^{-x})^7$

$\qquad = e^{7x} - 7e^{5x} + 21e^{3x} - 35e^x + 35e^{-x} - 21e^{-3x} + 7e^{-5x} - e^{-7x}$

14. $(a+b-c)^3 = [(a+b)-c]^3 = (a+b)^3 + 3(a+b)^2(-c) + \frac{3\cdot 2}{1\cdot 2}(a+b)(-c)^2 + (-c)^3$

$\qquad = a^3 + 3a^2b + 3ab^2 + b^3 - 3a^2c - 6abc - 3b^2c + 3ac^2 + 3bc^2 - c^3$

15. $(x^2+x-3)^3 = [x^2+(x-3)]^3 = (x^2)^3 + 3(x^2)^2(x-3) + \frac{3\cdot 2}{1\cdot 2}(x^2)(x-3)^2 + (x-3)^3$

$\qquad = x^6 + (3x^5 - 9x^4) + (3x^4 - 18x^3 + 27x^2) + (x^3 - 9x^2 + 27x - 27)$

$\qquad = x^6 + 3x^5 - 6x^4 - 17x^3 + 18x^2 + 27x - 27$

In Problems 16-21 write the indicated term of each expansion, using the formula

$$r\text{th term of } (a+x)^n = \frac{n(n-1)(n-2)\ldots(n-r+2)}{(r-1)!}\, a^{n-r+1} x^{r-1}.$$

16. Sixth term of $(x+y)^{15}$.　　　$n = 15,\ r = 6,\ n-r+2 = 11,\ r-1 = 5,\ n-r+1 = 10$

$\qquad$ 6th term $= \dfrac{15\cdot 14\cdot 13\cdot 12\cdot 11}{1\cdot 2\cdot 3\cdot 4\cdot 5} x^{10} y^5 = 3003x^{10}y^5$

17. Fifth term of $(a-\sqrt{b})^9$.　　　$n = 9,\ r = 5,\ n-r+2 = 6,\ r-1 = 4,\ n-r+1 = 5$

$\qquad$ 5th term $= \dfrac{9\cdot 8\cdot 7\cdot 6}{1\cdot 2\cdot 3\cdot 4} a^5(-\sqrt{b})^4 = 126a^5 b^2$

18. Fourth term of $(x^2-y^2)^{11}$.　　$n = 11,\ r = 4,\ n-r+2 = 9,\ r-1 = 3,\ n-r+1 = 8$

$\qquad$ 4th term $= \dfrac{11\cdot 10\cdot 9}{1\cdot 2\cdot 3}(x^2)^8(-y^2)^3 = -165x^{16}y^6$

19. Ninth term of $(\frac{x}{2}+\frac{1}{x})^{12}$.　　　$n = 12,\ r = 9,\ n-r+2 = 5,\ r-1 = 8,\ n-r+1 = 4$

$\qquad$ 9th term $= \dfrac{12\cdot 11\cdot 10\cdot 9\cdot 8\cdot 7\cdot 6\cdot 5}{1\cdot 2\cdot 3\cdot 4\cdot 5\cdot 6\cdot 7\cdot 8}(\frac{x}{2})^4(\frac{1}{x})^8 = \dfrac{495}{16x^4}$

20. Eighteenth term of $(1-\frac{1}{x})^{20}$. $n = 20,\ r = 18,\ n-r+2 = 4,\ r-1 = 17,\ n-r+1 = 3$

$\qquad$ 18th term $= \dfrac{20\cdot 19\cdot 18\cdot 17\ldots 4}{1\cdot 2\cdot 3\cdot 4\ldots 17}(-\frac{1}{x})^{17} = -\dfrac{20\cdot 19\cdot 18}{1\cdot 2\cdot 3x^{17}} = -\dfrac{1140}{x^{17}}$

21. Middle (4th) term of $(x^{1/3}-\frac{1}{2}x^{-2})^6$. $n = 6,\ r = 4,\ n-r+2 = 4,\ r-1 = 3,\ n-r+1 = 3$

$\qquad$ 4th term $= \dfrac{6\cdot 5\cdot 4}{1\cdot 2\cdot 3}(x^{1/3})^3(-\frac{1}{2}x^{-2})^3 = 20(x)(-\frac{1}{8}x^{-6}) = -\dfrac{5}{2x^5}$

22. Find the term involving x^2 in the expansion of $(x^3+\frac{a}{x})^{10}$.

$\qquad$ From $(x^3)^{10-r+1}(x^{-1})^{r-1} = x^2$ we obtain $3(10-r+1) - 1(r-1) = 2$ or $r = 8$.

$\qquad$ For the 8th term: $n = 10,\ r = 8,\ n-r+2 = 4,\ r-1 = 7,\ n-r+1 = 3$.

$\qquad$ 8th term $= \dfrac{10\cdot 9\cdot 8\cdot 7\cdot 6\cdot 5\cdot 4}{1\cdot 2\cdot 3\cdot 4\cdot 5\cdot 6\cdot 7}(x^3)^3(\frac{a}{x})^7 = 120a^7x^2$

23. Find the term independent of x in the expansion of $(x^2-\frac{1}{x})^9$.

From $\ (x^2)^{9-r+1}(x^{-1})^{r-1} \ = \ x^0 \ $ we obtain $\ \ 2(9-r+1) - 1(r-1) = 0 \ \ $ or $\ \ r = 7$.

For the 7th term: $\ \ n = 9, \ \ r = 7, \ \ n-r+2 = 4, \ \ r-1 = 6, \ \ n-r+1 = 3$.

$$\text{7th term} \ = \ \frac{9 \cdot 8 \cdot 7 \cdot 6 \cdot 5 \cdot 4}{1 \cdot 2 \cdot 3 \cdot 4 \cdot 5 \cdot 6}(x^2)^3 \, (-x^{-1})^6 \ = \ 84$$

24. Evaluate $(1.03)^{10}$ to five significant figures.

$$(1.03)^{10} \ = \ (1+.03)^{10} \ = \ 1 + 10(.03) + \frac{10 \cdot 9}{1 \cdot 2}(.03)^2 + \frac{10 \cdot 9 \cdot 8}{1 \cdot 2 \cdot 3}(.03)^3 + \frac{10 \cdot 9 \cdot 8 \cdot 7}{1 \cdot 2 \cdot 3 \cdot 4}(.03)^4 + \cdots$$

$$= \ 1 + .3 + .0405 + .00324 + .00017 + \cdots \ = \ 1.3439$$

Note that all 11 terms of the expansion of $(.03+1)^{10}$ would be required in order to evaluate $(1.03)^{10}$.

25. Evaluate $(.99)^{15}$ to four decimal places.

$$(.99)^{15} \ = \ (1-.01)^{15} \ = \ 1 + 15(-.01) + \frac{15 \cdot 14}{1 \cdot 2}(-.01)^2 + \frac{15 \cdot 14 \cdot 13}{1 \cdot 2 \cdot 3}(-.01)^3$$

$$+ \ \frac{15 \cdot 14 \cdot 13 \cdot 12}{1 \cdot 2 \cdot 3 \cdot 4}(-.01)^4 + \cdots$$

$$= \ 1 - .15 + .0105 - .000455 + .000014 - \cdots \ = \ .8601$$

26. Find the sum of the coefficients in the expansion of $a)$ $(1+x)^{10}$, $\ b)$ $(1-x)^{10}$.

$a)$ If $\ 1, c_1, c_2, \ldots, c_{10} \ $ are the coefficients, we have the identity

$$(1 + x)^{10} \ = \ 1 + c_1 x + c_2 x^2 + \ldots + c_{10} x^{10}. \ \ \ \ \text{Let } x = 1.$$

Then $\ \ (1 + 1)^{10} \ = \ 1 + c_1 + c_2 + \ldots + c_{10} \ = \ $ sum of coefficients $\ = \ 2^{10} \ = \ 1024$.

$b)$ Let $x = 1$. Then $\ (1 - x)^{10} \ = \ (1 - 1)^{10} \ = \ 0 \ = \ $ sum of coefficients.

EXPANSION FOR FRACTIONAL AND NEGATIVE EXPONENTS.

In Problems 27-28 write the first four terms of the expansion.

27. $(a - x)^{1/3} \ = \ a^{1/3} + \frac{1}{3}a^{-2/3}(-x) + \frac{(1/3)(-2/3)}{1 \cdot 2 \cdot}a^{-5/3}(-x)^2 + \frac{(1/3)(-2/3)(-5/3)}{1 \cdot 2 \cdot 3}a^{-8/3}(-x)^3 + \cdots$

$$= \ a^{1/3} - \frac{1}{3}a^{-2/3}x - \frac{1}{9}a^{-5/3}x^2 - \frac{5}{81}a^{-8/3}x^3 - \cdots \ \ \ \ \ \ |x| < |a|$$

28. $(1 + x)^{3/2} \ = \ 1 + \frac{3}{2}x + \frac{(3/2)(1/2)}{1 \cdot 2}x^2 + \frac{(3/2)(1/2)(-1/2)}{1 \cdot 2 \cdot 3}x^3 + \cdots \ \ \ \ \ \ |x| < 1$

$$= \ 1 + \frac{3}{2}x + \frac{3}{8}x^2 - \frac{1}{16}x^3 + \cdots$$

29. Evaluate $\sqrt{26}$ to six significant figures.

$$\sqrt{26} = (5^2 + 1)^{1/2} = (5^2)^{1/2} + \frac{1}{2}(5^2)^{-1/2} + \frac{(1/2)(-1/2)}{1 \cdot 2}(5^2)^{-3/2}$$

$$+ \frac{(1/2)(-1/2)(-3/2)}{1 \cdot 2 \cdot 3}(5^2)^{-5/2} + \dots$$

$$= 5 + \frac{1}{2}(\frac{1}{5}) - \frac{1}{8}(\frac{1}{5^3}) + \frac{1}{16}(\frac{1}{5^5}) - \dots = 5 + \frac{1}{10} - \frac{1}{1000} + \frac{1}{50,000} - \dots$$

$$= 5 + .1 - .001 + .00002 - \dots = 5.09902$$

Note. It is incorrect to evaluate $\sqrt{26}$ by expanding $(1 + 5^2)^{1/2}$. Why?

30. Evaluate $\sqrt[3]{998}$ to six significant figures.

$$\sqrt[3]{998} = (10^3 - 2)^{1/3} = (10^3)^{1/3} + \frac{1}{3}(10^3)^{-2/3}(-2) + \frac{(1/3)(-2/3)}{1 \cdot 2}(10^3)^{-5/3}(-2)^2 + \dots$$

$$= 10 - \frac{2}{3}(\frac{1}{10^2}) - \frac{4}{9}(\frac{1}{10^5}) - \dots = 10 - .006667 - .000004 - \dots = 9.99333$$

31. Write the sixth term of the expansion of $(1 - x^2)^{1/2}$.

$$n = 1/2, \quad r = 6, \quad n - r + 2 = -7/2, \quad r - 1 = 5, \quad n - r + 1 = -9/2$$

$$\text{6th term} = \frac{(1/2)(-1/2)(-3/2)(-5/2)(-7/2)}{1 \cdot 2 \cdot 3 \cdot 4 \cdot 5}(-x^2)^5 = -\frac{7}{256}x^{10}$$

In Problems 32-36 write the first four terms of the expansion.

32. $(a + x)^{-1} = a^{-1} + (-1)a^{-2}x + \frac{(-1)(-2)}{1 \cdot 2}a^{-3}x^2 + \frac{(-1)(-2)(-3)}{1 \cdot 2 \cdot 3}a^{-4}x^3 + \dots$

$\qquad = a^{-1} - a^{-2}x + a^{-3}x^2 - a^{-4}x^3 + \dots \qquad\qquad$ valid for $|x| < |a|$

33. $(1 - x)^{-2} = 1 + (-2)(-x) + \frac{(-2)(-3)}{1 \cdot 2}(-x)^2 + \frac{(-2)(-3)(-4)}{1 \cdot 2 \cdot 3}(-x)^3 + \dots$

$\qquad = 1 + 2x + 3x^2 + 4x^3 + \dots \qquad\qquad$ valid for $|x| < 1$

34. $(1 - x)^{-4} = 1 + (-4)(-x) + \frac{(-4)(-5)}{1 \cdot 2}(-x)^2 + \frac{(-4)(-5)(-6)}{1 \cdot 2 \cdot 3}(-x)^3 + \dots$

$\qquad = 1 + 4x + 10x^2 + 20x^3 + \dots$

35. $(2 + x)^{-3} = 2^{-3} + (-3)(2^{-4})x + \frac{(-3)(-4)}{1 \cdot 2}(2^{-5})x^2 + \frac{(-3)(-4)(-5)}{1 \cdot 2 \cdot 3}(2^{-6})x^3 + \dots$

$\qquad = \frac{1}{8} - \frac{3}{16}x + \frac{3}{16}x^2 - \frac{5}{32}x^3 + \dots \qquad\qquad$ valid for $|x| < 2$

36. $(9 + x)^{-1/2} = 9^{-1/2} + (-1/2)(9^{-3/2})x + \frac{(-1/2)(-3/2)}{1 \cdot 2}(9^{-5/2})x^2 \qquad$ valid for $|x| < 9$

$\qquad + \frac{(-1/2)(-3/2)(-5/2)}{1 \cdot 2 \cdot 3}(9^{-7/2})x^3 + \dots = \frac{1}{3} - \frac{x}{54} + \frac{x^2}{648} - \frac{5x^3}{34,992} + \dots$

37. Write the fifth term of the expansion of $(\sqrt{x/y} - \sqrt{y/x})^{-4}$.

$n = -4, \quad r = 5, \quad n - r + 2 = -7, \quad r - 1 = 4, \quad n - r + 1 = -8$

$$\text{5th term} = \frac{(-4)(-5)(-6)(-7)}{1 \cdot 2 \cdot 3 \cdot 4}\left(\frac{x^{1/2}}{y^{1/2}}\right)^{-8}\left(-\frac{y^{1/2}}{x^{1/2}}\right)^{4} = 35x^{-6}y^{6}$$

38. Write the first four terms of the expansion of $(1 + \frac{1}{n})^{n}$.

$$(1 + \frac{1}{n})^{n} = 1 + n(\frac{1}{n}) + \frac{n(n-1)}{1 \cdot 2}(\frac{1}{n})^{2} + \frac{n(n-1)(n-2)}{1 \cdot 2 \cdot 3}(\frac{1}{n})^{3} + \cdots$$

$$= 1 + 1 + \frac{1}{2!}\frac{n(n-1)}{n^{2}} + \frac{1}{3!}\frac{n(n-1)(n-2)}{n^{3}} + \cdots$$

SUPPLEMENTARY PROBLEMS

39. Evaluate each of the following.

a) $\dfrac{6!}{3!}$ b) $\dfrac{8!}{5!}$ c) $\dfrac{10!}{6!\,4!}$ d) $\dfrac{12!}{7!\,3!\,2!}$ e) $\dfrac{(2n+1)!}{(2n-1)!}$ f) $\dfrac{n!}{(n-3)!\,3!}$ g) $\dfrac{(n-1)!\,4!}{(n+2)!}$

40. Expand by the binomial formula.

a) $(x + \frac{1}{2})^{6}$ c) $(y + 3)^{4}$ e) $(x^{2} - y^{3})^{4}$ g) $(\frac{x}{2} + \frac{3}{y})^{4}$

b) $(x - 2)^{5}$ d) $(x + \frac{1}{x})^{5}$ f) $(a - 2b)^{6}$ h) $(y^{1/2} + y^{-1/2})^{6}$

41. Write the indicated term in the expansion of each of the following.

a) Fifth term of $(a - b)^{7}$

b) Seventh term of $(x^{2} - \frac{1}{x})^{9}$

c) Middle term of $(y - \frac{1}{y})^{8}$

d) Eighth term of $(\frac{x^{2}}{2} - 2y)^{16}$

e) Seventh term of $(a - \frac{1}{\sqrt{a}})^{10}$

f) Sixteenth term of $(2 - 1/x)^{18}$

g) Sixth term of $(x^{2} - 2y)^{11}$

h) Eleventh term of $(x + \frac{1}{\sqrt{x}})^{14}$

42. Find the term independent of x in the expansion of $(\sqrt{x} + \frac{1}{3x^{2}})^{10}$.

43. Find the term involving x^{3} in the expansion of $(x^{2} + \frac{1}{x})^{12}$.

44. Evaluate $(0.98)^{6}$ correct to five decimal places.

45. Evaluate $(1.1)^{10}$ correct to the nearest hundredth.

46. Write the first four terms of the expansion of each of the following.

a) $(1 + 2x)^{-1}$ c) $(a + x)^{-1/2}$ e) $(1 + x)^{1/3}$ g) $(4 + x)^{1/2}$

b) $(1 - y)^{-3}$ d) $(1 + r)^{-6}$ f) $(1 - 3y)^{-1/3}$ h) $(1 + \frac{1}{s})^{-2/3}$

47. Evaluate to the indicated accuracy.

 a) $\sqrt{17}$; four significant figures   *c)* $\sqrt[5]{34}$; nearest hundredth

 b) $\sqrt[3]{28}$; three decimal places   *d)* $\sqrt[5]{1.01}$; five decimal places

48. Write the sixth term of the expansion of $(x + \frac{2}{x})^{-5}$.

ANSWERS TO SUPPLEMENTARY PROBLEMS.

39. *a)* 120   *b)* 336   *c)* 210   *d)* 7920   *e)* $2n(2n+1)$ *f)* $\dfrac{n(n-1)(n-2)}{6}$ *g)* $\dfrac{24}{n(n+1)(n+2)}$

40. *a)* $x^6 + 3x^5 + \dfrac{15}{4}x^4 + \dfrac{5}{2}x^3 + \dfrac{15}{16}x^2 + \dfrac{3}{16}x + \dfrac{1}{64}$

 b) $x^5 - 10x^4 + 40x^3 - 80x^2 + 80x - 32$

 c) $y^4 + 12y^3 + 54y^2 + 108y + 81$

 d) $x^5 + 5x^3 + 10x + \dfrac{10}{x} + \dfrac{5}{x^3} + \dfrac{1}{x^5}$ *e)* $x^8 - 4x^6y^3 + 6x^4y^6 - 4x^2y^9 + y^{12}$

 f) $a^6 - 12a^5b + 60a^4b^2 - 160a^3b^3 + 240a^2b^4 - 192ab^5 + 64b^6$

 g) $\dfrac{x^4}{16} + \dfrac{3x^3}{2y} + \dfrac{27x^2}{2y^2} + \dfrac{54x}{y^3} + \dfrac{81}{y^4}$ *h)* $y^3 + 6y^2 + 15y + 20 + 15y^{-1} + 6y^{-2} + y^{-3}$

41. *a)* $35a^3b^4$ *c)* 70   *e)* $210a$ *g)* $-14784x^{12}y^5$

 b) 84   *d)* $-2860x^{18}y^7$ *f)* $-\dfrac{6528}{x^{15}}$ *h)* $\dfrac{1001}{x}$

42. 5   **43.** $792x^3$ **44.** 0.88584   **45.** 2.59

46. *a)* $1 - 2x + 4x^2 - 8x^3 + \cdots$ $|x| < \frac{1}{2}$ *e)* $1 + \dfrac{x}{3} - \dfrac{x^2}{9} + \dfrac{5x^3}{81} - \cdots$

 b) $1 + 3y + 6y^2 + 10y^3 + \cdots$ $|y| < 1$ *f)* $1 + y + 2y^2 + \dfrac{14}{3}y^3 + \cdots$

 c) $a^{-1/2} - \dfrac{1}{2}a^{-3/2}x + \dfrac{3}{8}a^{-5/2}x^2 - \dfrac{5}{16}a^{-7/2}x^3 + \cdots$

 d) $1 - 6r + 21r^2 - 56r^3 + \cdots$ *g)* $2 + \dfrac{x}{4} - \dfrac{x^2}{64} + \dfrac{x^3}{512} - \cdots$

          h) $1 - \dfrac{2}{3s} + \dfrac{5}{9s^2} - \dfrac{40}{81s^3} + \cdots$ $|s| > 1$

47. *a)* 4.123   *b)* 3.037   *c)* 2.02   *d)* 1.00199

48. $-\dfrac{4032}{x^{15}}$

CHAPTER 18

Mathematical Induction

MATHEMATICAL INDUCTION is the process of proving a general theorem or formula from particular cases. There are two distinct steps in the proof.

1) Show by actual substitution that the proposed theorem or formula is true for some one positive integral value of n, as $n = 1$, or $n = 2$, etc.

2) Assume that the theorem or formula is true for $n = k$. Then prove that it is true for $n = k + 1$.

SOLVED PROBLEMS

1. Prove by mathematical induction that, for all positive integral values of n,

$$1 + 2 + 3 + \ldots + n = \frac{n(n+1)}{2}.$$

Step 1. The formula is true for $n = 1$, since $1 = \frac{1(1+1)}{2} = 1$.

Step 2. Assume that the formula is true for $n = k$. Then, adding $(k+1)$ to both sides,

$$1 + 2 + 3 + \ldots + k + (k+1) = \frac{k(k+1)}{2} + (k+1) = \frac{(k+1)(k+2)}{2}$$

which is the value of $\frac{n(n+1)}{2}$ when $(k+1)$ is substituted for n.

Hence if the formula is true for $n = k$, we have proved it to be true for $n = k + 1$. But the formula holds for $n = 1$; hence it holds for $n = 1 + 1 = 2$. Then, since it holds for $n = 2$, it holds for $n = 2 + 1 = 3$, and so on. Thus the formula is true for all positive integral values of n.

2. Prove by mathematical induction that the sum of n terms of an arithmetic progression a, $a + d$, $a + 2d$, $\ldots$ is $\frac{n}{2}[2a + (n-1)d]$, that is

$$a + (a+d) + (a+2d) + \ldots + [a + (n-1)d] = \frac{n}{2}[2a + (n-1)d].$$

Step 1. The formula holds for $n = 1$, since $a = \frac{1}{2}[2a + (1-1)d] = a$.

Step 2. Assume that the formula holds for $n = k$. Then

$$a + (a+d) + (a+2d) + \ldots + [a + (k-1)d] = \frac{k}{2}[2a + (k-1)d].$$

Add the $(k+1)$th term, which is $(a+kd)$, to both sides of the latter equation.   Then

$$a + (a+d) + (a+2d) + \ldots + [a+(k-1)d] + (a+kd) \;\; = \;\; \frac{k}{2}[2a+(k-1)d] + (a+kd).$$

The right hand side of this equation $= ka + \dfrac{k^2 d}{2} - \dfrac{kd}{2} + a + kd \;\; = \;\; \dfrac{k^2 d + kd + 2ka + 2a}{2}$

$$= \;\; \frac{kd(k+1) + 2a(k+1)}{2} \;\; = \;\; \frac{k+1}{2}(2a+kd)$$

which is the value of $\dfrac{n}{2}[2a+(n-1)d]$ when n is replaced by $(k+1)$.

Hence if the formula is true for $n = k$, we have proved it to be true for $n = k+1$. But the formula holds for $n = 1$; hence it holds for $n = 1+1 = 2$. Then, since it holds for $n = 2$, it holds for $n = 2+1 = 3$, and so on. Thus the formula is true for all positive integral values of n.

3. Prove by mathematical induction that, for all positive integral values of n,
$$1^2 + 2^2 + 3^2 + \ldots + n^2 \;\; = \;\; \frac{n(n+1)(2n+1)}{6}.$$

Step 1.   The formula is true for $n = 1$, since $1^2 = \dfrac{1(1+1)(2+1)}{6} = 1$.

Step 2.   Assume that the formula is true for $n = k$.   Then
$$1^2 + 2^2 + 3^2 + \ldots + k^2 \;\; = \;\; \frac{k(k+1)(2k+1)}{6}.$$

Add the $(k+1)$th term, which is $(k+1)^2$, to both sides of this equation.   Then
$$1^2 + 2^2 + 3^2 + \ldots + k^2 + (k+1)^2 \;\; = \;\; \frac{k(k+1)(2k+1)}{6} + (k+1)^2.$$

The right hand side of this equation $= \dfrac{k(k+1)(2k+1) + 6(k+1)^2}{6}$

$$= \;\; \frac{(k+1)\left[(2k^2+k)+(6k+6)\right]}{6} \;\; = \;\; \frac{(k+1)(k+2)(2k+3)}{6}$$

which is the value of $\dfrac{n(n+1)(2n+1)}{6}$ when n is replaced by $(k+1)$.

Hence if the formula is true for $n = k$, it is true for $n = k+1$.   But the formula holds for $n = 1$; hence it holds for $n = 1+1 = 2$.   Then, since it holds for $n = 2$, it holds for $n = 2+1 = 3$, and so on.   Thus the formula is true for all positive integral values of n.

4. Prove by mathematical induction that, for all positive integral values of n,
$$\frac{1}{1\cdot 3} + \frac{1}{3\cdot 5} + \frac{1}{5\cdot 7} + \ldots + \frac{1}{(2n-1)(2n+1)} \;\; = \;\; \frac{n}{2n+1}.$$

Step 1.   The formula is true for $n = 1$, since $\dfrac{1}{(2-1)(2+1)} = \dfrac{1}{2+1} = \dfrac{1}{3}$.

Step 2.   Assume that the formula is true for $n = k$.   Then
$$\frac{1}{1\cdot 3} + \frac{1}{3\cdot 5} + \frac{1}{5\cdot 7} + \ldots + \frac{1}{(2k-1)(2k+1)} \;\; = \;\; \frac{k}{2k+1}.$$

Add the $(k+1)$th term, which is $\dfrac{1}{(2k+1)(2k+3)}$, to both sides of the above equation. Then

$$\frac{1}{1\cdot 3} + \frac{1}{3\cdot 5} + \frac{1}{5\cdot 7} + \dots + \frac{1}{(2k-1)(2k+1)} + \frac{1}{(2k+1)(2k+3)} = \frac{k}{2k+1} + \frac{1}{(2k+1)(2k+3)}.$$

The right hand side of this equation $= \dfrac{k(2k+3)+1}{(2k+1)(2k+3)} = \dfrac{k+1}{2k+3}$, which is the value of $\dfrac{n}{2n+1}$ when n is replaced by $(k+1)$.

Hence if the formula is true for $n = k$, it is true for $n = k+1$. But the formula holds for $n = 1$; hence it holds for $n = 1+1 = 2$. Then, since it holds for $n = 2$, it holds for $n = 2+1 = 3$, and so on. Thus the formula is true for all positive integral values of n.

5. Prove by mathematical induction that $a^{2n} - b^{2n}$ is divisible by $a+b$ when n is any positive integer.

Step 1. The theorem is true for $n = 1$, since $a^2 - b^2 = (a+b)(a-b)$.

Step 2. Assume that the theorem is true for $n = k$. Then

$$a^{2k} - b^{2k} \text{ is divisible by } a+b.$$

We must show that $a^{2k+2} - b^{2k+2}$ is divisible by $a+b$. From the identity

$$a^{2k+2} - b^{2k+2} = a^2(a^{2k} - b^{2k}) + b^{2k}(a^2 - b^2)$$

it follows that $a^{2k+2} - b^{2k+2}$ is divisible by $a+b$ if $a^{2k} - b^{2k}$ is.

Hence if the theorem is true for $n = k$, we have proved it to be true for $n = k+1$. But the theorem holds for $n = 1$; hence it holds for $n = 1+1 = 2$. Then, since it holds for $n = 2$, it holds for $n = 2+1 = 3$, and so on. Thus the theorem is true for all positive integral values of n.

6. Prove the binomial formula

$$(a+x)^n = a^n + na^{n-1}x + \frac{n(n-1)}{2!}a^{n-2}x^2 + \dots + \frac{n(n-1)\dots(n-r+2)}{(r-1)!}a^{n-r+1}x^{r-1} + \dots + x^n$$

for positive integral values of n.

Step 1. The formula is true for $n = 1$.

Step 2. Assume the formula is true for $n = k$. Then

$$(a+x)^k = a^k + ka^{k-1}x + \frac{k(k-1)}{2!}a^{k-2}x^2 + \dots + \frac{k(k-1)\dots(k-r+2)}{(r-1)!}a^{k-r+1}x^{r-1} + \dots + x^k.$$

Multiply both sides by $a+x$. The multiplication on the right may be written

$$a^{k+1} + ka^k x + \frac{k(k-1)}{2!}a^{k-1}x^2 + \dots + \frac{k(k-1)\dots(k-r+2)}{(r-1)!}a^{k-r+2}x^{r-1} + \dots + ax^k$$

$$+ a^k x + ka^{k-1}x^2 + \dots + \frac{k(k-1)\dots(k-r+3)}{(r-2)!}a^{k-r+2}x^{r-1} + \dots + x^{k+1}.$$

Since $\dfrac{k(k-1)\dots(k-r+2)}{(r-1)!}a^{k-r+2}x^{r-1} + \dfrac{k(k-1)\dots(k-r+3)}{(r-2)!}a^{k-r+2}x^{r-1}$

$$= \frac{k(k-1)\ldots(k-r+3)}{(r-2)!}a^{k-r+2}x^{r-1}\left\{\frac{k-r+2}{r-1}+1\right\} = \frac{(k+1)k(k-1)\ldots(k-r+3)}{(r-1)!}a^{k-r+2}x^{r-1},$$

the product may be written

$$(a+x)^{k+1} = a^{k+1} + (k+1)a^kx + \ldots + \frac{(k+1)k(k-1)\ldots(k-r+3)}{(r-1)!}a^{k-r+2}x^{r-1} + \ldots + x^{k+1}$$

which is the binomial formula with n replaced by $k+1$.

Hence if the formula is true for $n = k$, it is true for $n = k+1$. But the formula holds for $n = 1$; hence it holds for $n = 1+1 = 2$, and so on. Thus the formula is true for all positive integral values of n.

SUPPLEMENTARY PROBLEMS

Prove each of the following by mathematical induction. In each case n is a positive integer.

7. $1 + 3 + 5 + \ldots + (2n-1) = n^2$

8. $1 + 3 + 3^2 + \ldots + 3^{n-1} = \frac{3^n - 1}{2}$

9. $1^3 + 2^3 + 3^3 + \ldots + n^3 = \frac{n^2(n+1)^2}{4}$

10. $a + ar + ar^2 + \ldots + ar^{n-1} = \frac{a(r^n-1)}{r-1}$, $r \neq 1$

11. $\frac{1}{1\cdot 2} + \frac{1}{2\cdot 3} + \frac{1}{3\cdot 4} + \ldots + \frac{1}{n(n+1)} = \frac{n}{n+1}$

12. $1\cdot 3 + 2\cdot 3^2 + 3\cdot 3^3 + \ldots + n\cdot 3^n = \frac{(2n-1)3^{n+1} + 3}{4}$

13. $\frac{1}{2\cdot 5} + \frac{1}{5\cdot 8} + \frac{1}{8\cdot 11} + \ldots + \frac{1}{(3n-1)(3n+2)} = \frac{n}{6n+4}$

14. $\frac{1}{1\cdot 2\cdot 3} + \frac{1}{2\cdot 3\cdot 4} + \frac{1}{3\cdot 4\cdot 5} + \ldots + \frac{1}{n(n+1)(n+2)} = \frac{n(n+3)}{4(n+1)(n+2)}$

15. $a^n - b^n$ is divisible by $a-b$, for n = positive integer.

16. $a^{2n-1} + b^{2n-1}$ is divisible by $a+b$, for n = positive integer.

CHAPTER 19

Inequalities

DEFINITIONS. An *inequality* is a statement that one real quantity or expression is greater or less than another real quantity or expression.

The following indicate the meaning of inequality signs.

1) $a > b$ means "a is greater than b", (or $a - b$ is a positive number).

2) $a < b$ means "a is less than b", (or $a - b$ is a negative number).

3) $a \geq b$ means "a is greater than or equal to b."

4) $a \leq b$ means "a is less than or equal to b."

5) $0 < a < 2$ means "a is greater than zero but less than 2."

6) $-2 \leq x < 2$ means "x is greater than or equal to -2 but less than 2."

An *absolute inequality* is true for all real values of the letters involved. For example, $(a - b)^2 > -1$ holds for all real values of a and b, since the square of any real number is positive or zero.

A *conditional inequality* holds only for particular values of the letters involved. Thus $x - 5 > 3$ is true only when x is greater than 8.

The inequalities $a > b$ and $c > d$ have the *same sense*. The inequalities $a > b$ and $x < y$ have *opposite sense*.

PRINCIPLES OF INEQUALITIES.

1) The sense of an inequality is unchanged if each side is increased or decreased by the same real number. It follows that any term may be transposed from one side of an inequality to the other, provided the sign of the term is changed.

Thus if $a > b$, then $a + c > b + c$, and $a - c > b - c$, and $a - b > 0$.

2) The sense of an inequality is unchanged if each side is multiplied or divided by the same positive number.

Thus if $a > b$ and $k > 0$, then $ka > kb$ and $\dfrac{a}{k} > \dfrac{b}{k}$.

3) The sense of an inequality is reversed if each side is multiplied or divided by the same negative number.

Thus if $a > b$ and $k < 0$, then $ka < kb$ and $\dfrac{a}{k} < \dfrac{b}{k}$.

4) If $a > b$ and a, b, n are positive, then $a^n > b^n$ but $a^{-n} < b^{-n}$.

Examples. $5 > 4$; then $5^3 > 4^3$ or $125 > 64$, but $5^{-3} < 4^{-3}$ or $\dfrac{1}{125} < \dfrac{1}{64}$.

$16 > 9$; then $16^{\frac{1}{2}} > 9^{\frac{1}{2}}$ or $4 > 3$, but $16^{-\frac{1}{2}} < 9^{-\frac{1}{2}}$ or $\dfrac{1}{4} < \dfrac{1}{3}$.

5) If $a > b$ and $c > d$, then $(a+c) > (b+d)$.

6) If $a > b > 0$ and $c > d > 0$, then $ac > bd$.

SOLVED PROBLEMS

1. If $a > b$ and $c > d$, prove that $a+c > b+d$.

Since $(a-b)$ and $(c-d)$ are each positive, $(a-b)+(c-d)$ is positive.

Hence $(a-b)+(c-d) > 0$, $(a+c)-(b+d) > 0$ and $(a+c) > (b+d)$.

2. Find the fallacy.
  a) Let $a = 3$, $b = 5$; then $\qquad\qquad\qquad a < b$
  b) Multiply by a: $\qquad\qquad\qquad\qquad\quad a^2 < ab$
  c) Subtract b^2: $\qquad\qquad\qquad\qquad a^2 - b^2 < ab - b^2$
  d) Factor: $\qquad\qquad\qquad\quad (a+b)(a-b) < b(a-b)$
  e) Divide by $a-b$: $\qquad\qquad\qquad\qquad a+b < b$
  f) Substitute $a = 3$, $b = 5$: $\qquad\qquad\qquad 8 < 5$

There is nothing wrong with steps a), b), c), d). The error is made in step e) where the inequality is divided by $a-b$, a negative number, without reversing the inequality sign.

3. Find the values of x for which each of the following inequalities holds.

a) $4x + 5 > 2x + 9$.   We have $4x - 2x > 9 - 5$, $2x > 4$ and $x > 2$.

b) $\dfrac{x}{2} - \dfrac{1}{3} < \dfrac{2x}{3} + \dfrac{1}{2}$. Multiplying by 6, we obtain

$$3x - 2 < 4x + 3, \quad 3x - 4x < 2 + 3, \quad -x < 5, \quad x > -5.$$

c) $x^2 < 16$.

Method 1. $x^2 - 16 < 0$, $(x-4)(x+4) < 0$. The product of the factors $(x-4)$ and $(x+4)$ is negative. Two cases are possible.

1) $x - 4 > 0$ and $x + 4 < 0$ simultaneously. Thus $x > 4$ and $x < -4$. This is impossible, as x cannot be both greater than 4 and less than -4 simultaneously.

2) $x - 4 < 0$ and $x + 4 > 0$ simultaneously. Thus $x < 4$ and $x > -4$. This is possible if and only if $-4 < x < 4$. Hence $-4 < x < 4$.

Method 2. $(x^2)^{\frac{1}{2}} < (16)^{\frac{1}{2}}$. Now $(x^2)^{\frac{1}{2}} = x$ if $x \geqq 0$, and $(x^2)^{\frac{1}{2}} = -x$ if $x \leqq 0$.

If $x \geqq 0$, $(x^2)^{\frac{1}{2}} < (16)^{\frac{1}{2}}$ may be written $x < 4$. Hence $0 \leqq x < 4$.

If $x \leqq 0$, $(x^2)^{\frac{1}{2}} < (16)^{\frac{1}{2}}$ may be written $-x < 4$ or $x > -4$. Hence $-4 < x \leqq 0$.

Thus $0 \leqq x < 4$ and $-4 < x \leqq 0$, or $-4 < x < 4$.

4. Prove that $a^2 + b^2 > 2ab$ if a and b are real and unequal numbers.

If $a^2 + b^2 > 2ab$, then $a^2 - 2ab + b^2 > 0$ or $(a - b)^2 > 0$. This last statement is true since the square of any real number different from zero is positive.

The above provides a clue as to the method of proof. Starting with $(a - b)^2 > 0$ which we know to be true if $a \neq b$, we obtain $a^2 - 2ab + b^2 > 0$ or $a^2 + b^2 > 2ab$.

Note that the proof is essentially a reversal of the steps in the first paragraph.

5. Prove that the sum of any positive number and its reciprocal is never less than 2.

We must prove that $(a + 1/a) \geqq 2$ if $a > 0$.

If $(a + 1/a) \geqq 2$, then $a^2 + 1 \geqq 2a$, $a^2 - 2a + 1 \geqq 0$, and $(a - 1)^2 \geqq 0$ which is true.

To prove the theorem we start with $(a - 1)^2 \geqq 0$, which is known to be true.

Then $a^2 - 2a + 1 \geqq 0$, $a^2 + 1 \geqq 2a$ and $a + 1/a \geqq 2$ upon division by a.

6. Show that $a^2 + b^2 + c^2 > ab + bc + ca$ for all real values of a, b, c unless $a = b = c$.

Since $a^2 + b^2 > 2ab$, $b^2 + c^2 > 2bc$, $c^2 + a^2 > 2ca$ (see Problem 4), we have by addition

$$2(a^2 + b^2 + c^2) > 2(ab + bc + ca) \quad \text{or} \quad a^2 + b^2 + c^2 > ab + bc + ca.$$

(If $a = b = c$, then $a^2 + b^2 + c^2 = ab + bc + ca$.)

7. If $a^2 + b^2 = 1$ and $c^2 + d^2 = 1$, show that $ac + bd < 1$.

$a^2 + c^2 > 2ac$ and $b^2 + d^2 > 2bd$; hence by addition

$$(a^2 + b^2) + (c^2 + d^2) > 2ac + 2bd \quad \text{or} \quad 2 > 2ac + 2bd, \text{ i.e. } 1 > ac + bd.$$

8. Prove that $x^3 + y^3 > x^2 y + y^2 x$, if x and y are real, positive and unequal numbers.

If $x^3 + y^3 > x^2 y + y^2 x$, then $(x + y)(x^2 - xy + y^2) > xy(x + y)$. Dividing by $x + y$, which is positive,

$$x^2 - xy + y^2 > xy \quad \text{or} \quad x^2 - 2xy + y^2 > 0, \text{ i.e. } (x - y)^2 > 0 \text{ which is true if } x \neq y.$$

The steps are reversible and supply the proof. Starting with $(x - y)^2 > 0$, $x \neq y$, obtain

$$x^2 - xy + y^2 > xy.$$

Multiplying both sides by $x + y$, we have $(x + y)(x^2 - xy + y^2) > xy(x + y)$ or $x^3 + y^3 > x^2 y + y^2 x$.

9. Prove that $a^n + b^n > a^{n-1}b + ab^{n-1}$, provided a and b are positive and unequal, and $n > 1$.

If $a^n + b^n > a^{n-1}b + ab^{n-1}$, then $(a^n - a^{n-1}b) - (ab^{n-1} - b^n) > 0$ or

$$a^{n-1}(a - b) - b^{n-1}(a - b) > 0, \text{ i.e. } (a^{n-1} - b^{n-1})(a - b) > 0.$$

This is true since the factors are both positive or both negative.

Reversing the steps, which are reversible, provides the proof.

10. Prove that $a^3 + \dfrac{1}{a^3} > a^2 + \dfrac{1}{a^2}$ if $a > 0$ and $a \neq 1$.

Multiplying both sides of the inequality by a^3 (which is positive since $a > 0$), we have

$$a^6 + 1 > a^5 + a, \quad a^6 - a^5 - a + 1 > 0 \quad \text{and} \quad (a^5 - 1)(a - 1) > 0.$$

If $a > 1$ both factors are positive, while if $0 < a < 1$ both factors are negative. In either case the product is positive. (If $a = 1$ the product is zero.)

Reversal of the steps provides the proof.

11. If a,b,c,d are positive numbers and $\dfrac{a}{b} > \dfrac{c}{d}$, prove that $\dfrac{a+c}{b+d} > \dfrac{c}{d}$.

   Method 1. If $\dfrac{a+c}{b+d} > \dfrac{c}{d}$, then multiplying by $d(b+d)$ we obtain

   $(a+c)d > c(b+d)$, $ad + cd > bc + cd$, $ad > bc$ and, dividing by bd, $\dfrac{a}{b} > \dfrac{c}{d}$ which is given as true. Reversing the steps provides the proof.

   Method 2. Since $\dfrac{a}{b} > \dfrac{c}{d}$, then $\dfrac{a}{b} + \dfrac{c}{b} > \dfrac{c}{d} + \dfrac{c}{b}$, $\dfrac{a+c}{b} > \dfrac{c(b+d)}{bd}$ and $\dfrac{a+c}{b+d} > \dfrac{c}{d}$.

12. Prove: *a)* $x^2 - y^2 > x - y$ if $x + y > 1$ and $x > y$

    b) $x^2 - y^2 < x - y$ if $x + y > 1$ and $x < y$.

   a) Since $x > y$, $x - y > 0$. Multiplying both sides of $x + y > 1$ by the positive number $x - y$,
   $$(x+y)(x-y) > (x-y) \quad \text{or} \quad x^2 - y^2 > x - y.$$

   b) Since $x < y$, $x - y < 0$. Multiplying both sides of $x + y > 1$ by the negative number $x - y$ reverses the sense of the inequality; thus
   $$(x+y)(x-y) < (x-y) \quad \text{or} \quad x^2 - y^2 < x - y.$$

13. The arithmetic mean of two numbers a and b is $\dfrac{a+b}{2}$, the geometric mean is $\sqrt{ab}$, and the harmonic mean is $\dfrac{2ab}{a+b}$. Prove that $\dfrac{a+b}{2} > \sqrt{ab} > \dfrac{2ab}{a+b}$ if a and b are positive and unequal.

   a) If $\dfrac{a+b}{2} > \sqrt{ab}$, then $(a+b)^2 > (2\sqrt{ab})^2$, $a^2 + 2ab + b^2 > 4ab$, $a^2 - 2ab + b^2 > 0$ and $(a-b)^2 > 0$ which is true if $a \neq b$. Reversing the steps, we have $\dfrac{a+b}{2} > \sqrt{ab}$.

   b) If $\sqrt{ab} > \dfrac{2ab}{a+b}$, then $ab > \dfrac{4a^2 b^2}{(a+b)^2}$, $(a+b)^2 > 4ab$ and $(a-b)^2 > 0$ which is true if $a \neq b$. Reversing the steps, we have $\sqrt{ab} > \dfrac{2ab}{a+b}$.

   From *a)* and *b)*, $\dfrac{a+b}{2} > \sqrt{ab} > \dfrac{2ab}{a+b}$.

14. Find the values of x for which *a)* $x^2 - 7x + 12 = 0$, *b)* $x^2 - 7x + 12 > 0$, *c)* $x^2 - 7x + 12 < 0$.

   a) $x^2 - 7x + 12 = (x-3)(x-4) = 0$ when $x = 3$ or 4.

   b) $x^2 - 7x + 12 > 0$ or $(x-3)(x-4) > 0$ when $(x-3) > 0$ and $(x-4) > 0$ simultaneously, or when $(x-3) < 0$ and $(x-4) < 0$ simultaneously.

   $(x-3) > 0$ and $(x-4) > 0$ simultaneously when $x > 3$ and $x > 4$, i.e. when $x > 4$.

   $(x-3) < 0$ and $(x-4) < 0$ simultaneously when $x < 3$ and $x < 4$, i.e. when $x < 3$.

   Hence $x^2 - 7x + 12 > 0$ is satisfied when $x > 4$ or $x < 3$.

   c) $x^2 - 7x + 12 < 0$ or $(x-3)(x-4) < 0$ when $(x-3) > 0$ and $(x-4) < 0$ simultaneously, or when $(x-3) < 0$ and $(x-4) > 0$ simultaneously.

   $(x-3) > 0$ and $(x-4) < 0$ simultaneously when $x > 3$ and $x < 4$, i.e. when $3 < x < 4$.

   $(x-3) < 0$ and $(x-4) > 0$ simultaneously when $x < 3$ and $x > 4$, which is absurd.

   Hence $x^2 - 7x + 12 < 0$ is satisfied when $3 < x < 4$.

15. Determine graphically the range of values of x defined
    by

$$a) \ x^2 + 2x - 3 = 0$$
$$b) \ x^2 + 2x - 3 > 0$$
$$c) \ x^2 + 2x - 3 < 0.$$

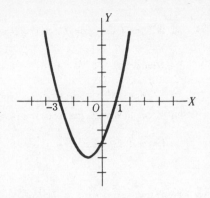

The figure shows the graph of the function defined
by $y = x^2 + 2x - 3$.   From the graph it is clear that

$$a) \ y = 0 \ \text{when} \ x = 1, \ x = -3$$
$$b) \ y > 0 \ \text{when} \ x > 1 \ \text{or} \ x < -3$$
$$c) \ y < 0 \ \text{when} \ -3 < x < 1.$$

SUPPLEMENTARY PROBLEMS

16. If $a > b$, prove that $a - c > b - c$ where c is any real number.

17. If $a > b$ and $k > 0$, prove that $ka > kb$.

18. Find the values of x for which each of the following inequalities holds.

   $a) \ 2(x+3) > 3(x-1) + 6$ $b) \ \dfrac{x}{4} + \dfrac{2}{3} < \dfrac{2x}{3} - \dfrac{1}{6}$ $c) \ \dfrac{1}{x} + \dfrac{3}{4x} > \dfrac{7}{8}$ $d) \ x^2 > 9$

19. For what values of a will $(a+3) < 2(2a+1)$?

20. Prove that $\frac{1}{2}(a^2 + b^2) \geqq ab$ for all real values of a and b, the equality holding if and only
    if $a = b$.

21. Prove that $\dfrac{1}{x} + \dfrac{1}{y} > \dfrac{2}{x+y}$ if x and y are positive and $x \neq y$.

22. Prove that $\dfrac{x^2 + y^2}{x+y} < x+y$ if $x > 0$, $y > 0$.

23. Prove that $xy + 1 \geqq x + y$ if $x \geqq 1$ and $y \geqq 1$ or if $x \leqq 1$ and $y \leqq 1$.

24. If $a > 0$, $a \neq 1$ and n is any positive integer, prove that $a^{n+1} + \dfrac{1}{a^{n+1}} > a^n + \dfrac{1}{a^n}$.

25. Show that $\sqrt{2} + \sqrt{6} < \sqrt{3} + \sqrt{5}$.

26. Determine the values of x for which each of the following inequalities holds.

   $a) \ x^2 + 2x - 24 > 0$ $b) \ x^2 - 6 < x$ $c) \ 3x^2 - 2x < 1$ $d) \ 3x + \dfrac{1}{x} > \dfrac{7}{2}$

27. Determine graphically the range of values of x for which $a) \ x^2 - 3x - 4 > 0$, $b) \ 2x^2 - 5x + 2 < 0$.

ANSWERS TO SUPPLEMENTARY PROBLEMS.

18. $a) \ x < 3$ $b) \ x > 2$ $c) \ 0 < x < 2$ $d) \ x < -3$ or $x > 3$

19. $a > \dfrac{1}{3}$

26. $a) \ x > 4$ or $x < -6$ $b) \ -2 < x < 3$ $c) -\dfrac{1}{3} < x < 1$ $d) \ x > \dfrac{2}{3}$ or $0 < x < \dfrac{1}{2}$

27. $a) \ x > 4$ or $x < -1$ $b) \ \dfrac{1}{2} < x < 2$

CHAPTER 20

Polar Form of Complex Numbers

A COMPLEX NUMBER is an expression of the form $a + bi$ where a and b are real numbers and $i = \sqrt{-1}$. For a discussion of the operations of addition, subtraction, multiplication and division of complex numbers see Chapter 8.

GRAPHICAL REPRESENTATION OF COMPLEX NUMBERS.

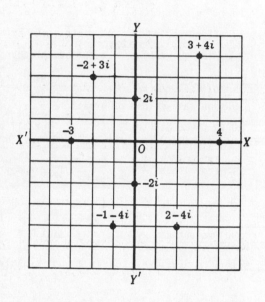

Employing rectangular coordinate axes, the complex number $x + yi$ is represented by, or corresponds to, the point whose coordinates are (x, y). For example:

To represent the complex number $3 + 4i$, measure off 3 units distance along $X'X$ and to the right of O, and then up 4 units distance.

To represent the number $-2 + 3i$, measure off 2 units distance along $X'X$ and to the left of O, and then up 3 units distance.

To represent the number $-1 - 4i$, measure off 1 unit distance along $X'X$ and to the left of O, and then down 4 units distance.

To represent the number $2 - 4i$, measure off 2 units distance along $X'X$ and to the right of O, and then down 4 units distance.

Pure imaginary numbers (as $2i$, $-2i$) are represented by points on the line $Y'Y$. Real numbers (as 4, -3) are represented by points on the line $X'X$.

POLAR FORM OF COMPLEX NUMBERS.

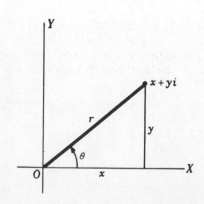

In the figure, $x = r \cos \theta$ and $y = r \sin \theta$.

Then $\qquad x + yi = r(\cos \theta + i \sin \theta)$.

The form $r(\cos \theta + i \sin \theta)$ is the polar form, and the form $x + yi$ is the rectangular form, of the same complex number.

The length $r = \sqrt{x^2 + y^2}$ is always positive and is called the *modulus* or *absolute value* of the complex number. The angle θ is called the *amplitude* or *argument* of the complex number.

MULTIPLICATION OF COMPLEX NUMBERS IN POLAR FORM.

The modulus of the product of two complex numbers is the product of their moduli, and the amplitude of the product is the sum of their amplitudes.

$$r_1(\cos \theta_1 + i \sin \theta_1) \cdot r_2(\cos \theta_2 + i \sin \theta_2) = r_1 r_2[\cos(\theta_1 + \theta_2) + i \sin(\theta_1 + \theta_2)]$$

Thus $4(\cos 45° + i \sin 45°)\, 7(\cos 30° + i \sin 30°) = 28(\cos 75° + i \sin 75°)$.

DIVISION OF COMPLEX NUMBERS IN POLAR FORM.

The modulus of the quotient of two complex numbers is the modulus of the dividend divided by the modulus of the divisor, and the amplitude of the quotient is the amplitude of the dividend minus the amplitude of the divisor.

$$\frac{r_1(\cos \theta_1 + i \sin \theta_1)}{r_2(\cos \theta_2 + i \sin \theta_2)} = \frac{r_1}{r_2}[\cos(\theta_1 - \theta_2) + i \sin(\theta_1 - \theta_2)]$$

For example, $\dfrac{6(\cos 82° + i \sin 82°)}{2(\cos 50° + i \sin 50°)} = 3(\cos 32° + i \sin 32°)$.

DE MOIVRE'S THEOREM. The nth power of $r(\cos \theta + i \sin \theta)$ is

$$[r(\cos \theta + i \sin \theta)]^n = r^n(\cos n\theta + i \sin n\theta).$$

This relation is De Moivre's Theorem and holds for any real value of the exponent. For example, if the exponent is a fraction $1/n$,

$$[r(\cos \theta + i \sin \theta)]^{1/n} = r^{1/n}(\cos \frac{\theta}{n} + i \sin \frac{\theta}{n}).$$

ROOTS OF COMPLEX NUMBERS IN POLAR FORM. If k is any integer,

$$\cos \theta = \cos(\theta + k\,360°) \quad \text{and} \quad \sin \theta = \sin(\theta + k\,360°).$$

Then $(x + yi)^{1/n} = [r(\cos \theta + i \sin \theta)]^{1/n}$

$$= \{r[\cos(\theta + k\,360°) + i \sin(\theta + k\,360°)]\}^{1/n}$$

$$= r^{1/n}[\cos \frac{\theta + k\,360°}{n} + i \sin \frac{\theta + k\,360°}{n}].$$

Any number (real or complex), except zero, has n distinct nth roots. To obtain the n nth roots of the complex number $x + yi$, or $r(\cos \theta + i \sin \theta)$, let k take on the successive values $0, 1, 2, 3, \ldots, n-1$ in the above formula.

SOLVED PROBLEMS

GRAPHICAL ADDITION AND SUBTRACTION OF COMPLEX NUMBERS.

1. Perform the indicated operations both algebraically and graphically:
   a) $(2 + 6i) + (5 + 3i)$,   b) $(-4 + 2i) - (3 + 5i)$.

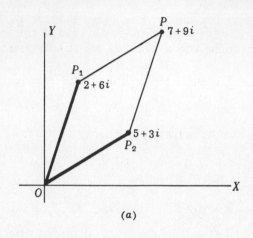

(a)

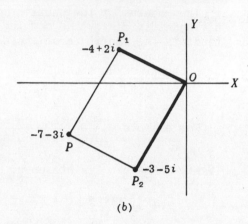

(b)

a) Algebraically. $(2 + 6i) + (5 + 3i) = 7 + 9i$

   Graphically.   Represent the two complex numbers by the points P_1 and P_2 respectively, as shown in Figure (a) above.   Connect P_1 and P_2 with the origin O.   Complete the parallelogram having OP_1 and OP_2 as adjacent sides.   The vertex P (point $7 + 9i$) represents the sum of the two given complex numbers.

b) Algebraically. $(-4 + 2i) - (3 + 5i) = -7 - 3i$

   Graphically. $(-4 + 2i) - (3 + 5i) = (-4 + 2i) + (-3 - 5i)$. We now proceed to add $(-4 + 2i)$ with $(-3 - 5i)$.

   Represent the two complex numbers $(-4 + 2i)$ and $(-3 - 5i)$ by the points P_1 and P_2 respectively, as shown in Figure (b) above.   Connect P_1 and P_2 with the origin O.   Complete the parallelogram having OP_1 and OP_2 as adjacent sides.   The vertex P (point $-7 - 3i$) represents the subtraction $(-4 + 2i) - (3 + 5i)$.

POLAR FORM OF COMPLEX NUMBERS.

   Find the polar form of each of the following complex numbers.

2. $2 + 2i$

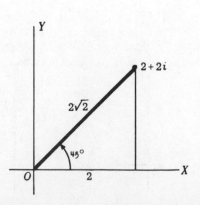

   Amplitude or argument, $\theta = \tan^{-1} \dfrac{2}{2} = \tan^{-1} 1 = 45°$.

   Modulus or absolute value, $r = \sqrt{2^2 + 2^2} = \sqrt{8} = 2\sqrt{2}$.

   Hence $2 + 2i = r(\cos \theta + i \sin \theta)$

   $= 2\sqrt{2}(\cos 45° + i \sin 45°)$.

3. $1 + \sqrt{3}\,i$

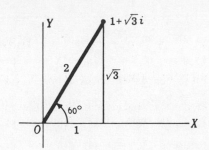

Amplitude or argument, $\theta = \cos^{-1}\dfrac{1}{2} = 60°$.

Modulus or absolute value, $r = \sqrt{3 + 1} = 2$.

Hence $\quad 1 + \sqrt{3}\,i = r(\cos\theta + i\sin\theta)$

$\qquad\qquad = 2(\cos 60° + i\sin 60°)$.

4. $-3 + 3i$

$\theta = 180° - 45° = 135°, \quad r = \sqrt{9 + 9} = 3\sqrt{2}; \quad$ then $\quad -3 + 3i = 3\sqrt{2}\,(\cos 135° + i\sin 135°)$.

5. $-1 - \sqrt{3}\,i$

$\theta = 180° + 60° = 240°, \quad r = \sqrt{3 + 1} = 2; \quad$ then $\quad -1 - \sqrt{3}\,i = 2(\cos 240° + i\sin 240°)$.

6. $\sqrt{3} - i$

$\theta = 360° - 30° = 330°, \quad r = \sqrt{3 + 1} = 2; \quad$ then $\quad \sqrt{3} - i = 2(\cos 330° + i\sin 330°)$.

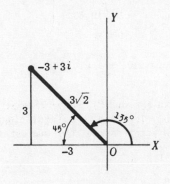

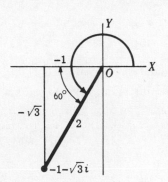

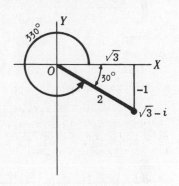

| Problem 4 | Problem 5 | Problem 6 |

7. Find the polar form of the following complex numbers: $\quad a)\ 5,\quad b)\ 2i,\quad c)\ -4,\quad d)\ -4i$.

$a)\ 5. \qquad \theta = 0°, \quad r = 5\,;\quad$ then $\quad 5 = 5(\cos 0° + i\sin 0°)$.

$b)\ 2i. \qquad \theta = 90°, \quad r = 2\,;\quad$ then $\quad 2i = 2(\cos 90° + i\sin 90°)$.

$c)\ -4. \qquad \theta = 180°, \quad r = 4\,;\quad$ then $\quad -4 = 4(\cos 180° + i\sin 180°)$.

$d)\ -4i. \qquad \theta = 270°, \quad r = 4\,;\quad$ then $\quad -4i = 4(\cos 270° + i\sin 270°)$.

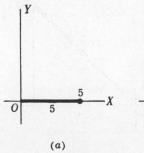

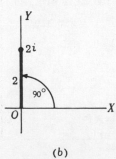

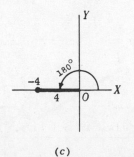

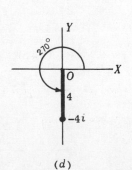

| (a) | (b) | (c) | (d) |

8. Write the following complex numbers in the rectangular form $(a + bi)$.

a) $2(\cos 30° + i \sin 30°)$ $= 2(\frac{1}{2}\sqrt{3} + \frac{1}{2}i) = \sqrt{3} + i$

b) $6(\cos 60° + i \sin 60°)$ $= 6(\frac{1}{2} + \frac{1}{2}\sqrt{3}i) = 3 + 3\sqrt{3}i$

c) $10(\cos 45° + i \sin 45°)$ $= 10(\frac{1}{2}\sqrt{2} + \frac{1}{2}\sqrt{2}i) = 5\sqrt{2} + 5\sqrt{2}i$

d) $3(\cos 90° + i \sin 90°)$ $= 3(0 + i) = 3i$

e) $2(\cos 150° + i \sin 150°)$ $= 2(-\frac{1}{2}\sqrt{3} + \frac{1}{2}i) = -\sqrt{3} + i$

f) $8(\cos 240° + i \sin 240°)$ $= 8(-\frac{1}{2} - \frac{1}{2}\sqrt{3}i) = -4 - 4\sqrt{3}i$

g) $6(\cos 315° + i \sin 315°)$ $= 6(\frac{1}{2}\sqrt{2} - \frac{1}{2}\sqrt{2}i) = 3\sqrt{2} - 3\sqrt{2}i$

h) $4(\cos 720° + i \sin 720°)$ $= 4(\cos 0° + i \sin 0°) = 4(1 + 0) = 4$

POLAR FORM PRODUCTS AND QUOTIENTS.

9. Perform the indicated operations and express the results in rectangular form.

a) $[4(\cos 20° + i \sin 20°)][3(\cos 25° + i \cos 25°)] = 12(\cos 45° + i \sin 45°)$
$$= 12(\frac{1}{2}\sqrt{2} + \frac{1}{2}\sqrt{2}i) = 6\sqrt{2} + 6\sqrt{2}i$$

b) $[2(\cos 18° + i \sin 18°)][5(\cos 42° + i \sin 42°)] = 10(\cos 60° + i \sin 60°)$
$$= 10(\frac{1}{2} + \frac{1}{2}\sqrt{3}i) = 5 + 5\sqrt{3}i$$

c) $[3(\cos 80° + i \sin 80°)][6(\cos 130° + i \sin 130°)] = 18(\cos 210° + i \sin 210°)$
$$= 18(-\frac{1}{2}\sqrt{3} - \frac{1}{2}i) = -9\sqrt{3} - 9i$$

d) $\dfrac{12(\cos 54° + i \sin 54°)}{3(\cos 24° + i \sin 24°)} = 4(\cos 30° + i \sin 30°) = 4(\frac{1}{2}\sqrt{3} + \frac{1}{2}i) = 2\sqrt{3} + 2i$

e) $\dfrac{4\sqrt{2}(\cos 45° + i \sin 45°)}{2(\cos 315° + i \sin 315°)} = 2\sqrt{2}(\cos -270° + i \sin -270°)$
$$= 2\sqrt{2}(\cos 90° + i \sin 90°) = 2\sqrt{2}(0 + i) = 2\sqrt{2}i$$

10. Determine the indicated product or quotient both analytically and graphically.

a) $(\sqrt{3} + i)(-\sqrt{3} + 3i)$

Analytically: $(\sqrt{3} + i)(-\sqrt{3} + 3i) = -3 + 3\sqrt{3}i - \sqrt{3}i - 3 = -6 + 2\sqrt{3}i$.

Graphically: $P_1 = \sqrt{3} + i = 2(\cos 30° + i \sin 30°)$,

$P_2 = -\sqrt{3} + 3i = 2\sqrt{3}(\cos 120° + i \sin 120°)$, and

$P = P_1 \cdot P_2 = 4\sqrt{3}(\cos 150° + i \sin 150°)$.

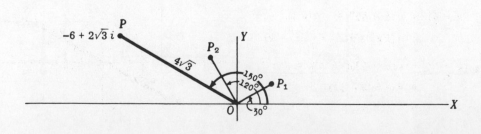

b) $\dfrac{2 - 2\sqrt{3}\,i}{1 + i}$

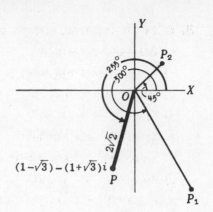

Analytically: $\dfrac{2 - 2\sqrt{3}\,i}{1 + i} \cdot \dfrac{1 - i}{1 - i} = \dfrac{2 - 2i - 2\sqrt{3}\,i - 2\sqrt{3}}{2}$

$\qquad\qquad\qquad = (1 - \sqrt{3}) - (1 + \sqrt{3})\,i.$

Graphically: $P_1 = 2 - 2\sqrt{3}\,i = 4(\cos 300° + i \sin 300°),$

$\qquad\qquad P_2 = \sqrt{2}\,(\cos 45° + i \sin 45°),$ and

$\qquad\qquad P = \dfrac{P_1}{P_2} = 2\sqrt{2}\,(\cos 255° + i \sin 255°).$

POWERS OF COMPLEX NUMBERS.

11. Determine the indicated powers of the following complex numbers and express the results in rectangular form.

a) $[4(\cos 15° + i \sin 15°)]^2 = 4^2(\cos 30° + i \sin 30°) = 16(\tfrac{1}{2}\sqrt{3} + \tfrac{1}{2}i) = 8\sqrt{3} + 8i$

b) $[3(\cos 45° + i \sin 45°)]^4 = 3^4(\cos 180° + i \sin 180°) = 81(-1 + 0i) = -81$

c) $[2(\cos 80° + i \sin 80°)]^6 = 2^6(\cos 480° + i \sin 480°)$

$\qquad\qquad\qquad\qquad\qquad = 64(\cos 120° + i \sin 120°)$

$\qquad\qquad\qquad\qquad\qquad = 64(-\tfrac{1}{2} + \tfrac{1}{2}\sqrt{3}\,i) = -32 + 32\sqrt{3}\,i$

d) $(\cos 30° + i \sin 30°)^3 = 1^3(\cos 90° + i \sin 90°) = 1(0 + i) = i$

e) $(1 + i)^3 = [\sqrt{2}\,(\cos 45° + i \sin 45°)]^3 = 2\sqrt{2}\,(\cos 135° + i \sin 135°)$

$\qquad\qquad\qquad\qquad\qquad\qquad = 2\sqrt{2}\,(-\tfrac{1}{2}\sqrt{2} + \tfrac{1}{2}\sqrt{2}\,i) = -2 + 2i$

f) $(1 - i)^4 = [\sqrt{2}\,(\cos 315° + i \sin 315°)]^4 = 4(\cos 1260° + i \sin 1260°)$

$\qquad\qquad\qquad\qquad = 4(\cos 180° + i \sin 180°) = 4(-1 + 0i) = -4$

g) $(\tfrac{1}{2}\sqrt{3} + \tfrac{1}{2}i)^{100} = [1(\cos 30° + i \sin 30°)]^{100} = \cos 3000° + i \sin 3000°$

$\qquad\qquad\qquad\qquad = \cos 120° + i \sin 120° = -\tfrac{1}{2} + \tfrac{1}{2}\sqrt{3}\,i$

ROOTS OF COMPLEX NUMBERS.

12. Obtain all the indicated roots $(z_1, z_2, \text{etc.})$ and represent them graphically.

a) $\sqrt{16(\cos 60° + i \sin 60°)} = 4(\cos \dfrac{60° + k\,360°}{2} + i \sin \dfrac{60° + k\,360°}{2})$

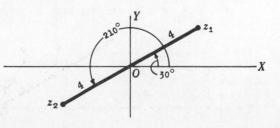

$\qquad z_1\ (k = 0) = 4(\cos 30° + i \sin 30°)$

$\qquad\qquad\qquad = 4(\tfrac{1}{2}\sqrt{3} + \tfrac{1}{2}i) = 2\sqrt{3} + 2i$

$\qquad z_2\ (k = 1) = 4(\cos 210° + i \sin 210°)$

$\qquad\qquad\qquad = 4(-\tfrac{1}{2}\sqrt{3} - \tfrac{1}{2}i)$

$\qquad\qquad\qquad = -2\sqrt{3} - 2i$

b) $\sqrt[5]{32(\cos 50° + i \sin 50°)} = 2(\cos \dfrac{50° + k\,360°}{5} + i \sin \dfrac{50° + k\,360°}{5})$

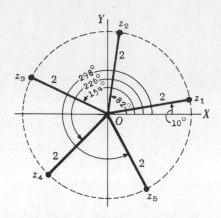

$z_1\ (k = 0) = 2(\cos 10° + i \sin 10°)$

$z_2\ (k = 1) = 2(\cos 82° + i \sin 82°)$

$z_3\ (k = 2) = 2(\cos 154° + i \sin 154°)$

$z_4\ (k = 3) = 2(\cos 226° + i \sin 226°)$

$z_5\ (k = 4) = 2(\cos 298° + i \sin 298°)$

Note that the roots z_1, z_2, z_3, z_4, z_5 lie on a circle whose radius (2) is the modulus of each complex number, and the angle between successive roots is $\dfrac{360°}{5} = 72°$.

c) $\sqrt[3]{8} = 8^{1/3} = [8(\cos 0° + i \sin 0°)]^{1/3} = 2(\cos \dfrac{0° + k\,360°}{3} + i \sin \dfrac{0° + k\,360°}{3})$

$z_1\ (k = 0) = 2(\cos 0° + i \sin 0°) = 2(1 + 0i) = 2$

$z_2\ (k = 1) = 2(\cos 120° + i \sin 120°) = 2(-\tfrac{1}{2} + \tfrac{1}{2}\sqrt{3}i) = -1 + \sqrt{3}i$

$z_3\ (k = 2) = 2(\cos 240° + i \sin 240°) = 2(-\tfrac{1}{2} - \tfrac{1}{2}\sqrt{3}i) = -1 - \sqrt{3}i$

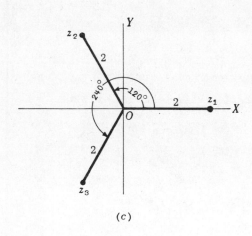

(c)

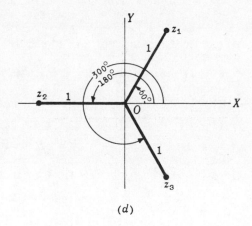

(d)

d) $\sqrt[3]{-1} = (-1)^{1/3} = [1(\cos 180° + i \sin 180°)]^{1/3} = \cos \dfrac{180° + k\,360°}{3} + i \sin \dfrac{180° + k\,360°}{3}$

$z_1\ (k = 0) = \cos 60° + i \sin 60° = \tfrac{1}{2} + \tfrac{1}{2}\sqrt{3}i$

$z_2\ (k = 1) = \cos 180° + i \sin 180° = -1$

$z_3\ (k = 2) = \cos 300° + i \sin 300° = \tfrac{1}{2} - \tfrac{1}{2}\sqrt{3}i$

13. Find all the indicated roots.
   a) Fourth roots of 1
   b) Cube roots of $1-i$
   c) Cube roots of $-8i$
   d) Fourth roots of $2 + 2\sqrt{3}i$

a) $1^{1/4} = [1(\cos 0° + i \sin 0°)]^{1/4} = \cos \dfrac{0° + k\,360°}{4} + i \sin \dfrac{0° + k\,360°}{4}$

$z_1\ (k = 0) = \cos 0° + i \sin 0° = 1 + 0i = 1$ $z_3\ (k = 2) = \cos 180° + i \sin 180° = -1$

$z_2\ (k = 1) = \cos 90° + i \sin 90° = 0 + i = i$ $z_4\ (k = 3) = \cos 270° + i \sin 270° = -i$

b) $(1 - i)^{1/3} = [\sqrt{2}(\cos 315° + i \sin 315°)]^{1/3} = \sqrt[6]{2}(\cos \dfrac{315° + k\,360°}{3} + i \sin \dfrac{315° + k\,360°}{3})$

$z_1\ (k = 0) = \sqrt[6]{2}(\cos 105° + i \sin 105°)$

$z_2\ (k = 1) = \sqrt[6]{2}(\cos 225° + i \sin 225°)$

$z_3\ (k = 2) = \sqrt[6]{2}(\cos 345° + i \sin 345°)$

c) $(-8i)^{1/3} = [8(\cos 270° + i \sin 270°)]^{1/3} = 2(\cos \dfrac{270° + k\,360°}{3} + i \sin \dfrac{270° + k\,360°}{3})$

$z_1\ (k = 0) = 2(\cos 90° + i \sin 90°) = 2(0 + i) = 2i$

$z_2\ (k = 1) = 2(\cos 210° + i \sin 210°) = 2(-\tfrac{1}{2}\sqrt{3} - \tfrac{1}{2}i) = -\sqrt{3} - i$

$z_3\ (k = 2) = 2(\cos 330° + i \sin 330°) = 2(\tfrac{1}{2}\sqrt{3} - \tfrac{1}{2}i) = \sqrt{3} - i$

d) $(2 + 2\sqrt{3}i)^{1/4} = [4(\cos 60° + i \sin 60°)]^{1/4} = \sqrt{2}(\cos \dfrac{60° + k\,360°}{4} + i \sin \dfrac{60° + k\,360°}{4})$

$z_1\ (k = 0) = \sqrt{2}(\cos 15° + i \sin 15°)$ $z_3\ (k = 2) = \sqrt{2}(\cos 195° + i \sin 195°)$

$z_2\ (k = 1) = \sqrt{2}(\cos 105° + i \sin 105°)$ $z_4\ (k = 3) = \sqrt{2}(\cos 285° + i \sin 285°)$

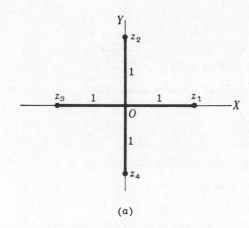

(a)

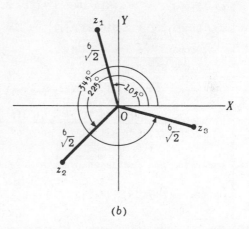

(b)

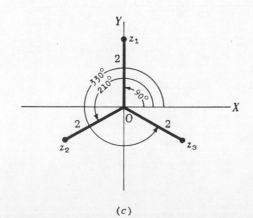

(c)

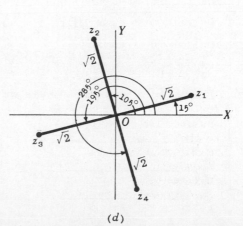

(d)

SUPPLEMENTARY PROBLEMS

14. Perform the indicated operations both algebraically and graphically.

   a) $(3+4i) + (4+3i)$ c) $(4-3i) - (-2+i)$
   b) $(2-i) + (-4+5i)$ d) $(-2+2i) - (-2-i)$

15. Write the polar form of each of the following complex numbers.

   a) $3 - 3i$ c) $4 - 4\sqrt{3}\,i$ e) $6\sqrt{3} + 6i$ g) 2 i) $-2i$
   b) $-\sqrt{3} + i$ d) -5 f) $-4 - 4i$ h) $i\sqrt{3}$ j) $-1 + i$

16. Write the following complex numbers in the rectangular form $(a + bi)$.

   a) $4(\cos 45° + i \sin 45°)$ f) $10\sqrt{2}(\cos 225° + i \sin 225°)$

   b) $12(\cos 30° + i \sin 30°)$ g) $2(\cos 300° + i \sin 300°)$

   c) $6(\cos 120° + i \sin 120°)$ h) $5(\cos 360° + i \sin 360°)$

   d) $8(\cos 180° + i \sin 180°)$ i) $8(\cos 90° + i \sin 90°)$

   e) $3(\cos 270° + i \sin 270°)$ j) $16(\cos 210° + i \sin 210°)$

17. Perform the indicated operations and express the results in rectangular form.

   a) $[3(\cos 15° + i \sin 15°)][2(\cos 75° + i \sin 75°)]$

   b) $[4(\cos 40° + i \sin 40°)][5(\cos 20° + i \sin 20°)]$

   c) $[2(\cos 100° + i \sin 100°)][4(\cos 50° + i \sin 50°)]$

   d) $[6(\cos 25° + i \sin 25°)][3(\cos 290° + i \sin 290°)]$

   e) $\dfrac{20(\cos 83° + i \sin 83°)}{5(\cos 23° + i \sin 23°)}$ f) $\dfrac{6\sqrt{3}(\cos 40° + i \sin 40°)}{3(\cos 190° + i \sin 190°)}$

   g) $[2(\cos 12° + i \sin 12°)][3(\cos 84° + i \sin 84°)][5(\cos 24° + i \sin 24°)]$

   h) $\dfrac{12(\cos 16° + i \sin 16°)}{[3(\cos 44° + i \sin 44°)][2(\cos 62° + i \sin 62°)]}$

18. Determine the indicated product or quotient both analytically and graphically.

   a) $(-1 + \sqrt{3}\,i)(2\sqrt{3} - 2i)$ b) $\dfrac{4 - 4i}{\sqrt{3} - i}$

19. Find the indicated powers of the complex numbers and express the results in rectangular form.

   a) $[2(\cos 75° + i \sin 75°)]^2$ d) $[4(\cos 20° + i \sin 20°)]^3$ g) $(\tfrac{1}{2}\sqrt{3} - \tfrac{1}{2}i)^{10}$
   b) $[5(\cos 30° + i \sin 30°)]^3$ e) $(-1 + i)^6$
   c) $[\sqrt{2}(\cos 36° + i \sin 36°)]^5$ f) $(1 + \sqrt{3}i)^7$ h) $(\tfrac{1}{2}\sqrt{2} + \tfrac{1}{2}\sqrt{2}i)^{30}$

20. Determine all the indicated roots and represent them graphically.

   a) $\sqrt{4(\cos 120° + i \sin 120°)}$ e) $\sqrt[3]{1 + i}$ i) $\sqrt[3]{\sqrt{3} - i}$

   b) $\sqrt[4]{81(\cos 180° + i \sin 180°)}$ f) $\sqrt[5]{-32}$

   c) $\sqrt[3]{8(\cos 60° + i \sin 60°)}$ g) $\sqrt[3]{1}$ j) $\sqrt[5]{2 - 2\sqrt{3}i}$

   d) $\sqrt[5]{32(\cos 200° + i \sin 200°)}$ h) $\sqrt[4]{16i}$

ANSWERS TO SUPPLEMENTARY PROBLEMS.

14. *a*) $7 + 7i$ *b*) $-2 + 4i$ *c*) $6 - 4i$ *d*) $3i$

15. *a*) $3\sqrt{2}(\cos 315° + i \sin 315°)$ *f*) $4\sqrt{2}(\cos 225° + i \sin 225°)$

   b) $2(\cos 150° + i \sin 150°)$ *g*) $2(\cos 0° + i \sin 0°)$

   c) $8(\cos 300° + i \sin 300°)$ *h*) $\sqrt{3}(\cos 90° + i \sin 90°)$

   d) $5(\cos 180° + i \sin 180°)$ *i*) $2(\cos 270° + i \sin 270°)$

   e) $12(\cos 30° + i \sin 30°)$ *j*) $\sqrt{2}(\cos 135° + i \sin 135°)$

16. *a*) $2\sqrt{2} + 2\sqrt{2}i$ *c*) $-3 + 3\sqrt{3}i$ *e*) $-3i$ *g*) $1 - \sqrt{3}i$ *i*) $8i$

   b) $6\sqrt{3} + 6i$ *d*) -8 *f*) $-10 - 10i$ *h*) 5 *j*) $-8\sqrt{3} - 8i$

17. *a*) $6i$ *c*) $-4\sqrt{3} + 4i$ *e*) $2 + 2\sqrt{3}i$ *g*) $-15 + 15\sqrt{3}i$

   b) $10 + 10\sqrt{3}i$ *d*) $9\sqrt{2} - 9\sqrt{2}i$ *f*) $-3 - \sqrt{3}i$ *h*) $-2i$

18. *a*) $8i$ *b*) $\sqrt{3} + 1 + (1 - \sqrt{3})i$

19. *a*) $-2\sqrt{3} + 2i$ *c*) $-4\sqrt{2}$ *e*) $8i$ *g*) $\frac{1}{2} + \frac{1}{2}\sqrt{3}i$

   b) $125i$ *d*) $32 + 32\sqrt{3}i$ *f*) $64 + 64\sqrt{3}i$ *h*) $-i$

20. *a*) $2(\cos 60° + i \sin 60°)$, $2(\cos 240° + i \sin 240°)$

   b) $3(\cos 45° + i \sin 45°)$, $3(\cos 135° + i \sin 135°)$, $3(\cos 225° + i \sin 225°)$,
      $3(\cos 315° + i \sin 315°)$

   c) $2(\cos 20° + i \sin 20°)$, $2(\cos 140° + i \sin 140°)$, $2(\cos 260° + i \sin 260°)$

   d) $2(\cos 40° + i \sin 40°)$, $2(\cos 112° + i \sin 112°)$, $2(\cos 184° + i \sin 184°)$,
      $2(\cos 256° + i \sin 256°)$, $2(\cos 328° + i \sin 328°)$

   e) $\sqrt[6]{2}(\cos 15° + i \sin 15°)$, $\sqrt[6]{2}(\cos 135° + i \sin 135°)$, $\sqrt[6]{2}(\cos 255° + i \sin 255°)$

   f) $2(\cos 36° + i \sin 36°)$, $2(\cos 108° + i \sin 108°)$, $2(\cos 180° + i \sin 180°)$,
      $2(\cos 252° + i \sin 252°)$, $2(\cos 324° + i \sin 324°)$

   g) $\cos 0° + i \sin 0°$, $\cos 120° + i \sin 120°$, $\cos 240° + i \sin 240°$

   h) $2(\cos 22.5° + i \sin 22.5°)$, $2(\cos 112.5° + i \sin 112.5°)$, $2(\cos 202.5° + i \sin 202.5°)$,
      $2(\cos 292.5° + i \sin 292.5°)$

   i) $\sqrt[3]{2}(\cos 110° + i \sin 110°)$, $\sqrt[3]{2}(\cos 230° + i \sin 230°)$, $\sqrt[3]{2}(\cos 350° + i \sin 350°)$

   j) $\sqrt[5]{4}(\cos 60° + i \sin 60°)$, $\sqrt[5]{4}(\cos 132° + i \sin 132°)$, $\sqrt[5]{4}(\cos 204° + i \sin 204°)$,
      $\sqrt[5]{4}(\cos 276° + i \sin 276°)$, $\sqrt[5]{4}(\cos 348° + i \sin 348°)$

CHAPTER 21

Theory of Equations

A RATIONAL INTEGRAL EQUATION of degree n in the variable x is an equation which can be written in the form

$$a_0 x^n + a_1 x^{n-1} + a_2 x^{n-2} + \ldots + a_{n-1} x + a_n = 0, \qquad a_0 \neq 0$$

where n is a positive integer and a_0, a_1, a_2, $\ldots$, a_{n-1}, a_n are constants. The equation can be written in the p-form

$$x^n + p_1 x^{n-1} + p_2 x^{n-2} + \ldots + p_{n-1} x + p_n = 0$$

upon division by $a_0 \neq 0$ so that the coefficient of the highest power of x is 1.

Thus $4x^3 - 2x^2 + 3x - 5 = 0$, $x^2 - \sqrt{2}x + \frac{1}{4} = 0$ and $x^4 + \sqrt{-3}x - 8 = 0$ are rational integral equations in x of degree 3, 2 and 4 respectively. Note that in each equation the exponents of x are positive and integral, and the coefficients are constants (real or complex numbers).

In this chapter, only rational integral equations are considered.

A POLYNOMIAL of degree n in the variable x is a function of x which can be written in the form

$$f(x) = a_0 x^n + a_1 x^{n-1} + a_2 x^{n-2} + \ldots + a_{n-1} x + a_n, \qquad a_0 \neq 0$$

where n is a positive integer and a_0, a_1, a_2, $\ldots$, a_{n-1}, a_n are constants. Then $f(x) = 0$ is a rational integral equation of degree n in x.

If $f(x) = 3x^3 + x^2 + 5x - 6$, then $f(-2) = 3(-2)^3 + (-2)^2 + 5(-2) - 6 = -36$.

If $f(x) = x^2 + 2x - 8$, then $f(\sqrt{5}) = 5 + 2\sqrt{5} - 8 = 2\sqrt{5} - 3$.

Any value of x which makes $f(x)$ vanish is called a *root* of the equation $f(x) = 0$. Thus 2 is a root of the equation $f(x) = 3x^3 - 2x^2 - 5x - 6 = 0$ since $f(2) = 24 - 8 - 10 - 6 = 0$.

REMAINDER THEOREM. If r is any constant and if a polynomial $f(x)$ is divided by $(x - r)$, the remainder is $f(r)$.

For example, if $f(x) = 2x^3 - 3x^2 - x + 8$ is divided by $x + 1$, then $r = -1$ and the remainder $= f(-1) = -2 - 3 + 1 + 8 = 4$. That is,

$$\frac{2x^3 - 3x^2 - x + 8}{x + 1} = P(x) + \frac{4}{x + 1}, \quad \text{where } P(x) \text{ is a polynomial in } x.$$

FACTOR THEOREM. If r is a root of the equation $f(x) = 0$, i.e. if $f(r) = 0$, then $(x - r)$ is a factor of $f(x)$. Conversely if $(x - r)$ is a factor of $f(x)$, then r is a root of $f(x) = 0$, or $f(r) = 0$.

Thus $1, -2, -3$ are the three roots of the equation $f(x) = x^3 + 4x^2 + x - 6 = 0$ since $f(1) = f(-2) = f(-3) = 0$. Then $(x - 1)$, $(x + 2)$ and $(x + 3)$ are factors of $x^3 + 4x^2 + x - 6$.

SYNTHETIC DIVISION is a simplified method of dividing a polynomial $f(x)$ by $x - r$, where r is any assigned number. By this method we determine values of the coefficients of the quotient and the value of the remainder can readily be determined.

FUNDAMENTAL THEOREM OF ALGEBRA. Every rational integral equation $f(x) = 0$ has at least one root, real or complex.

Thus $x^7 - 3x^5 + 2 = 0$ has at least one root.

But $f(x) = \sqrt{x} + 3 = 0$ has no root, since no number r exists such that $f(r) = 0$. Since this equation is not rational, the fundamental theorem does not apply.

NUMBER OF ROOTS OF AN EQUATION. Every rational integral equation $f(x) = 0$ of the nth degree has exactly n roots.

Thus $2x^3 + 5x^2 - 14x - 8 = 0$ has exactly 3 roots, namely $2, -\frac{1}{2}, -4$.

Some of the n roots may be equal. Thus the equation of the sixth degree $(x - 2)^3 (x - 5)^2 (x + 4) = 0$ has 2 as a triple root, 5 as a double root, and -4 as a single root; i.e. the six roots are $2, 2, 2, 5, 5, -4$.

IDENTICALLY EQUAL POLYNOMIALS. If two polynomials of degree n in the same variable x are equal for more than n values of x, the coefficients of like powers of x are equal and the two polynomials are identically equal.

For example, the equation $5x^2 - 2x + 3 = Ax^2 + (B + C)x + 7(B - C)$ is an identity if $A = 5$, $B + C = -2$, $7(B - C) = 3$.

COMPLEX AND IRRATIONAL ROOTS.

1) If a complex number $a + bi$ is a root of the rational integral equation $f(x) = 0$ with *real coefficients*, then the conjugate complex number $a - bi$ is also a root.

It follows that every rational integral equation of odd degree with real coefficients has at least one real root.

2) If the rational integral equation $f(x) = 0$ with *rational coefficients* has $a + \sqrt{b}$ as a root, where a and b are rational and $\sqrt{b}$ is irrational, then $a - \sqrt{b}$ is also a root.

RATIONAL ROOTS.

1) If b/c, a rational fraction in lowest terms, is a root of the equation

$$a_0 x^n + a_1 x^{n-1} + a_2 x^{n-2} + \ldots + a_{n-1} x + a_n = 0, \qquad a_0 \neq 0$$

with integral coefficients, then b is a factor of a_n and c is a factor of a_0.

Thus if b/c is a rational root of $6x^3 + 5x^2 - 3x - 2 = 0$, the values of b are limited to the factors of 2, which are ± 1, ± 2; and the values of c are limited to the factors of 6, which are ± 1, ± 2, ± 3, ± 6. Hence the only possible rational roots are ± 1, ± 2, $\pm 1/2$, $\pm 1/3$, $\pm 1/6$, $\pm 2/3$.

2) It follows that if an equation $f(x) = 0$ has integral coefficients and is in the p-form (in which the coefficient of the highest power of x is 1),

$$x^n + p_1 x^{n-1} + p_2 x^{n-2} + \ldots + p_{n-1} x + p_n = 0,$$

then any rational root of $f(x) = 0$ is an integer and a factor of p_n.

Thus the rational roots, if any, of $x^3 + 2x^2 - 11x - 12 = 0$ are limited to the integral factors of 12, which are ± 1, ± 2, ± 3, ± 4, ± 6, ± 12.

GRAPHICAL PROCEDURE FOR FINDING REAL ROOTS. If $f(x) = 0$ is a rational integral equation with real coefficients, then approximate values of the real roots of $f(x) = 0$ may be found by obtaining the graph of $y = f(x)$ and determining the values of x at the points where the graph intersects the x-axis ($y = 0$). Fundamental in this procedure is the fact that if $f(a)$ and $f(b)$ have opposite signs then $f(x) = 0$ has at least one root between $x = a$ and $x = b$. This fact is based on the continuity of the graph of $y = f(x)$ when $f(x)$ is a polynomial with real coefficients.

UPPER AND LOWER LIMITS FOR THE ROOTS. A number a is called an *upper limit* or *upper bound* for the real roots of $f(x) = 0$ if no root is greater than a. A number b is called a *lower limit* or *lower bound* for the real roots of $f(x) = 0$ if no root is less than b. The following theorem is useful in determining upper and lower limits.

Let $f(x) = a_0 x^n + a_1 x^{n-1} + \ldots + a_n = 0$, where $a_0, a_1, \ldots, a_n$ are real and $a_0 > 0$. Then:

1) If upon synthetic division of $f(x)$ by $x - a$, where $a \geqq 0$, all of the numbers obtained in the third row are positive or zero, then a is an upper limit for all the real roots of $f(x) = 0$.

2) If upon synthetic division of $f(x)$ by $x - b$, where $b \leqq 0$, all of the numbers obtained in the third row are alternately positive and negative (or zero), then b is a lower limit for all the real roots of $f(x) = 0$.

DESCARTES' RULE OF SIGNS.

If the terms of a polynomial $f(x)$ with real coefficients are arranged in order of descending powers of x, a *variation of sign* occurs when two consecutive terms differ in sign. For example, $x^3 - 2x^2 + 3x - 12$ has 3 variations of sign, and $2x^7 - 6x^5 - 4x^4 + x^2 - 2x + 4$ has 4 variations of sign.

Descartes' Rule of Signs: The number of positive roots of $f(x) = 0$ is either equal to the number of variations of sign of $f(x)$ or is less than that number by an even integer. The number of negative roots of $f(x) = 0$ is either equal to the number of variations of sign of $f(-x)$ or is less than that number by an even integer.

Thus in $f(x) = x^9 - 2x^5 + 2x^2 - 3x + 12 = 0$ there are 4 variations of sign of $f(x)$; hence the number of positive roots of $f(x) = 0$ are 4, $(4-2)$ or $(4-4)$. Since $f(-x) = (-x)^9 - 2(-x)^5 + 2(-x)^2 - 3(-x) + 12 = -x^9 + 2x^5 + 2x^2 + 3x + 12 = 0$ has one variation of sign, then $f(x) = 0$ has exactly one negative root. Hence there are 4, 2 or 0 positive roots, 1 negative root, and at least $9 - (4+1) = 4$ complex roots. (There are 4, 6 or 8 complex roots. Why?)

TRANSFORMATION OF EQUATIONS.

MULTIPLYING EACH ROOT BY A CONSTANT. To obtain an equation each of whose roots is k times the corresponding root of a given equation, multiply the second term of the given equation by k, the third term by k^2, the fourth term by k^3, etc., taking account of terms with zero coefficients, if any.

Thus the equation whose roots are double those of $x^3 - 3x^2 - 10x + 24 = 0$ is $x^3 - 3x^2(2) - 10x(2^2) + 24(2^3) = 0$ or $x^3 - 6x^2 - 40x + 192 = 0$.

CHANGING THE SIGN OF EACH ROOT. To obtain an equation each of whose roots is the negative of the corresponding roots of a given equation, change the signs of the odd degree terms of the given equation.

Thus since 1, $-\frac{1}{2}$, -2 are the roots of $2x^3 + 3x^2 - 3x - 2 = 0$, then -1, $\frac{1}{2}$, 2 are the roots of $-2x^3 + 3x^2 + 3x - 2 = 0$ or $2x^3 - 3x^2 - 3x + 2 = 0$.

DIMINISHING EACH ROOT BY A CONSTANT. To obtain an equation each of whose roots is h less than the corresponding root of a given equation $f(x) = 0$, divide $f(x)$ by $(x-h)$; the remainder is the coefficient of the last term of the required equation. Then divide the quotient thus obtained by $(x-h)$, and the remainder is the next-to-the-last term of the required equation, etc. This work is best done by synthetic division. (See Problem 48.)

HORNER'S METHOD is a process by which the irrational roots of integral rational equations may be approximated to any desired accuracy. (See Problems 57-59.)

RELATION BETWEEN ROOTS AND COEFFICIENTS. In the equation in the p-form, in which the coefficient of the first term is 1,

$$x^n + p_1 x^{n-1} + p_2 x^{n-2} + p_3 x^{n-3} + \ldots + p_{n-1} x + p_n = 0,$$

the following relations exist between coefficients and roots:

1) $-p_1$ = sum of roots;

2) p_2 = sum of products of roots taken two at a time;

3) $-p_3$ = sum of products of roots taken three at a time; etc.;

4) $(-1)^n p_n$ = product of all the roots.

Thus if $x^3 - 6x^2 - 7x - 8 = 0$ and the roots are x_1, x_2, x_3, then $x_1 + x_2 + x_3 = -(-6) = 6$, $x_1 x_2 + x_2 x_3 + x_3 x_1 = -7$, $x_1 x_2 x_3 = -(-8) = 8$.

SOLVED PROBLEMS

REMAINDER AND FACTOR THEOREMS.

1. Prove the remainder theorem: If a polynomial $f(x)$ is divided by $(x-r)$ the remainder is $f(r)$.

   In the division of $f(x)$ by $(x-r)$, let $Q(x)$ be the quotient and R, a constant, the remainder.

   By definition $f(x) = (x-r)Q(x) + R$, an identity for all values of x. Letting $x = r$, $f(r) = R$.

2. Determine the remainder R after each of the following divisions.

   a) $(2x^3 + 3x^2 - 18x - 4) \div (x - 2)$. $R = f(2) = 2(2^3) + 3(2^2) - 18(2) - 4 = -12$

   b) $(x^4 - 3x^3 + 5x + 8) \div (x + 1)$. $R = f(-1) = (-1)^4 - 3(-1)^3 + 5(-1) + 8 = 1 + 3 - 5 + 8 = 7$

   c) $(4x^3 + 5x^2 - 1) \div (x + \frac{1}{2})$. $R = f(-\frac{1}{2}) = 4(-\frac{1}{2})^3 + 5(-\frac{1}{2})^2 - 1 = -\frac{1}{4}$

   d) $(x^3 - 2x^2 + x - 4) \div x$. $R = f(0) = -4$

   e) $(\frac{8}{27}x^3 - \frac{4}{9}x^2 + x - \frac{3}{2}) \div (2x - 3)$. $R = f(\frac{3}{2}) = \frac{8}{27}(\frac{3}{2})^3 - \frac{4}{9}(\frac{3}{2})^2 + \frac{3}{2} - \frac{3}{2} = 0$

   f) $(x^8 - x^5 - x^3 + 1) \div (x + \sqrt{-1})$. $R = f(-i) = (-i)^8 - (-i)^5 - (-i)^3 + 1 = i^8 + i^5 + i^3 + 1$
   $$= 1 + i - i + 1 = 2$$

3. Prove the factor theorem: If r is a root of the equation $f(x) = 0$, then $(x-r)$ is a factor of $f(x)$; and conversely if $(x-r)$ is a factor of $f(x)$, then r is a root of $f(x) = 0$.

   In the division of $f(x)$ by $(x-r)$, let $Q(x)$ be the quotient and R, a constant, the remainder. Then $f(x) = (x-r)Q(x) + R$ or $f(x) = (x-r)Q(x) + f(r)$ by the remainder theorem.

   If r is a root of $f(x) = 0$, then $f(r) = 0$. Hence $f(x) = (x-r)Q(x)$, or $(x-r)$ is a factor of $f(x)$.

   Conversely if $(x-r)$ is a factor of $f(x)$, then the remainder in the division of $f(x)$ by $(x-r)$ is zero. Hence $f(r) = 0$, i.e. r is a root of $f(x) = 0$.

4. Show that $(x-3)$ is a factor of the polynomial $f(x) = x^4 - 4x^3 - 7x^2 + 22x + 24$.

   $f(3) = 81 - 108 - 63 + 66 + 24 = 0$.   Hence $(x-3)$ is a factor of $f(x)$, 3 is a *zero* of the polynomial $f(x)$, and 3 is a *root* of the equation $f(x) = 0$.

5. a) Is -1 a root of the equation $f(x) = x^3 - 7x - 6 = 0$?

   b) Is 2 a root of the equation $f(y) = y^4 - 2y^2 - y + 7 = 0$?

   c) Is $2i$ a root of the equation $f(z) = 2z^3 + 3z^2 + 8z + 12 = 0$?

   a) $f(-1) = -1 + 7 - 6 = 0$.   Hence -1 is a root of the equation $f(x) = 0$, and $[x - (-1)] = x + 1$ is a factor of the polynomial $f(x)$.

   b) $f(2) = 16 - 8 - 2 + 7 = 13$.   Hence 2 is not a root of $f(y) = 0$, and $(y - 2)$ is not a factor of $y^4 - 2y^2 - y + 7$.

   c) $f(2i) = 2(2i)^3 + 3(2i)^2 + 8(2i) + 12 = -16i - 12 + 16i + 12 = 0$.   Hence $2i$ is a root of $f(z) = 0$, and $(z - 2i)$ is a factor of the polynomial $f(z)$.

6. Prove that $x-a$ is a factor of x^n-a^n, if n is any positive integer.

$f(x) = x^n - a^n$; then $f(a) = a^n - a^n = 0$. Since $f(a) = 0$, $x-a$ is a factor of $x^n - a^n$.

7. a) Show that $x^5 + a^5$ is exactly divisible by $x+a$.
   b) What is the remainder when $y^6 + a^6$ is divided by $y+a$?

a) $f(x) = x^5 + a^5$; then $f(-a) = (-a)^5 + a^5 = -a^5 + a^5 = 0$.
   Since $f(-a) = 0$, $x^5 + a^5$ is exactly divisible by $x+a$.

b) $f(y) = y^6 + a^6$. Remainder $= f(-a) = (-a)^6 + a^6 = a^6 + a^6 = 2a^6$.

8. Show that $x+a$ is a factor of $x^n - a^n$ when n is an even positive integer, but not a factor when n is an odd positive integer. Assume $a \neq 0$.

$f(x) = x^n - a^n$.

When n is even, $f(-a) = (-a)^n - a^n = a^n - a^n = 0$. Since $f(-a) = 0$, $x+a$ is a factor of $x^n - a^n$ when n is even.

When n is odd, $f(-a) = (-a)^n - a^n = -a^n - a^n = -2a^n$. Since $f(-a) \neq 0$, $x^n - a^n$ is not exactly divisible by $x+a$ when n is odd, (the remainder being $-2a^n$).

9. Find the values of p for which a) $2x^3 - px^2 + 6x - 3p$ is exactly divisible by $x+2$,
   b) $(x^4 - p^2x + 3 - p) \div (x-3)$ has a remainder of 4.

a) The remainder is $2(-2)^3 - p(-2)^2 + 6(-2) - 3p = -16 - 4p - 12 - 3p = -28 - 7p = 0$. Then $p = -4$.

b) The remainder is $3^4 - p^2(3) + 3 - p = 84 - 3p^2 - p = 4$.
   Then $3p^2 + p - 80 = 0$, $(p-5)(3p+16) = 0$ and $p = 5, -16/3$.

SYNTHETIC DIVISION.

By synthetic division, determine the quotient and remainder in the following.

10. $(3x^5 - 4x^4 - 5x^3 - 8x + 25) \div (x-2)$

$$\begin{array}{r} 3 - 4 - 5 + 0 - 8 + 25 \underline{|2} \\ 6 + 4 - 2 - 4 - 24 \\ \hline 3 + 2 - 1 - 2 - 12 + 1 \end{array}$$

Quotient: $3x^4 + 2x^3 - x^2 - 2x - 12$
Remainder: 1

The top row of figures gives the coefficients of the dividend, with zero as the coefficient of the missing power of x ($0x^2$). The 2 at the extreme right is the second term of the divisor with the sign changed (since the coefficient of x in the divisor is 1).

The first coefficient of the top row, 3, is written first in the third row and then multiplied by the 2 of the divisor. The product 6 is placed first in the second row and added to the -4 above it to give 2, which is the next figure in the third row. This 2 is then multiplied by the 2 of the divisor. The product 4 is placed in the second row and added to the -5 above it to give the -1 in the third row; etc. The last figure in the third row is the Remainder, while all the figures to its left constitute the coefficients of the Quotient.

Since the dividend is of the 5th degree and the divisor of the 1st degree, the Quotient is of the 4th degree.

The answer may be written: $3x^4 + 2x^3 - x^2 - 2x - 12 + \dfrac{1}{x-2}$.

11. $(x^4 - 2x^3 - 24x^2 + 15x + 50) \div (x + 4)$

$$\begin{array}{r} 1 - 2 - 24 + 15 + 50 \, \underline{|-4} \\ \underline{-4 + 24 - 0 - 60} \\ 1 - 6 + 0 + 15 - 10 \end{array}$$

Answer: $x^3 - 6x^2 + 15 - \dfrac{10}{x+4}$

12. $(2x^4 - 17x^2 - 4) \div (x + 3)$

$$\begin{array}{r} 2 + 0 - 17 + 0 - 4 \, \underline{|-3} \\ \underline{-6 + 18 - 3 + 9} \\ 2 - 6 + 1 - 3 + 5 \end{array}$$

Answer: $2x^3 - 6x^2 + x - 3 + \dfrac{5}{x+3}$

13. $(4x^3 - 10x^2 + x - 1) \div (x - 1/2)$

$$\begin{array}{r} 4 - 10 + 1 - 1 \, \underline{|1/2} \\ \underline{+ 2 - 4 - 3/2} \\ 4 - 8 - 3 - 5/2 \end{array}$$

Answer: $4x^2 - 8x - 3 - \dfrac{5}{2x-1}$

14. Given $f(x) = x^3 - 6x^2 - 2x + 40$,   compute a) $f(-5)$ and b) $f(4)$ using synthetic division.

a)
$$\begin{array}{r} 1 - 6 - 2 + 40 \, \underline{|-5} \\ \underline{- 5 + 55 - 265} \\ 1 - 11 + 53 - 225 \end{array}$$

$f(-5) = -225$

b)
$$\begin{array}{r} 1 - 6 - 2 + 40 \, \underline{|4} \\ \underline{+ 4 - 8 - 40} \\ 1 - 2 - 10 + 0 \end{array}$$

$f(4) = 0$

DEPRESSED EQUATIONS.

15. Given that one root of $x^3 + 2x^2 - 23x - 60 = 0$ is 5,   solve the equation.

$$\begin{array}{r} 1 + 2 - 23 - 60 \, \underline{|5} \\ \underline{+ 5 + 35 + 60} \\ 1 + 7 + 12 + 0 \end{array}$$

Divide $x^3 + 2x^2 - 23x - 60$ by $x - 5$.

The depressed equation is $x^2 + 7x + 12 = 0$ whose roots are $-3, -4$.

The three roots are $5, -3, -4$.

16. Two roots of $x^4 - 2x^2 - 3x - 2 = 0$ are -1 and 2. Solve the equation.

$$\begin{array}{r} 1 + 0 - 2 - 3 - 2 \, \underline{|-1} \\ \underline{- 1 + 1 + 1 + 2} \\ 1 - 1 - 1 - 2 + 0 \end{array}$$

Divide $x^4 - 2x^2 - 3x - 2$ by $x + 1$.

The first depressed equation is $x^3 - x^2 - x - 2 = 0$.

$$\begin{array}{r} 1 - 1 - 1 - 2 \, \underline{|2} \\ \underline{+ 2 + 2 + 2} \\ 1 + 1 + 1 + 0 \end{array}$$

Divide $x^3 - x^2 - x - 2$ by $x - 2$.

The second depressed equation is $x^2 + x + 1 = 0$ whose roots are

$$-\frac{1}{2} \pm \frac{1}{2} i \sqrt{3}.$$

The four roots are $-1, \, 2, \, -\dfrac{1}{2} \pm \dfrac{1}{2} i \sqrt{3}.$

TO FORM AN EQUATION WITH GIVEN ROOTS.

17. Determine the roots of each of the following equations.

a) $(x-1)^2(x+2)(x+4) = 0.$ Ans.   1 as a double root, -2, -4

b) $(2x+1)(3x-2)^3(2x-5) = 0.$ $-1/2$, $2/3$ as a triple root, $5/2$

c) $x^3(x^2-2x-15) = 0.$ 0 as a triple root, 5, -3

d) $(x+1+\sqrt{3})(x+1-\sqrt{3})(x-6) = 0.$ $(-1-\sqrt{3})$, $(-1+\sqrt{3})$, 6

e) $[(x-i)(x+i)]^3(x+1)^2 = 0.$ $\pm i$ as triple roots, -1 as a double root

f) $3(x+m)^4(5x-n)^2 = 0.$ $-m$ as a quadruple root, $n/5$ as a double root

18. Write the equation having only the following roots.

a) 5, 1, -3;   b) 2, $-1/4$, $-1/2$;   c) ± 2, $2 \pm \sqrt{3}$;   d) 0, $1 \pm 5i$.

a) $(x-5)(x-1)(x+3) = 0$ or $x^3 - 3x^2 - 13x + 15 = 0.$

b) $(x-2)(x+\frac{1}{4})(x+\frac{1}{2}) = 0$ or $x^3 - \frac{5x^2}{4} - \frac{11x}{8} - \frac{1}{4} = 0$ or $8x^3 - 10x^2 - 11x - 2 = 0$ which has integral coefficients.

c) $(x-2)(x+2)[x-(2-\sqrt{3})][x-(2+\sqrt{3})] = (x^2-4)[(x-2)+\sqrt{3}][(x-2)-\sqrt{3}]$

    $= (x^2-4)[(x-2)^2-3] = (x^2-4)(x^2-4x+1) = 0,$ or $x^4 - 4x^3 - 3x^2 + 16x - 4 = 0.$

d) $x[x-(1+5i)][x-(1-5i)] = x[(x-1)-5i][(x-1)+5i] = x[(x-1)^2+25]$

    $= x(x^2-2x+26) = 0,$ or $x^3 - 2x^2 + 26x = 0.$

19. Form the equation with integral coefficients having only the following roots.

a) 1, $\frac{1}{2}$, $-\frac{1}{3}$;   b) 0, $\frac{3}{4}$, $\frac{2}{3}$, -1;   c) $\pm 3i$, $\pm\frac{1}{2}\sqrt{2}$;   d) 2 as a triple root, -1.

a) $(x-1)(2x-1)(3x+1) = 0$ or $6x^3 - 7x^2 + 1 = 0$

b) $x(4x-3)(3x-2)(x+1) = 0$ or $12x^4 - 5x^3 - 11x^2 + 6x = 0$

c) $(x-3i)(x+3i)(x-\frac{1}{2}\sqrt{2})(x+\frac{1}{2}\sqrt{2}) = (x^2+9)(x^2-\frac{1}{2}) = 0,$ $(x^2+9)(2x^2-1) = 0,$

    or $2x^4 + 17x^2 - 9 = 0$

d) $(x-2)^3(x+1) = 0$ or $x^4 - 5x^3 + 6x^2 + 4x - 8 = 0$

IDENTICALLY EQUAL POLYNOMIALS.

20. Find the values of A and B for which the equation $A(2x+3) + B(x-4) = 3x + 10$ is an identity.

    Write the equation as $(2A+B)x + 3A - 4B = 3x + 10.$
    This is an identity if and only if $2A+B = 3$, $3A-4B = 10$. Solving, $A = 2$ and $B = -1$.

21. Find the values of A,B,C for which $A(x-3)(x-1) + B(x+1)(x-1) + C(x+1)(x-3) = 6x - 10$ is an identity.

    Write the equation as $A(x^2-4x+3) + B(x^2-1) + C(x^2-2x-3) = 6x - 10$

or $(A+B+C)x^2 + (-4A-2C)x + 3A - B - 3C = 6x - 10.$ This is an iden-

tity if and only if $A+B+C = 0$, $-4A-2C = 6$, $3A-B-3C = -10$. Solving, $A = -2$, $B = 1$, $C = 1$.

COMPLEX AND IRRATIONAL ROOTS.

22. Each given number is a root of an equation with *real coefficients*. What other number is a root?   *a)* $2i$,   *b)* $-3+2i$,   *c)* $-3-i\sqrt{2}$.

     a) $-2i$,   *b)* $-3-2i$,   *c)* $-3+i\sqrt{2}$

23. Each given number is a root of an equation with *rational coefficients*. What other number is a root?   *a)* $-\sqrt{7}$,   *b)* $-4+2\sqrt{3}$,   *c)* $5-\frac{1}{2}\sqrt{2}$.

     a) $\sqrt{7}$,   *b)* $-4-2\sqrt{3}$,   *c)* $5+\frac{1}{2}\sqrt{2}$

24. Criticize the validity of each of the following conclusions.

     a) $x^3+7x-6i=0$ has $x=i$ as a root; hence $x=-i$ is a root.

     b) $x^3+(1-2\sqrt{3})x^2+(5-2\sqrt{3})x+5=0$ has $\sqrt{3}-i\sqrt{2}$ as a root; hence $\sqrt{3}+i\sqrt{2}$ is a root.

     c) $x^4+(1-2\sqrt{2})x^3+(4-2\sqrt{2})x^2+(3-4\sqrt{2})x+1=0$ has $x=-1+\sqrt{2}$ as a root; hence $x=-1-\sqrt{2}$ is a root.

     a) $x=-i$ is not necessarily a root, since not all the coefficients of the given equation are *real*. By substitution it is, in fact, found that $x=-i$ is not a root.

     b) The conclusion is valid, since the given equation has real coefficients.

     c) $x=-1-\sqrt{2}$ is not necessarily a root, since not all the coefficients of the given equation are *rational*. By substitution it is found that $x=-1-\sqrt{2}$ is not a root.

25. Write the equation of lowest degree with real coefficients having 2 and $1-3i$ as two of its roots.

$$(x-2)[x-(1-3i)][x-(1+3i)] = (x-2)(x^2-2x+10) = 0 \quad \text{or} \quad x^3-4x^2+14x-20=0$$

26. Form the equation of lowest degree with rational coefficients having $-1+\sqrt{5}$ and -6 as two of its roots.

$$[x-(-1+\sqrt{5})][x-(-1-\sqrt{5})](x+6) = (x^2+2x-4)(x+6) = 0 \quad \text{or} \quad x^3+8x^2+8x-24=0$$

27. Form the quartic equation with rational coefficients having as two of its roots   *a)* $-5i$ and $\sqrt{6}$,   *b)* $2+i$ and $1-\sqrt{3}$.

     a) $(x+5i)(x-5i)(x-\sqrt{6})(x+\sqrt{6}) = (x^2+25)(x^2-6) = 0$ or $x^4+19x^2-150=0$

     b) $[x-(2+i)][x-(2-i)][x-(1-\sqrt{3})][x-(1+\sqrt{3})] = (x^2-4x+5)(x^2-2x-2) = 0$ or

$$x^4-6x^3+11x^2-2x-10=0$$

28. Find the four roots of $x^4+2x^2+1=0$.

$$x^4+2x^2+1 = (x^2+1)^2 = [(x+i)(x-i)]^2 = 0. \quad \text{The roots are} \quad i, \; i, \; -i, \; -i.$$

29. Solve $x^4-3x^3+5x^2-27x-36=0$, given that one root is a pure imaginary number of the form bi where b is real.

     Substituting bi for x, $b^4+3b^3i-5b^2-27bi-36=0$.

     Equating real and imaginary parts to zero:

$$b^4-5b^2-36=0, \quad (b^2-9)(b^2+4)=0 \quad \text{and} \quad b=\pm3 \quad \text{since } b \text{ is real;}$$

$$3b^3-27b=0, \quad 3b(b^2-9)=0 \quad \text{and} \quad b=0, \pm3.$$

The common solution is $b = \pm 3$; hence two roots are $\pm 3i$ and $(x - 3i)(x + 3i) = x^2 + 9$ is a factor of $x^4 - 3x^3 + 5x^2 - 27x - 36$. By division the other factor is $x^2 - 3x - 4 = (x - 4)(x + 1)$, and the other two roots are $4, -1$.

The four roots are $\pm 3i, 4, -1$.

30. Form the equation of lowest degree with *rational coefficients*, one of whose roots is:
    a) $\sqrt{3} - \sqrt{2}$, b) $\sqrt{2} + \sqrt{-1}$.

    a) Let $x = \sqrt{3} - \sqrt{2}$.

    Squaring both sides, $x^2 = 3 - 2\sqrt{6} + 2 = 5 - 2\sqrt{6}$ and $x^2 - 5 = -2\sqrt{6}$.

    Squaring again, $x^4 - 10x^2 + 25 = 24$ and $x^4 - 10x^2 + 1 = 0$.

    b) Let $x = \sqrt{2} + \sqrt{-1}$.

    Squaring both sides, $x^2 = 2 + 2\sqrt{-2} - 1 = 1 + 2\sqrt{-2}$ and $x^2 - 1 = 2\sqrt{-2}$.

    Squaring again, $x^4 - 2x^2 + 1 = -8$ and $x^4 - 2x^2 + 9 = 0$.

31. a) Write the equation of lowest degree with *constant* (real or complex) coefficients having the roots 2 and $1 - 3i$.   Compare with Problem 25 above.

    b) Write the equation of lowest degree with *real* coefficients having the roots -6 and $-1 + \sqrt{5}$. Compare with Problem 26 above.

    a) $(x - 2)[x - (1 - 3i)] = 0$ or $x^2 - 3(1 - i)x + 2 - 6i = 0$

    b) $(x + 6)[x - (-1 + \sqrt{5})] = 0$ or $x^2 + (7 - \sqrt{5})x - 6(\sqrt{5} - 1) = 0$

RATIONAL ROOTS.

32. Obtain the rational roots, if any, of each of the following equations.

    a) $x^4 - 2x^2 - 3x - 2 = 0$

    The rational roots are limited to the integral factors of 2, which are $\pm 1, \pm 2$.

    Testing these values for x in order, $+1, -1, +2, -2$, by synthetic division or by substitution, we find that the only rational roots are -1 and 2.

    b) $x^3 - x - 6 = 0$

    The rational roots are limited to the integral factors of 6, which are $\pm 1, \pm 2, \pm 3, \pm 6$.

    Testing these values for x in order, $+1, -1, +2, -2, +3, -3, +6, -6$, the only rational root obtained is 2.

    c) $2x^3 + x^2 - 7x - 6 = 0$

    If b/c (in lowest terms) is a rational root, the only possible values of b are $\pm 1, \pm 2, \pm 3, \pm 6$; and the only possible values of c are $\pm 1, \pm 2$. Hence the possible rational roots are limited to the following numbers: $\pm 1, \pm 2, \pm 3, \pm 6, \pm 1/2, \pm 3/2$.

    Testing these values for x, we obtain $-1, 2, -3/2$ as the rational roots.

    d) $2x^4 + x^2 + 2x - 4 = 0$

    If b/c is a rational root, the values of b are limited to $\pm 1, \pm 2, \pm 4$; and the values of c are limited to $\pm 1, \pm 2$. Hence the possible rational roots are limited to the numbers $\pm 1, \pm 2, \pm 4, \pm 1/2$.

    Testing these values for x, we find that there are no rational roots.

33. Solve the equation $x^3 - 2x^2 - 31x + 20 = 0$.

Any rational root of this equation is an integral factor of 20. Then the possibilities for rational roots are: $\pm 1, \pm 2, \pm 4, \pm 5, \pm 10, \pm 20$.

Testing these values for x by synthetic division, we find that -5 is a root.

$$
\begin{array}{r}
1 - 2 - 31 + 20 \underline{|-5} \\
 - 5 + 35 - 20 \\
\hline
1 - 7 + 4 + 0
\end{array}
$$

The depressed equation $x^2 - 7x + 4 = 0$ has irrational roots $7/2 \pm \sqrt{33}/2$.

Hence the three roots of the given equation are -5, $7/2 \pm \sqrt{33}/2$.

34. Solve the equation $2x^4 - 3x^3 - 7x^2 - 8x + 6 = 0$.

If b/c is a rational root, the only possible values of b are $\pm 1, \pm 2, \pm 3, \pm 6$; and the only possible values of c are $\pm 1, \pm 2$. Hence the possibilities for rational roots are $\pm 1, \pm 2, \pm 3, \pm 6, \pm 1/2, \pm 3/2$.

Testing these values for x by synthetic division, we find that 3 is a root.

$$
\begin{array}{r}
2 - 3 - 7 - 8 + 6 \underline{|3} \\
 + 6 + 9 + 6 - 6 \\
\hline
2 + 3 + 2 - 2 + 0
\end{array}
$$

The first depressed equation $2x^3 + 3x^2 + 2x - 2 = 0$ is tested and $1/2$ is obtained as a root.

The second depressed equation $2x^2 + 4x + 4 = 0$ or $x^2 + 2x + 2 = 0$ has the complex roots $-1 \pm i$.

$$
\begin{array}{r}
2 + 3 + 2 - 2 \underline{|1/2} \\
 + 1 + 2 + 2 \\
\hline
2 + 4 + 4 + 0
\end{array}
$$

The four roots are $3, 1/2, -1 \pm i$.

35. Prove that $\sqrt{3} + \sqrt{2}$ is an irrational number.

Let $x = \sqrt{3} + \sqrt{2}$; then $x^2 = (\sqrt{3} + \sqrt{2})^2 = 3 + 2\sqrt{6} + 2 = 5 + 2\sqrt{6}$ and $x^2 - 5 = 2\sqrt{6}$.

Squaring again, $x^4 - 10x^2 + 25 = 24$ or $x^4 - 10x^2 + 1 = 0$. The only possible rational roots of this equation are ± 1. Testing these values, we find that there is no rational root. Hence $x = \sqrt{3} + \sqrt{2}$ is irrational.

GRAPHICAL PROCEDURE FOR FINDING REAL ROOTS.

36. Graph $f(x) = x^3 + x - 3$. From the graph determine
a) the number of positive, negative and complex roots of $x^3 + x - 3 = 0$,
b) an approximate value for any real root of $x^3 + x - 3 = 0$ accurate to two decimal places.

x	-3	-2	-1	0	1	2	3	4
$f(x)$	-33	-13	-5	-3	-1	7	27	65

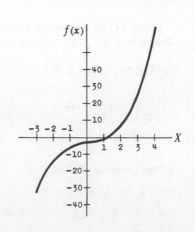

a) From the graph it is seen that there is one positive and no negative real root.

Hence there are two conjugate complex roots.

b) The approximate value of the real root is $1+$. Note that $f(x)$ changes sign between $x = 1$ and $x = 2$.

It appears that the root is closer to 1.0 than to 2.0. To obtain greater accuracy, construct a more accurate graph for values of x between 1 and 2.

x	1.0	1.1	1.2	1.3
$f(x)$	-1	$-.569$	$-.072$	$+.497$

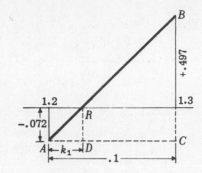

Since $f(x)$ changes sign between $x = 1.2$ and $x = 1.3$, the root lies between these values. The line AB in the figure is an approximation to the actual curve joining points A and B. The root is approximately located at R and its value is $1.2 + k_1$. From similar triangles ABC and ARD we have

$$\frac{k_1}{.1} = \frac{.072}{.072 + .497} = \frac{.072}{.569} = .1+$$

or $k_1 = .01+$. Hence the root is $1.21+$.

This process of locating the root by assuming a straight line joining A and B is called *linear interpolation*.

x	1.21	1.22
$f(x)$	$-.0184$	$+.0358$

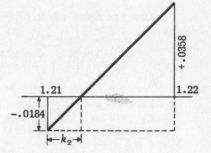

The root is approximately $1.21 + k_2$. By interpolation,

$$\frac{k_2}{.01} = \frac{.0184}{.0184 + .0358} = \frac{.0184}{.0542} = .3+$$

or $k_2 = .003+$. Hence the root is $1.21 + .003$ or $1.213+$.

Thus the root is 1.21 accurate to two decimal places.

UPPER AND LOWER LIMITS OF REAL ROOTS.

37. Find upper and lower limits of the real roots of *a*) $x^3 - 3x^2 + 5x + 4 = 0$, *b*) $x^3 + x^2 - 6 = 0$.

a) The possible rational roots are $\pm 1, \pm 2, \pm 4$.

Testing for Upper Limit.

```
1 - 3 + 5 + 4|1          1 - 3 + 5 + 4|2          1 - 3 + 5 + 4|3
  + 1 - 2 + 3              + 2 - 2 + 6              + 3 + 0 + 15
1 - 2 + 3 + 7          1 - 1 + 3 + 10          1 + 0 + 5 + 19
```

Since all the numbers in the third row of the synthetic division of $f(x)$ by $x - 3$ are positive (or zero), an upper limit of the roots is 3, i.e. no root is greater than 3.

Testing for Lower Limit.

```
1 - 3 + 5 + 4|-1
  - 1 + 4 - 9
1 - 4 + 9 - 5
```

Since the numbers in the third row are alternately positive and negative, -1 is a lower limit of the roots, i.e. no root is less than -1.

b) The possible rational roots are $\pm 1, \pm 2, \pm 3, \pm 6$.

Testing for Upper Limit.

```
1 + 1 + 0 - 6|1          1 + 1 + 0 - 6|2
  + 1 + 2 + 2              + 2 + 6 + 12
1 + 2 + 2 - 4          1 + 3 + 6 + 6
```

Hence 2 is an upper limit of the roots.

Testing for Lower Limit.

$$1 + 1 + 0 - 6\underline{|-1}$$
$$\underline{\quad - 1 - 0 + 0\quad}$$
$$1 + 0 + 0 - 6$$

Since all the numbers of the third row are alternately positive and negative (or zero), a lower limit to the roots is –1.

38. Determine the rational roots of $4x^3 + 15x - 36 = 0$ and thus solve the equation completely.

The possible rational roots are ±1, ±2, ±3, ±4, ±6, ±9, ±12, ±18, ±36, ±1/2, ±3/2, ±9/2, ±1/4, ±3/4, ±9/4. To avoid testing all of these possibilities, find upper and lower limits of the roots.

Testing for Upper Limit.

$$4 + 0 + 15 - 36\underline{|1} \qquad 4 + 0 + 15 - 36\underline{|2}$$
$$\underline{\quad + 4 + \ \ 4 + 19\quad} \qquad \underline{\quad + 8 + 16 + 62\quad}$$
$$4 + 4 + 19 - 17 \qquad\quad 4 + 8 + 31 + 26$$

Hence no (real) root is greater than or equal to 2.

Testing for Lower Limit.

$$4 + 0 + 15 - 36\underline{|-1}$$
$$\underline{\quad - 4 + \ \ 4 - 19\quad}$$
$$4 - 4 + 19 - 55$$

Hence no real root is less than or equal to –1.

The only possible rational roots greater than –1 and less than 2 are +1, ±1/2, ±3/2, ±1/4, ±3/4. Testing these we find that 3/2 is the only rational root.

$$4 + 0 + 15 - 36\underline{|3/2}$$
$$\underline{\quad + 6 + \ \ 9 + 36\quad}$$
$$4 + 6 + 24 + \ \ 0$$

The other roots are solutions of $4x^2 + 6x + 24 = 0$ or $2x^2 + 3x + 12 = 0$, i.e. $x = -\dfrac{3}{4} \pm \dfrac{\sqrt{87}}{4}i$.

DESCARTES' RULE OF SIGNS.

39. Employing Descartes' Rule of Signs, what may be inferred as to the number of positive, negative and complex roots of the following equations?

a) $2x^3 + 3x^2 - 13x + 6 = 0$ 　　　*d)* $2x^4 + 7x^2 + 6 = 0$ 　　　*g)* $x^6 + x^3 - 1 = 0$

b) $x^4 - 2x^2 - 3x - 2 = 0$ 　　　*e)* $x^4 - 3x^2 - 4 = 0$ 　　　*h)* $x^6 - 3x^2 - 4x + 1 = 0$

c) $x^2 - 2x + 7 = 0$ 　　　　　　*f)* $x^3 + 3x - 14 = 0$

a) There are 2 variations of sign in $f(x) = 2x^3 + 3x^2 - 13x + 6$. There is 1 variation of sign in $f(-x) = -2x^3 + 3x^2 + 13x + 6$. Hence there are at most 2 positive roots and 1 negative root.

The roots may be: (1) 2 positive, 1 negative, 0 complex; or (2) 0 positive, 1 negative, 2 complex. (Complex roots occur in conjugate pairs.)

b) There is 1 variation of sign in $f(x) = x^4 - 2x^2 - 3x - 2$, and 3 variations of sign in $f(-x) = x^4 - 2x^2 + 3x - 2$. Hence there are at most 1 positive root and 3 negative roots.

The roots may be: (1) 1 positive, 3 negative, 0 complex;
　　　　　　　or (2) 1 positive, 1 negative, 2 complex.

c) There are 2 variations of sign in $f(x) = x^2 - 2x + 7$, and no variation of sign in $f(-x) = x^2 + 2x + 7$.

Hence the roots may be: (1) 2 positive, 0 negative, 0 complex;
　　　　　　　　　or (2) 0 positive, 0 negative, 2 complex.

d) Neither $f(x) = 2x^4 + 7x^2 + 6$ nor $f(-x) = 2x^4 + 7x^2 + 6$ has a variation of sign. Hence all 4 roots are complex, since $f(0) \neq 0$.

e) There is 1 variation of sign in $f(x) = x^4 - 3x^2 - 4 = 0$, and 1 variation of sign in $f(-x) = x^4 - 3x^2 - 4$.

  Hence the roots are: 1 positive, 1 negative, 2 complex.

f) There is 1 variation of sign in $f(x) = x^3 + 3x - 14$, and no variation in $f(-x) = -x^3 - 3x - 14$.

  Hence the roots are: 1 positive, 2 complex.

g) There is 1 variation of sign in $f(x) = x^6 + x^3 - 1$, and 1 variation in $f(-x) = x^6 - x^3 - 1$.

  Hence the roots are: 1 positive, 1 negative, 4 complex.

h) There are 2 variations of sign in $f(x) = x^6 - 3x^2 - 4x + 1$, and 2 variations of sign in $f(-x) = x^6 - 3x^2 + 4x + 1$.

  Hence the roots may be:
      (1) 2 positive, 2 negative, 2 complex;   (3) 0 positive, 2 negative, 4 complex;
      (2) 2 positive, 0 negative, 4 complex;   (4) 0 positive, 0 negative, 6 complex.

40. Determine the nature of the roots of $x^n - 1 = 0$ when n is a positive integer and *a*) n is even, *b*) n is odd.

a) $f(x) = x^n - 1$ has 1 variation of sign, and $f(-x) = x^n - 1$ has 1 variation of sign.
  Hence the roots are: 1 positive, 1 negative, $(n-2)$ complex.

b) $f(x) = x^n - 1$ has 1 variation of sign, and $f(-x) = -x^n - 1$ has no variation of sign.
  Hence the roots are: 1 positive, 0 negative, $(n-1)$ complex.

41. Obtain the rational roots, if any, of each equation, making use of Descartes' rule of signs.
  a) $x^3 - x^2 + 3x - 27 = 0$,   *b*) $x^3 + 2x + 12 = 0$,   *c*) $2x^5 + x - 66 = 0$,   *d*) $3x^4 + 7x^2 + 6 = 0$.

a) By Descartes' rule of signs, the equation has 3 or 1 positive roots and no negative root. Hence the rational roots are limited to positive integral factors of 27, which are 1, 3, 9, 27.

  Testing these values for x, the only rational root obtained is 3.

b) By the rule of signs, the equation has no positive root and 1 negative root. Hence the rational roots are limited to the negative integral factors of 12, i.e. to $-1, -2, -3, -4, -6, -12$.

  Testing these values for x, we obtain -2 as the only rational root.

c) By the rule of signs, the equation has 1 positive root and no negative root. Hence the rational roots are limited to positive rational numbers of the form b/c where b is limited to integral factors of 66 and c is limited to integral factors of 2. The possible rational roots are then $1, 2, 3, 6, 11, 22, 33, 66, 1/2, 3/2, 11/2, 33/2$.

  Testing these values for x, we obtain 2 as the only rational root.

d) The equation has no real root, since neither $f(x) = 3x^4 + 7x^2 + 6$ nor $f(-x) = 3x^4 + 7x^2 + 6$ has a variation of sign and $f(0) \neq 0$.

  Hence all of its four roots are complex.

MULTIPLYING EACH ROOT BY A CONSTANT.

42. Form an equation in y whose roots equal twice those of the equation $x^3 - 6x^2 + 11x - 6 = 0$.

Any root y of the required equation is equal to twice the corresponding root x of the given equation. Hence $y = 2x$ or $x = y/2$, and the required equation is

$$(\tfrac{y}{2})^3 - 6(\tfrac{y}{2})^2 + 11(\tfrac{y}{2}) - 6 = 0 \quad \text{or} \quad y^3 - 2(6y^2) + 2^2(11y) - 2^3(6) = 0 \quad \text{or} \quad y^3 - 12y^2 + 44y - 48 = 0.$$

Check: The roots of $x^3 - 6x^2 + 11x - 6 = 0$ are $1, 2, 3$.
         The roots of $y^3 - 12y^2 + 44y - 48 = 0$ are $2, 4, 6$.

43. Form equations in y whose roots are equal to those of the given equations multiplied by the numbers in parentheses.

a) $x^3 - x^2 - 7x + 3 = 0$ (2)   d) $32x^4 - 2x - 1 = 0$ (4)

b) $x^3 - 19x - 30 = 0$ (3)   e) $2x^5 + 3x^2 + 1 = 0$ (-3)

c) $x^4 + \dfrac{3x^2}{5} - \dfrac{4}{25} = 0$ (5)   f) $x^3 - 12x^2 - 16x + 192 = 0$ (1/4)

a) $y^3 - 2(y^2) - 2^2(7y) + 2^3(3) = 0$ or $y^3 - 2y^2 - 28y + 24 = 0$

b) $y^3 + 3(0y^2) - 3^2(19y) - 3^3(30) = 0$ or $y^3 - 171y - 810 = 0$

c) $y^4 + 5(0y^3) + 5^2(\dfrac{3y^2}{5}) + 5^3(0y) - 5^4(\dfrac{4}{25}) = 0$ or $y^4 + 15y^2 - 100 = 0$

d) $32y^4 - 4^3(2y) - 4^4 = 0$ or $32y^4 - 128y - 256 = 0$ or $y^4 - 4y - 8 = 0$

e) $2y^5 + (-3)^3(3y^2) + (-3)^5(1) = 0$ or $2y^5 - 81y^2 - 243 = 0$

f) $y^3 - \dfrac{12y^2}{4} - \dfrac{16y}{4^2} + \dfrac{192}{4^3} = 0$ or $y^3 - 3y^2 - y + 3 = 0$

44. Obtain equations in y whose roots are equal to those of the given equations multiplied by the smallest positive integer which will make all the coefficients integers and the coefficient of the highest power unity.

a) $x^3 - x^2 - \dfrac{3x}{2} + \dfrac{9}{8} = 0$,   b) $x^4 - 3x^3 - \dfrac{x}{27} + \dfrac{1}{9} = 0$,   c) $8x^3 + 18x^2 + x - 6 = 0$,   d) $5x^3 + 3 = 0$

a) The equation in y has roots twice those of the given equation.

$$y^3 - 2(y^2) - 2^2(\tfrac{3y}{2}) + 2^3(\tfrac{9}{8}) = 0 \quad \text{or} \quad y^3 - 2y^2 - 6y + 9 = 0$$

b) The equation in y has roots three times those of the given equation.

$$y^4 - 3(3y^3) - 3^3(\tfrac{y}{27}) + 3^4(\tfrac{1}{9}) = 0 \quad \text{or} \quad y^4 - 9y^3 - y + 9 = 0$$

c) Divide each term of the given equation by 8: $x^3 + \dfrac{9x^2}{4} + \dfrac{x}{8} - \dfrac{3}{4} = 0.$

The equation in y has roots four times those of the given equation.

$$y^3 + 4(\tfrac{9y^2}{4}) + 4^2(\tfrac{y}{8}) - 4^3(\tfrac{3}{4}) = 0 \quad \text{or} \quad y^3 + 9y^2 + 2y - 48 = 0$$

d) Divide each term of the given equation by 5: $x^3 + \dfrac{3}{5} = 0.$

The equation in y has roots five times those of the given equation.

$$y^3 + 5^3(3/5) = 0 \quad \text{or} \quad y^3 + 75 = 0$$

45. Solve the equation $54x^3 - 9x^2 - 12x - 4 = 0$.

Dividing by 54, the equation in the p-form is $x^3 - \dfrac{x^2}{6} - \dfrac{2x}{9} - \dfrac{2}{27} = 0$.

Multiply the roots by 6 to obtain an equation in y with integral coefficients.

$$y^3 - 6(\dfrac{y^2}{6}) - 6^2(\dfrac{2y}{9}) - 6^3(\dfrac{2}{27}) = 0 \quad \text{or} \quad y^3 - y^2 - 8y - 16 = 0$$

The rational roots of the equation in y are limited to integral factors of 16, which are ±1, ±2, ±4, ±8, ±16. Testing these values by synthetic division, 4 is obtained as a root of the equation in y. The depressed equation, $y^2 + 3y + 4 = 0$, has roots $-\dfrac{3}{2} \pm \dfrac{i\sqrt{7}}{2}$.

$$
\begin{array}{r}
1 - 1 - 8 - 16 \underline{|4} \\
+ 4 + 12 + 16 \\
\hline
1 + 3 + 4 + 0
\end{array}
$$

The roots of the equation in y divided by 6 give the roots of the equation in x, i.e. $x = y/6$.

Hence the required roots are $\dfrac{2}{3}$, $-\dfrac{1}{4} \pm \dfrac{i\sqrt{7}}{12}$.

46. Solve the equation $64x^4 - 32x^3 + 4x^2 - 8x - 3 = 0$.

Dividing by 64, the equation in the p-form is $x^4 - \dfrac{x^3}{2} + \dfrac{x^2}{16} - \dfrac{x}{8} - \dfrac{3}{64} = 0$.

Multiply the roots by 4 to obtain an equation in y with integral coefficients.

$$y^4 - 4(\dfrac{y^3}{2}) + 4^2(\dfrac{y^2}{16}) - 4^3(\dfrac{y}{8}) - 4^4(\dfrac{3}{64}) = 0 \quad \text{or} \quad y^4 - 2y^3 + y^2 - 8y - 12 = 0$$

The rational roots of the equation in y are limited to integral factors of 12, which are ±1, ±2, ±3, ±4, ±6, ±12. Testing these values by synthetic division, -1 is obtained as a root of the equation in y.

$$
\begin{array}{r}
1 - 2 + 1 - 8 - 12 \underline{|-1} \\
- 1 + 3 - 4 + 12 \\
\hline
1 - 3 + 4 - 12 + 0
\end{array}
$$

The first depressed equation is $y^3 - 3y^2 + 4y - 12 = 0$ which by the rule of signs has no negative root, and its rational roots are limited to the positive integral factors of 12, which are 1, 2, 3, 4, 6, 12. When these values are tested, 3 is obtained as a root. The second depressed equation, $y^2 + 4 = 0$, has roots $\pm2i$.

$$
\begin{array}{r}
1 - 3 + 4 - 12 \underline{|3} \\
+ 3 + 0 + 12 \\
\hline
1 + 0 + 4 + 0
\end{array}
$$

The roots of the transformed equation in y are -1, 3, $\pm2i$. Hence the roots of the given equation in x are $-1/4$, $3/4$ and $\pm i/2$, since $x = y/4$.

CHANGING THE SIGN OF EACH ROOT.

47. Write equations in y whose roots are numerically equal but opposite in sign to the roots of the following equations.

a) $x^3 + 7x^2 + 11x + 5 = 0$,   b) $x^4 + 3x^2 - x - 27 = 0$,   c) $2x^5 - 10x^4 - 3x + 15 = 0$

The required equations in y are obtained by changing the signs of the odd degree terms in the given equations and replacing x by y, or by letting $x = -y$.

a) $-y^3 + 7y^2 - 11y + 5 = 0$ or $y^3 - 7y^2 + 11y - 5 = 0$

The roots of $x^3 + 7x^2 + 11x + 5 = 0$ are $-5, -1, -1$; of $y^3 - 7y^2 + 11y - 5 = 0$ are $5, 1, 1$.

b) $y^4 + 3y^2 + y - 27 = 0$

c) $-2y^5 - 10y^4 + 3y + 15 = 0$ or $2y^5 + 10y^4 - 3y - 15 = 0$

DIMINISHING EACH ROOT BY A CONSTANT.

48. Form equations in y whose roots are equal to those of the given equations diminished by the numbers in parentheses.

a) $2x^3 - 17x^2 + 26x + 45 = 0$ (3)

b) $x^3 - x^2 - 17x - 15 = 0$ (–3)

c) $x^4 - 12x - 5 = 0$ (2)

d) $x^3 + 8x^2 - 2 = 0$ (0.4)

a)
```
2 - 17 + 26 + 45 |3
    +  6 - 33 - 21
─────────────────────
2 - 11 -  7 + 24
    +  6 - 15
─────────────────────
2 -  5  -22
    +  6
─────────────────────
2 +  1
```

The required equation is

$$2y^3 + y^2 - 22y + 24 = 0.$$

The equation in x has roots 5, 9/2, –1.
The equation in y has roots 2, 3/2, –4.

Otherwise: Let $y = x - 3$ or $x = y + 3$
            and obtain the same result.

b)
```
1 - 1 - 17 - 15 |-3
    - 3 + 12 + 15
─────────────────────
1 - 4 -  5 +  0
    - 3 + 21
─────────────────────
1 - 7 + 16
    - 3
─────────────────────
1 -10
```

The required equation is

$$y^3 - 10y^2 + 16y = 0.$$

The equation in x has roots –3, –1, 5.
The equation in y has roots 0, 2, 8.

Note that diminishing the roots by –3 is the same as increasing the roots by 3.

c)
```
1 + 0 + 0 - 12 - 5 |2
    + 2 + 4 +  8 - 8
──────────────────────
1 + 2 + 4 -  4 -13
    + 2 + 8 + 24
──────────────────────
1 + 4 + 12 + 20
    + 2 + 12
──────────────────────
1 + 6 + 24
    + 2
──────────────────────
1 + 8
```

The required equation is

$$y^4 + 8y^3 + 24y^2 + 20y - 13 = 0.$$

d)
```
1 + 8   + 0    - 2      |0.4
    + 0.4 + 3.36 + 1.344
──────────────────────────
1 + 8.4 + 3.36 - 0.656
    + 0.4 + 3.52
──────────────────────────
1 + 8.8 + 6.88
    + 0.4
──────────────────────────
1 + 9.2
```

The required equation is

$$y^3 + 9.2y^2 + 6.88y - 0.656 = 0.$$

RELATION BETWEEN ROOTS AND COEFFICIENTS.

49. Employing the relations between roots and coefficients, write the equation whose roots are 1, 3, –2, –4.

The equation in the p-form is $x^4 + p_1x^3 + p_2x^2 + p_3x + p_4 = 0.$

p_1 = –(sum of roots) = –(1 + 3 – 2 – 4) = 2.

p_2 = +(sum of products of roots taken two at a time)
    = (1)(3) + (1)(–2) + (1)(–4) + (3)(–2) + (3)(–4) + (–2)(–4) = –13.

p_3 = –(sum of products of roots taken three at a time)
    = –[(1)(3)(–2) + (1)(3)(–4) + (1)(–2)(–4) + (3)(–2)(–4)] = –14.

p_4 = +(product of all the roots) = (1)(3)(–2)(–4) = 24.

The required equation is $x^4 + 2x^3 - 13x^2 - 14x + 24 = 0.$

50. Given $x^3 - 8x^2 + 9x + k = 0$. Determine the integral value of k if one of the roots is twice another.

Let the roots be $a, 2a, b$.

Sum of roots = $-(-8) = 8 = a + 2a + b$ or   (1) $b = 8 - 3a$.
Sum of products of roots two at a time = $9 = a(2a) + a(b) + 2a(b)$ or   (2) $2a^2 + 3ab = 9$.
Product of all the roots = $-k = a(2a)(b)$ or   (3) $k = -2a^2 b$.

The pertinent solution of (1) and (2) is $a = 3$, $b = -1$. Substituting in (3), $k = 18$.

51. Transform $2x^4 + 8x^3 + 5x^2 - 3x + 6 = 0$ into an equation in y in which the third degree term is missing.

First write the equation in the p-form to obtain the relationship between the roots and coefficients.

Sum of roots = $-8/2 = -4$. The sum of the roots of the required equation in y must be zero, or 4 more than the sum of the roots of the given equation. This is done by increasing each of the four roots of the given equation by 1, i.e. $y = x + 1$.

The required equation is $2y^4 - 7y^2 + 3y + 8 = 0$.

$$
\begin{array}{rrrrr|r}
2 & +8 & +5 & -3 & +6 & \underline{-1} \\
  & -2 & -6 & +1 & +2 & \\
\hline
2 & +6 & -1 & -2 & +8 & \\
  & -2 & -4 & +5 & & \\
\hline
2 & +4 & -5 & +3 & & \\
  & -2 & -2 & & & \\
\hline
2 & +2 & -7 & & & \\
  & -2 & & & & \\
\hline
2 & +0 & & & & \\
\end{array}
$$

52. Determine the sum of the squares of the roots of $x^3 - 2x^2 - 23x + k = 0$.

Let the roots be a, b, c. Since $(a + b + c)^2 = a^2 + b^2 + c^2 + 2(ab + bc + ca)$, we have
$a^2 + b^2 + c^2 = (a + b + c)^2 - 2(ab + bc + ca) = 2^2 - 2(-23) = 50$.

53. Two roots of the incomplete equation $3x^3 - 17x^2 + \cdots = 0$ are $2, 4$. Find the third root and complete the equation.

Let the equation be $3x^3 - 17x^2 + hx + k = 0$ and its roots $2, 4, r$.

Sum of roots = $17/3 = 2 + 4 + r$; then $r = -1/3$.

Sum of products of roots two at a time = $h/3 = 2(4) + 2(-1/3) + 4(-1/3) = 18/3$; then $h = 18$.

Product of roots = $-k/3 = 2(4)(-1/3) = -8/3$; then $k = 8$.

Thus $r = -1/3$, and the complete equation is $3x^3 - 17x^2 + 18x + 8 = 0$.

Another Method. Sum of roots = $17/3 = 2 + 4 + r$, $r = -1/3$. Thus the three roots are $2, 4, -1/3$. An equation having only these three roots is $(x - 2)(x - 4)(3x + 1) = 0$ or $3x^3 - 17x^2 + 18x + 8 = 0$.

54. Given $x^3 - 9x + k = 0$, where k is a constant. Find k if
a) one root is the negative of another,
b) there is a double root,
c) the three roots are in geometric progression,
d) the three roots are in arithmetic progression,
e) one root is $\sqrt[3]{9} + \sqrt[3]{3}$.

a) Let the roots be $a, -a, b$.
   Then $a + (-a) + b = 0$, since the second degree term is missing. Hence $b = 0$.
   Product of roots = $a(-a)(b) = 0 = -k$. Hence $k = 0$.

b) Let the roots be a, a, b.
   Sum of roots = $a + a + b = 0$. (1)
   Product of roots taken two at a time = $2ab + a^2 = -9$. (2)
   Product of roots = $a^2 b = -k$. (3)

Solving (1) and (2) simultaneously, $a = \pm\sqrt{3}$, $b = \mp 2\sqrt{3}$. Substituting in (3), $k = \pm 6\sqrt{3}$.

c) Let the roots be a/r, a, ar.

Then $a/r + a + ar = a(1/r + 1 + r) = 0$, $a^2/r + a^2 + a^2r = a^2(1/r + 1 + r) = -9$, $a^3 = -k$.
But neither a nor $(1/r + 1 + r)$ can be zero. Why? Hence no value of k exists for which the roots are in geometric progression.

d) Let the roots be $a - d$, a, $a + d$.

Sum of roots $= a - d + a + a + d = 3a = 0$ or $a = 0$. Hence $k = -(a - d)(a)(a + d) = 0$.

e) Substituting in the given equation, $(\sqrt[3]{9} + \sqrt[3]{3})^3 - 9(\sqrt[3]{9} + \sqrt[3]{3}) + k = 0$.

Expanding, $9 + 9\sqrt[3]{9} + 9\sqrt[3]{3} + 3 - 9(\sqrt[3]{9} + \sqrt[3]{3}) + k = 0$ and $k = -12$.

55. If the roots of $2x^3 - 5x^2 + 6x - 1 = 0$ are a, b, c, form the equation whose roots are $\frac{1}{a}, \frac{1}{b}, \frac{1}{c}$.

Let $y = 1/x$. Then $2(1/y)^3 - 5(1/y)^2 + 6(1/y) - 1 = 0$ or $y^3 - 6y^2 + 5y - 2 = 0$ has roots which are reciprocals of the roots of the given equation. (Compare the coefficients of the given and required equations.)

56. The roots of $x^3 - 2x^2 + 3x - 4 = 0$ are a, b, c. Find the value of each of the following.

a) $a + b + c$ c) abc e) $a^3 + b^3 + c^3$

b) $ab + bc + ca$ d) $a^2 + b^2 + c^2$ f) $\frac{1}{a} + \frac{1}{b} + \frac{1}{c}$

a) 2 (sum of roots)

b) 3 (sum of products of roots taken two at a time)

c) 4 (product of roots)

d) $a^2 + b^2 + c^2 = (a + b + c)^2 - 2(ab + bc + ca) = 2^2 - 2(3) = -2$

e) $a^3 - 2a^2 + 3a - 4 = 0$, $b^3 - 2b^2 + 3b - 4 = 0$ and $c^3 - 2c^2 + 3c - 4 = 0$.

By addition, $a^3 + b^3 + c^3 - 2(a^2 + b^2 + c^2) + 3(a + b + c) - 12 = 0$.

Then $a^3 + b^3 + c^3 = 2(-2) - 3(2) + 12 = 2$.

f) $abc(\frac{1}{a} + \frac{1}{b} + \frac{1}{c}) = bc + ac + ab$.   Then $4(\frac{1}{a} + \frac{1}{b} + \frac{1}{c}) = 3$ and $\frac{1}{a} + \frac{1}{b} + \frac{1}{c} = \frac{3}{4}$.

Another Method. The equation whose roots are the reciprocals of the roots of the given equation is $4y^3 - 3y^2 + 2y - 1 = 0$.

The sum of the roots of this equation is $\frac{3}{4}$. Hence $\frac{1}{a} + \frac{1}{b} + \frac{1}{c} = \frac{3}{4}$.

HORNER'S METHOD FOR IRRATIONAL ROOTS.

57. Find a positive root of $x^3 - 2x^2 - 2x - 7 = 0$ correct to three decimal places.

a) By Descartes' rule of signs, this equation has one positive root.

b) Plot the graph of $f(x) = x^3 - 2x^2 - 2x - 7$ from $x = 0$ to $x = 4$.

Since $f(x)$ changes in sign between $x = 3$ and $x = 4$, there is a root between $x = 3$ and $x = 4$, and the first figure of the root is 3. This fact is confirmed by the graph.

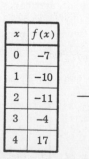

x	$f(x)$
0	−7
1	−10
2	−11
3	−4
4	17

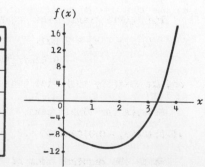

c) Transform the given equation by diminishing its roots by 3, i.e. $x_1 = x - 3$.

```
1 - 2 -  2 - 7 |3
    + 3 +  3 + 3
  1 + 1 +  1 - 4
      + 3 + 12
  1 + 4 + 13
      + 3
  1 + 7
```

This gives $x_1^3 + 7x_1^2 + 13x_1 - 4 = 0$ which has a root between 0 and 1.

d) Evaluating $f_1(x_1) = x_1^3 + 7x_1^2 + 13x_1 - 4$ for successive tenths, we find it is negative for $x_1 = 0.2$ (−1.112) and positive for $x_1 = 0.3$ (0.557).

```
1 + 7    + 13     - 4      |0.2
    + 0.2 + 1.44 + 2.888
1 + 7.2 + 14.44 - 1.112
```

Hence the positive root of $f_1(x_1) = 0$ is 0.2+ (and the root of the given equation is 3.2+).

```
1 + 7    + 13     - 4      |0.3
    + 0.3 + 2.19 + 4.557
1 + 7.3 + 15.19 + 0.557
```

e) Diminish the roots of $f_1(x_1) = 0$ by 0.2 ($x_2 = x_1 - 0.2$).

```
1 + 7    + 13     - 4      |0.2
    + 0.2 + 1.44 + 2.888
1 + 7.2 + 14.44 - 1.112
    + 0.2 + 1.48
1 + 7.4 + 15.92
    + 0.2
1 + 7.6
```

The second transformed equation is

$$f_2(x_2) = x_2^3 + 7.6x_2^2 + 15.92x_2 - 1.112 = 0$$

and its positive root lies between 0 and 0.1.

To approximate this root to the nearest hundredth, solve the linear part. This gives $15.92x_2 - 1.112 = 0$, or $x_2 = 0.06+$, which indicates that the positive root of $f_2(x_2) = 0$ lies between 0.06 and 0.07 (and the root of the given equation is 3.26+).

f) Diminish the roots of $f_2(x_2) = 0$ by 0.06.

```
1 + 7.6  + 15.92   - 1.112    |0.06
    + 0.06 + 0.4596 + 0.982776
1 + 7.66 + 16.3796 - 0.129224
     + 0.06 + 0.4632
1 + 7.72 + 16.8428
     + 0.06
1 + 7.78
```

The third transformed equation is

$$f_3(x_3) = x_3^3 + 7.78x_3^2 + 16.8428x_3 - 0.129224 = 0$$

and its positive root lies between 0 and 0.01.

Solving the linear part,

$$16.8428x_3 - 0.129224 = 0 \quad \text{or} \quad x_3 = 0.007+.$$

This indicates that the positive root of $f_3(x_3) = 0$ lies between 0.007 and 0.008 (and the root of the given equation is 3.267+).

g) Diminish the roots of $f_3(x_3) = 0$ by 0.007.

```
1 + 7.78  + 16.8428    - 0.129224    |0.007
    + 0.007 + 0.054509 + 0.118281163
1 + 7.787 + 16.897309 - 0.010942837
     + 0.007 + 0.054558
1 + 7.794 + 16.951867
     + 0.007
1 + 7.801
```

The fourth transformed equation is

$$f_4(x_4) = x_4^3 + 7.801x_4^2 + 16.951867x_4 - 0.010942837 = 0$$

and its positive root lies between 0 and 0.001.

Solving the linear part,

$$16.951867x_4 - 0.010942837 = 0 \quad \text{or} \quad x_4 = 0.0006+.$$

Hence the required root of the given equation is 3.2676+, or 3.268 correct to three decimal places.

58. Find a real root of $x^3 + 3x + 8 = 0$ correct to two decimal places.

Let $f(x) = x^3 + 3x + 8$; then $f(-x) = -x^3 - 3x + 8$.

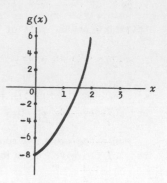

$g(x)$

a) By Descartes' rule of signs, $f(x) = 0$ has 1 real root which is negative and $f(-x) = 0$ has 1 real root which is positive.

b) Find the positive root of $-f(-x) = x^3 + 3x - 8 = 0$. This root with its sign changed is the required negative root of $f(x) = 0$.

c) Let $g(x) = x^3 + 3x - 8$. The positive root of $f(-x) = 0$ is the same as the positive root of $g(x) = 0$.

x	0	1	2
$g(x)$	–8	–4	6

Since $g(x)$ changes sign between $x = 1$ and $x = 2$, the first figure of the root of $g(x) = 0$ is 1.

d) Diminish the roots of $g(x) = 0$ by 1.

This gives $x_1^3 + 3x_1^2 + 6x_1 - 4 = 0$ which has a root between 0 and 1.

```
1 + 0 + 3 - 8 |1
    + 1 + 1 + 4
1 + 1 + 4 - 4
    + 1 + 2
1 + 2 + 6
    + 1
1 + 3
```

e) Evaluating $g_1(x_1) = x_1^3 + 3x_1^2 + 6x_1 - 4$ for successive tenths, we find that it is negative for $x_1 = 0.5$ and positive for $x_1 = 0.6$.

Hence the positive root of $g_1(x_1) = 0$ is 0.5+ (and the root of $g(x) = 0$ is 1.5+).

f) Diminish the roots of $g_1(x_1) = 0$ by 0.5.

The second transformed equation is

$$g_2(x_2) = x_2^3 + 4.5x_2^2 + 9.75x_2 - 0.125 = 0$$

and its positive root lies between 0 and 0.1.

```
1 + 3    + 6     - 4      |0.5
    + 0.5 + 1.75 + 3.875
1 + 3.5 + 7.75 - 0.125
    + 0.5 + 2
1 + 4    + 9.75
    + 0.5
1 + 4.5
```

To approximate this root to the nearest hundredth, solve the linear part. This gives $9.75x_2 - 0.125 = 0$ or $x_2 = 0.01+$, which indicates that the positive root of $g_2(x_2) = 0$ lies between 0.01 and 0.02 (and the root of $g(x) = 0$ is 1.51+).

g) Diminish the roots of $g_2(x_2) = 0$ by 0.01.

The third transformed equation is

$$g_3(x_3) = x_3^3 + 4.53x_3^2 + 9.8403x_3 - 0.027049 = 0$$

and its positive root lies between 0 and 0.01.

```
1 + 4.5  + 9.75    - 0.125      |0.01
    + 0.01 + 0.0451 + 0.097951
1 + 4.51 + 9.7951 - 0.027049
    + 0.01 + 0.0452
1 + 4.52 + 9.8403
    + 0.01
1 + 4.53
```

Solving the linear part, $9.8403x_3 - 0.027049 = 0$ or $x_3 = 0.002+$. This indicates that the root of $g_3(x_3) = 0$ is 0.002+, and the root of $g(x) = 0$ is 1.512+.

Hence the required root of $x^3 + 3x + 8 = 0$ correct to two decimal places is –1.51.

59. Find $\sqrt[3]{3}$ correct to three decimal places.

Let $x = \sqrt[3]{3}$. Then $x^3 = 3$ and $f(x) = x^3 - 3 = 0$.

By Descartes' rule of signs, $f(x) = 0$ has one positive root $(\sqrt[3]{3})$ and two imaginary roots.

Since $f(1)$ is $-$ and $f(2)$ is $+$, the first figure of the root is 1.

$\begin{array}{l} 1 + 0 + 0 - 3 \lfloor 1 \\ \underline{+ 1 + 1 + 1} \\ 1 + 1 + 1 - 2 \\ \underline{+ 1 + 2} \\ 1 + 2 + 3 \\ \underline{+ 1} \\ 1 + 3 \end{array}$	Diminish the roots of $f(x) = x^3 - 3 = 0$ by 1.

Then $f_1(x_1) = x_1^3 + 3x_1^2 + 3x_1 - 2 = 0$.

$\begin{array}{l} 1 + 3 \quad + 3 \quad - 2 \quad \lfloor 0.4 \\ \underline{+ 0.4 + 1.36 + 1.744} \\ 1 + 3.4 + 4.36 - 0.256 \\ \underline{+ 0.4 + 1.52} \\ 1 + 3.8 + 5.88 \\ \underline{+ 0.4} \\ 1 + 4.2 \end{array}$	Since $f_1(0.4)$ is $-$ and $f_1(0.5)$ is $+$, diminish the roots by 0.4.

$5.88x_2 - 0.256 = 0$, approximately.

Hence $x_2 = 0.04$, approximately.

$\begin{array}{l} 1 + 4.2 \quad + 5.88 \quad - 0.256 \lfloor 0.04 \\ \underline{+ 0.04 + 0.1696 + 0.2420} \\ 1 + 4.24 + 6.0496 - 0.0140 \\ \underline{+ 0.04 + 0.1712} \\ 1 + 4.28 + 6.2208 \\ \underline{+ 0.04} \\ 1 + 4.32 \end{array}$

$6.2208x_3 - 0.0140 = 0$, approximately.

Hence $x_3 = 0.002$, approximately.

$\begin{array}{l} 1 + 4.32 \quad + 6.2208 - 0.0140 \lfloor 0.002 \\ \underline{+ 0.002 + 0.0086 + 0.0125} \\ 1 + 4.322 + 6.2294 - 0.0015 \\ \underline{+ 0.002 + 0.0086} \\ 1 + 4.324 + 6.2380 \\ \underline{+ 0.002} \\ 1 + 4.326 \end{array}$

$(\dfrac{0.0015}{6.2380} = 0.0002$, approximately$)$

Hence $\sqrt[3]{3} = 1.442$, correct to three decimal places.

SUPPLEMENTARY PROBLEMS

60. If $f(x) = 2x^3 - x^2 - x + 2$, find a) $f(0)$, b) $f(2)$, c) $f(-1)$, d) $f(\frac{1}{2})$, e) $f(\sqrt{2})$.

61. Determine the remainder in each of the following.

 a) $(2x^5 - 7) \div (x + 1)$ d) $(4y^3 + y + 27) \div (2y + 3)$

 b) $(x^3 + 3x^2 - 4x + 2) \div (x - 2)$ e) $(x^{12} + x^6 + 1) \div (x - \sqrt{-1})$

 c) $(3x^3 + 4x - 4) \div (x - \frac{1}{2})$ f) $(2x^{33} + 35) \div (x + 1)$

62. Prove that $x + 3$ is a factor of $x^3 + 7x^2 + 10x - 6$ and that $x = -3$ is a root of the equation $x^3 + 7x^2 + 10x - 6 = 0$.

63. Determine which of the following numbers are roots of the equation $y^4 + 3y^3 + 12y - 16 = 0$:
a) 2, b) -4, c) 3, d) 1, e) 2i.

64. Find the values of k for which a) $4x^3 + 3x^2 - kx + 6k$ is exactly divisible by $x + 3$,
b) $x^5 + 4kx - 4k^2 = 0$ has the root $x = 2$.

65. By synthetic division determine the quotient and remainder in each of the following.

 a) $(2x^3 + 3x^2 - 4x - 2) \div (x + 1)$ c) $(y^6 - 3y^5 + 4y - 5) \div (y + 2)$

 b) $(3x^5 + x^3 - 4) \div (x - 2)$ d) $(4x^3 + 6x^2 - 2x + 3) \div (2x + 1)$

66. If $f(x) = 2x^4 - 3x^3 + 4x - 4$, compute $f(2)$ and $f(-3)$ using synthetic division.

67. Given that one root of $x^3 - 7x - 6 = 0$ is -1, find the other two roots.

68. Show that $2x^4 - x^3 - 3x^2 - 31x - 15 = 0$ has roots 3, -½. Find the other roots.

69. Find the roots of each equation.

 a) $(x + 3)^2 (x - 2)^3 (x + 1) = 0$ c) $(x^2 + 3x + 2)(x^2 - 4x + 5) = 0$

 b) $4x^4 (x + 2)^4 (x - 1) = 0$ d) $(y^2 + 4)^2 (y + 1)^2 = 0$

70. Form equations with integral coefficients having only the following roots.
a) 2, -3, -½   b) 0, -4, 2/3, 1   c) ±3i, double root 2   d) -1±2i, 2±i

71. Form an equation whose only roots are $1 \pm \sqrt{2}$, $-1 \pm i\sqrt{3}$.

72. Find the values of A and B for which the equation $A(2x - 3) + B(x - 2) = x$ is an identity.

73. Find the values of A, B and C for which the following equation is an identity:
$A(x - 1)(x - 2) + B(x + 2)(x - 2) + C(x + 2)(x - 1) = x^2 - 5x - 2$.

74. Write the equation of lowest possible degree with integral coefficients having the given roots.
a) 1, 0, i b) 2 + i c) $-1 + \sqrt{3}$, 1/3   d) -2, $i\sqrt{3}$ e) $\sqrt{2}$, i f) $i/2$, 6/5

75. In the equation $x^3 + ax^2 + bx + a = 0$, a and b are real numbers. If $x = 2 + i$ is a root of the equation, find a and b.

76. Write an equation of lowest degree with integral coefficients having $\sqrt{2} - 1$ as a double root.

77. Write an equation of lowest degree with integral coefficients having $\sqrt{3} + 2i$ as a root.

78. Solve each equation, given the indicated root.

    a) $x^4 + x^3 - 12x^2 + 32x - 40 = 0$; $1 - i\sqrt{3}$ c) $x^3 - 5x^2 + 6 = 0$; $3 - \sqrt{3}$

    b) $6x^4 - 11x^3 + x^2 + 33x - 45 = 0$; $1 + i\sqrt{2}$ d) $x^4 - 4x^3 + 6x^2 - 16x + 8 = 0$; $2i$

79. Obtain the rational roots, if any, of each equation.

    a) $x^4 + 2x^3 - 4x^2 - 5x - 6 = 0$ c) $2x^4 - x^3 + 2x^2 - 2x - 4 = 0$

    b) $4x^3 - 3x + 1 = 0$ d) $3x^3 + x^2 - 12x - 4 = 0$

80. Solve each equation.

    a) $x^3 - x^2 - 9x + 9 = 0$ d) $4x^4 + 8x^3 - 5x^2 - 2x + 1 = 0$

    b) $2x^3 - 3x^2 - 11x + 6 = 0$ e) $5x^4 + 3x^3 + 8x^2 + 6x - 4 = 0$

    c) $3x^3 + 2x^2 + 2x - 1 = 0$ f) $3x^5 + 2x^4 - 15x^3 - 10x^2 + 12x + 8 = 0$

81. Prove that a) $\sqrt{5} - \sqrt{2}$ and b) $\sqrt[5]{2}$ are irrational numbers.

82. Graph $f(x) = 2x^3 - 3x^2 + 12x - 16$. From the graph determine a) the number of positive, negative and complex roots of $2x^3 - 3x^2 + 12x - 16 = 0$, b) an approximate value, using interpolation, for any real zero of $f(x)$ accurate to two decimal places.

83. Locate graphically between two successive integers the real roots of $x^4 - 3x^2 - 6x - 2 = 0$. Find the least positive root of the equation accurate to two decimal places.

84. Find upper and lower limits for the real roots of each equation.

    a) $x^3 - 3x^2 + 2x - 4 = 0$ b) $2x^4 + 5x^2 - 6x - 14 = 0$

85. Find the rational roots of $2x^3 - 5x^2 + 4x + 24 = 0$ and thus solve the equation completely.

86. Using Descartes' rule of signs, what may be inferred as to the number of positive, negative and complex roots of the following equations?

    a) $2x^3 + 3x^2 + 7 = 0$ c) $x^5 + 4x^3 - 3x^2 - x + 12 = 0$

    b) $3x^3 - x^2 + 2x - 1 = 0$ d) $x^5 - 3x - 2 = 0$

87. Given the equation $3x^4 - x^3 + x^2 - 5 = 0$, determine a) the maximum number of positive roots, b) the minimum number of positive roots, c) the exact number of negative roots, d) the maximum number of complex roots.

88. Given the equation $5x^3 + 2x - 4 = 0$, how many roots are a) negative, b) real ?

89. Does the equation $x^6 + 4x^4 + 3x^2 + 16 = 0$ have a) 4 complex and 2 real roots, b) 4 real and 2 complex roots, c) 6 complex roots, d) 6 real roots?

90. a) How many positive roots has the equation $x^6 - 7x^2 - 11 = 0$?

    b) How many complex roots has the equation $x^7 + x^4 - x^2 - 3 = 0$?

    c) Show that $x^6 + 2x^3 + 3x - 4 = 0$ has exactly 4 complex roots.

    d) Show that $x^4 + x^3 - x^2 - 1 = 0$ has only one negative root.

91. Transform each equation into another whose roots shall be those of the given equation but changed as indicated.

    a) $2x^3 - x^2 + 6x - 3 = 0$; multiplied by 2

    b) $2x^4 - 5 = 0$; divided by 3

c) $x^3 - 2x^2 + 5x + 7 = 0$; multiplied by -2

d) $x^3 - 20x^2 + 500x - 4000 = 0$; divided by 10

e) $3x^4 + 2x^3 - 5x^2 + 4x - 2 = 0$; with signs changed

f) $x^3 + 3x^2 - 2x + 5 = 0$; diminished by 2

g) $x^3 + 3x^2 - 2x + 1 = 0$; increased by 2

h) $x^3 + 3x^2 + 3x + 2 = 0$; increased by 1

92. Show that the equation $ax^3 + bx^2 + cx + d = 0$ may be transformed into an equation with the x^2 term missing by increasing each of its roots by $b/3a$. Thus transform $x^3 + 6x^2 + 3x + 8 = 0$ into an equation in which the x^2 term is missing.

93. Transform each equation into another with integral coefficients in which the coefficient of the term of highest degree is unity.

a) $2x^3 - 3x^2 - 1 = 0$ b) $5x^3 + 2x - 4 = 0$ c) $2x^2 - 1 = 0$

94. Solve completely each equation.

a) $8x^3 - 20x^2 + 14x - 3 = 0$ c) $4x^3 + 5x^2 + 2x - 6 = 0$

b) $8x^4 - 14x^3 - 9x^2 + 11x - 2 = 0$ d) $2x^4 - x^3 - 23x^2 + 18x + 18 = 0$

95. Given the equation $5x^3 + 2x - 4 = 0$, find *a)* the sum of the roots, *b)* the product of the roots, *c)* the sum of the products of the roots taken two at a time.

96. Given the equation $3x^6 - x^2 - 6 = 0$, find *a)* the sum of the products of the roots taken three at a time, *b)* the sum of the products of the roots taken four at a time, *c)* the product of the roots.

97. Find the sum and product of all the roots of $2x^3 - 3x^2 + bx - 6 = 0$.

98. Find the constants a and b in the equation $ax^3 - 6x^2 + 2ax - 3b = 0$ if the sum of its roots is 3 and the product of its roots is 6.

99. Given $x^3 + 3x^2 - 16x + k = 0$. Determine the value of k if one of the roots is twice another and the three roots are integers.

100. The equation $x^3 + kx + 16 = 0$ has two equal roots. Find the real value of k.

101. Two of the roots of $2x^3 - x^2 + cx + 4 = 0$ are equal in absolute value but opposite in sign. Find the three roots.

102. What must be the value of k in the equation $x^3 - 3x^2 - 6x + k = 0$ if the roots are in arithmetic progression?

103. Find the values of k for which two of the roots of $x^3 - 3x + k = 0$ are equal.

104. Find the roots of $2x^3 + (k+2)x^2 + (2k-2)x + 1 - k = 0$, given that their sum is ½.

105. Transform $3x^4 - 12x^3 + 7x + 5 = 0$ into an equation in y in which the third degree term is missing.

106. The roots of the equation $2x^3 + 3x^2 + 4x + 2 = 0$ are a, b, c. Find the value of :

a) $a + b + c$ c) abc e) $a^3 + b^3 + c^3$ f) $\dfrac{1}{a} + \dfrac{1}{b} + \dfrac{1}{c}$ g) $\dfrac{1}{ab} + \dfrac{1}{bc} + \dfrac{1}{ca}$.

b) $ab + bc + ca$ d) $a^2 + b^2 + c^2$

107. Find by Horner's Method the indicated root of each equation to the specified accuracy.

    a) $2x^3 + 3x^2 - 9x - 7 = 0$; positive root, to the nearest tenth

    b) $x^3 + 9x^2 + 27x - 50 = 0$; positive root, to the nearest hundredth

    c) $x^3 - 3x^2 - 3x + 18 = 0$; negative root, to the nearest tenth

    d) $x^3 + 6x^2 + 9x + 17 = 0$; negative root, to the nearest tenth

    e) $x^5 + x^4 - 27x^3 - 83x^2 + 50x + 162 = 0$; root between 5 and 6, to the nearest hundredth

    f) $x^4 - 3x^3 + x^2 - 7x + 12 = 0$; root between 1 and 2, to the nearest hundredth

108. In finding the maximum deflection of a beam of given length loaded in a certain way, it is necessary to solve the equation $4x^3 - 150x^2 + 1500x - 2871 = 0$. Find correct to the nearest tenth the root of the equation lying between 2 and 3.

109. The length of a rectangular box is twice its width, and its depth is one foot greater than its width. If its volume is 64 cubic feet, find its width to the nearest tenth of a foot.

110. Find $\sqrt[3]{20}$ correct to the nearest hundredth.

ANSWERS TO SUPPLEMENTARY PROBLEMS.

60. *a*) 2   *b*) 12   *c*) 0   *d*) 3/2   *e*) $3\sqrt{2}$

61. *a*) –9   *b*) 14   *c*) –13/8   *d*) 12   *e*) 1   *f*) 33

63. –4, 1 and $2i$ are roots

64. *a*) $k = 9$ *b*) $k = 4, -2$

65. *a*) $2x^2 + x - 5 + \dfrac{3}{x+1}$ *c*) $y^5 - 5y^4 + 10y^3 - 20y^2 + 40y - 76 + \dfrac{147}{y+2}$

    b) $3x^4 + 6x^3 + 13x^2 + 26x + 52 + \dfrac{100}{x-2}$ *d*) $2x^2 + 2x - 2 + \dfrac{5}{2x+1}$

66. 12, 227

67. 3, –2

68. $-1 \pm 2i$

69. *a*) double root –3, triple root 2, –1   *c*) –1, –2, $2 \pm i$
    b) quadruple root 0, quadruple root –2, 1   *d*) double roots $\pm 2i$, double root –1

70. *a*) $2x^3 + 3x^2 - 11x - 6 = 0$ *c*) $x^4 - 4x^3 + 13x^2 - 36x + 36 = 0$
    b) $3x^4 + 7x^3 - 18x^2 + 8x = 0$ *d*) $x^4 - 2x^3 + 2x^2 - 10x + 25 = 0$

71. $x^4 - x^2 - 10x - 4 = 0$

72. $A = 2, B = -3$

73. $A = 1, B = 2, C = -2$

74. *a*) $x^4 - x^3 + x^2 - x = 0$ *d*) $x^3 + 2x^2 + 3x + 6 = 0$

    b) $x^2 - 4x + 5 = 0$ *e*) $x^4 - x^2 - 2 = 0$

    c) $3x^3 + 5x^2 - 8x + 2 = 0$ *f*) $20x^3 - 24x^2 + 5x - 6 = 0$

75. $a = -5, b = 9$

76. $x^4 + 4x^3 + 2x^2 - 4x + 1 = 0$ 77. $x^4 + 2x^2 + 49 = 0$

78. *a*) $1 \pm i\sqrt{3}, -5, 2$ *b*) $1 \pm i\sqrt{2}, -5/3, 3/2$ *c*) $3 \pm \sqrt{3}, -1$ *d*) $\pm 2i, 2 \pm \sqrt{2}$

79. *a*) –3, 2   *b*) 1/2, 1/2, –1   *c*) no rational root   *d*) –1/3, ±2

80. *a*) 1, ±3   *c*) 1/3, $-\frac{1}{2} \pm \frac{1}{2}\sqrt{3}i$ *e*) –1, 2/5, $\pm\sqrt{2}i$

     b) 3, –2, 1/2   *d*) $\pm\frac{1}{2}$, $-1\pm\sqrt{2}$ *f*) ±1, ±2, –2/3

82. 1.37

83. Positive root between 2 and 3; negative root between –1 and 0; pos. root = 2.41 approx.

84. *a*) Upper limit 3, lower limit –1   *b*) Upper limit 2, lower limit –2

85. –3/2, $2\pm 2i$

86. *a*) 1 negative, 2 complex
     b) 3 positive or 1 positive, 2 complex
     c) 1 negative, 2 positive, 2 complex   or   1 negative, 4 complex
     d) 1 positive, 2 negative, 2 complex   or   1 positive, 4 complex

87. *a*) 3   *b*) 1   *c*) 1   *d*) 2   88. *a*) none   *b*) one   89. *c*)

90. *a*) one   *b*) four or six

91. *a*) $x^3 - x^2 + 12x - 12 = 0$ *d*) $x^3 - 2x^2 + 5x - 4 = 0$ *g*) $x^3 - 3x^2 - 2x + 9 = 0$

     b) $162x^4 - 5 = 0$ *e*) $3x^4 - 2x^3 - 5x^2 - 4x - 2 = 0$

     c) $x^3 + 4x^2 + 20x - 56 = 0$ *f*) $x^3 + 9x^2 + 22x + 21 = 0$ *h*) $x^3 + 1 = 0$

92. $x^3 - 9x + 18 = 0$

93. *a*) $x^3 - 3x^2 - 4 = 0$ *b*) $x^3 + 10x - 100 = 0$ *c*) $x^2 - 2 = 0$

94. *a*) 1/2, 1/2, 3/2   *b*) 2, –1, 1/4, 1/2   *c*) 3/4, $-1\pm i$ *d*) 3, 3/2, $-2\pm\sqrt{2}$

95. *a*) 0   *b*) 4/5   *c*) 2/5   96. *a*) 0   *b*) –1/3   *c*) –2

97. *a*) Sum = 3/2, product = 3   98. *a* = 2, *b* = 4   99. 12

100. –12   101. ±2, 1/2   102. 8   103. ±2   104. ±2, 1/2

105. $3y^4 - 18y^2 - 17y + 3 = 0$

106. *a*) –3/2   *b*) 2   *c*) –1   *d*) –7/4   *e*) 21/8   *f*) –2   *g*) 3/2

107. *a*) 1.9   *b*) 1.25   *c*) –2.2   *d*) –4.9   *e*) 5.77   *f*) 1.38

108. 2.5

109. 2.9 ft

110. 2.71

CHAPTER 22

Logarithms

DEFINITION OF A LOGARITHM. If $b^x = N$, where N is a positive number and b is a positive number different from 1, then the exponent x is the logarithm of N to the base b and is written $x = \log_b N$.

Examples. 1) Since $3^2 = 9$, then 2 is the logarithm of 9 to the base 3, i.e. $2 = \log_3 9$.

2) $\log_2 8$ is that number x to which the base 2 must be raised in order to yield 8, i.e. $2^x = 8$, $x = 3$. Hence $\log_2 8 = 3$.

Both $b^x = N$ and $x = \log_b N$ are equivalent relationships; $b^x = N$ is called the *exponential form* and $x = \log_b N$ the *logarithmic form* of the relationship. As a consequence, corresponding to *laws of exponents* there are *laws of logarithms*.

LAWS OF LOGARITHMS.

I. The logarithm of the product of two positive numbers M and N is equal to the sum of the logarithms of the numbers, i.e.,

$$\log_b MN = \log_b M + \log_b N.$$

II. The logarithm of the quotient of two positive numbers M and N is equal to the difference of the logarithms of the numbers, i.e.,

$$\log_b \frac{M}{N} = \log_b M - \log_b N.$$

III. The logarithm of the pth power of a positive number M is equal to p multiplied by the logarithm of the number, i.e.,

$$\log_b M^p = p \log_b M.$$

Examples. 1) $\log_2 3(5) = \log_2 3 + \log_2 5$

2) $\log_{10} \frac{17}{24} = \log_{10} 17 - \log_{10} 24$

3) $\log_7 5^3 = 3 \log_7 5$

$\log_{10} \sqrt[3]{2} = \log_{10} 2^{1/3} = \frac{1}{3} \log_{10} 2$

COMMON LOGARITHMS

COMMON LOGARITHMS. The system of logarithms whose base is 10 is called the common or Briggsian system. When the base is omitted, it is understood that base 10 is to be used. Thus $\log 25 = \log_{10} 25$.

Consider the following table.

Number N	.0001	.001	.01	.1	1	10	100	1000	10,000
Exponential form of N	10^{-4}	10^{-3}	10^{-2}	10^{-1}	10^0	10^1	10^2	10^3	10^4
$\log N$	-4	-3	-2	-1	0	1	2	3	4

It is obvious that $10^{1.5377}$ will give some number greater than 10 (which is 10^1) but smaller than 100 (which is 10^2). Actually, $10^{1.5377} = 34.49$; hence $\log 34.49 = 1.5377$.

The digit before the decimal point is the *characteristic* of the log, and the decimal fraction part is the *mantissa* of the log. In the above example, the characteristic is 1 and the mantissa is .5377.

The mantissa of the log of a number is found in tables, ignoring the decimal point of the number. Each mantissa in the tables is understood to have a decimal point preceding it, and the mantissa is always considered positive.

THE CHARACTERISTIC is determined by inspection from the number itself according to the following rules.

1) For a number greater than 1, the characteristic is positive and is one *less* than the number of digits before the decimal point. For example:

Number	5297	348	900	34.8	60	5.764	3
Characteristic	3	2	2	1	1	0	0

2) For a number less than 1, the characteristic is negative and is one *more* than the number of zeros immediately following the decimal point. The negative sign of the characteristic is written in either of these two ways: *a*) above the characteristic as $\bar{1}$, $\bar{2}$, etc.; *b*) as 9 -10, 8 -10, etc. Thus the characteristic of 0.3485 is $\bar{1}$ or 9 -10, of 0.0513 is $\bar{2}$ or 8 -10, of 0.0024 is $\bar{3}$ or 7 -10.

TO FIND THE COMMON LOGARITHM OF A POSITIVE NUMBER using the table of logarithms in the Appendix.

Suppose it is required to find the log of the number 728. In the table of logarithms glance down the N column to 72, then horizontally to the right to column 8 and note the figures 8621 which is the required mantissa. Since the characteristic is 2, $\log 728 = 2.8621$. (This means that $728 = 10^{2.8621}$.)

The mantissa for $\log 72.8$, for $\log 7.28$, for $\log 0.728$, for $\log 0.0728$, etc., is .8621, but the characteristics differ. Thus:

log 728 = 2.8621   log 0.728 = $\overline{1}$.8621 or 9.8621-10

log 72.8 = 1.8621   log 0.0728 = $\overline{2}$.8621 or 8.8621-10

log 7.28 = 0.8621   log 0.00728 = $\overline{3}$.8621 or 7.8621-10

When the number contains four digits, interpolate using the method of proportional parts.

Example. Find log 4.638.

      The characteristic is 0. The mantissa is found as follows.

$$\begin{array}{r}
\text{Mantissa of log } 4640 = .6665 \\
\text{Mantissa of log } 4630 = \underline{.6656} \\
\text{Tabular difference} = .0009
\end{array}$$

.8 × tabular difference = .00072 or .0007 to four decimal places.

Mantissa of log 4638 = .6656 + .0007 = .6663 to four digits.

Hence log 4.638 = 0.6663.

The mantissa for log 4638, for log 463.8, for log 46.38, etc., is .6663, but the characteristics differ. Thus:

log 4638 = 3.6663   log 0.4638   = $\overline{1}$.6663 or 9.6663-10

log 463.8 = 2.6663   log 0.04638   = $\overline{2}$.6663 or 8.6663-10

log 46.38 = 1.6663   log 0.004638 = $\overline{3}$.6663 or 7.6663-10

log 4.638 = 0.6663   log 0.0004638 = $\overline{4}$.6663 or 6.6663-10

THE ANTILOGARITHM is the number corresponding to a given logarithm. "The antilog of 3" means "the number whose log is 3"; that number is obviously 1000.

      Example 1. Given log N = 1.9058, find N.

        In the table the mantissa .9058 corresponds to the number 805. Since the characteristic of log N is 1, the number must have two digits before the decimal point; thus N = 80.5 (or antilog 1.9058 = 80.5).

      2. If log N = 7.8657-10, find N.

        In the table the mantissa .8657 corresponds to the number 734. Since the characteristic is 7 −10, the number must have two zeros immediately following the decimal point; thus N = 0.00734 (or antilog 7.8657-10 = 0.00734).

      3. Given log N = 9.3842-10, find N.

        Since the mantissa .3842 is not found in the tables, interpolation must be used.

Mantissa of log 2430 = .3856	Given mantissa = .3842
Mantissa of log 2420 = .3838	Next smaller mantissa = .3838
Tabular difference = .0018	Difference = .0004

Then $2420 + \dfrac{4}{18}(2430 - 2420) = 2422$ to four digits, and $N = 0.2422$.

THE COLOGARITHM of a positive number is the logarithm of its reciprocal.

Thus $\text{colog } N = \log \dfrac{1}{N} = \log 1 - \log N = -\log N$,   since $\log 1 = 0$.

Cologarithms are often used in computations involving division where instead of subtracting a logarithm one adds its cologarithm.

For example, $\log \dfrac{56}{73} = \log 56 - \log 73 = \log 56 + \text{colog } 73$.

SOLVED PROBLEMS

EXPONENTIAL AND LOGARITHMIC FORMS.

1. Express each of the following exponential forms in logarithmic form:

   a) $p^q = r$,   b) $2^3 = 8$,   c) $4^2 = 16$,   d) $3^{-2} = \dfrac{1}{9}$,   e) $8^{-2/3} = \dfrac{1}{4}$.

   a) $q = \log_p r$,   b) $3 = \log_2 8$,   c) $2 = \log_4 16$,   d) $-2 = \log_3 \dfrac{1}{9}$,   e) $-\dfrac{2}{3} = \log_8 \dfrac{1}{4}$

2. Express each of the following logarithmic forms in exponential form:

   a) $\log_5 25 = 2$,   b) $\log_2 64 = 6$,   c) $\log_{1/4} \dfrac{1}{16} = 2$,   d) $\log_a a^3 = 3$,   e) $\log_r 1 = 0$.

   a) $5^2 = 25$,   b) $2^6 = 64$,   c) $(\dfrac{1}{4})^2 = \dfrac{1}{16}$,   d) $a^3 = a^3$,   e) $r^0 = 1$

3. Determine the value of each of the following.

   a) $\log_4 64$.   Let $\log_4 64 = x$; then $4^x = 64 = 4^3$ and $x = 3$.

   b) $\log_3 81$.   Let $\log_3 81 = x$; then $3^x = 81 = 3^4$ and $x = 4$.

   c) $\log_{1/2} 8$.   Let $\log_{1/2} 8 = x$; then $(\dfrac{1}{2})^x = 8$, $(2^{-1})^x = 2^3$, $2^{-x} = 2^3$ and $x = -3$.

   d) $\log \sqrt[3]{10} = x$, $10^x = \sqrt[3]{10} = 10^{1/3}$, $x = 1/3$

   e) $\log_5 125\sqrt{5} = x$, $5^x = 125\sqrt{5} = 5^3 \cdot 5^{1/2} = 5^{7/2}$, $x = 7/2$

4. Solve each of the following equations.

   a) $\log_3 x = 2$, $3^2 = x$, $x = 9$

   b) $\log_4 y = -\dfrac{3}{2}$, $4^{-3/2} = y$, $y = \dfrac{1}{8}$

   c) $\log_x 25 = 2$, $x^2 = 25$, $x = \pm 5$. Since bases are positive, the solution is $x = 5$.

   d) $\log_y \dfrac{9}{4} = -\dfrac{2}{3}$, $y^{-2/3} = \dfrac{9}{4}$, $y^{2/3} = \dfrac{4}{9}$, $y = (\dfrac{4}{9})^{3/2} = \dfrac{8}{27}$ is the required solution.

   e) $\log(3x^2 + 2x - 4) = 0$, $10^0 = 3x^2 + 2x - 4$, $3x^2 + 2x - 5 = 0$, $x = 1, -5/3$

LAWS OF LOGARITHMS.

5. Prove the laws of logarithms.

Let $M = b^x$ and $N = b^y$; then $x = \log_b M$ and $y = \log_b N$.

I. Since $MN = b^x \cdot b^y = b^{x+y}$, then $\log_b MN = x + y = \log_b M + \log_b N$.

II. Since $\dfrac{M}{N} = \dfrac{b^x}{b^y} = b^{x-y}$, then $\log_b \dfrac{M}{N} = x - y = \log_b M - \log_b N$.

III. Since $M^p = (b^x)^p = b^{px}$, then $\log_b M^p = px = p \log_b M$.

6. Express each of the following as an algebraic sum of logarithms, using the laws I, II, III.

a) $\log_b UVW = \log_b (UV)W = \log_b UV + \log_b W = \log_b U + \log_b V + \log_b W$

b) $\log_b \dfrac{UV}{W} = \log_b UV - \log_b W = \log_b U + \log_b V - \log_b W$

c) $\log \dfrac{XYZ}{PQ} = \log XYZ - \log PQ = \log X + \log Y + \log Z - (\log P + \log Q)$

$$= \log X + \log Y + \log Z - \log P - \log Q$$

d) $\log \dfrac{U^2}{V^3} = \log U^2 - \log V^3 = 2 \log U - 3 \log V$

e) $\log \dfrac{U^2 V^3}{W^4} = \log U^2 V^3 - \log W^4 = \log U^2 + \log V^3 - \log W^4$

$$= 2 \log U + 3 \log V - 4 \log W$$

f) $\log \dfrac{U^{1/2}}{V^{2/3}} = \log U^{1/2} - \log V^{2/3} = \dfrac{1}{2} \log U - \dfrac{2}{3} \log V$

g) $\log_e \dfrac{\sqrt{x^3}}{\sqrt[4]{y^3}} = \log_e \dfrac{x^{3/2}}{y^{3/4}} = \log_e x^{3/2} - \log_e y^{3/4} = \dfrac{3}{2} \log_e x - \dfrac{3}{4} \log_e y$

h) $\log \sqrt[4]{a^2 b^{-3/4} c^{1/3}} = \dfrac{1}{4} \left\{ 2 \log a - \dfrac{3}{4} \log b + \dfrac{1}{3} \log c \right\}$

$$= \dfrac{1}{2} \log a - \dfrac{3}{16} \log b + \dfrac{1}{12} \log c$$

7. Given that $\log 2 = .3010$, $\log 3 = .4771$, $\log 5 = .6990$, $\log 7 = .8451$ (all base 10) accurate to four decimal places, evaluate the following.

a) $\log 105 = \log (3 \cdot 5 \cdot 7) = \log 3 + \log 5 + \log 7 = .4771 + .6990 + .8451 = 2.0212$

b) $\log 108 = \log (2^2 \cdot 3^3) = 2 \log 2 + 3 \log 3 = 2(.3010) + 3(.4771) = 2.0333$

c) $\log \sqrt[3]{72} = \log \sqrt[3]{3^2 \cdot 2^3} = \log (3^{2/3} \cdot 2) = \dfrac{2}{3} \log 3 + \log 2 = .6191$

d) $\log 2.4 = \log \dfrac{24}{10} = \log \dfrac{3 \cdot 2^3}{10} = \log 3 + 3 \log 2 - \log 10$

$$= .4771 + 3(.3010) - 1 = .3801$$

e) $\log .0081 = \log \dfrac{81}{10^4} = \log 81 - \log 10^4 = \log 3^4 - \log 10^4$

$$= 4 \log 3 - 4 \log 10 = 4(.4771) - 4 = -2.0916 \text{ or } 7.9084-10$$

Note. In exponential form this means $10^{-2.0916} = .0081$.

8. Express each of the following as a single logarithm (base is 10 unless otherwise indicated).

a) $\log 2 - \log 3 + \log 5 = \log \dfrac{2}{3} + \log 5 = \log \dfrac{2}{3}(5) = \log \dfrac{10}{3}$

b) $3 \log 2 - 4 \log 3 = \log 2^3 - \log 3^4 = \log \dfrac{2^3}{3^4} = \log \dfrac{8}{81}$

c) $\dfrac{1}{2} \log 25 - \dfrac{1}{3} \log 64 + \dfrac{2}{3} \log 27 = \log 25^{1/2} - \log 64^{1/3} + \log 27^{2/3}$

$$= \log 5 - \log 4 + \log 9 = \log \dfrac{5}{4} + \log 9 = \log \dfrac{5}{4}(9) = \log \dfrac{45}{4}$$

d) $\log 5 - 1 = \log 5 - \log 10 = \log \dfrac{5}{10} = \log \dfrac{1}{2}$

e) $2 \log 3 + 4 \log 2 - 3 = \log 3^2 + \log 2^4 - 3 \log 10 = \log 9 + \log 16 - \log 10^3$

$$= \log (9 \cdot 16) - \log 10^3 = \log \dfrac{9 \cdot 16}{10^3} = \log .144$$

f) $3 \log_a b - \dfrac{1}{2} \log_a c = \log_a b^3 + \log_a c^{-1/2} = \log_a (b^3 c^{-1/2})$

9. In each of the following equations, solve for the indicated letter in terms of the other quantities.

a) $\log_2 x = y + c \; : x.$ $\qquad x = 2^{y+c}$

b) $\log a = 2 \log b \; : a.$ $\qquad \log a = \log b^2, \;\; a = b^2$

c) $\log_e I = \log_e I_0 - t \; : I.$ $\qquad \log_e I = \log_e I_0 - t \log_e e = \log_e I_0 + \log_e e^{-t}$

$$= \log_e I_0 e^{-t}, \qquad I = I_0 e^{-t}$$

d) $2 \log x + 3 \log y = 4 \log z - 2 \; : y.$

Solving for $\log y$, $\quad 3 \log y = 4 \log z - 2 - 2 \log x \quad$ and

$\log y = \dfrac{4}{3} \log z - \dfrac{2}{3} - \dfrac{2}{3} \log x = \log z^{4/3} + \log 10^{-2/3} + \log x^{-2/3} = \log z^{4/3} 10^{-2/3} x^{-2/3}.$

Hence $y = 10^{-2/3} x^{-2/3} z^{4/3}.$

e) $\log (x+3) = \log x + \log 3 \; : x.$ $\qquad \log (x+3) = \log 3x, \;\; x+3 = 3x, \;\; x = 3/2$

COMMON LOGARITHMS.

10. Determine the characteristic of the common logarithm of each of the following numbers.

a) 57   c) 5.63   e) 982.5   g) 186,000   i) .7314   k) .0071
b) 57.4   d) 35.63   f) 7824   h) .71   j) .0325   l) .0003

a) 1   c) 0   e) 2   g) 5   i) 9 – 10   k) 7 – 10
b) 1   d) 1   f) 3   h) 9 – 10   j) 8 – 10   l) 6 – 10

11. Verify each of the following logarithms.

a) log 87.2   = 1.9405   h) log 6.753   = 0.8295   (8293 + 2)
b) log 37300 = 4.5717   i) log 183.2   = 2.2630   (2625 + 5)
c) log 753   = 2.8768   j) log 43.15   = 1.6350   (6345 + 5)
d) log 9.21   = 0.9643   k) log 876,400 = 5.9427   (9425 + 2)
e) log .382   = 9.5821 – 10   l) log .2548   = 9.4062 – 10 (4048 + 14)
f) log .00159 = 7.2014 – 10   m) log .04372 = 8.6407 – 10 (6405 + 2)
g) log .0256 = 8.4082 – 10   n) log .009848 = 7.9933 – 10 (9930 + 3)

12. Verify each of the following.

a) Antilog 3.8531   = 7130   h) Antilog 2.6715   = 469.3   (3/9 × 10 = 3 approx.)
b) Antilog 1.4997   = 31.6   i) Antilog 4.1853   = 15320   (6/28 × 10 = 2 approx.)
c) Antilog 9.8267 – 10 = .671   j) Antilog 0.9245   = 8.404   (2/5 × 10 = 4)
d) Antilog 7.7443 – 10 = .00555   k) Antilog $\overline{1}$.6089 = .4064   (4/11 × 10 = 4 approx.)
e) Antilog 0.1875   = 1.54   l) Antilog 8.8907 – 10 = .07775 (3/6 × 10 = 5)
f) Antilog $\overline{2}$.3927 = .0247   m) Antilog 1.2000 = 15.85   (13/27 × 10 = 5 approx.)
g) Antilog 4.9360   = 86300   n) Antilog 7.2409 – 10 = .001742 (4/25 × 10 = 2 approx.)

13. Write each of the following numbers as a power of 10: a) 893, b) 0.358.

a) We require x such that $10^x = 893$. Then $x = \log 893 = 2.9509$ and $893 = 10^{2.9509}$.

b) We require x such that $10^x = 0.358$.
   Then $x = \log 0.358 = 9.5539 - 10 = -0.4461$ and $0.358 = 10^{-0.4461}$.

Calculate each of the following using logarithms.

14. $P = 3.81 \times 43.4$

$\log P = \log 3.81 + \log 43.4$

$$
\begin{array}{rl}
\log 3.81 & = 0.5809 \\
(+) \log 43.4 & = 1.6375 \\
\hline
\log P & = 2.2184
\end{array}
$$
Hence $P = $ antilog $2.2184 = 165.3$.

Note the exponential significance of the computation. Thus

$$3.81 \times 43.4 = 10^{0.5809} \times 10^{1.6375}$$

$$= 10^{0.5809 + 1.6375} = 10^{2.2184} = 165.3.$$

15. $P = 73.42 \times .00462 \times .5143$ $\log P = \log 73.42 + \log .00462 + \log .5143$

$$
\begin{array}{rll}
\log 73.42 & = & 1.8658 \\
(+)\ \log .00462 & = & 7.6646 - 10 \\
(+)\ \log .5143 & = & 9.7112 - 10 \\
\hline
\log P & = & 19.2416 - 20 = 9.2416 - 10. \qquad \text{Hence } P = .1744.
\end{array}
$$

16. $P = \dfrac{784.6 \times .0431}{28.23}$ $\log P = \log 784.6 + \log .0431 - \log 28.23$

$$
\begin{array}{rll}
\log 784.6 & = & 2.8947 \\
(+)\ \log .0431 & = & 8.6345 - 10 \\
\hline
 & & 11.5292 - 10 \\
(-)\ \log 28.23 & = & 1.4507 \\
\hline
\log P & = & 10.0785 - 10 = .0785 \\
P & = & 1.198
\end{array}
$$

17. $P = \dfrac{.4932 \times 653.7}{.07213 \times 8456}$

Numerator N			Denominator D		
$\log .4932$	$=$	$9.6930 - 10$	$\log .07213$	$=$	$8.8581 - 10$
$(+)\ \log 653.7$	$=$	2.8154	$(+)\ \log 8456$	$=$	3.9272
$\log N$	$=$	$12.5084 - 10$	$\log D$	$=$	$12.7853 - 10$
$(-)\ \log D$	$=$	2.7853		$=$	2.7853
$\log P$	$=$	$9.7231 - 10$			
P	$=$	$.5286$			

18. $P = (7.284)^5$

$\log P = 5 \log 7.284 = 5(.8623) = 4.3115$ and $P = 20490.$

19. $P = \dfrac{(63.28)^3 (.00843)^2 (.4623)}{(412.3)(2.184)^5}$

$\log P = 3 \log 63.28 + 2 \log .00843 + \log .4623 - (\log 412.3 + 5 \log 2.184)$

Numerator N				Denominator D		
$3 \log 63.28$	$=$	$3(1.8013)$	$= 5.4039$	$\log 412.3$	$=$	2.6152
$(+)\ 2 \log .00843$	$=$	$2(7.9258 - 10)$	$= 15.8516 - 20$	$(+)\ 5 \log 2.184$	$=$	1.6965
$(+)\ \ \ \log .4623$			$= 9.6649 - 10$	$\log D$	$=$	4.3117
$\log N$	$=$	$30.9204 - 30$				
$(-)\ \log D$	$=$	4.3117				
$\log P$	$=$	$26.6087 - 30 = 6.6087 - 10$				
P	$=$	$.0004062$ (or 4.062×10^{-4})				

20. $P = \sqrt[5]{.8532}$

$\log P = \dfrac{1}{5} \log .8532 = \dfrac{1}{5}(9.9310 - 10) = \dfrac{1}{5}(49.9310 - 50) = 9.9862 - 10$ and $P = .9688$.

21. $P = \dfrac{(78.41)^3 \sqrt{142.3}}{\sqrt[4]{.1562}}$ $\qquad\qquad \log P = 3 \log 78.41 + \dfrac{1}{2} \log 142.3 - \dfrac{1}{4} \log .1562$.

Numerator N

$$3 \log 78.41 = 3(1.8944) = 5.6832$$
$$(+) \tfrac{1}{2} \log 142.3 = \tfrac{1}{2}(2.1532) = 1.0766$$
$$\log N = \overline{6.7598} = 16.7598 - 10$$
$$(-) \log D = 9.7984 - 10$$
$$\log P = \overline{6.9614}$$
$$P = 9,150,000 \ \text{ or } \ 9.15 \times 10^6$$

Denominator D

$$\tfrac{1}{4} \log .1562 = \tfrac{1}{4}(9.1937 - 10)$$
$$= \tfrac{1}{4}(39.1937 - 40)$$
$$\log D = 9.7984 - 10$$

22. The period T of a simple pendulum of length l is given by the formula $T = 2\pi\sqrt{l/g}$, where g is the acceleration due to gravity. Find T (in seconds) if $l = 281.3$ cm and $g = 981.0$ cm/sec^2. Take $2\pi = 6.283$.

$$T = 2\pi\sqrt{\dfrac{l}{g}} = 6.283 \sqrt{\dfrac{281.3}{981.0}}$$

$$\log T = \log 6.283 + \tfrac{1}{2}(\log 281.3 - \log 981.0)$$

$$\log 6.283 = \qquad\quad = 0.7982$$
$$(+) \tfrac{1}{2} \log 281.3 = \tfrac{1}{2}(2.4492) = 1.2246$$
$$\overline{2.0228}$$
$$(-) \tfrac{1}{2} \log 981.0 = \tfrac{1}{2}(2.9917) = 1.4959$$
$$\log T = 0.5269$$
$$T = 3.365 \text{ seconds}$$

COLOGARITHMS.

23. Find a) colog 42.36, b) colog .8536.

a) $\text{colog } 42.36 = \log \dfrac{1}{42.36} = \log 1 - \log 42.36$

$$\log 1 = 10.0000 - 10$$
$$(-) \log 42.36 = 1.6270$$
$$\text{colog } 42.36 = \overline{8.3730 - 10}$$

b) $\text{colog } .8536 = \log \dfrac{1}{.8536} = \log 1 - \log .8536$

$$\log 1 = 10.0000 - 10$$
$$(-) \log .8536 = 9.9313 - 10$$
$$\text{colog } .8536 = \overline{0.0687}$$

24. Prove that a) colog MN = colog M + colog N and b) colog M^p = p colog M,   where $M, N > 0$.

    a) colog MN = $\log \dfrac{1}{MN}$ = $\log \dfrac{1}{M} + \log \dfrac{1}{N}$ = colog M + colog N

    b) colog M^p = $\log \dfrac{1}{M^p}$ = $p \log \dfrac{1}{M}$ = p colog M

25. Use cologarithms to evaluate $P = \dfrac{(372.1)(.0862)}{(4.315)(.7460)}$.

    $\log P$ = $\log 372.1 + \log .0862 - \log 4.315 - \log .7460$

$$
\begin{array}{rll}
\log 372.1 &= 2.5706 & \\
(+) \quad \log .0862 &= 8.9355 - 10 & \\
(+) \ \text{colog } 4.315 &= 9.3650 - 10 & (\log 4.315 = 0.6350) \\
(+) \ \text{colog } .7460 &= 0.1273 & (\log .7460 = 9.8727 - 10) \\
\hline
\log P &= 20.9984 - 20 = 0.9984 & \\
P &= 9.963 &
\end{array}
$$

26. Use cologarithms to calculate $P = \dfrac{\sqrt{.8730}}{\sqrt[4]{37.31} \ (4.863)^3}$.

$$
\begin{array}{rlll}
\tfrac{1}{2} \log .8730 &= \tfrac{1}{2}(9.9410 - 10) &= \tfrac{1}{2}(19.9410 - 20) &= 9.9705 - 10 \\
(+) \ \tfrac{1}{4} \text{ colog } 37.31 &= \tfrac{1}{4}(8.4282 - 10) &= \tfrac{1}{4}(38.4282 - 40) &= 9.6071 - 10 \\
(+) \ 3 \text{ colog } 4.863 &= 3(9.3131 - 10) &= 27.9393 - 30 &= 7.9393 - 10 \\
\hline
& & \log P = 27.5169 - 30 &= 7.5169 - 10 \\
& & P = .003288 & \text{or} \ 3.288 \times 10^{-3}
\end{array}
$$

EXPONENTIAL EQUATIONS.

27. Solve for x: $5^{2x+2} = 3^{5x-1}$.

    Taking logarithms, $(2x + 2) \log 5$ = $(5x - 1) \log 3$.

    Then $2x \log 5 - 5x \log 3$ = $- \log 3 - 2 \log 5,$

                         $x(2 \log 5 - 5 \log 3)$ = $- \log 3 - 2 \log 5,$

    and $x = \dfrac{\log 3 + 2 \log 5}{5 \log 3 - 2 \log 5} = \dfrac{0.4771 + 2(0.6990)}{5(0.4771) - 2(0.6990)} = \dfrac{1.8751}{0.9875}$.

$$
\begin{array}{rl}
\log 1.875 &= 10.2730 - 10 \\
(-) \ \log .9875 &= 9.9946 - 10 \\
\hline
\log x &= 0.2784 \\
x &= 1.898
\end{array}
$$

SUPPLEMENTARY PROBLEMS

28. Evaluate: a) $\log_2 32$, b) $\log \sqrt[4]{10}$, c) $\log_3 1/9$, d) $\log_{1/4} 16$, e) $\log_e e^x$, f) $\log_8 4$.

29. Solve each equation for the unknown.

   a) $\log_2 x = 3$ c) $\log_x 8 = -3$ e) $\log_4 x^3 = 3/2$

   b) $\log y = -2$ d) $\log_3 (2x + 1) = 1$ f) $\log_{(x-1)} (4x - 4) = 2$

30. Express as an algebraic sum of logarithms.

   a) $\log \dfrac{U^3 V^2}{W^5}$ b) $\log \sqrt{\dfrac{2x^3 y}{z^7}}$ c) $\log_e \sqrt[3]{x^{1/2} y^{-1/2}}$ d) $\log \dfrac{xy^{-3/2} z^3}{a^2 b^{-4}}$

31. Solve each equation for the indicated letter in terms of the other quantities.

   a) $2 \log x = \log 16$; x c) $\log_3 F = \log_3 4 - 2 \log_3 x$; F

   b) $3 \log y + 2 \log 2 = \log 32$; y d) $\log_e (30 - U) = \log_e 30 - 2t$; U

32. Prove that if a and b are positive and $\neq 1$, $(\log_a b)(\log_b a) = 1$.

33. Prove that $10^{\log N} = N$ where $N > 0$.

34. Determine the characteristic of the common logarithm of each number.

   a) 248 c) .024 e) .0006 g) 1.06 i) 4 k) 237.63 m) 7,000,000
   b) 2.48 d) .162 f) 18.36 h) 6000 j) 40.60 l) 146.203 n) .000007

35. Find the common logarithm of each number.

   a) 237 c) 1.26 e) .086 g) 10400 i) .000000728 k) 23.70 m) 1
   b) 28.7 d) .263 f) .007 h) .00607 j) 6,000,000 l) 6.03 n) 1000

36. Find the antilogarithm of each of the following.

   a) 2.8802 c) 0.6946 e) 8.3160 − 10 g) 4.6618 i) $\overline{1}.9484$
   b) 1.6590 d) $\overline{2}.9042$ f) 7.8549 − 10 h) 0.4216 j) 9.8344 − 10

37. Find the common logarithm of each number by interpolation.

   a) 1463 c) 86.27 e) .6041 g) 1.006 i) 460.3
   b) 810.6 d) 8.106 f) .04622 h) 300.6 j) .003001

38. Find the antilogarithm of each of the following by interpolation.

   a) 2.9060 c) 1.6600 e) 3.7045 g) 2.2500 i) $\overline{1}.4700$
   b) $\overline{1}.4860$ d) $\overline{1}.9840$ f) 8.9266 − 10 h) 0.8003 j) 1.2925

39. Write each number as a power of 10: a) 45.4, b) .005278.

40. Evaluate.

   a) $(42.8)(3.26)(8.10)$

   b) $\dfrac{(.148)(47.6)}{284}$

   c) $\dfrac{(1.86)(86.7)}{(2.87)(1.88)}$

   d) $\dfrac{2453}{(67.2)(8.55)}$

   e) $\dfrac{5608}{(.4536)(11000)}$

   f) $\dfrac{(3.92)^3 (72.16)}{\sqrt[4]{654}}$

   g) $3.14 \sqrt{11.65/32}$

   h) $\sqrt{\dfrac{906}{(3.142)(14.6)}}$

   i) $\sqrt{\dfrac{(1600)(310.6)^2}{7290}}$

   j) $\sqrt[3]{\dfrac{(5.52)(2610)}{(7.36)(3.142)}}$

41. Solve the following hydraulics equation: $\dfrac{20.0}{14.7} = (\dfrac{.0613}{x})^{1.32}$.

42. The formula $D = \sqrt[3]{\dfrac{W}{.5236(A-G)}}$ gives the diameter of a spherical balloon required to lift a weight W. Find D if $A = .0807$, $G = .0056$ and $W = 1250$.

43. Given the formula $T = 2\pi\sqrt{l/g}$, find l if $T = 2.75$, $\pi = 3.142$ and $g = 32.16$.

44. Solve for x.

   a) $3^x = 243$ c) $2^{x+2} = 64$ e) $x^{-3/4} = 8$ g) $7x^{-1/2} = 4$ i) $5^{x-2} = 1$

   b) $5^x = 1/125$ d) $x^{-2} = 16$ f) $x^{-2/3} = 1/9$ h) $3^x = 1$ j) $2^{2x+3} = 1$

45. Solve each exponential equation: a) $4^{2x-1} = 5^{x+2}$, b) $3^{x-1} = 4 \cdot 5^{1-3x}$.

46. Find a) colog 58.3, b) colog .07312.

47. Use cologarithms to evaluate a) $\dfrac{(28.3)(471.5)}{(55.21)(3.142)}$, b) $\sqrt{\dfrac{(4.36)(2.143)^3}{(.0258)^2}}$.

ANSWERS TO SUPPLEMENTARY PROBLEMS.

28. a) 5   b) 1/4   c) –2   d) –2   e) x f) 2/3

29. a) 8   b) .01   c) 1/2   d) 1   e) 2   f) 5

30. a) $3 \log U + 2 \log V - 5 \log W$ c) $\dfrac{1}{6} \log_e x - \dfrac{1}{6} \log_e y$

   b) $\dfrac{1}{2} \log 2 + \dfrac{3}{2} \log x + \dfrac{1}{2} \log y - \dfrac{7}{2} \log z$ d) $\log x - \dfrac{3}{2} \log y + 3 \log z - 2 \log a + 4 \log b$

31. a) 4   b) 2   c) $F = 4/x^2$ d) $U = 30(1 - e^{-2t})$

34. a) 2   c) $\bar{2}$ e) $\bar{4}$ g) 0   i) 0   k) 2   m) 6

   b) 0   d) $\bar{1}$ f) 1   h) 3   j) 1   l) 2   n) $\bar{6}$

35. a) 2.3747   c) 0.1004   e) $\bar{2}.9345$ g) 4.0170   i) $\bar{7}.8621$ k) 1.3747   m) 0.0000

   b) 1.4579   d) $\bar{1}.4200$ f) 7.8451-10   h) $\bar{3}.7832$ j) 6.7782   l) 0.7803   n) 3.0000

36. a) 759   c) 4.95   e) .0207   g) 45,900   i) .888

   b) 45.6   d) .0802   f) .00716   h) 2.64   j) .683

37. a) 3.1653   c) 1.9359   e) $\bar{1}.7811$ g) 0.0026   i) 2.6631

   b) 2.9088   d) 0.9088   f) 8.6648-10   h) 2.4780   j) 7.4773-10

38. a) 805.4   c) 45.71   e) 5064   g) 177.8   i) .2951

   b) .3062   d) .9638   f) .08445   h) 6.314   j) 19.61

39. a) $10^{1.6571}$ b) $10^{-2.2776}$

40. a) 1130   c) 29.9   e) 1.124   g) 1.90   i) 145.5   41. .0486   42. 31.7   43. 6.16

   b) .0248   d) 4.27   f) 860   h) 4.44   j) 8.54

44. a) 5   c) 4   e) 1/16   g) 49/16   i) 2

   b) –3   d) ±1/4   f) ±27   h) 0   j) –3/2

45. a) 3.958   b) .6907

46. a) 8.2343 – 10   b) 1.1360

47. a) 76.93   b) 254.0

CHAPTER 23

Interest and Annuities

INTEREST is money paid by an individual or organization for the use of a sum of money called the *principal*. This interest is usually paid at the ends of specified equal intervals of time, for example annually, semiannually, or quarterly. The sum of the principal and the interest is called the *amount*.

THE RATE OF INTEREST is the ratio of the interest charged in one unit of time to the principal. The unit of time is taken to be one year unless otherwise specified. This rate of interest is generally expressed as a percentage.

For example, if the principal is $100 and the interest is $2 per year then the interest rate is $2/100 = .02 = 2\%$.

SIMPLE INTEREST is interest computed on the original principal for the time during which the money is being used. The simple interest I on the principal P for t years at a rate of interest r per year is given by

$$I = Prt$$

and the amount A (principal P plus interest I) is given by

$$A = P(1 + rt).$$

For example, if an individual borrows $800 at 4% to be paid in 2½ years, the interest is $I = 800(.04)(2\tfrac{1}{2}) = \$80$ and the amount due at the end of 2½ years is $A = \$880$.

COMPOUND INTEREST. Suppose the interest due at the end of the first of a specified number of equal intervals of time is added to the original principal and that this amount acts as a second principal for the second interval, the process being continued for a given time. In such case we say that the interest has been *compounded* or converted into principal. The sum by which the original principal has been increased for a given number of intervals of time is the *compound interest* for that time. The sum of the compound interest and original principal is called the *compound amount*. The successive equal intervals of time during which interest is compounded is called the *conversion period* or *interest period* and is usually either three months or six months or one year; in such cases interest is said to be compounded quarterly, semiannually, or annually respectively.

Interest rates are usually quoted on a yearly basis even though the compound interest is computed for each conversion period. The rate of interest quoted as a yearly rate is called the *nominal rate*.

Thus if the nominal rate is 4% compounded quarterly, the conversion period is 3 months and the interest rate is ¼(4%) = 1% for each conversion period.

221

If P is the original principal, i the rate of interest per conversion period and n the number of conversion periods, then the compound amount A at the end of these n conversion periods is given by

$$A = P(1 + i)^n.$$

The compound interest is $I = A - P$.

Example. A man invests \$1000 at 6% compounded semiannually. Find the compound amount A and compound interest I after 2 years.

$P = \$1000$, $i = \frac{1}{2}(6\%) = 3\% = .03$, $n = 4$ (since each conversion period is $\frac{1}{2}$ year and there are 4 such periods in 2 years). Then

$$A = 1000(1 + .03)^4 = 1000(1.03)^4 = \$1125.50$$

and $\qquad I = A - P = \$1125.50 - \$1000 = \$125.50.$

THE PRESENT VALUE of a given sum of money which is due at the end of a specified period of time is that principal which, when invested at interest, will accumulate to the given sum in the specified time. If the given sum of money due after n conversion periods is A and if the interest rate per period is i, then the present value P of the amount A is

$$P = \frac{A}{(1 + i)^n} = A(1 + i)^{-n}.$$

Example. Find the present value of \$1000 due in 3 years at 6% compounded quarterly.

$A = \$1000$, $i = \frac{1}{4}(6\%) = 1.5\% = .015$, $n = 3(4) = 12$. Thus

$$P = \frac{1000}{(1 + .015)^{12}} = 1000(1.015)^{-12} = \$836.39.$$

To discount amount A for n conversion periods at interest rate i per period is to determine the present value P of A, n periods before A is due. The difference $A - P$ is called the *discount* on A.

Thus in the above example the discount is $A - P = \$1000 - \$836.39 = \$163.61.$

AN ANNUITY is a sequence of equal periodic payments. The length of time between two successive payments is called the *payment period* or *payment interval*. The length of time between the beginning of the first payment period and the end of the last payment period is called the *term* of the annuity. The sum of payments made in one year is the *annual rent*. We shall consider payments made at the end of each payment period.

For example in an annuity of \$200 payable semiannually for ten years, the payment period or interval is 6 months, the term is 10 years and the annual rent is \$400.

THE AMOUNT OF AN ANNUITY is the sum of the compound amounts which would be obtained if each payment when due were kept at interest until the end of the specified term.

An annuity in which the periodic payment at the end of each of n equal periods is one dollar and whose interest rate per period is i has an amount given by

$$s_{\overline{n}|i} = \frac{(1 + i)^n - 1}{i}$$

where $s_{\overline{n}|i}$ is read "s angle n at i".

The amount of an annuity in which the periodic payment is R dollars is

$$S_{\overline{n}|i} = R\, s_{\overline{n}|i}$$

THE PRESENT VALUE OF AN ANNUITY is the sum of the present values of all the payments. An annuity in which the periodic payment at the end of each of n equal periods is one dollar and whose interest rate per period is i has a present value given by

$$a_{\overline{n}|i} = \frac{1 - (1 + i)^{-n}}{i}.$$

The present value of an annuity in which the periodic payment is R dollars is given by

$$A_{\overline{n}|i} = R\, a_{\overline{n}|i}.$$

SOLVED PROBLEMS

SIMPLE INTEREST.

1. A man borrows $400 for 2 years at a simple interest rate of 3%. Find the amount required to repay the loan at the end of 2 years.

   Interest $I = Prt = 400(.03)(2) = \$24$. Amount A = principal P + interest I = $424.

2. Find the interest I and amount A for a) $600 for 8 months (2/3 yr) at 4%,
   b) $1562.60 for 3 years, 4 months (10/3 yr) at 3½%.

   a) $I = Prt = 600(.04)(2/3) = \$16$. $A = P + I = \$616$.
   b) $I = Prt = 1562.60(.035)(10/3) = \$182.30$. $A = P + I = \$1744.90$.

3. What principal invested at 4% for 5 years will amount to $1200?

   $$A = P(1 + rt) \quad \text{or} \quad P = \frac{A}{1 + rt} = \frac{1200}{1 + (.04)(5)} = \frac{1200}{1.2} = \$1000.$$

   The principal of $1000 is called the present value of $1200. This means that $1200 to be paid 5 *years from now* is worth $1000 *now* (the interest rate being 4%).

4. What rate of interest will yield $1000 on a principal of $800 in 5 years?

   $$A = P(1 + rt) \quad \text{or} \quad r = \frac{A - P}{Pt} = \frac{1000 - 800}{800(5)} = .05 \text{ or } 5\%.$$

5. A man wishes to borrow $200. He goes to the bank where he is told that the interest rate is 5%, interest payable in advance, and that the $200 is to be paid back at the end of one year. What interest rate is he actually paying?

   The simple interest on $200 for 1 year at 5% is $I = 200(.05)(1) = \$10$. Thus he receives

$200 - $10 = $190. \quad Since he must pay back $200 after a year, $P = \$190$, $A = \$200$, $t = 1$ year.

Thus $r = \dfrac{A - P}{Pt} = \dfrac{200 - 190}{190(1)} = .0526$,   i.e. the effective interest rate is 5.26%.

6. A merchant borrows $4000 under the condition that he pay at the end of every 3 months $200 on the principal plus the simple interest of 6% on the principal outstanding at the time. Find the total amount he must pay.

Since $4000 is to be paid (excluding interest) at the rate of $200 every 3 months, it will take $\dfrac{4000}{200(4)} = 5$ years, i.e. 20 payments.

Interest paid at 1st payment (for first 3 months) $= 4000(.06)(\tfrac{1}{4}) = \$60.00.$
Interest paid at 2nd payment $= 3800(.06)(\tfrac{1}{4}) = \$57.00.$
Interest paid at 3rd payment $= 3600(.06)(\tfrac{1}{4}) = \$54.00.$
$\qquad\qquad\qquad\vdots\qquad\qquad\qquad\qquad\qquad\vdots\qquad\qquad\vdots$
Interest paid at 20th payment $= 200(.06)(\tfrac{1}{4}) = \$\ 3.00.$

The total interest is $60 + 57 + 54 + \ldots + 9 + 6 + 3$ an arithmetic progression having sum given by $S = \dfrac{n}{2}(a + l)$ where a = 1st term, l = last term, n = number of terms.

Then $S = \dfrac{20}{2}(60 + 3) = \$630$, and the total amount he must pay is $4630.

7. A man wants to receive $800 immediately and pay it back in 1 year. The bank charges a simple discount of 6% payable at once. How much must he borrow?

Let x dollars = amount borrowed.
Interest on x dollars at 6% for 1 year $= x(.06)(1) = .06x.$
The bank pays $x - .06x = .94x = 800$ from which $x = 800/.94 = \$851.06.$

8. A mortgage debt in the amount D dollars is to be discharged by k payments per year, in each payment the sum to be paid is p dollars on the principal plus simple interest at rate r on the principal outstanding at the time. Develop a formula for the total amount of money to be paid.

Since k payments per year are made,

$\qquad$ interest paid at 1st payment $\qquad = D(r)(1/k) \qquad = Dr/k$
$\qquad$ interest paid at 2nd payment $\qquad = (D - p)(r)(1/k) \quad = (D - p)r/k$
$\qquad$ interest paid at 3rd payment $\qquad = (D - 2p)(r)(1/k) \quad = (D - 2p)r/k$
$\qquad\qquad\qquad\qquad\vdots\qquad\qquad\qquad\qquad\qquad\vdots$
$\qquad$ interest paid at nth (last) payment $= [D - (n - 1)p](r)(1/k) = [D - (n - 1)p]r/k.$

Number of payments × principal paid per payment = total principal.

Then $np = D$ or $n = \dfrac{D}{p}$ (assumed equal to an integer).

Total interest $= \dfrac{Dr}{k} + \dfrac{(D - p)r}{k} + \dfrac{(D - 2p)r}{k} + \ldots + \dfrac{[D - (n - 1)p]r}{k}$, an arithmetic progression with sum

$$\frac{n}{2}(a + l) = \frac{n}{2}\left\{\frac{Dr}{k} + \frac{[D - (n - 1)p]r}{k}\right\} = \frac{Dr(D + p)}{2pk} \quad \text{using } n = \frac{D}{p}.$$

The total amount in dollars to be paid $= D + \dfrac{Dr(D + p)}{2pk}.$

Problem 6 is a special case where $D = 4000$, $k = 4$, $p = 200$, $r = .06$.

COMPOUND INTEREST.

9. What will $500 deposited in a bank amount to in 2 years if interest is compounded semiannually at 2%?

Method 1. Without formula.

At end of 1st half year, interest = $500(.02)(\frac{1}{2})$ = \$5.00.
At end of 2nd half year, interest = $505(.02)(\frac{1}{2})$ = \$5.05.
At end of 3rd half year, interest = $510.05(.02)(\frac{1}{2})$ = \$5.10.
At end of 4th half year, interest = $515.15(.02)(\frac{1}{2})$ = \$5.15.

Total interest = \$20.30.   Total amount = \$520.30.

Method 2. Using formula.

P = \$500, i = rate per period = $.02/2$ = $.01$, n = number of periods = 4.

$A = P(1 + i)^n = 500(1.01)^4 = 500(1.0406) = \$520.30.$

Note. $(1.01)^4$ may be evaluated by the binomial formula, logarithms or tables.

10. Find the compound interest and amount of $2800 in 8 years at 5% compounded quarterly.

$A = P(1+i)^n = 2800(1 + .05/4)^{32} = 2800(1.0125)^{32} = 2800(1.4881) = \$4166.68.$

Interest $= A - P = \$4166.68 - \$2800 = \$1366.68.$

11. A man expects to receive $2000 in 10 years. How much is that money worth now considering interest at 6% compounded quarterly? What is the discount?

We are asked for the present value P which will amount to A = \$2000 in 10 years.

$$A = P(1+i)^n \quad \text{or} \quad P = \frac{A}{(1+i)^n} = \frac{2000}{(1.015)^{40}} = \$1102.52 \quad \text{using tables.}$$

The discount is $2000 - \$1102.52 = \$897.48.$

12. What rate of interest compounded annually is the same as the rate of interest of 6% compounded semiannually?

Amount from principal P in 1 year at rate $r = P(1 + r)$.

Amount from principal P in 1 year at rate 6% compounded semiannually $= P(1 + .03)^2$.

The amounts are equal if $P(1+r) = P(1.03)^2$, $1+r = (1.03)^2$, $r = .0609$ or 6.09%.

The rate of interest i per year compounded a given number of times per year is called the *nominal rate*. The rate of interest r which if compounded annually would result in the same amount of interest is called the *effective rate*. In this example, 6% compounded semiannually is the nominal rate and 6.09% is the effective rate.

13. Derive a formula for the effective rate in terms of the nominal rate.

Let r = effective interest rate,
    i = interest rate per annum compounded k times per year, i.e. nominal rate.

Amount from principal P in 1 year at rate $r = P(1 + r)$.

Amount from principal P in 1 year at rate i compounded k times per year $= P(1 + i/k)^k$.

The amounts are equal if $P(1 + r) = P(1 + i/k)^k$.

Hence $r = (1 + i/k)^k - 1.$

ANNUITIES.

14. Find the amount of an annuity of $100 per year at the end of each year for 5 years at 3% compounded annually.

Method 1. Without formula.

Compound amount of 1st payment after 4 years $= 100(1.03)^4 = \$112.55.$
Compound amount of 2nd payment after 3 years $= 100(1.03)^3 = \$109.27.$
Compound amount of 3rd payment after 2 years $= 100(1.03)^2 = \$106.09.$
Compound amount of 4th payment after 1 year $= 100(1.03)^1 = \$103.00.$
Compound amount of 5th payment after 0 years $= 100 \qquad = \underline{\$100.00.}$
$$\text{Amount of annuity} = \$530.91.$$

Method 2. Using formula.

$$S_{\overline{n}|i} = S_{\overline{5}|.03} = R\{\frac{(1+i)^n - 1}{i}\} = 100\{\frac{(1+.03)^5 - 1}{.03}\} = 100(5.3091) = \$530.91 \text{ (tables)}$$

15. Determine the present value of an annuity of $100 per year at the end of each year for 5 years at 3% compounded annually.

Method 1. Without formula.

1st payment has present value $= 100/(1.03)^1 = 100(1.03)^{-1} = \$ 97.09.$
2nd payment has present value $= 100/(1.03)^2 = 100(1.03)^{-2} = \$ 94.26.$
3rd payment has present value $= 100/(1.03)^3 = 100(1.03)^{-3} = \$ 91.51.$
4th payment has present value $= 100/(1.03)^4 = 100(1.03)^{-4} = \$ 88.85.$
5th payment has present value $= 100/(1.03)^5 = 100(1.03)^{-5} = \underline{\$ 86.26.}$
$$\text{Present value of annuity} \doteq \$457.97.$$

Method 2. Using formula.

$$A_{\overline{n}|i} = A_{\overline{5}|.03} = R\{\frac{1-(1+i)^{-n}}{i}\} = 100\{\frac{1-(1+.03)^{-5}}{.03}\} = 100(4.5797) = \$457.97 \text{ (tables)}$$

16. Compute the amount and present value of an annuity of $120 at the end of each 3 months for 12 years at 6% compounded quarterly.

$R = \$120, \quad n = 4(12) = 48, \quad i = \frac{1}{4}(6\%) = 1.5\% = .015.$

$$\text{Amount } S_{\overline{48}|.015} = 120\,s_{\overline{48}|.015} = 120\{\frac{(1.015)^{48} - 1}{.015}\} = 120(69.5652) = \$8347.82$$

$$\text{Present value } A_{\overline{48}|.015} = 120\,a_{\overline{48}|.015} = 120\{\frac{1-(1.015)^{-48}}{.015}\} = 120(34.0426)$$
$$= \$4085.11$$

17. What equal amounts must a person invest at the end of each year in order to have $20,000 in 20 years if interest is 3% compounded annually?

Let R dollars = amount invested at the end of each year. The yearly payments of R dollars constitute an annuity with amount $S_{\overline{n}|i} = \$20,000, n = 20, i = 3\% = .03.$

$$S_{\overline{n}|i} = R\{\frac{(1+i)^n - 1}{i}\} = R\,s_{\overline{n}|i} \quad \text{or} \quad R = \frac{S_{\overline{n}|i}}{s_{\overline{n}|i}} = \frac{20,000}{s_{\overline{20}|.03}} = \frac{20,000}{26.8704} = \$744.31$$

18. What equal amounts should a company deposit every 6 months into a fund (called a sinking fund) which pays 4% compounded semiannually, in order that it may be able to replace at the end of 15 years some equipment worth $12,000, assuming the old equipment to be worthless then?

Let R dollars = amount deposited at the end of each 6 months. The amount in the sinking fund after 15 years is the amount of the annuity of 30 deposits made into the fund.

$$S_{\overline{n}|i} = \$12,000, \quad n = 30, \quad i = \tfrac{1}{2}(.04) = .02 \quad \text{and} \quad R = \frac{S_{\overline{n}|i}}{s_{\overline{n}|i}} = \frac{12,000}{s_{\overline{30}|.02}} = \frac{12,000}{40.5681} = \$295.80.$$

19. A mortgage debt in the amount of $8000 is to be discharged (amortized) in 6 years by equal payments made at the end of each year, the interest rate being 5% compounded annually. What annual payments must be made? What is the total interest paid?

Let annual payment = R dollars. The annual payments constitute an annuity whose present value $A_{\overline{n}|i}$ = $8000, $n = 6$, $i = 5\% = .05$.

$$A_{\overline{n}|i} = R\left\{\frac{1-(1+i)^{-n}}{i}\right\} = R\,a_{\overline{n}|i} \quad \text{or} \quad R = \frac{A_{\overline{n}|i}}{a_{\overline{n}|i}} = \frac{8000}{a_{\overline{6}|.05}} = \frac{8000}{5.0757} = \$1576.14.$$

The total principal paid is $8000 and the total amount paid = 6(1576.14) = $9456.84. The total interest = $9456.84 - $8000 = $1456.84.

20. A house may be purchased by the cash payment of $1000 at the present time and $300 paid at the end of each 3 months for 10 years. What is a fair cash price to pay for the house at the present time, assuming that money is worth 4% compounded quarterly?

The present value of an annuity of $300 at the end of each 3 months for 10 years at 4% compounded quarterly is

$$A_{\overline{40}|.01} = 300\,a_{\overline{40}|.01} = 300(32.8347) = \$9850.41.$$

Thus a fair cash price is $1000 + $9850.41 = $10,850.41.

21. Determine the amount and present value of an annuity in which n payments of R dollars are made at the beginning of each payment period at an interest rate of i per payment period.

The amount and present value of an annuity in which n payments are made at the beginning of each payment period are the same as the respective amount and present value of an annuity in which n payments of $R(1+i)$ dollars are made at the end of each payment period. Thus the

amount of the annuity = $R(1+i)\left\{\dfrac{(1+i)^n - 1}{i}\right\}$ and the present value = $R(1+i)\left\{\dfrac{1-(1+i)^{-n}}{i}\right\}$.

22. Use the formulas of Problem 21 to compute the amount and present value of an annuity in which $200 is paid at the beginning of each 6 months for 20 years at 2% compounded semiannually.

$R = 200$, $n = 2(20) = 40$, $i = \tfrac{1}{2}(2\%) = 1\% = .01$.

Amount of annuity = $200(1+.01)s_{\overline{40}|.01} = 202(48.8864) = \$9875.05.$

Present value of annuity = $200(1+.01)a_{\overline{40}|.01} = 202(32.8347) = \$6632.61.$

SUPPLEMENTARY PROBLEMS

23. Find the simple interest and amount for a) \$1200 for 4 months at 3%, b) \$1800 for 16 months at $2\frac{1}{2}$%, c) \$612.30 for 2 years and 8 months at 4%.

24. A man wishes to borrow D dollars. He goes to a bank where he is told that the interest rate is r%, interest payable in advance, and that the D dollars are to be paid back at the end of one year. Show that the effective interest rate is $\dfrac{100r}{100-r}$ %. What is the effective interest rate if $r = 6$?

25. A mortgage debt of \$3000 is to be discharged in semiannual payments; in each payment the sum to be paid is \$500 on the principal plus simple interest at rate $5\frac{1}{2}$% on the principal outstanding at the time. What is the total amount of money to be paid?

26. A man wishes to receive D dollars immediately and pay it back in 1 year. If the bank charges a simple discount rate r% payable at once, show that the amount he must borrow is \$ $\dfrac{100D}{100-r}$.

27. Find the compound interest and amount for a) \$2500 for 3 years at 4% compounded quarterly, b) 4800 for 6 years at 5% compounded annually, c) \$7350 for 10 years at 3% compounded semiannually.

28. How much money should be deposited in a fund in order to receive \$2000 in 6 years if the fund pays 6% interest compounded a) quarterly, b) semiannually?

29. What is the present value of \$5000 due in 8 years if the interest rate is 4% compounded a) quarterly, b) semiannually, c) annually? Find the corresponding discounts.

30. Find the present value of a) \$1500 due in 3 years if the interest rate is 6% compounded quarterly, b) \$5450.20 due in 10 years if the interest rate is 5% compounded semiannually.

31. What effective rate of interest corresponds to the nominal rate a) 4% compounded semiannually, b) 4% compounded quarterly?

32. Compute the amount and present value of each of the following annuities:
    a) \$400 at the end of each 6 months for 8 years at 4% compounded semiannually,
    b) \$1000 at the end of each year for 20 years at 5% compounded annually,
    c) \$250 at the end of each 3 months for 12 years at 6% compounded quarterly.

33. A man wishes to invest equal sums of money at the end of each 3 months so that in 10 years he may have \$5000. What must each investment be if interest is at 4% compounded quarterly?

34. A man owes \$10,000 on his house. He agrees to pay the debt in 10 years by equal payments to be made at the end of each 6 months, the interest rate being 6% compounded semiannually. Find a) his semiannual payments, b) the total amound paid at the end of 10 years.

35. A machine may be purchased by a down payment of \$1500 and \$500 paid at the end of each 6 months for 5 years. Assuming money to be worth 5% compounded semiannually, determine a fair cash price to offer for the machine at the present time.

ANSWERS TO SUPPLEMENTARY PROBLEMS.

23. a) \$12, \$1212 b) \$60, \$1860 c) \$65.31, \$677.61   25. \$3288.75
27. a) \$317.00, \$2817.00 b) \$1632.48, \$6432.48 c) \$2549.72, \$9899.72
28. a) \$1399.08 b) \$1402.76
29. a) \$3636.50, \$1363.50 b) \$3642.25, \$1357.75 c) \$3653.45, \$1346.55
30. a) \$1254.59 b) \$3326.09
31. a) 4.04% b) 4.06%
32. a) \$7455.72, \$5431.08 b) \$33,066.00, \$12,462.20 c) \$17,391.30, \$8510.65
33. \$102.28
34. a) \$672.16 b) \$13,443.20
35. \$5876.05

CHAPTER 24

Permutations and Combinations

FUNDAMENTAL PRINCIPLE. If one thing can be done in m different ways and, when it is done in any one of these ways, a second thing can be done in n different ways, then the two things in succession can be done in mn different ways.

For example, if there are 3 candidates for governor and 5 for mayor, then the two offices may be filled in $3 \cdot 5 = 15$ ways.

A PERMUTATION is an arrangement of all or part of a number of things in a definite order.

For example, the permutations of the three letters a, b, c taken all at a time are abc, acb, bca, bac, cba, cab. The permutations of the three letters a, b, c taken two at a time are ab, ac, ba, bc, ca, cb.

A COMBINATION is a grouping or selection of all or part of a number of things without reference to the arrangement of the things selected.

Thus the combinations of the three letters a, b, c taken 2 at a time are ab, ac, bc. Note that ab and ba are 1 combination but 2 permutations of the letters a, b.

FACTORIAL NOTATION. The following identities indicate the meaning of *factorial n*, or $n!$.

$$2! = 1 \cdot 2 = 2, \quad 3! = 1 \cdot 2 \cdot 3 = 6, \quad 4! = 1 \cdot 2 \cdot 3 \cdot 4 = 24$$

$$5! = 1 \cdot 2 \cdot 3 \cdot 4 \cdot 5 = 120, \quad n! = 1 \cdot 2 \cdot 3 \ldots n, \quad (r-1)! = 1 \cdot 2 \cdot 3 \ldots (r-1)$$

Note. $0! = 1$ by definition.

THE SYMBOL $_nP_r$ represents the number of permutations (arrangements, orders) of n things taken r at a time.

Thus $_8P_3$ denotes the number of permutations of 8 things taken 3 at a time, and $_5P_5$ denotes the number of permutations of 5 things taken 5 at a time.

Note. The symbol $P(n, r)$ having the same meaning as $_nP_r$ is sometimes used.

PERMUTATIONS OF n DIFFERENT THINGS TAKEN r AT A TIME.

$$_nP_r = n(n-1)(n-2)\ldots(n-r+1) = \frac{n!}{(n-r)!}$$

When $r = n$, $\quad _nP_r = {_nP_n} = n(n-1)(n-2)\ldots 1 = n!$.

229

Thus $_5P_1 = 5$, $_5P_2 = 5 \cdot 4 = 20$, $_5P_3 = 5 \cdot 4 \cdot 3 = 60$, $_5P_4 = 5 \cdot 4 \cdot 3 \cdot 2 = 120$,

$_5P_5 = 5! = 5 \cdot 4 \cdot 3 \cdot 2 \cdot 1 = 120$, $_{10}P_7 = 10 \cdot 9 \cdot 8 \cdot 7 \cdot 6 \cdot 5 \cdot 4 = 604,800$.

For example, the number of ways in which 4 persons can take their places in a cab having 6 seats is $_6P_4 = 6 \cdot 5 \cdot 4 \cdot 3 = 360$.

PERMUTATIONS WITH SOME THINGS ALIKE, TAKEN ALL AT A TIME. The number of permutations P of n things taken all at a time, of which n_1 are alike, n_2 others are alike, n_3 others are alike, etc., is

$$P = \frac{n!}{n_1! \, n_2! \, n_3! \, \ldots} \qquad \text{where} \quad n_1 + n_2 + n_3 + \ldots = n.$$

For example, the number of ways 3 dimes and 7 quarters can be distributed among 10 boys, each to receive one coin, is $\dfrac{10!}{3! \, 7!} = \dfrac{10 \cdot 9 \cdot 8}{1 \cdot 2 \cdot 3} = 120$.

CIRCULAR PERMUTATIONS. The number of ways of arranging n different objects around a circle is $(n-1)!$ ways.

Thus 10 persons may be seated at a round table in $(10-1)! = 9!$ ways.

THE SYMBOL $_nC_r$ represents the number of combinations (selections, groups) of n things taken r at a time.

Thus $_9C_4$ denotes the number of combinations of 9 things taken 4 at a time.

Note. The symbol $C(n, r)$ having the same meaning as $_nC_r$ is sometimes used.

COMBINATIONS OF n DIFFERENT THINGS TAKEN r AT A TIME.

$$_nC_r = \frac{_nP_r}{r!} = \frac{n!}{r! \, (n-r)!} = \frac{n(n-1)(n-2)\ldots(n-r+1)}{r!}$$

For example, the number of handshakes that may be exchanged among a party of 12 students if each student shakes hands once with each other student is

$$_{12}C_2 = \frac{12!}{2! \, (12-2)!} = \frac{12!}{2! \, 10!} = \frac{12 \cdot 11}{1 \cdot 2} = 66.$$

The following formula is very useful in simplifying calculations:

$$_nC_r = {_nC_{n-r}}.$$

This formula indicates that the number of selections of r out of n things is the same as the number of selections of $n-r$ out of n things.

Examples. $_5C_1 = \dfrac{5}{1} = 5$, $_5C_2 = \dfrac{5 \cdot 4}{1 \cdot 2} = 10$, $_5C_5 = \dfrac{5!}{5!} = 1$

$$_9C_7 = {_9C_{9-7}} = {_9C_2} = \frac{9 \cdot 8}{1 \cdot 2} = 36, \qquad _{25}C_{22} = {_{25}C_3} = \frac{25 \cdot 24 \cdot 23}{1 \cdot 2 \cdot 3} = 2300$$

Note that in each case the numerator and denominator have the same number of factors.

COMBINATIONS OF DIFFERENT THINGS TAKEN ANY NUMBER AT A TIME. The total number of combinations C of n different things taken $1, 2, 3, \ldots, n$ at a time is

$$C = 2^n - 1.$$

For example, a man has in his pocket a quarter, a dime, a nickel, and a penny. The total number of ways he can draw a sum of money from his pocket is $2^4 - 1 = 15$.

SOLVED PROBLEMS

PERMUTATIONS.

1. Evaluate $_{20}P_2$, $_8P_5$, $_7P_5$, $_7P_7$.

$_{20}P_2 = 20 \cdot 19 = 380$ $_7P_5 = 7 \cdot 6 \cdot 5 \cdot 4 \cdot 3 = 2520$

$_8P_5 = 8 \cdot 7 \cdot 6 \cdot 5 \cdot 4 = 6720$ $_7P_7 = 7! = 7 \cdot 6 \cdot 5 \cdot 4 \cdot 3 \cdot 2 \cdot 1 = 5040$

2. Find n if $a)$ $7 \cdot {_nP_3} = 6 \cdot {_{n+1}P_3}$, $b)$ $3 \cdot {_nP_4} = {_{n-1}P_5}$.

$a)$ $7n(n-1)(n-2) = 6(n+1)(n)(n-1)$.

Since $n \neq 0, 1$ we may divide by $n(n-1)$ to obtain $7(n-2) = 6(n+1)$, $n = 20$.

$b)$ $3n(n-1)(n-2)(n-3) = (n-1)(n-2)(n-3)(n-4)(n-5)$.

Since $n \neq 1, 2, 3$ we may divide by $(n-1)(n-2)(n-3)$ to obtain

$$3n = (n-4)(n-5), \qquad n^2 - 12n + 20 = 0, \qquad (n-10)(n-2) = 0.$$

Thus $n = 10$.

3. A student has a choice of 5 foreign languages and 4 sciences. In how many ways can he choose 1 language and 1 science?

He can choose a language in 5 ways, and with each of these choices there are 4 ways of choosing a science.

Hence the required number of ways $= 5 \cdot 4 = 20$ ways.

4. In how many ways can 2 different prizes be awarded among 10 contestants if both prizes $a)$ may not be given to the same person, $b)$ may be given to the same person?

$a)$ The first prize can be awarded in 10 different ways and, when it is awarded, the second prize can be given in 9 ways, since both prizes may not be given to the same contestant.
Hence the required number of ways $= 10 \cdot 9 = 90$ ways.

$b)$ The first prize can be awarded in 10 ways, and the second prize also in 10 ways, since both prizes may be given to the same contestant.
Hence the required number of ways $= 10 \cdot 10 = 100$ ways.

5. In how many ways can 5 letters be mailed if there are 3 mailboxes available?

Each of the 5 letters may be mailed in any of the 3 mailboxes.
Hence the required number of ways $= 3 \cdot 3 \cdot 3 \cdot 3 \cdot 3 = 3^5 = 243$ ways.

6. There are 4 candidates for president of a club, 6 for vice-president and 2 for secretary. In how many ways can these three positions be filled?

A president may be selected in 4 ways, a vice-president in 6 ways, and a secretary in 2 ways. Hence the required number of ways $= 4 \cdot 6 \cdot 2 = 48$ ways.

7. In how many different orders may 5 persons be seated in a row?

The first person may take any one of 5 seats, and after the first person is seated, the second person may take any one of the remaining 4 seats, etc. Hence the required number of orders $= 5 \cdot 4 \cdot 3 \cdot 2 \cdot 1 = 120$ orders.

Otherwise. Number of orders = number of arrangements of 5 persons taken all at a time
$$= {}_5P_5 = 5! = 5 \cdot 4 \cdot 3 \cdot 2 \cdot 1 = 120 \text{ orders.}$$

8. In how many ways can 7 books be arranged on a shelf?

Number of ways = number of arrangements of 7 books taken all at a time
$$= {}_7P_7 = 7! = 7 \cdot 6 \cdot 5 \cdot 4 \cdot 3 \cdot 2 \cdot 1 = 5040 \text{ ways.}$$

9. Twelve different pictures are available, of which 4 are to be hung in a row. In how many ways can this be done?

The first place may be occupied by any one of 12 pictures, the second place by any one of 11, the third place by any one of 10, and the fourth place by any one of 9.

Hence the required number of ways $= 12 \cdot 11 \cdot 10 \cdot 9 = 11,880$ ways.

Otherwise. Number of ways = number of arrangements of 12 pictures taken 4 at a time
$$= {}_{12}P_4 = 12 \cdot 11 \cdot 10 \cdot 9 = 11,880 \text{ ways.}$$

10. It is required to seat 5 men and 4 women in a row so that the women occupy the even places. How many such arrangements are possible?

The men may be seated in ${}_5P_5$ ways, and the women in ${}_4P_4$ ways. Each arrangement of the men may be associated with each arrangement of the women.

Hence the required number of arrangements $= {}_5P_5 \cdot {}_4P_4 = 5! \, 4! = 120 \cdot 24 = 2880$.

11. In how many orders can 7 different pictures be hung in a row so that 1 specified picture is *a*) at the center, *b*) at either end?

a) Since 1 given picture is to be at the center, 6 pictures remain to be arranged in a row. Hence the number of orders $= {}_6P_6 = 6! = 720$ orders.

b) After the specified picture is hung in any one of 2 ways, the remaining 6 can be arranged in ${}_6P_6$ ways.
Hence the number of orders $= 2 \cdot {}_6P_6 = 1440$ orders.

12. In how many ways can 9 different books be arranged on a shelf so that *a*) 3 of the books are always together, *b*) 3 of the books are never all 3 together?

a) The specified 3 books can be arranged among themselves in ${}_3P_3$ ways. Since the specified 3 books are always together, they may be considered as 1 thing. Then together with the other 6 books (things) we have a total of 7 things which can be arranged in ${}_7P_7$ ways.
Total number of ways $= {}_3P_3 \cdot {}_7P_7 = 3! \, 7! = 6 \cdot 5040 = 30,240$ ways.

b) Number of ways in which 9 books can be arranged on a shelf if there are no restrictions $= 9! = 362,880$ ways.
Number of ways in which 9 books can be arranged on a shelf when 3 specified books are always together (from *a*) above) $= 3! \, 7! = 30,240$ ways.

Hence the number of ways in which 9 books can be arranged on a shelf so that 3 specified books are never all 3 together = $362,880 - 30,240 = 332,640$ ways.

13. In how many ways can n men be seated in a row so that 2 particular men will not be next to each other?

With no restrictions, n men may be seated in a row in $_nP_n$ ways. If 2 of the n men must always sit next to each other, the number of arrangements = $2!(_{n-1}P_{n-1})$.

Hence the number of ways n men can be seated in a row if 2 particular men may never sit together = $_nP_n - 2(_{n-1}P_{n-1}) = n! - 2(n-1)! = n(n-1)! - 2(n-1)! = (n-2)\cdot(n-1)!$.

14. Six different biology books, 5 different chemistry books and 2 different physics books are to be arranged on a shelf so that the biology books stand together, the chemistry books stand together and the physics books stand together. How many such arrangements are possible?

The biology books can be arranged among themselves in 6! ways, the chemistry books in 5! ways, the physics books in 2! ways, and the three groups in 3! ways.

Required number of arrangements = $6!\,5!\,2!\,3! = 1,036,800$.

15. Determine the number of different words of 5 letters each that can be formed with the letters of the word *chromate* a) if each letter is used not more than once, b) if each letter may be repeated in any arrangement. (These words need not have meaning.)

a) Number of words = arrangements of 8 different letters taken 5 at a time
$$= {_8P_5} = 8\cdot7\cdot6\cdot5\cdot4 = 6720 \text{ words.}$$

b) Number of words = $8\cdot8\cdot8\cdot8\cdot8 = 8^5 = 32,768$ words.

16. How many numbers may be formed by using 4 out of the 5 digits $1,2,3,4,5$ a) if the digits must not be repeated in any number, b) if they may be repeated? If the digits must not be repeated, how many of the 4-digit numbers c) begin with 2, d) end with 25?

a) Numbers formed = $_5P_4 = 5\cdot4\cdot3\cdot2 = 120$ numbers.

b) Numbers formed = $5\cdot5\cdot5\cdot5 = 5^4 = 625$ numbers.

c) Since the first digit of each number is specified, there remain 4 digits to be arranged in 3 places.
Numbers formed = $_4P_3 = 4\cdot3\cdot2 = 24$ numbers.

d) Since the last two digits of every number are specified, there remain 3 digits to be arranged in 2 places.
Numbers formed = $_3P_2 = 3\cdot2 = 6$ numbers.

17. How many 4-digit numbers may be formed with the 10 digits $0,1,2,3,\ldots,9$ if each digit is used only once in each number? How many of these numbers are odd?

a) The first place may be filled by any one of the 10 digits except 0, i.e. by any one of 9 digits. The 9 digits remaining may be arranged in the 3 other places in $_9P_3$ ways.
Numbers formed = $9\cdot{_9P_3} = 9(9\cdot8\cdot7) = 4536$ numbers.

b) The last place may be filled by any one of the 5 odd digits, $1,3,5,7,9$. The first place may be filled by any one of the 8 digits, i.e. by the remaining 4 odd digits and the even digits, $2,4,6,8$. The 8 remaining digits may be arranged in the 2 middle positions in $_8P_2$ ways.
Numbers formed = $5\cdot8\cdot{_8P_2} = 5\cdot8\cdot8\cdot7 = 2240$ odd numbers.

18. a) How many 5-digit numbers can be formed from the 10 digits $0,1,2,3,\ldots,9$, repetitions allowed? How many of these numbers b) begin with 40, c) are even, d) are divisible by 5?

a) The first place may be filled by any one of 9 digits (any of the 10 except 0). Each of the other 4 places may be filled by any one of the 10 digits whatever.

Numbers formed = $9 \cdot 10 \cdot 10 \cdot 10 \cdot 10$ = $9 \cdot 10^4$ = 90,000 numbers.

b) The first 2 places may be filled in 1 way, by 40. The other 3 places may be filled by any one of the 10 digits whatever.

Numbers formed = $1 \cdot 10 \cdot 10 \cdot 10$ = 10^3 = 1000 numbers.

c) The first place may be filled in 9 ways, and the last place in 5 ways (0,2,4,6,8). Each of the other 3 places may be filled by any one of the 10 digits whatever.

Even numbers = $9 \cdot 10 \cdot 10 \cdot 10 \cdot 5$ = 45,000 numbers.

d) The first place may be filled in 9 ways, the last place in 2 ways (0,5), and the other 3 places in 10 ways each.

Numbers divisible by 5 = $9 \cdot 10 \cdot 10 \cdot 10 \cdot 2$ = 18,000 numbers.

19. How many numbers between 3000 and 5000 can be formed by using the 7 digits 0,1,2,3,4,5,6 if each digit must not be repeated in any number?

Since the numbers are between 3000 and 5000, they consist of 4 digits. The first place may be filled in 2 ways, i.e. by digits 3,4. Then the remaining 6 digits may be arranged in the 3 other places in $_6P_3$ ways.

Numbers formed = $2 \cdot {_6P_3}$ = $2(6 \cdot 5 \cdot 4)$ = 240 numbers.

20. From 11 novels and 3 dictionaries, 4 novels and 1 dictionary are to be selected and arranged on a shelf so that the dictionary is always in the middle. How many such arrangements are possible?

The dictionary may be chosen in 3 ways. The number of arrangements of 11 novels taken 4 at a time is $_{11}P_4$.

Required number of arrangements = $3 \cdot {_{11}P_4}$ = $3(11 \cdot 10 \cdot 9 \cdot 8)$ = 23,760.

21. How many signals can be made with 5 different flags by raising them any number at a time?

Signals may be made by raising the flags 1,2,3,4, and 5 at a time. Hence the total number of signals is

$$_5P_1 + {_5P_2} + {_5P_3} + {_5P_4} + {_5P_5} = 5 + 20 + 60 + 120 + 120 = 325 \text{ signals.}$$

22. Compute the sum of the 4-digit numbers which can be formed with the four digits 2,5,3,8 if each digit is used only once in each arrangement.

The number of arrangements, or numbers, is $_4P_4$ = 4! = $4 \cdot 3 \cdot 2 \cdot 1$ = 24.

The sum of the digits = $2 + 5 + 3 + 8 = 18$, and each digit will occur 24/4 = 6 times each in the units', tens', hundreds', and thousands' positions. Hence the sum of all the numbers formed is

$$1(6 \cdot 18) + 10(6 \cdot 18) + 100(6 \cdot 18) + 1000(6 \cdot 18) = 119,988.$$

23. *a*) How many arrangements can be made from the letters of the word *cooperator* when all are taken at a time? How many of such arrangements *b*) have the three *o*'s together, *c*) begin with the two *r*'s ?

a) The word *cooperator* consists of 10 letters: 3 *o*'s, 2 *r*'s, and 5 different letters.

Number of arrangements = $\dfrac{10!}{3! \, 2!}$ = $\dfrac{10 \cdot 9 \cdot 8 \cdot 7 \cdot 6 \cdot 5 \cdot 4 \cdot 3 \cdot 2 \cdot 1}{(1 \cdot 2 \cdot 3)(1 \cdot 2)}$ = 302,400.

b) Consider the 3 *o*'s as 1 letter. Then we have 8 letters of which 2 *r*'s are alike.

Number of arrangements = $\dfrac{8!}{2!}$ = 20,160.

c) The number of arrangements of the remaining 8 letters, of which 3 *o*'s are alike, = 8!/3!
= 6720.

24. There are 3 copies each of 4 different books. In how many different ways can they be arranged on a shelf?

There are $3 \cdot 4 = 12$ books of which 3 are alike, 3 others alike, etc.

Number of arrangements $= \dfrac{(3 \cdot 4)!}{3!\,3!\,3!\,3!} = \dfrac{12!}{(3!)^4} = 369,600$.

25. *a*) In how many ways can 5 persons be seated at a round table?
b) In how many ways can 8 persons be seated at a round table if 2 particular persons must always sit together?

a) Let 1 of them be seated anywhere. Then the 4 persons remaining can be seated in 4! ways. Hence there are 4! = 24 ways of arranging 5 persons in a circle.

b) Consider the two particular persons as one person. Since there are 2! ways of arranging 2 persons among themselves and 6! ways of arranging 7 persons in a circle, the required number of ways $= 2!\,6! = 2 \cdot 720 = 1440$ ways.

26. In how many ways can 4 men and 4 women be seated at a round table if each woman is to be between two men?

Consider that the men are seated first. Then the men can be arranged in 3! ways, and the women in 4! ways.

Required number of circular arrangements $= 3!\,4! = 144$.

27. By stringing together 9 differently colored beads, how many different bracelets can be made?

There are 8! arrangements of the beads on the bracelet, but half of these can be obtained from the other half simply by turning the bracelet over.

Hence there are $\frac{1}{2}(8!) = 20,160$ different bracelets.

COMBINATIONS.

28. In each case, find *n*: *a*) $_nC_{n-2} = 10$,   *b*) $_nC_{15} = _nC_{11}$,   *c*) $_nP_4 = 30 \cdot {}_nC_5$.

a) $_nC_{n-2} = {}_nC_2 = \dfrac{n(n-1)}{2!} = \dfrac{n^2 - n}{2} = 10$, $n^2 - n - 20 = 0$, $n = 5$

b) $_nC_r = {}_nC_{n-r}$, $_nC_{15} = {}_nC_{n-11}$, $15 = n - 11$, $n = 26$

c) $30 \cdot {}_nC_5 = 30\left(\dfrac{_nP_5}{5!}\right) = \dfrac{30 \cdot {}_nP_4 \cdot (n-4)}{5!}$

Then $_nP_4 = \dfrac{30 \cdot {}_nP_4 \cdot (n-4)}{5!}$, $1 = \dfrac{30(n-4)}{120}$, $n = 8$.

29. Given $_nP_r = 3024$ and $_nC_r = 126$, find *r*.

$_nP_r = r!({}_nC_r)$, $r! = \dfrac{_nP_r}{_nC_r} = \dfrac{3024}{126} = 24$, $r = 4$

30. How many different sets of 4 students can be chosen out of 17 qualified students to represent a school in a mathematics contest?

Number of sets = number of combinations of 4 out of 17 students

$$= {}_{17}C_4 = \frac{17 \cdot 16 \cdot 15 \cdot 14}{1 \cdot 2 \cdot 3 \cdot 4} = 2380 \text{ sets of 4 students.}$$

31. In how many ways can 5 styles be selected out of 8 styles?

Number of ways = number of combinations of 5 out of 8 styles

$$= {}_8C_5 = {}_8C_3 = \frac{8 \cdot 7 \cdot 6}{1 \cdot 2 \cdot 3} = 56 \text{ ways.}$$

32. In how many ways can 12 books be divided between A and B so that one may get 9 and the other 3 books?

In each separation of 12 books into 9 and 3, A may get the 9 and B the 3, or A may get the 3 and B the 9.

Hence the number of ways $= 2 \cdot {}_{12}C_9 = 2 \cdot {}_{12}C_3 = 2(\frac{12 \cdot 11 \cdot 10}{1 \cdot 2 \cdot 3}) = 440 \text{ ways.}$

33. Determine the number of different triangles which can be formed by joining the six vertices of a hexagon, the vertices of each triangle being on the hexagon.

Number of triangles = number of combinations of 3 out of 6 points

$$= {}_6C_3 = \frac{6 \cdot 5 \cdot 4}{1 \cdot 2 \cdot 3} = 20 \text{ triangles.}$$

34. How many angles less than $180°$ are formed by 12 straight lines which terminate in a point, if no two of them are in the same straight line?

Number of angles = number of combinations of 2 out of 12 lines

$$= {}_{12}C_2 = \frac{12 \cdot 11}{1 \cdot 2} = 66 \text{ angles.}$$

35. How many diagonals has an octagon?

Lines formed = number of combinations of 2 out of 8 corners (points) $= {}_8C_2 = \frac{8 \cdot 7}{2} = 28.$

Since 8 of these 28 lines are the sides of the octagon, the number of diagonals = 20.

36. How many parallelograms are formed by a set of 4 parallel lines intersecting another set of 7 parallel lines?

Each combination of 2 lines out of 4 can intersect each combination of 2 lines out of 7 to form a parallelogram.

Number of parallelograms $= {}_4C_2 \cdot {}_7C_2 = 6 \cdot 21 = 126 \text{ parallelograms.}$

37. There are 10 points in a plane. No three of these points are in a straight line, except 4 points which are all in the same straight line. How many straight lines can be formed by joining the 10 points?

Number of lines formed if no 3 of the 10 points were in a straight line $= {}_{10}C_2 = \frac{10 \cdot 9}{2} = 45.$

Number of lines formed by 4 points, no 3 of which are collinear $= {}_4C_2 = \frac{4 \cdot 3}{2} = 6.$

Since the 4 points are collinear, they form 1 line instead of 6 lines.

Required number of lines = 45 – 6 + 1 = 40 lines.

38. In how many ways can 3 men be selected out of 15 men
a) if 1 of the men is to be included in every selection,
b) if 2 of the men are to be excluded from every selection,
c) if 1 is always included and 2 are always excluded?

a) Since 1 is always included, we must select 2 out of 14 men.

Hence the number of ways = $_{14}C_2$ = $\dfrac{14 \cdot 13}{2}$ = 91 ways.

b) Since 2 are always excluded, we must select 3 out of 13 men.

Hence the number of ways = $_{13}C_3$ = $\dfrac{13 \cdot 12 \cdot 11}{3!}$ = 286 ways.

c) Number of ways = $_{15-1-2}C_{3-1}$ = $_{12}C_2$ = $\dfrac{12 \cdot 11}{2}$ = 66 ways.

39. An organization has 25 members, 4 of whom are doctors. In how many ways can a committee of 3 members be selected so as to include at least 1 doctor?

Total number of ways in which 3 can be selected out of 25 = $_{25}C_3$.

Number of ways in which 3 can be selected so that no doctor is included = $_{25-4}C_3$ = $_{21}C_3$.

Then the number of ways in which 3 members can be selected so that at least 1 doctor is included = $_{25}C_3 - {_{21}C_3}$ = $\dfrac{25 \cdot 24 \cdot 23}{3!}$ – $\dfrac{21 \cdot 20 \cdot 19}{3!}$ = 970 ways.

40. From 6 chemists and 5 biologists, a committee of 7 is to be chosen so as to include 4 chemists. In how many ways can this be done?

Each selection of 4 out of 6 chemists can be associated with each selection of 3 out of 5 biologists.

Hence the number of ways = $_6C_4 \cdot {_5C_3}$ = $_6C_2 \cdot {_5C_2}$ = 15·10 = 150 ways.

41. Given 8 consonants and 4 vowels, how many 5-letter words can be formed, each word consisting of 3 different consonants and 2 different vowels?

The 3 different consonants can be selected in $_8C_3$ ways, the 2 different vowels in $_4C_2$ ways, and the 5 different letters (3 consonants, 2 vowels) can be arranged among themselves in $_5P_5$ = 5! ways.

Hence the number of words = $_8C_3 \cdot {_4C_2} \cdot 5!$ = 56·6·120 = 40,320.

42. From 7 capitals, 3 vowels and 5 consonants, how many words of 4 letters each can be formed if each word begins with a capital and contains at least 1 vowel, all the letters of each word being different?

The first letter, or capital, may be selected in 7 ways.

The remaining 3 letters may be
a) 1 vowel and 2 consonants, which may be selected in $_3C_1 \cdot {_5C_2}$ ways,
b) 2 vowels and 1 consonant, which may be selected in $_3C_2 \cdot {_5C_1}$ ways, and
c) 3 vowels, which may be selected in $_3C_3$ = 1 way.
Each of these selections of 3 letters may be arranged among themselves in $_3P_3$ = 3! ways.

Hence the number of words = $7 \cdot 3!({_3C_1} \cdot {_5C_2} + {_3C_2} \cdot {_5C_1} + 1)$

$$= 7 \cdot 6(3 \cdot 10 + 3 \cdot 5 + 1) = 1932 \text{ words.}$$

43. A has 3 maps and B has 9 maps. Determine the number of ways in which they can exchange maps if each keeps his initial number of maps.

   A can exchange 1 map with B in $_3C_1 \cdot {_9}C_1$ = $3 \cdot 9$ = 27 ways.

   A can exchange 2 maps with B in $_3C_2 \cdot {_9}C_2$ = $3 \cdot 36$ = 108 ways.

   A can exchange 3 maps with B in $_3C_3 \cdot {_9}C_3$ = $1 \cdot 84$ = 84 ways.

   Total number of ways = $27 + 108 + 84$ = 219 ways.

   Another Method. Consider that A and B put their maps together. Then the problem is to find the number of ways A can select 3 maps out of 12, not including the selection by A of his original three maps. Hence, $_{12}C_3 - 1 = \dfrac{12 \cdot 11 \cdot 10}{1 \cdot 2 \cdot 3} - 1 = 219$ ways.

44. a) In how many ways can 12 books be divided among 3 students so that each receives 4 books?
   b) In how many ways can 12 books be divided into 3 groups of 4 each?

   a) The first student can select 4 out of 12 books in $_{12}C_4$ ways.
   The second student can select 4 of the remaining 8 books in $_8C_4$ ways.
   The third student can select 4 of the remaining 4 books in 1 way.

   Number of ways = $_{12}C_4 \cdot {_8}C_4 \cdot 1$ = $495 \cdot 70 \cdot 1$ = 34,650 ways.

   b) The 3 groups could be distributed among the students in $3! = 6$ ways.

   Hence the number of groups = $\dfrac{34,650}{3!}$ = 5775 groups.

45. In how many ways can a person choose 1 or more of 4 electrical appliances?

   Each appliance may be dealt with in 2 ways, as it can be chosen or not chosen. Since each of the 2 ways of dealing with an appliance is associated with 2 ways of dealing with each of the other appliances, the number of ways of dealing with the 4 appliances = $2 \cdot 2 \cdot 2 \cdot 2 = 2^4$ ways. But 2^4 ways includes the case in which no appliance is chosen.

   Hence the required number of ways = $2^4 - 1$ = $16 - 1$ = 15 ways.

   Another Method. The appliances may be chosen singly, in twos, etc. Hence the required number of ways = $_4C_1 + {_4}C_2 + {_4}C_3 + {_4}C_4$ = $4 + 6 + 4 + 1$ = 15 ways.

46. How many different sums of money can be drawn from a wallet containing one bill each of 1, 2, 5, 10, 20 and 50 dollars?

   Number of sums = $2^6 - 1$ = 63 sums.

47. In how many ways can 2 or more ties be selected out of 8 ties?

   One or more ties may be selected in $(2^8 - 1)$ ways. But since 2 or more must be chosen, the required number of ways = $2^8 - 1 - 8$ = 247 ways.

   Another Method. 2, 3, 4, 5, 6, 7, or 8 ties may be selected in

   $_8C_2 + {_8}C_3 + {_8}C_4 + {_8}C_5 + {_8}C_6 + {_8}C_7 + {_8}C_8$ = $_8C_2 + {_8}C_3 + {_8}C_4 + {_8}C_3 + {_8}C_2 + {_8}C_1 + 1$
   $= 28 + 56 + 70 + 56 + 28 + 8 + 1$ = 247 ways.

48. There are available 5 different green dyes, 4 different blue dyes, and 3 different red dyes. How many selections of dyes can be made, taking at least 1 green and 1 blue dye?

   The green dyes can be chosen in $(2^5 - 1)$ ways, the blue dyes in $(2^4 - 1)$ ways, and the red dyes in 2^3 ways.

   Number of selections = $(2^5 - 1)(2^4 - 1)(2^3)$ = $31 \cdot 15 \cdot 8$ = 3720 selections.

SUPPLEMENTARY PROBLEMS

PERMUTATIONS.

49. Evaluate: $_{16}P_3$, $_7P_4$, $_5P_5$, $_{12}P_1$.

50. Find n if $a)$ $10 \cdot {}_nP_2 = {}_{n+1}P_4$, $b)$ $3 \cdot {}_{2n+4}P_3 = 2 \cdot {}_{n+4}P_4$.

51. In how many ways can six people be seated on a bench?

52. With four signal flags of different colors, how many different signals can be made by displaying two flags one above the other?

53. With six signal flags of different colors, how many different signals can be made by displaying three flags one above the other?

54. In how many ways can a club consisting of 12 members choose a president, a secretary and a treasurer?

55. If no two books are alike, in how many ways can 2 red, 3 green and 4 blue books be arranged on a shelf so that all the books of the same color are together?

56. There are 4 hooks on a wall. In how many ways can 3 coats be hung on them, one coat on a hook?

57. How many two digit numbers can be formed with the digits 0,3,5,7 if no repetition in any of the numbers is allowed?

58. How many even numbers of two different digits can be formed from the digits 3,4,5,6,8 ?

59. How many three digit numbers can be formed from the digits 1,2,3,4,5 if no digit is repeated in any number?

60. How many numbers of three digits each can be written with the digits 1,2,...,9 if no digit is repeated in any number?

61. How many three digit numbers can be formed from the digits 3,4,5,6,7 if digits are allowed to be repeated?

62. How many odd numbers of three digits each can be formed without the repetition of any digit in a number, from the digits $a)$ 1,2,3,4, $b)$ 1,2,4,6,8 ?

63. How many even numbers of four different digits each can be formed from the digits 3,5,6,7,9?

64. How many different numbers of 5 digits each can be formed from the digits 2,3,5,7,9 if no digit is repeated?

65. How many integers are there between 100 and 1000 in which no digit is repeated?

66. How many integers greater than 300 and less than 1000 can be made with the digits 1,2,3,4,5 if no digit is repeated in any number?

67. How many numbers between 100 and 1000 can be written with the digits 0,1,2,3,4 if no digit is repeated in any number?

68. How many four digit numbers greater than 2000 can be formed with the digits 1,2,3,4 if repetitions $a)$ are not allowed, $b)$ are allowed ?

69. How many of the arrangements of the letters of the word *logarithm* begin with a vowel and end with a consonant?

70. In a telephone system four different letters P,R,S,T and the four digits 3,5,7,8 are used. Find the maximum number of "telephone numbers" the system can have if each consists of a letter followed by a four digit number in which the digits may be repeated.

71. In how many ways can 3 girls and 3 boys be seated in a row, if no two girls and no two boys are to occupy adjacent seats?

72. How many telegraphic characters could be made by using three dots and two dashes in each character?

73. In how many ways can three dice fall?

74. How many fraternities can be named with the 24 letters of the Greek alphabet if each has three letters and none is repeated in any name?

75. How many signals can be shown with 8 flags of which 2 are red, 3 white and 3 blue, if they are all strung up on a vertical pole at once?

76. In how many ways can 4 men and 4 women sit at a round table so that no two men are adjacent?

77. How many different arrangements are possible with the factors of the term $a^2b^4c^3$ written at full length?

78. In how many ways can 9 different prizes be awarded to two students so that one receives 3 and the other 6?

79. How many different radio stations can be named with 3 different letters of the alphabet? How many with 4 different letters in which W must come first?

COMBINATIONS.

80. In each case find n: a) $4 \cdot {}_nC_2 = {}_{n+2}C_3$, b) ${}_{n+2}C_n = 45$, c) ${}_nC_{12} = {}_nC_8$.

81. If $5 \cdot {}_nP_3 = 24 \cdot {}_nC_4$, find n.

82. Evaluate a) ${}_7C_7$, b) ${}_5C_3$, c) ${}_7C_2$, d) ${}_7C_5$, e) ${}_7C_6$, f) ${}_8C_7$, g) ${}_8C_5$, h) ${}_{100}C_{98}$.

83. How many straight lines are determined by a) 6, b) n points, no three of which lie in the same straight line?

84. How many chords are determined by seven points on a circle?

85. A student is allowed to choose 5 questions out of 9. In how many ways can he choose them?

86. How many different sums of money can be formed by taking two of the following: a cent, a nickel, a dime, a quarter, a half-dollar?

87. How many different sums of money can be formed from the coins of Problem 86?

88. A baseball league is made up of 6 teams. If each team is to play each of the other teams a) twice, b) three times, how many games will be played?

89. How many different committees of two men and one woman can be formed from a) 7 men and 4 women, b) 5 men and 3 women?

90. In how many ways can 5 colors be selected out of 8 different colors including red, blue and green
a) if blue and green are always to be included,
b) if red is always excluded,
c) if red and blue are always included but green excluded?

91. From 5 physicists, 4 chemists and 3 mathematicians a committee of 6 is to be chosen so as to include 3 physicists, 2 chemists and 1 mathematician. In how many ways can this be done?

92. In Problem 91, in how many ways can the committee of 6 be chosen so that
a) 2 members of the committee are mathematicians,
b) at least 3 members of the committee are physicists?

93. How many words of 2 vowels and 3 consonants may be formed (considering any set a word) from the letters of the word a) *stenographic*, b) *facetious*?

94. In how many ways can a picture be colored if 7 different colors are available for use?

95. In how many ways can 8 women form a committee if at least 3 women are to be on the committee?

96. A box contains 7 red cards, 6 white cards and 4 blue cards. How many selections of three cards can be made so that *a*) all three are red, *b*) none are red?

97. How many baseball nines can be chosen from 13 candidates if *A,B,C,D* are the only candidates for two positions and can play no other position?

98. How many different committees including 3 Democrats and 2 Republicans can be chosen from 8 Republicans and 10 Democrats?

99. At a meeting, after everyone had shaken hands once with everyone else, it was found that 45 handshakes were exchanged. How many were at the meeting?

100. Find the number of *a*) combinations and *b*) permutations of four letters each that can be made from the letters of the word TENNESSEE.

ANSWERS TO SUPPLEMENTARY PROBLEMS.

49. 3360, 840, 120, 12

50. *a*) 4, *b*) 6

51. 720

52. 12

53. 120

54. 1320

55. 1728

56. 24

57. 9

58. 12

59. 60

60. 504

61. 125

62. *a*) 12, *b*) 12

63. 24

64. 120

65. 648

66. 36

67. 48

68. *a*) 18, *b*) 192

69. 90,720

70. 1024

71. 72

72. 10

73. 216

74. 12,144

75. 560

76. 144

77. 1260

78. 168

79. 15,600; 13,800

80. *a*) 2, 7 *b*) 8 *c*) 20

81. 8

82. *a*) 1 *b*) 10 *c*) 21
d) 21 e) 7 *f*) 8
g) 56 h) 4950

83. *a*) 15 b) $\dfrac{n(n-1)}{2}$

84. 21

85. 126

86. 10

87. 31

88. *a*) 30, *b*) 45

89. *a*) 84, *b*) 30

90. *a*) 20, b) 21, c) 10

91. 180

92. *a*) 378, b) 462

93. *a*) 40,320, *b*) 4800

94. 127

95. 219

96. *a*) 35, *b*) 120

97. 216

98. 3360

99. 10

100. *a*) 17, *b*) 163

CHAPTER 25

Probability

DEFINITION. Suppose that an event can happen in h ways and fail to happen in f ways, all these $h + f$ ways supposed equally likely. Then the probability of the occurrence of the event (called its success) is $p = \dfrac{h}{h + f} = \dfrac{h}{n}$, and the probability of the non-occurrence of the event (called its failure) is $q = \dfrac{f}{h + f} = \dfrac{f}{n}$, where $n = h + f$.

It follows that $p + q = 1$, $p = 1 - q$, and $q = 1 - p$.

The odds in favor of the occurrence of the event are $h : f$ or h/f; the odds against its happening are $f : h$ or f/h.

If p is the probability that an event will occur, the odds in favor of its happening are $p : q = p : (1 - p)$ or $p/(1 - p)$; the odds against its happening are $q : p = (1 - p) : p$ or $(1 - p)/p$.

INDEPENDENT EVENTS. Two or more events are said to be independent if the occurrence or non-occurrence of any one of them does not affect the probabilities of occurrence of any of the others.

Thus if a coin is tossed four times and it turns up head each time, the fifth toss may be head or tail and is not influenced by the previous tosses.

The probability that two or more independent events will happen is equal to the product of their separate probabilities.

Thus the probability of getting head on both the fifth and sixth tosses is $\dfrac{1}{2}(\dfrac{1}{2}) = \dfrac{1}{4}$.

DEPENDENT EVENTS. Two or more events are said to be dependent if the occurrence or non-occurrence of one of the events affects the probabilities of occurrence of any of the others.

Consider that two or more events are dependent. If p_1 is the probability of a first event, p_2 the probability that after the first has happened the second will occur, p_3 the probability that after the first and second have happened the third will occur, etc., then the probability that all events will happen in the given order is the product $p_1 \cdot p_2 \cdot p_3 \cdots$.

For example, a box contains 3 white balls and 2 black balls. If a ball is drawn at random, the probability that it is black is $\dfrac{2}{3 + 2} = \dfrac{2}{5}$. If this ball

242

is not replaced and a second ball is drawn, the probability that it also is black is $\frac{1}{3+1} = \frac{1}{4}$. Thus the probability that both will be black is $\frac{2}{5}(\frac{1}{4}) = \frac{1}{10}$.

MUTUALLY EXCLUSIVE EVENTS. Two or more events are said to be mutually exclusive if the occurrence of any one of them excludes the occurrence of the others.

The probability of occurrence of some one of two or more mutually exclusive events is the *sum* of the probabilities of the individual events.

MATHEMATICAL EXPECTATION. If p is the probability that a person will receive a sum of money m, the value of his expectation is $p \cdot m$.

Thus if the probability of your winning a $10 prize is 1/5, your expectation is $\frac{1}{5}(\$10) = \$2$.

REPEATED TRIALS. If p is the probability that an event will happen in any single trial and $q = 1 - p$ is the probability that it will fail to happen in any single trial, then the probability of its happening *exactly* r times in n trials is $_nC_r p^r q^{n-r}$. (See Problems 22-23.)

The probability that an event will happen *at least* r times in n trials is

$$p^n + {}_nC_1 p^{n-1}q + {}_nC_2 p^{n-2}q^2 + \ldots + {}_nC_r p^r q^{n-r}.$$

This expression is the sum of the first $n - r + 1$ terms of the binomial expansion of $(p+q)^n$. (See Problems 24-26.)

SOLVED PROBLEMS

1. One ball is drawn at random from a box containing 3 red balls, 2 white balls, and 4 blue balls. Determine the probability p that it is a) red, b) not red, c) white, d) red or blue.

a) $p = \dfrac{\text{ways of drawing 1 out of 3 red balls}}{\text{ways of drawing 1 out of } (3+2+4) \text{ balls}} = \dfrac{3}{3+2+4} = \dfrac{3}{9} = \dfrac{1}{3}$

b) $p = 1 - \dfrac{1}{3} = \dfrac{2}{3}$ c) $p = \dfrac{2}{9}$ d) $p = \dfrac{3+4}{9} = \dfrac{7}{9}$

2. One bag contains 4 white balls and 2 black balls; another bag contains 3 white balls and 5 black balls. If one ball is drawn from each bag, determine the probability p that a) both are white, b) both are black, c) 1 is white and 1 is black.

a) $p = (\dfrac{4}{4+2})(\dfrac{3}{3+5}) = \dfrac{1}{4}$ b) $p = (\dfrac{2}{4+2})(\dfrac{5}{3+5}) = \dfrac{5}{24}$

c) Probability that first ball is white and second black $= \dfrac{4}{6}(\dfrac{5}{8}) = \dfrac{5}{12}$.

Probability that first ball is black and second white $= \dfrac{2}{6}(\dfrac{3}{8}) = \dfrac{1}{8}$.

These are mutually exclusive; hence the required probability $p = \dfrac{5}{12} + \dfrac{1}{8} = \dfrac{13}{24}$.

Another Method. $p = 1 - \dfrac{1}{4} - \dfrac{5}{24} = \dfrac{13}{24}$.

3. Determine the probability of throwing a total of 8 in a single throw with two dice, each of whose faces are numbered from 1 to 6.

Each of the faces of one die can be associated with any of the 6 faces of the other die; thus the total number of possible cases = $6 \cdot 6$ = 36 cases.

There are 5 ways of throwing an 8: 2,6; 3,5; 4,4; 5,3; 6,2.

Required probability = $\dfrac{\text{number of favorable cases}}{\text{possible number of cases}} = \dfrac{5}{36}$.

4. What is the probability of getting at least 1 ace in 2 throws of a die?

The probability of not getting an ace in any single throw = $1 - 1/6$ = 5/6.

The probability of not getting an ace in 2 throws = (5/6)(5/6) = 25/36.

Hence the probability of getting at least 1 ace in 2 throws = $1 - 25/36$ = 11/36.

5. The probability of A's winning a game of chess against B is 1/3. What is the probability that A will win at least 1 of a total of 3 games?

The probability of A's losing any single game = $1 - 1/3 = 2/3$, and the probability of his losing all 3 games = $(2/3)^3 = 8/27$.

Hence the probability of his winning at least 1 game = $1 - 8/27$ = 19/27.

6. Three cards are drawn from a pack of 52, each card being replaced before the next one is drawn. Compute the probability p that all are a) spades, b) aces, c) red cards.

A pack of 52 cards includes 13 spades, 4 aces, and 26 red cards.

a) $p = (\dfrac{13}{52})^3 = \dfrac{1}{64}$
   b) $p = (\dfrac{4}{52})^3 = \dfrac{1}{2197}$
   c) $p = (\dfrac{26}{52})^3 = \dfrac{1}{8}$

7. The odds are 23 to 2 against a person winning a \$500 prize. What is his mathematical expectation?

Expectation = probability of winning × sum of money = $(\dfrac{2}{23 + 2})(\$500)$ = \$40.

8. Nine tickets, numbered from 1 to 9, are in a box. If 2 tickets are drawn at random, determine the probability p that a) both are odd, b) both are even, c) one is odd and one is even, d) they are numbered 2,5.

There are 5 odd and 4 even numbered tickets.

a) $p = \dfrac{\text{number of selections of 2 out of 5 odd tickets}}{\text{number of selections of 2 out of 9 tickets}} = \dfrac{{}_5C_2}{{}_9C_2} = \dfrac{5}{18}$

b) $p = \dfrac{{}_4C_2}{{}_9C_2} = \dfrac{1}{6}$
   c) $p = \dfrac{{}_5C_1 \cdot {}_4C_1}{{}_9C_2} = \dfrac{5 \cdot 4}{36} = \dfrac{5}{9}$
   d) $p = \dfrac{{}_2C_2}{{}_9C_2} = \dfrac{1}{36}$

9. A bag contains 6 red, 4 white and 8 blue balls. If 3 balls are drawn at random, determine the probability p that a) all 3 are red, b) all 3 are blue, c) 2 are white and 1 is red, d) at least 1 is red, e) 1 of each color is drawn, f) the balls are drawn in the order red, white, blue.

a) $p = \dfrac{\text{number of selections of 3 out of 6 red balls}}{\text{number of selections of 3 out of 18 balls}} = \dfrac{_6C_3}{_{18}C_3} = \dfrac{5}{204}$

b) $p = \dfrac{_8C_3}{_{18}C_3} = \dfrac{7}{102}$ c) $p = \dfrac{_4C_2 \cdot _6C_1}{_{18}C_3} = \dfrac{3}{68}$

d) Probability that none is red $= \dfrac{_{(4+8)}C_3}{_{18}C_3} = \dfrac{_{12}C_3}{_{18}C_3} = \dfrac{55}{204}$.

Hence the probability that at least 1 is red $= 1 - \dfrac{55}{204} = \dfrac{149}{204}$.

e) $p = \dfrac{6 \cdot 4 \cdot 8}{_{18}C_3} = \dfrac{6 \cdot 4 \cdot 8}{18 \cdot 17 \cdot 16/6} = \dfrac{4}{17}$

f) $p = \dfrac{4}{17} \cdot \dfrac{1}{3!} = \dfrac{4}{17} \cdot \dfrac{1}{6} = \dfrac{2}{51}$ or $p = \dfrac{6 \cdot 4 \cdot 8}{_{18}P_3} = \dfrac{6 \cdot 4 \cdot 8}{18 \cdot 17 \cdot 16} = \dfrac{2}{51}$

10. Three cards are drawn from a pack of 52 cards. Determine the probability p that a) all are aces, b) all are aces and drawn in the order spade, club, diamond, c) all are spades, d) all are of the same suit, e) no two are of the same suit.

a) There are $_{52}C_3$ selections of 3 out of 52 cards, and $_4C_3$ selections of 3 out of 4 aces.

Hence $p = \dfrac{_4C_3}{_{52}C_3} = \dfrac{_4C_1}{_{52}C_3} = \dfrac{1}{5525}$.

b) There are $_{52}P_3$ orders of drawing 3 out of 52 cards, one of which is the given order.

Hence $p = \dfrac{1}{_{52}P_3} = \dfrac{1}{52 \cdot 51 \cdot 50} = \dfrac{1}{132,600}$.

c) There are $_{13}C_3$ selections of 3 out of 13 spades. Hence $p = \dfrac{_{13}C_3}{_{52}C_3} = \dfrac{11}{850}$.

d) There are 4 suits, each consisting of 13 cards. Hence there are 4 ways of selecting a suit, and $_{13}C_3$ ways of selecting 3 cards from a given suit.

Hence $p = \dfrac{4 \cdot _{13}C_3}{_{52}C_3} = \dfrac{22}{425}$.

e) There are $_4C_3 = _4C_1 = 4$ ways of selecting 3 out of 4 suits, and $13 \cdot 13 \cdot 13$ ways of selecting 1 card from each of 3 given suits.

Hence $p = \dfrac{4 \cdot 13 \cdot 13 \cdot 13}{_{52}C_3} = \dfrac{169}{425}$.

11. What is the probability that any two different cards of a well-shuffled deck of 52 cards will be together in the deck, if their suit is not considered?

Consider the probability that, for example, an ace and a king are together. There are 4 aces and 4 kings in a deck. Hence an ace can be chosen in 4 ways and when that is done, a king can be chosen in 4 ways. Thus an ace and then a king can be selected in $4 \cdot 4 = 16$ ways.

Similarly, a king and then an ace can be selected in 16 ways. Then an ace and a king can be together in $2 \cdot 16 = 32$ ways.

For every one way the combination (ace,king) occurs, the remaining 50 cards and the (ace, king) combination can be permuted in 51! ways. The number of favorable arrangements is thus 32(51!). Since the total number of arrangements of all the cards in the deck is 52!, the required probability is $\dfrac{32(51!)}{52!} = \dfrac{32}{52} = \dfrac{8}{13}$.

12. A man holds 2 of a total of 20 tickets in a lottery. If there are 2 winning tickets, determine the probability that he has a) both, b) neither, c) exactly one.

a) There are $_{20}C_2$ ways of selecting 2 out of 20 tickets.

Hence the probability of his winning both prizes $= \dfrac{1}{_{20}C_2} = \dfrac{1}{190}$.

Another Method. The probability of winning the first prize $= 2/20 = 1/10$.   After winning the first prize (he has 1 ticket left and there remain 19 tickets from which to choose the second prize) the probability of winning the second prize is 1/19.

Hence the probability of winning both prizes $= \dfrac{1}{10}\left(\dfrac{1}{19}\right) = \dfrac{1}{190}$.

b) There are 20 tickets, 18 of which are losers.

Hence the probability of winning neither prize $= \dfrac{_{18}C_2}{_{20}C_2} = \dfrac{153}{190}$.

Another Method. The probability of not winning the first prize $= 1 - 2/20 = 9/10$. If he does not win the first prize (he still has 2 tickets), the probability of not winning the second prize $= 1 - 2/19 = 17/19$.

Hence the probability of winning neither prize $= \dfrac{9}{10}\left(\dfrac{17}{19}\right) = \dfrac{153}{190}$.

c) Probability of winning exactly one prize

$\qquad = 1 -$ probability of winning neither $-$ probability of winning both

$\qquad = 1 - \dfrac{153}{190} - \dfrac{1}{190} = \dfrac{36}{190} = \dfrac{18}{95}$.

Another Method. Probability of winning first but not second prize $= \dfrac{2}{20}\left(\dfrac{18}{19}\right) = \dfrac{9}{95}$.

Probability of not winning first but winning second $= \dfrac{18}{20}\left(\dfrac{2}{19}\right) = \dfrac{9}{95}$.

Hence the probability of winning exactly 1 prize $= \dfrac{9}{95} + \dfrac{9}{95} = \dfrac{18}{95}$.

13. A box contains 7 tickets, numbered from 1 to 7 inclusive. If 3 tickets are drawn from the box, one at a time, determine the probability that they are alternately either odd, even, odd or even, odd, even.

The probability that the first drawn is odd (4/7), then the second even (3/6) and then the third odd (3/5) is $\dfrac{4}{7}\left(\dfrac{3}{6}\right)\left(\dfrac{3}{5}\right) = \dfrac{6}{35}$.

The probability that the first drawn is even (3/7), then the second odd (4/6) and then the third even (2/5) is $\dfrac{3}{7}\left(\dfrac{4}{6}\right)\left(\dfrac{2}{5}\right) = \dfrac{4}{35}$.

Hence the required probability $= \dfrac{6}{35} + \dfrac{4}{35} = \dfrac{2}{7}$.

Another Method. Possible orders of 7 numbers taken 3 at a time $= {_7}P_3 = 7 \cdot 6 \cdot 5 = 210.$

Orders where numbers are alternately odd, even, odd $= 4 \cdot 3 \cdot 3 = 36.$

Orders where numbers are alternately even, odd, even $= 3 \cdot 4 \cdot 2 = 24.$

Hence the required probability $= \dfrac{36 + 24}{210} = \dfrac{60}{210} = \dfrac{2}{7}.$

14. The probability that A can solve a given problem is 4/5, that B can solve it is 2/3 and that C can solve it is 3/7. If all three try, compute the probability that the problem will be solved.

The probability that A will fail to solve it $= 1 - 4/5 = 1/5$, that B will fail $= 1 - 2/3 = 1/3$, and that C will fail $= 1 - 3/7 = 4/7$.

The probability that all three fail $= \dfrac{1}{5}(\dfrac{1}{3})(\dfrac{4}{7}).$ Hence the probability that all three will not fail, i.e. that at least one will solve it, is $1 - \dfrac{1}{5}(\dfrac{1}{3})(\dfrac{4}{7}) = 1 - \dfrac{4}{105} = \dfrac{101}{105}.$

15. The probability that a certain man will be alive 25 years hence is 3/7, and the probability that his wife will be alive 25 years hence is 4/5. Determine the probability that, 25 years hence, *a*) both will be alive, *b*) at least one of them will be alive, *c*) only the man will be alive.

a) The probability that both will be alive $= \dfrac{3}{7}(\dfrac{4}{5}) = \dfrac{12}{35}.$

b) The probability that both will die within 25 years $= (1 - \dfrac{3}{7})(1 - \dfrac{4}{5}) = \dfrac{4}{7}(\dfrac{1}{5}) = \dfrac{4}{35}.$

Hence the probability that at least one will be alive $= 1 - \dfrac{4}{35} = \dfrac{31}{35}.$

c) The probability that the man will be alive $= 3/7$, and the probability that his wife will not be alive $= 1 - 4/5 = 1/5.$

Hence the probability that only the man will be alive $= \dfrac{3}{7}(\dfrac{1}{5}) = \dfrac{3}{35}.$

16. There are three candidates, A, B and C, for an office. The odds that A will win are 7 to 5, and the odds that B will win are 1 to 3. *a*) What is the probability that either A or B will win? *b*) What are the odds in favor of C?

a) Probability that A will win $= \dfrac{7}{7+5} = \dfrac{7}{12}$, that B will win $= \dfrac{1}{1+3} = \dfrac{1}{4}.$

Hence the probability that either A or B will win $= \dfrac{7}{12} + \dfrac{1}{4} = \dfrac{5}{6}.$

b) Probability that C will win $= 1 - \dfrac{5}{6} = \dfrac{1}{6}.$ Hence the odds in favor of C are 1 to 5.

17. One purse contains 5 dimes and 2 quarters, and a second purse contains 1 dime and 3 quarters. If a coin is taken from one of the two purses at random, what is the probability that it is a quarter?

The probability of selecting the first purse (1/2) and of then drawing a quarter from it (2/7) is (1/2)(2/7) = 1/7. The probability of selecting the second purse (1/2) and of then drawing a quarter from it (3/4) is (1/2)(3/4) = 3/8.

Hence the required probability $= \dfrac{1}{7} + \dfrac{3}{8} = \dfrac{29}{56}.$

18. A bag contains 2 white balls and 3 black balls. Four persons, A,B,C,D, in the order named, each draws one ball and does not replace it. The first to draw a white ball receives $10. Determine their expectations.

   A's probability of winning $= \dfrac{2}{5}$, and his expectation $= \dfrac{2}{5}(\$10) = \$4$.

   To find B's expectation: The probability that A fails $= 1 - 2/5 = 3/5$. If A fails, the bag contains 2 white and 2 black balls. Thus the probability that if A fails B will win $= 2/4 = 1/2$. Hence B's probability of winning $= (3/5)(1/2) = 3/10$, and his expectation is $3.

   To find C's expectation: The probability that A fails $= 3/5$, and the probability that B fails $= 1 - 1/2 = 1/2$. If A and B both fail, the bag contains 2 white balls and 1 black ball. Thus the probability that, if A and B both fail, C will win $= 2/3$. Hence C's probability of winning $= \dfrac{3}{5}(\dfrac{1}{2})(\dfrac{2}{3}) = \dfrac{1}{5}$, and his expectation $= \dfrac{1}{5}(\$10) = \$2$.

   If A, B and C fail, only white balls remain and D must win. Hence D's probability of winning $= \dfrac{3}{5}(\dfrac{1}{2})(\dfrac{1}{3})(\dfrac{1}{1}) = \dfrac{1}{10}$, and his expectation $= \dfrac{1}{10}(\$10) = \$1$.

   Check. $\$4 + \$3 + \$2 + \$1 = \$10$, and $\dfrac{2}{5} + \dfrac{3}{10} + \dfrac{1}{5} + \dfrac{1}{10} = 1$.

19. Eleven books, consisting of 5 engineering books, 4 mathematics books and 2 chemistry books, are placed on a shelf at random. What is the probability p that the books of each kind are all together?

   When the books of each kind are all together, the engineering books could be arranged in 5! ways, the mathematics books in 4! ways, the chemistry books in 2! ways, and the 3 groups in 3! ways.

   $$p = \frac{\text{ways in which books of each kind are together}}{\text{total number of ways of arranging 11 books}} = \frac{5!\,4!\,2!\,3!}{11!} = \frac{1}{1155}.$$

20. Five red blocks and 4 white blocks are placed at random in a row. What is the probability p that the extreme blocks are both red?

   Total possible arrangements of 5 red and 4 white blocks $= \dfrac{(5+4)!}{5!\,4!} = \dfrac{9!}{5!\,4!} = 126$.

   Arrangements where extreme blocks are both red $= \dfrac{(9-2)!}{(5-2)!\,4!} = \dfrac{7!}{3!\,4!} = 35$.

   Hence the required probability $p = \dfrac{35}{126} = \dfrac{5}{18}$.

21. One purse contains 6 copper coins and 1 silver coin; a second purse contains 4 copper coins. Five coins are drawn from the first purse and put into the second, and then 2 coins are drawn from the second and put into the first. Determine the probability that the silver coin is in a) the second purse, b) the first purse.

   Initially, the first purse contains 7 coins. When 5 coins are drawn from the first purse and put into the second, the probability that the silver coin is put into the second purse is 5/7, and the probability that it remains in the first purse is 2/7.

   The second purse now contains $5 + 4 = 9$ coins. Finally, after 2 of these 9 coins are put into the first purse, the probability that the silver coin is in the second purse $= \dfrac{5}{7}(\dfrac{7}{9}) = \dfrac{5}{9}$, and the probability that it is in the first purse $= \dfrac{2}{7} + \dfrac{5}{7}(\dfrac{2}{9}) = \dfrac{4}{9}$ (or $1 - \dfrac{5}{9} = \dfrac{4}{9}$).

22. Compute the probability that a single throw with 9 dice will result in exactly 2 aces.

The probability that a certain pair of the 9 dice thrown will yield aces $= \frac{1}{6}(\frac{1}{6}) = (\frac{1}{6})^2$.

The probability that the other 7 dice will not yield aces $= (1 - \frac{1}{6})^7 = (\frac{5}{6})^7$. Since $_9C_2$ different pairs may be selected from the 9 dice, the probability that exactly 1 pair will be aces $= {}_9C_2(\frac{1}{6})^2(\frac{5}{6})^7 = \frac{78,125}{279,936}$.

Or, by formula: Probability $= {}_nC_r p^r q^{n-r} = {}_9C_2(\frac{1}{6})^2(\frac{5}{6})^7 = \frac{78,125}{279,936}$

23. What is the probability of getting a 9 exactly once in 3 throws with a pair of dice?

A 9 can occur in 4 ways: 3,6; 4,5; 5,4; 6,3.

In any throw with a pair of dice, the probability of getting a 9 $= \frac{4}{6\cdot 6} = \frac{1}{9}$, and the probability of not getting a 9 $= 1 - \frac{1}{9} = \frac{8}{9}$. The probability that any given throw with a pair of dice is a 9 and that the other two throws are not $= (\frac{1}{9})(\frac{8}{9})^2$. Since there are $_3C_1 = 3$ different ways in which one throw is a 9 and the other two throws are not, the probability of throwing a 9 exactly once in 3 throws $= {}_3C_1(\frac{1}{9})(\frac{8}{9})^2 = \frac{64}{243}$.

Or, by formula: Probability $= {}_nC_r p^r q^{n-r} = {}_3C_1(\frac{1}{9})(\frac{8}{9})^2 = \frac{64}{243}$.

24. If the probability that the average freshman will not complete four years of college is 1/3, what is the probability p that of 4 freshmen at least 3 will complete four years of college?

Probability that 3 will complete and 1 will not $= {}_4C_3(\frac{2}{3})^3(\frac{1}{3}) = {}_4C_1(\frac{2}{3})^3(\frac{1}{3})$.

Probability that 4 will complete $= (\frac{2}{3})^4$. Hence $p = (\frac{2}{3})^4 + {}_4C_1(\frac{2}{3})^3(\frac{1}{3}) = \frac{16}{27}$.

Or, by formula: $p =$ first 2 $(n-r+1 = 4-3+1)$ terms of the expansion of $(\frac{2}{3} + \frac{1}{3})^4$

$$= (\frac{2}{3})^4 + {}_4C_1(\frac{2}{3})^3(\frac{1}{3}) = \frac{16}{81} + \frac{32}{81} = \frac{16}{27}.$$

25. A coin is tossed 6 times. What is the probability p of getting at least 3 heads? What are the odds in favor of getting at least 3 heads?

On each toss, probability of a head = probability of a tail = 1/2.

The probability that certain 3 of the 6 tosses will give heads $= (1/2)^3$. The probability that none of the other 3 tosses will be heads $= (1/2)^3$. Since $_6C_3$ different selections of 3 can be made from the 6 tosses, the probability that exactly 3 will be heads is

$$_6C_3(\frac{1}{2})^3(\frac{1}{2})^3 = {}_6C_3(\frac{1}{2})^6.$$

Similarly, the probability of exactly 4 heads $= {}_6C_4(1/2)^6 = {}_6C_2(1/2)^6$,

the probability of exactly 5 heads $= {}_6C_5(1/2)^6 = {}_6C_1(1/2)^6$,

the probability of exactly 6 heads $= (1/2)^6$.

Hence $\quad p \;=\; (\frac{1}{2})^6 \;+\; {}_6C_1(\frac{1}{2})^6 \;+\; {}_6C_2(\frac{1}{2})^6 \;+\; {}_6C_3(\frac{1}{2})^6$

$$= (\tfrac{1}{2})^6 (1 + {}_6C_1 + {}_6C_2 + {}_6C_3) \;=\; (\tfrac{1}{2})^6 (1 + 6 + 15 + 20) \;=\; \frac{21}{32}.$$

The odds in favor of getting at least 3 heads is 21:11 or 21/11.

Or, by formula: $\quad p \;=\; $ first 4 $(n - r + 1 = 6 - 3 + 1)$ terms of the expansion of $(\frac{1}{2} + \frac{1}{2})^6$

$$= (\tfrac{1}{2})^6 + {}_6C_1(\tfrac{1}{2})^6 + {}_6C_2(\tfrac{1}{2})^6 + {}_6C_3(\tfrac{1}{2})^6 \;=\; \frac{21}{32}.$$

26. Determine the probability p that in a family of 5 children there will be at least 2 boys and 1 girl. Assume that the probability of a male birth is 1/2.

The three favorable cases are: 2 boys, 3 girls; 3 boys, 2 girls; 4 boys, 1 girl.

$$p \;=\; (\tfrac{1}{2})^5 ({}_5C_2 + {}_5C_3 + {}_5C_4) \;=\; \frac{1}{32}(10 + 10 + 5) \;=\; \frac{25}{32}.$$

SUPPLEMENTARY PROBLEMS

27. Determine the probability that a digit chosen at random from the digits $1, 2, 3, \ldots, 9$ will be
   a) odd, b) even, c) a multiple of 3.

28. A coin is tossed three times. If H = head and T = tail, what is the probability of the tosses coming up in the order a) H TH, b) T HH, c) H HH?

29. If three coins are tossed, what is the probability of obtaining a) three heads, b) two heads and a tail?

30. Find the probability of throwing a total of 7 in a single throw with two dice.

31. What is the probability of throwing a total of 8 or 11 in a single throw with two dice?

32. A die is cast twice. What is the probability of getting a 4 or 5 on the first throw and a 2 or 3 on the second throw? What is the probability of not getting an ace on either throw?

33. What is the probability that a coin will turn up heads at least once in six tosses of a coin?

34. Five discs in a bag are numbered $1, 2, 3, 4, 5$. What is the probability that the sum of the numbers on three discs chosen at random is greater than 10?

35. Three balls are drawn at random from a box containing 5 red, 8 black and 4 white balls. Determine the probability that a) all three are white, b) two are black and one red, c) one of each color is selected.

36. Four cards are drawn from a pack of 52 cards. Find the probability that a) all are kings, b) two are kings and two are aces, c) all are of the same suit, d) all are clubs.

37. A man will win \$3.20 if in 5 tosses of a coin he gets either of the sequences HTHTH or THTHT where H = head and T = tail. Determine his expectation.

38. In a plane crash it was reported that three persons out of the total of twenty passengers were injured. Three newspapermen were in this plane. What is the probability that the three reported injured were the newspapermen?

39. A committee of three is to be chosen from a group consisting of 5 men and 4 women. If the selection is made at random, find the probability that a) all three are women, b) two are men.

40. Six persons seat themselves at a round table. What is the probability that two given persons are adjacent?

41. Four letters are chosen at random from the word MISSISSIPPI. Determine the probability that *a*) at least three I's are chosen, *b*) at most two consonants are chosen.

42. *A* and *B* alternately toss a coin. The first one to turn up a head wins. If no more than five tosses each are allowed for a single game, find the probability that the person who tosses first will win the game. What are the odds against *A*'s losing if he goes first?

43. Six red blocks and 4 white blocks are placed at random in a row. Find the probability that the two blocks in the middle are of the same color.

44. In 8 tosses of a coin determine the probability of *a*) exactly 4 heads, *b*) at least 2 tails, *c*) at most 5 heads, *d*) exactly 3 tails.

45. In 2 throws with a pair of dice determine the probability of getting *a*) an 11 exactly once, *b*) a 10 twice.

46. What is the probability of getting at least one 11 in 3 throws with a pair of dice?

47. In ten tosses of a coin, what is the probability of getting not less than 3 heads and not more than 6 heads?

ANSWERS TO SUPPLEMENTARY PROBLEMS.

27. *a*) 5/9   *b*) 4/9   *c*) 1/3

28. *a*) 1/8   *b*) 1/8   *c*) 1/8

29. *a*) 1/8   *b*) 3/8

30. 1/6

31. 7/36

32. 1/9,   25/36

33. 63/64

34. 1/5

35. *a*) $\dfrac{1}{170}$ *b*) $\dfrac{7}{34}$ *c*) $\dfrac{4}{17}$

36. *a*) $\dfrac{1}{270,725}$ *b*) $\dfrac{36}{270,725}$ *c*) $\dfrac{44}{4165}$ *d*) $\dfrac{11}{4165}$

37. 20 cents

38. 1/1140

39. *a*) 1/21   *b*) 10/21

40. 2/5

41. *a*) $\dfrac{29}{330}$ *b*) $\dfrac{31}{66}$

42. $\dfrac{21}{32}$,   21:11

43. $\dfrac{7}{15}$

44. *a*) $\dfrac{35}{128}$ *b*) $\dfrac{247}{256}$ *c*) $\dfrac{219}{256}$ *d*) $\dfrac{7}{32}$

45. *a*) $\dfrac{17}{162}$ *b*) $\dfrac{1}{144}$

46. $\dfrac{919}{5832}$

47. $\dfrac{99}{128}$

CHAPTER 26

Determinants and Systems of Linear Equations

DETERMINANTS OF SECOND ORDER. The symbol

$$\begin{vmatrix} a_1 & b_1 \\ a_2 & b_2 \end{vmatrix}$$

consisting of the four numbers a_1, b_1, a_2, b_2 arranged in two rows and two columns is called a *determinant of second order* or *determinant of order two*. The four numbers are called *elements* of the determinant. By definition,

$$\begin{vmatrix} a_1 & b_1 \\ a_2 & b_2 \end{vmatrix} = a_1 b_2 - b_1 a_2 .$$

Thus $\begin{vmatrix} 2 & 3 \\ -1 & -2 \end{vmatrix} = (2)(-2) - (3)(-1) = -4 + 3 = -1.$ Here the elements 2 and 3 are in the first row, the elements -1 and -2 are in the second row. Elements 2 and -1 are in the first column, and elements 3 and -2 are in the second column.

A determinant of order one is the number itself.

SYSTEMS OF TWO LINEAR EQUATIONS in two unknowns may be solved by use of second order determinants. Given the system of equations

$$\begin{cases} a_1 x + b_1 y = c_1 \\ a_2 x + b_2 y = c_2 \end{cases} \tag{1}$$

it is possible by any of the methods of Chapter 12 to obtain

$$x = \frac{c_1 b_2 - b_1 c_2}{a_1 b_2 - b_1 a_2}, \qquad y = \frac{a_1 c_2 - c_1 a_2}{a_1 b_2 - b_1 a_2} \qquad (a_1 b_2 - b_1 a_2 \neq 0).$$

These values for x and y may be written in terms of second order determinants as follows:

$$x = \frac{\begin{vmatrix} c_1 & b_1 \\ c_2 & b_2 \end{vmatrix}}{\begin{vmatrix} a_1 & b_1 \\ a_2 & b_2 \end{vmatrix}}, \qquad y = \frac{\begin{vmatrix} a_1 & c_1 \\ a_2 & c_2 \end{vmatrix}}{\begin{vmatrix} a_1 & b_1 \\ a_2 & b_2 \end{vmatrix}} \tag{2}$$

The form involving determinants is easy to remember if one keeps in mind the following:

1. The denominators in (2) are given by the determinant $\begin{vmatrix} a_1 & b_1 \\ a_2 & b_2 \end{vmatrix}$ in which the elements are the coefficients of x and y arranged as in the given equations (1). This determinant, usually denoted by Δ, is called the *determinant of the coefficients*.

2. The numerator in the solution for either unknown is the same as the determinant of the coefficients Δ with the exception that the column of coefficients of the unknown to be determined is replaced by the column of constants on the right side of equations (1).

Example. Solve the system $\begin{cases} 2x + 3y = 8 \\ x - 2y = -3 \end{cases}$.

The denominator for both x and y is $\Delta = \begin{vmatrix} 2 & 3 \\ 1 & -2 \end{vmatrix} = 2(-2) - 3(1) = -7.$

Then $x = \dfrac{\begin{vmatrix} 8 & 3 \\ -3 & -2 \end{vmatrix}}{-7} = \dfrac{8(-2) - 3(-3)}{-7} = 1$ and $y = \dfrac{\begin{vmatrix} 2 & 8 \\ 1 & -3 \end{vmatrix}}{-7} = \dfrac{2(-3) - 8(1)}{-7} = 2.$

The method of solution of linear equations by determinants is called *Cramer's Rule*.

DETERMINANTS OF THIRD ORDER.   The symbol

$$\begin{vmatrix} a_1 & b_1 & c_1 \\ a_2 & b_2 & c_2 \\ a_3 & b_3 & c_3 \end{vmatrix}$$

consisting of nine numbers arranged in three rows and three columns is called a *determinant of third order*. By definition, the value of this determinant is given by

$$a_1 b_2 c_3 + b_1 c_2 a_3 + c_1 a_2 b_3 - c_1 b_2 a_3 - a_1 c_2 b_3 - b_1 a_2 c_3$$

and is called the expansion of the determinant.

In order to remember this definition, the following scheme is given. Rewrite the first two columns on the right of the determinant as follows:

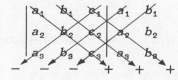

1. Form the products of the elements in each of the 3 diagonals shown which run down from left to right, and precede each of these 3 terms by a positive sign.

2. Form the products of the elements in each of the 3 diagonals shown which run down from right to left, and precede each of these 3 terms by a negative sign.

3. The algebraic sum of the six products of 1) and 2) is the required expansion of the determinant.

Example.   Expand $\begin{vmatrix} 3 & -2 & 2 \\ 6 & 1 & -1 \\ -2 & -3 & 2 \end{vmatrix}$.   Rewriting,

The value of the determinant is

$$(3)(1)(2) + (-2)(-1)(-2) + (2)(6)(-3) - (2)(1)(-2) - (3)(-1)(-3) - (-2)(6)(2) = -15.$$

CRAMER'S RULE for linear equations in 3 unknowns is a method of solving the following equations for x, y, z

$$\begin{cases} a_1 x + b_1 y + c_1 z = d_1 \\ a_2 x + b_2 y + c_2 z = d_2 \\ a_3 x + b_3 y + c_3 z = d_3 \end{cases} \tag{3}$$

by determinants. It is an extension of Cramer's rule for linear equations in two unknowns. If we solve equations (3) by the methods of Chapter 12, we obtain

$$x = \frac{d_1 b_2 c_3 + c_1 d_2 b_3 + b_1 c_2 d_3 - c_1 b_2 d_3 - b_1 d_2 c_3 - d_1 c_2 b_3}{a_1 b_2 c_3 + b_1 c_2 a_3 + c_1 a_2 b_3 - c_1 b_2 a_3 - b_1 a_2 c_3 - a_1 c_2 b_3}$$

$$y = \frac{a_1 d_2 c_3 + c_1 a_2 d_3 + d_1 c_2 a_3 - c_1 d_2 a_3 - d_1 a_2 c_3 - a_1 c_2 d_3}{a_1 b_2 c_3 + b_1 c_2 a_3 + c_1 a_2 b_3 - c_1 b_2 a_3 - b_1 a_2 c_3 - a_1 c_2 b_3}$$

$$z = \frac{a_1 b_2 d_3 + d_1 a_2 b_3 + b_1 d_2 a_3 - d_1 b_2 a_3 - b_1 a_2 d_3 - a_1 d_2 b_3}{a_1 b_2 c_3 + b_1 c_2 a_3 + c_1 a_2 b_3 - c_1 b_2 a_3 - b_1 a_2 c_3 - a_1 c_2 b_3}$$

These may be written in terms of determinants as follows

$$x = \frac{\begin{vmatrix} d_1 & b_1 & c_1 \\ d_2 & b_2 & c_2 \\ d_3 & b_3 & c_3 \end{vmatrix}}{\Delta} \qquad y = \frac{\begin{vmatrix} a_1 & d_1 & c_1 \\ a_2 & d_2 & c_2 \\ a_3 & d_3 & c_3 \end{vmatrix}}{\Delta} \qquad z = \frac{\begin{vmatrix} a_1 & b_1 & d_1 \\ a_2 & b_2 & d_2 \\ a_3 & b_3 & d_3 \end{vmatrix}}{\Delta} \tag{4}$$

where $\Delta = \begin{vmatrix} a_1 & b_1 & c_1 \\ a_2 & b_2 & c_2 \\ a_3 & b_3 & c_3 \end{vmatrix}$ is the determinant of the coefficients of x, y, z in equations (3) and is assumed not equal to zero.

The form involving determinants is easy to remember if one keeps in mind the following:

1. The denominators in (4) are given by the determinant Δ in which the elements are the coefficients of x, y and z arranged as in the given equations (3).

2. The numerator in the solution for any unknown is the same as the determinant of the coefficients Δ with the exception that the column of coefficients of the unknown to be determined is replaced by the column of constants on the right side of equations (3).

Example. Solve the system $\begin{cases} x + 2y - z = -3 \\ 3x + y + z = 4 \\ x - y + 2z = 6 \end{cases}$.

$$\Delta = \begin{vmatrix} 1 & 2 & -1 \\ 3 & 1 & 1 \\ 1 & -1 & 2 \end{vmatrix} = 2 + 2 + 3 + 1 + 1 - 12 = -3$$

$$x = \frac{\begin{vmatrix} -3 & 2 & -1 \\ 4 & 1 & 1 \\ 6 & -1 & 2 \end{vmatrix}}{-3} = \frac{-6 + 12 + 4 + 6 - 3 - 16}{-3} = \frac{-3}{-3} = 1$$

$$y = \frac{\begin{vmatrix} 1 & -3 & -1 \\ 3 & 4 & 1 \\ 1 & 6 & 2 \end{vmatrix}}{-3} = \frac{8 - 3 - 18 + 4 - 6 + 18}{-3} = \frac{3}{-3} = -1$$

$$z = \frac{\begin{vmatrix} 1 & 2 & -3 \\ 3 & 1 & 4 \\ 1 & -1 & 6 \end{vmatrix}}{-3} = \frac{6 + 8 + 9 + 3 + 4 - 36}{-3} = \frac{-6}{-3} = 2$$

SOLVED PROBLEMS

DETERMINANTS OF SECOND ORDER.

1. Evaluate the following determinants.

a) $\begin{vmatrix} 3 & 2 \\ 1 & 4 \end{vmatrix} = (3)(4) - (2)(1) = 12 - 2 = 10$

b) $\begin{vmatrix} 3 & -1 \\ 6 & -2 \end{vmatrix} = (3)(-2) - (-1)(6) = -6 + 6 = 0$

c) $\begin{vmatrix} 0 & 3 \\ 2 & -5 \end{vmatrix} = (0)(-5) - (3)(2) = 0 - 6 = -6$

d) $\begin{vmatrix} x & x^2 \\ y & y^2 \end{vmatrix} = xy^2 - x^2y$

e) $\begin{vmatrix} x + 2 & 2x + 5 \\ 3x - 1 & x - 3 \end{vmatrix} = (x+2)(x-3) - (2x+5)(3x-1) = -5x^2 - 14x - 1$

2. a) Show that if the rows and columns of a determinant of order two are interchanged the value of the determinant is the same.

b) Show that if the elements of one row (or column) are proportional respectively to the elements of the other row (or column), the determinant is equal to zero.

a) Let the determinant be $\begin{vmatrix} a_1 & b_1 \\ a_2 & b_2 \end{vmatrix} = a_1 b_2 - a_2 b_1$.

The determinant with rows and columns interchanged so that 1st row becomes 1st column and 2nd row becomes 2nd column is $\begin{vmatrix} a_1 & a_2 \\ b_1 & b_2 \end{vmatrix} = a_1 b_2 - a_2 b_1$.

b) The determinant with proportional rows is $\begin{vmatrix} a_1 & b_1 \\ ka_1 & kb_1 \end{vmatrix} = a_1 k b_1 - b_1 k a_1 = 0$.

3. Find the values of x for which $\begin{vmatrix} 2x-1 & 2x+1 \\ x+1 & 4x+2 \end{vmatrix} = 0$.

$$\begin{vmatrix} 2x-1 & 2x+1 \\ x+1 & 4x+2 \end{vmatrix} = (2x-1)(4x+2) - (2x+1)(x+1) = 6x^2 - 3x - 3 = 0.$$

Then $2x^2 - x - 1 = (x-1)(2x+1) = 0$ so that $x = 1, -1/2$.

LINEAR EQUATIONS IN TWO UNKNOWNS.

4. Solve for the unknowns in each of the following systems.

a) $\begin{cases} 4x + 2y = 5 \\ 3x - 4y = 1 \end{cases}$
$\quad x = \dfrac{\begin{vmatrix} 5 & 2 \\ 1 & -4 \end{vmatrix}}{\begin{vmatrix} 4 & 2 \\ 3 & -4 \end{vmatrix}} = \dfrac{-22}{-22} = 1, \quad y = \dfrac{\begin{vmatrix} 4 & 5 \\ 3 & 1 \end{vmatrix}}{\begin{vmatrix} 4 & 2 \\ 3 & -4 \end{vmatrix}} = \dfrac{-11}{-22} = \dfrac{1}{2}$

b) $\begin{cases} 3u + 2v = 18 \\ -5u - v = 12 \end{cases}$
$\quad u = \dfrac{\begin{vmatrix} 18 & 2 \\ 12 & -1 \end{vmatrix}}{\begin{vmatrix} 3 & 2 \\ -5 & -1 \end{vmatrix}} = \dfrac{-42}{7} = -6, \quad v = \dfrac{\begin{vmatrix} 3 & 18 \\ -5 & 12 \end{vmatrix}}{7} = \dfrac{126}{7} = 18$

c) $\begin{cases} 5x - 2y - 14 = 0 \\ 2x + 3y + 3 = 0 \end{cases}$ Rewrite as $\begin{cases} 5x - 2y = 14 \\ 2x + 3y = -3 \end{cases}$.

$$x = \dfrac{\begin{vmatrix} 14 & -2 \\ -3 & 3 \end{vmatrix}}{\begin{vmatrix} 5 & -2 \\ 2 & 3 \end{vmatrix}} = \dfrac{36}{19}, \quad y = \dfrac{\begin{vmatrix} 5 & 14 \\ 2 & -3 \end{vmatrix}}{19} = \dfrac{-43}{19}$$

5. Solve for x and y.

a) $\begin{cases} \dfrac{3x-2}{5} + \dfrac{7y+1}{10} = 10 \quad (1) \\ \dfrac{x+3}{2} - \dfrac{2y-5}{3} = 3 \quad (2) \end{cases}$

Multiply (1) by 10: $6x + 7y = 103$.

Multiply (2) by 6: $3x - 4y = -1$.

Then $\quad x = \dfrac{\begin{vmatrix} 103 & 7 \\ -1 & -4 \end{vmatrix}}{\begin{vmatrix} 6 & 7 \\ 3 & -4 \end{vmatrix}} = \dfrac{-405}{-45} = 9, \quad y = \dfrac{\begin{vmatrix} 6 & 103 \\ 3 & -1 \end{vmatrix}}{-45} = \dfrac{-315}{-45} = 7.$

$b)\begin{cases} \dfrac{2}{y+1} - \dfrac{3}{x+1} = 0 \quad (1) \\[3mm] \dfrac{2}{x-7} + \dfrac{3}{2y-3} = 0 \quad (2) \end{cases}$ Multiply (1) by $(x+1)(y+1)$: $2x - 3y = 1$.

Multiply (2) by $(x-7)(2y-3)$: $3x + 4y = 27$.

Then $x = \dfrac{\begin{vmatrix} 1 & -3 \\ 27 & 4 \end{vmatrix}}{\begin{vmatrix} 2 & -3 \\ 3 & 4 \end{vmatrix}} = \dfrac{85}{17} = 5,\quad y = \dfrac{\begin{vmatrix} 2 & 1 \\ 3 & 27 \end{vmatrix}}{17} = \dfrac{51}{17} = 3.$

6. Solve the following systems of equations.

$a)\begin{cases} \dfrac{3}{x} - \dfrac{6}{y} = \dfrac{1}{6} \\[3mm] \dfrac{2}{x} + \dfrac{3}{y} = \dfrac{1}{2} \end{cases}$ These are linear equations in $\dfrac{1}{x}$ and $\dfrac{1}{y}$.

Then $\dfrac{1}{x} = \dfrac{\begin{vmatrix} 1/6 & -6 \\ 1/2 & 3 \end{vmatrix}}{\begin{vmatrix} 3 & -6 \\ 2 & 3 \end{vmatrix}} = \dfrac{7/2}{21} = \dfrac{1}{6},\quad \dfrac{1}{y} = \dfrac{\begin{vmatrix} 3 & 1/6 \\ 2 & 1/2 \end{vmatrix}}{21} = \dfrac{7/6}{21} = \dfrac{1}{18}$ or $x = 6,\ y = 18.$

$b)\begin{cases} \dfrac{3}{2x} + \dfrac{8}{5y} = 3 \\[3mm] \dfrac{4}{3y} - \dfrac{1}{x} = 1 \end{cases}$ can be written $\begin{cases} \dfrac{3}{2}\left(\dfrac{1}{x}\right) + \dfrac{8}{5}\left(\dfrac{1}{y}\right) = 3 \\[3mm] -\left(\dfrac{1}{x}\right) + \dfrac{4}{3}\left(\dfrac{1}{y}\right) = 1 \end{cases}.$

Then $\dfrac{1}{x} = \dfrac{\begin{vmatrix} 3 & 8/5 \\ 1 & 4/3 \end{vmatrix}}{\begin{vmatrix} 3/2 & 8/5 \\ -1 & 4/3 \end{vmatrix}} = \dfrac{12/5}{18/5} = \dfrac{2}{3},\quad \dfrac{1}{y} = \dfrac{\begin{vmatrix} 3/2 & 3 \\ -1 & 1 \end{vmatrix}}{18/5} = \dfrac{9/2}{18/5} = \dfrac{5}{4}$ or $x = \dfrac{3}{2},\ y = \dfrac{4}{5}.$

DETERMINANTS OF THIRD ORDER.

7. Evaluate each of the following determinants.

$a)\begin{vmatrix} 3 & -2 & 2 \\ 1 & 4 & 5 \\ 6 & -1 & 2 \end{vmatrix}$ Repeat the first two columns: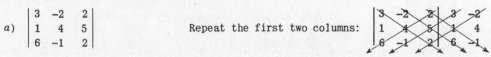

$(3)(4)(2) + (-2)(5)(6) + (2)(1)(-1) - (2)(4)(6) - (3)(5)(-1) - (-2)(1)(2) = -67$

$b)\begin{vmatrix} -1 & 2 & -3 \\ 5 & -3 & 2 \\ 1 & -1 & -3 \end{vmatrix} = 29$ $d)\begin{vmatrix} a & b & c \\ c & a & b \\ b & c & a \end{vmatrix} = a^3 + b^3 + c^3 - 3abc$

$c)\begin{vmatrix} 2 & 3 & 2 \\ 0 & -2 & 1 \\ -1 & 4 & 0 \end{vmatrix} = -15$ $e)\begin{vmatrix} (x-2) & (y+3) & (z-2) \\ -2 & 3 & 4 \\ 1 & -2 & 1 \end{vmatrix} = 11x + 6y + z - 6$

8. *a*) Show that if two rows (or two columns) of a third order determinant have their corresponding elements proportional, the value of the determinant is zero.

   b) Show that if the elements of any row (or column) are multiplied by any given constant and added to the corresponding elements of any other row (or column), the value of the determinant is unchanged.

a) We must show that $\begin{vmatrix} a_1 & b_1 & c_1 \\ ka_1 & kb_1 & kc_1 \\ a_3 & b_3 & c_3 \end{vmatrix} = 0$, where the elements in the first and second rows are proportional. This is shown by expansion of the determinant.

b) Let the given determinant be $\begin{vmatrix} a_1 & b_1 & c_1 \\ a_2 & b_2 & c_2 \\ a_3 & b_3 & c_3 \end{vmatrix}$. We must show that if k is any constant

$$\begin{vmatrix} a_1 & b_1 & c_1 \\ a_2 & b_2 & c_2 \\ a_3 + ka_2 & b_3 + kb_2 & c_3 + kc_2 \end{vmatrix} = \begin{vmatrix} a_1 & b_1 & c_1 \\ a_2 & b_2 & c_2 \\ a_3 & b_3 & c_3 \end{vmatrix}$$

where we have multiplied each of the elements in the second row of the given determinant by k and added to the corresponding elements in the third row. The result is proved by expanding each of the determinants and showing that they are equal.

LINEAR EQUATIONS IN THREE UNKNOWNS.

9. Solve the following systems of equations.

a) $\begin{cases} 2x + y - z = 5 \\ 3x - 2y + 2z = -3 \\ x - 3y - 3z = -2 \end{cases}$ Here $\Delta = \begin{vmatrix} 2 & 1 & -1 \\ 3 & -2 & 2 \\ 1 & -3 & -3 \end{vmatrix} = 42$ and

$$x = \frac{\begin{vmatrix} 5 & 1 & -1 \\ -3 & -2 & 2 \\ -2 & -3 & -3 \end{vmatrix}}{\Delta} = \frac{42}{42} = 1, \quad y = \frac{\begin{vmatrix} 2 & 5 & -1 \\ 3 & -3 & 2 \\ 1 & -2 & -3 \end{vmatrix}}{\Delta} = \frac{84}{42} = 2, \quad z = \frac{\begin{vmatrix} 2 & 1 & 5 \\ 3 & -2 & -3 \\ 1 & -3 & -2 \end{vmatrix}}{\Delta} = \frac{-42}{42} = -1.$$

b) $\begin{cases} x + 2z = 7 \\ 3x + y = 5 \\ 2y - 3z = -5 \end{cases}$ Write as $\begin{cases} x + 0y + 2z = 7 \\ 3x + y + 0z = 5 \\ 0x + 2y - 3z = -5 \end{cases}$. Then $\Delta = \begin{vmatrix} 1 & 0 & 2 \\ 3 & 1 & 0 \\ 0 & 2 & -3 \end{vmatrix} = 9$ and

$$x = \frac{\begin{vmatrix} 7 & 0 & 2 \\ 5 & 1 & 0 \\ -5 & 2 & -3 \end{vmatrix}}{\Delta} = \frac{9}{9} = 1, \quad y = \frac{\begin{vmatrix} 1 & 7 & 2 \\ 3 & 5 & 0 \\ 0 & -5 & -3 \end{vmatrix}}{\Delta} = \frac{18}{9} = 2, \quad z = \frac{\begin{vmatrix} 1 & 0 & 7 \\ 3 & 1 & 5 \\ 0 & 2 & -5 \end{vmatrix}}{\Delta} = \frac{27}{9} = 3.$$

10. The equations for the currents i_1, i_2, i_3 in a given electrical network are $\begin{cases} 3i_1 - 2i_2 + 4i_3 = 2 \\ i_1 + 3i_2 - 6i_3 = 8 \\ 2i_1 - i_2 - 2i_3 = 0 \end{cases}$. Find i_3.

$$i_3 = \frac{\begin{vmatrix} 3 & -2 & 2 \\ 1 & 3 & 8 \\ 2 & -1 & 0 \end{vmatrix}}{\begin{vmatrix} 3 & -2 & 4 \\ 1 & 3 & -6 \\ 2 & -1 & -2 \end{vmatrix}} = \frac{-22}{-44} = \frac{1}{2}$$

SUPPLEMENTARY PROBLEMS

11. Evaluate each of the following determinants.

a) $\begin{vmatrix} 4 & -3 \\ -1 & 2 \end{vmatrix}$ b) $\begin{vmatrix} -2 & 4 \\ -3 & 7 \end{vmatrix}$ c) $\begin{vmatrix} 2 & -1 \\ 4 & 0 \end{vmatrix}$ d) $\begin{vmatrix} -2x & -3y \\ 4x & -y \end{vmatrix}$ e) $\begin{vmatrix} a+b & a-b \\ a & -b \end{vmatrix}$ f) $\begin{vmatrix} 2x-1 & x+1 \\ x+2 & x-2 \end{vmatrix}$

12. Show that if the elements of one row (or column) of a second order determinant are multiplied by the same number, the determinant is multiplied by the number.

13. Solve for the unknowns in each of the following systems.

a) $\begin{cases} 5x + 2y = 4 \\ 2x - y = 7 \end{cases}$ b) $\begin{cases} 3r - 5s = -6 \\ 4r + 2s = 5 \end{cases}$ c) $\begin{cases} 28 + 4x + 5y = 0 \\ -3x + 4y + 10 = 0 \end{cases}$ d) $\begin{cases} 5x - 4y = 16 \\ 2x + 3y = -10 \end{cases}$

e) $\begin{cases} \dfrac{x-3}{3} + \dfrac{y+4}{5} = 7 \\ \dfrac{x+2}{7} - \dfrac{y-6}{2} = -3 \end{cases}$ f) $\begin{cases} \dfrac{3x+2y+1}{x+y} = 4 \\ \dfrac{5x+6y-7}{x+y} = 2 \end{cases}$ g) $\begin{cases} \dfrac{4}{x} + \dfrac{1}{y} = \dfrac{2}{5} \\ \dfrac{3}{x} - \dfrac{5}{y} = -\dfrac{1}{12} \end{cases}$ h) $\begin{cases} \dfrac{4}{3u} - \dfrac{3}{5v} = 1 \\ \dfrac{1}{u} - \dfrac{1}{v} = -\dfrac{1}{6} \end{cases}$

14. Evaluate each determinant.

a) $\begin{vmatrix} -2 & 1 & 2 \\ 3 & -1 & 3 \\ 1 & 3 & -2 \end{vmatrix}$ b) $\begin{vmatrix} 1 & 0 & -2 \\ 0 & -3 & 4 \\ -4 & 2 & -1 \end{vmatrix}$ c) $\begin{vmatrix} 3 & -1 & 4 \\ -2 & 1 & -3 \\ 1 & 3 & -2 \end{vmatrix}$ d) $\begin{vmatrix} x & y & z \\ -2 & 3 & 1 \\ 4 & 1 & 2 \end{vmatrix}$ e) $\begin{vmatrix} 1 & 1 & 1 \\ a & b & c \\ a^2 & b^2 & c^2 \end{vmatrix}$

15. For what value of k does $\begin{vmatrix} k+3 & 1 & -2 \\ 3 & -2 & 1 \\ -k & -3 & 3 \end{vmatrix} = 0$?

16. Show that if the elements of one row (or column) of a third order determinant are multiplied by the same number, the determinant is multiplied by the number.

17. Solve for the unknowns in each of the following systems.

a) $\begin{cases} 3x + y - 2z = 1 \\ 2x + 3y - z = 2 \\ x - 2y + 2z = -10 \end{cases}$ b) $\begin{cases} u + 2v - 3w = -7 \\ 2u - v + w = 5 \\ 3u - v + 2w = 8 \end{cases}$ c) $\begin{cases} 2x + 3y = -2 \\ 5y - 2z = 4 \\ 3z + 4x = -7 \end{cases}$

18. Solve for the indicated unknown.

a) $\begin{cases} 3i_1 + i_2 + 2i_3 = 0 \\ i_1 + 2i_2 - 3i_3 = 5 \\ 2i_1 - i_2 + i_3 = -1 \end{cases}$ for i_2 b) $\begin{cases} 1/x + 2/y + 1/z = 1/2 \\ 4/x + 2/y - 3/z = 2/3 \\ 3/x - 4/y + 4/z = 1/3 \end{cases}$ for x

ANSWERS TO SUPPLEMENTARY PROBLEMS.

11. a) 5   b) –2   c) 4   d) $14xy$ e) $-a^2 - b^2$ f) $x^2 - 8x$

13. a) $x = 2$, $y = -3$ d) $x = 8/23$, $y = -82/23$ g) $x = 12$, $y = 15$
    b) $r = 1/2$, $s = 3/2$ e) $x = 12$, $y = 16$ h) $u = 2/3$, $v = 3/5$
    c) $x = -2$, $y = -4$ f) $x = 5$, $y = -2$

14. a) 43   b) 19   c) 0   d) $5x + 8y - 14z$ e) $bc^2 - cb^2 + a^2c - ac^2 + ab^2 - ba^2$

15. All values of k.

17. a) $x = -2$, $y = 1$, $z = -3$ b) $u = 1$, $v = -1$, $w = 2$ c) $x = -4$, $y = 2$, $z = 3$

18. a) $i_2 = 0.8$ b) $x = 6$

CHAPTER 27

Determinants of Order n

AN INVERSION of the arrangement of the positive integers occurs whenever one integer precedes a smaller integer.

For example, in $4,3,1,5,2$ the integer 4 precedes 3, 1 and 2, the integer 3 precedes 1 and 2, and 5 precedes 2; hence there are 6 inversions.

Similarly, an inversion of the arrangement of letters in alphabetical order occurs whenever one letter precedes another letter which occurs earlier in alphabetical order. Thus $bdca$ has b preceding a, d preceding c and a, and c preceding a; hence there are 4 inversions.

DETERMINANTS OF ORDER n. The symbol

$$\begin{vmatrix} a_1 & b_1 & c_1 & \ldots & k_1 \\ a_2 & b_2 & c_2 & \ldots & k_2 \\ a_3 & b_3 & c_3 & \ldots & k_3 \\ \ldots & \ldots & \ldots & \ldots & \ldots \\ \ldots & \ldots & \ldots & \ldots & \ldots \\ a_n & b_n & c_n & \ldots & k_n \end{vmatrix}$$

consisting of n^2 numbers (called elements) arranged in n rows and n columns, is called a *determinant of order n*. This symbol is an abbreviation for the algebraic sum of all possible products, each consisting of n factors, where:

1) Each product has as factors one and only one element from each row and each column. There will be $n!$ such products.

2) Each product has associated with it a plus or minus sign according as the number of inversions of the subscripts is even or odd, after the letters in the product have been written in the order of appearance in the first row of the determinant.

The algebraic sum thus obtained is called the *expansion* or *value* of the determinant. Each product in the expansion with its associated sign is called a *term* in the expansion of the determinant.

Sometimes an nth order determinant is written

$$\begin{vmatrix} a_{11} & a_{12} & a_{13} & \ldots & a_{1n} \\ a_{21} & a_{22} & a_{23} & \ldots & a_{2n} \\ a_{31} & a_{32} & a_{33} & \ldots & a_{3n} \\ \ldots & \ldots & \ldots & \ldots & \ldots \\ \ldots & \ldots & \ldots & \ldots & \ldots \\ a_{n1} & a_{n2} & a_{n3} & \ldots & a_{nn} \end{vmatrix}$$

In this notation each element is characterized by two subscripts, the first indicating the *row* in which the element appears, the second indicating the *column* in which the element appears. Thus a_{23} is the element in the 2nd row and 3rd column whereas a_{32} is the element in the 3rd row and 2nd column.

The *principal diagonal* of a determinant consists of the elements in the determinant which lie in a straight line from the upper left-hand corner to the lower right-hand corner.

PROPERTIES OF DETERMINANTS.

I. Interchanging corresponding rows and columns of a determinant does not change the value of the determinant. Thus any theorem proved true for rows holds for columns, and conversely.

Example.
$$\begin{vmatrix} a_1 & b_1 & c_1 \\ a_2 & b_2 & c_2 \\ a_3 & b_3 & c_3 \end{vmatrix} = \begin{vmatrix} a_1 & a_2 & a_3 \\ b_1 & b_2 & b_3 \\ c_1 & c_2 & c_3 \end{vmatrix}$$

II. If each element in a row (or column) is zero, the value of the determinant is zero.

Example.
$$\begin{vmatrix} a_1 & 0 & c_1 \\ a_2 & 0 & c_2 \\ a_3 & 0 & c_3 \end{vmatrix} = 0$$

III. Interchanging any two rows (or columns) reverses the sign of the determinant.

Example.
$$\begin{vmatrix} a_1 & b_1 & c_1 \\ a_2 & b_2 & c_2 \\ a_3 & b_3 & c_3 \end{vmatrix} = - \begin{vmatrix} a_3 & b_3 & c_3 \\ a_2 & b_2 & c_2 \\ a_1 & b_1 & c_1 \end{vmatrix}$$

IV. If two rows (or columns) of a determinant are identical, the value of the determinant is zero.

Example.
$$\begin{vmatrix} a_1 & b_1 & a_1 \\ a_2 & b_2 & a_2 \\ a_3 & b_3 & a_3 \end{vmatrix} = 0$$

V. If each of the elements in a row (or column) of a determinant is multiplied by the same number p, the value of the determinant is multiplied by p.

Example.
$$\begin{vmatrix} pa_1 & b_1 & c_1 \\ pa_2 & b_2 & c_2 \\ pa_3 & b_3 & c_3 \end{vmatrix} = p \begin{vmatrix} a_1 & b_1 & c_1 \\ a_2 & b_2 & c_2 \\ a_3 & b_3 & c_3 \end{vmatrix}$$

VI. If each element of a row (or column) of a determinant is expressed as the sum of two (or more) terms, the determinant can be expressed as the sum of two (or more) determinants.

Example.
$$\begin{vmatrix} a_1+a_1' & b_1 & c_1 \\ a_2+a_2' & b_2 & c_2 \\ a_3+a_3' & b_3 & c_3 \end{vmatrix} = \begin{vmatrix} a_1 & b_1 & c_1 \\ a_2 & b_2 & c_2 \\ a_3 & b_3 & c_3 \end{vmatrix} + \begin{vmatrix} a_1' & b_1 & c_1 \\ a_2' & b_2 & c_2 \\ a_3' & b_3 & c_3 \end{vmatrix}$$

VII. If to each element of a row (or column) of a determinant is added m times the corresponding element of any other row (or column), the value of the determinant is not changed.

Example.
$$\begin{vmatrix} a_1+mb_1 & b_1 & c_1 \\ a_2+mb_2 & b_2 & c_2 \\ a_3+mb_3 & b_3 & c_3 \end{vmatrix} = \begin{vmatrix} a_1 & b_1 & c_1 \\ a_2 & b_2 & c_2 \\ a_3 & b_3 & c_3 \end{vmatrix}$$

These properties may be proved for the special cases of second and third order determinants by using the methods of expansion of Chapter 26. For proofs of the general cases, see the solved problems below.

THE MINOR of an element in a determinant of order n is the determinant of order $n-1$ obtained by removing the row and the column which contain the given element.

For example, the minor of b_3 in the 4th order determinant

$$\begin{vmatrix} a_1 & b_1 & c_1 & d_1 \\ a_2 & b_2 & c_2 & d_2 \\ a_3 & b_3 & c_3 & d_3 \\ a_4 & b_4 & c_4 & d_4 \end{vmatrix}$$

is obtained by crossing out the row and column containing b_3, as shown, and writing the resulting determinant of order 3, namely

$$\begin{vmatrix} a_1 & c_1 & d_1 \\ a_2 & c_2 & d_2 \\ a_4 & c_4 & d_4 \end{vmatrix}.$$

The minor of an element is denoted by capital letters. Thus the minor corresponding to the element b_3 is denoted by B_3.

THE VALUE OF A DETERMINANT may be obtained in terms of minors as follows:

1) Choose any row (or column).

2) Multiply each element in the row (or column) by its corresponding minor preceded by a sign which is *plus* or *minus* according as the sum of the column number and row number is *even* or *odd*. The minor of an element with the attached sign is called the *cofactor* of the element.

3) Add algebraically the products obtained in 2).

For example, let us expand the determinant

$$\begin{vmatrix} a_1 & b_1 & c_1 & d_1 \\ a_2 & b_2 & c_2 & d_2 \\ a_3 & b_3 & c_3 & d_3 \\ a_4 & b_4 & c_4 & d_4 \end{vmatrix}$$

by the elements in the third row. The minors of a_3, b_3, c_3, d_3 are A_3, B_3, C_3, D_3 respectively. The sign corresponding to the element a_3 is + since it appears in the 1st column and 3rd row and $1+3=4$ is even. Similarly, the signs associated with the elements b_3, c_3, d_3 are $-,+,-$ respectively. Thus the value of the determinant is

$$a_3A_3 - b_3B_3 + c_3C_3 - d_3D_3 .$$

Property VII is useful in producing zeros in a given row or column. This property coupled with the expansion in terms of minors makes for easy determination of the value of a determinant.

CRAMER'S RULE for the solution of n simultaneous linear equations in n unknowns is exactly analogous to the rule given in the preceding chapter for the case $n = 2$ and $n = 3$.

Given n linear equations in n unknowns $x_1, x_2, x_3, \ldots, x_n$

$$
\begin{aligned}
a_{11}x_1 + a_{12}x_2 + a_{13}x_3 + \ldots + a_{1n}x_n &= r_1 \\
a_{21}x_1 + a_{22}x_2 + a_{23}x_3 + \ldots + a_{2n}x_n &= r_2 \\
&\cdots \\
a_{n1}x_1 + a_{n2}x_2 + a_{n3}x_3 + \ldots + a_{nn}x_n &= r_n.
\end{aligned}
\tag{1}
$$

Let Δ be the determinant of the coefficients of $x_1, x_2, x_3, \ldots, x_n$, i.e.,

$$
\Delta =
\begin{vmatrix}
a_{11} & a_{12} & a_{13} & \cdots & a_{1n} \\
a_{21} & a_{22} & a_{23} & \cdots & a_{2n} \\
\cdots & \cdots & \cdots & \cdots & \cdots \\
\cdots & \cdots & \cdots & \cdots & \cdots \\
a_{n1} & a_{n2} & a_{n3} & \cdots & a_{nn}
\end{vmatrix}
$$

Denote by Δ_k the determinant Δ with the kth column (which corresponds to the coefficients of the unknown x_k) replaced by the column of the coefficients on the right-hand side of (1). Then

$$
x_1 = \frac{\Delta_1}{\Delta}, \quad x_2 = \frac{\Delta_2}{\Delta}, \quad x_3 = \frac{\Delta_3}{\Delta}, \quad \cdots \qquad \text{provided } \Delta \neq 0.
$$

If $\Delta \neq 0$ there is one and only one solution.

If $\Delta = 0$ the system of equations may or may not have solutions.

Equations having no simultaneous solution are called *inconsistent*, otherwise they are consistent. If $\Delta = 0$ and at least one of the determinants Δ_1, $\Delta_2, \ldots, \Delta_n \neq 0$, the given system is inconsistent. If $\Delta = \Delta_1 = \Delta_2 = \ldots = \Delta_n = 0$, the system may or may not be consistent.

Equations having an infinite number of simultaneous solutions are called *dependent*. If a system of equations is dependent then $\Delta = 0$ and all of the determinants $\Delta_1, \Delta_2, \ldots, \Delta_n = 0$. The converse, however, is not always true.

HOMOGENEOUS LINEAR EQUATIONS. If $r_1, r_2, \ldots, r_n$ in equations (1) are all zero, the system is said to be *homogeneous*. In this case $\Delta_1 = \Delta_2 = \Delta_3 = \ldots = \Delta_n = 0$ and the following theorem is true.

THEOREM. A necessary and sufficient condition that n homogeneous linear equations in n unknowns have solutions other than the trivial solution (where all the unknowns equal zero) is that the determinant of the coefficients, $\Delta = 0$.

A SYSTEM OF m EQUATIONS IN n UNKNOWNS may or may not have simultaneous solutions.

1) If $m > n$, the unknowns in n of the given equations may be obtained. If these values satisfy the remaining $m-n$ equations the system is consistent, otherwise it is inconsistent.

2) If $m < n$, then m of the unknowns may be determined in terms of the remaining $n-m$ unknowns.

SOLVED PROBLEMS

INVERSIONS.

1. Determine the number of inversions in each of the following groupings.

    a) 3, 1, 2   3 precedes 1 and 2. There are 2 inversions.

    b) 4, 2, 3, 1   4 precedes 2, 3, 1; 2 precedes 1; 3 precedes 1. There are 5 inversions.

    c) 5, 1, 4, 3, 2   5 precedes 1, 4, 3, 2; 4 precedes 3, 2; 3 precedes 2. There are 7 inversions.

    d) *b*, *a*, *c*   *b* precedes *a*. There is 1 inversion.

    e) *b*, *a*, *e*, *d*, *c*   *b* precedes *a*; *e* precedes *d*, *c*; *d* precedes *c*. There are 4 inversions.

2. *a*) What would be the number of inversions in the subscripts of $b_1 d_3 c_2 a_4$ when the letters are in alphabetical order?

    b) What would be the number of inversions in the letters of $b_1 d_3 c_2 a_4$ when the subscripts are arranged in natural order?

    a) Write $a_4 b_1 c_2 d_3$. The subscripts 4, 1, 2, 3 have 3 inversions.
    b) Write $b_1 c_2 d_3 a_4$. The letters *bcda* have 3 inversions.

    The fact that there are 3 inversions in *a*) and *b*) is not merely a coincidence. There are the same number of inversions with regard to subscripts when letters are in alphabetical order as inversions with regard to letters when subscripts are in natural order.

3. Write the expansion of the determinant $\begin{vmatrix} a_1 & b_1 & c_1 \\ a_2 & b_2 & c_2 \\ a_3 & b_3 & c_3 \end{vmatrix}$ by use of inversions.

    The expansion consists of terms of the form *abc* with all possible arrangements of the subscripts, the sign before the product being determined by the number of inversions in the subscripts, a plus sign if there is an even number of inversions, a minus sign if there is an odd number of inversions.

    There are 6 terms in the expansion. These terms are:

    $a_1 b_2 c_3$ 0 inversions, sign + $a_2 b_3 c_1$ 2 inversions, sign +
    $a_1 b_3 c_2$ 1 inversion,   sign − $a_3 b_2 c_1$ 3 inversions, sign −
    $a_2 b_1 c_3$ 1 inversion,   sign − $a_3 b_1 c_2$ 2 inversions, sign + .

    The required expansion is: $a_1 b_2 c_3 - a_1 b_3 c_2 - a_2 b_1 c_3 + a_2 b_3 c_1 - a_3 b_2 c_1 + a_3 b_1 c_2$.

4. Determine the signs associated with the terms $d_1 a_3 c_2 b_4$ and $a_3 c_2 d_4 b_1$ in the expansion of the determinant

$$\begin{vmatrix} a_1 & b_1 & c_1 & d_1 \\ a_2 & b_2 & c_2 & d_2 \\ a_3 & b_3 & c_3 & d_3 \\ a_4 & b_4 & c_4 & d_4 \end{vmatrix}$$

    $d_1 a_3 c_2 b_4$ written $a_3 b_4 c_2 d_1$ has 5 inversions in the subscripts. The associated sign is −.

    $a_3 c_2 d_4 b_1$ written $a_3 b_1 c_2 d_4$ has 2 inversions in the subscripts. The associated sign is + .

PROPERTIES OF DETERMINANTS.

5. Prove Property III: If two rows (or columns) are interchanged, the sign of the determinant is changed.

Case 1. Rows are adjacent.

Interchanging two adjacent rows results in the interchange of two adjacent subscripts in each term of the expansion. Thus the number of inversions of subscripts is either increased by one or decreased by one. Hence the sign of each term is changed and so the sign of the determinant is changed.

Case 2. Rows are not adjacent.

Suppose there are k rows between the ones to be interchanged. It will then take k interchanges of the adjacent rows to bring the upper row to the row just above the lower row, one more to interchange them, and k more interchanges to bring the lower row up to where the upper row was. This involves a total of $k + 1 + k = 2k + 1$ interchanges. Since $2k + 1$ is odd, there is an odd number of changes in sign and the result is the same as a single change in sign.

6. Prove Property IV: If two rows (or columns) are identical, the determinant has value zero.

Let D be the value of the determinant. By Property III, interchange of the two identical rows should change the value to $-D$. Since the determinants are the same, $D = -D$ or $D = 0$.

7. Prove Property V: If each of the elements of a row (or column) are multiplied by the same number p, the value of the determinant is multiplied by p.

Each term of the determinant contains one and only one element from the row multiplied by p and thus each term has factor p. This factor is therefore common to all the terms of the expansion and so the determinant is multiplied by p.

8. Prove Property VI: If each element of a row (or column) of a determinant is expressed as the sum of two (or more) terms, the determinant can be expressed as the sum of two (or more) determinants.

For the case of third order determinants we must show that

$$\begin{vmatrix} a_1 + a_1' & b_1 & c_1 \\ a_2 + a_2' & b_2 & c_2 \\ a_3 + a_3' & b_3 & c_3 \end{vmatrix} = \begin{vmatrix} a_1 & b_1 & c_1 \\ a_2 & b_2 & c_2 \\ a_3 & b_3 & c_3 \end{vmatrix} + \begin{vmatrix} a_1' & b_1 & c_1 \\ a_2' & b_2 & c_2 \\ a_3' & b_3 & c_3 \end{vmatrix} .$$

Each term in the expansion of the determinant on the left equals the sum of the two corresponding terms in the determinants on the right, e.g., $(a_2 + a_2')b_3c_1 = a_2b_3c_1 + a_2'b_3c_1$. Thus the property holds for third order determinants. The method of proof holds in the general case.

9. Prove Property VII: If to each element of a row (or column) of a determinant is added m times the corresponding element of any other row (or column), the value of the determinant is not changed.

For the case of a third order determinant we must show that

$$\begin{vmatrix} a_1 + mb_1 & b_1 & c_1 \\ a_2 + mb_2 & b_2 & c_2 \\ a_3 + mb_3 & b_3 & c_3 \end{vmatrix} = \begin{vmatrix} a_1 & b_1 & c_1 \\ a_2 & b_2 & c_2 \\ a_3 & b_3 & c_3 \end{vmatrix} .$$

By Property VI the right hand side may be written

$$
\begin{vmatrix} a_1 & b_1 & c_1 \\ a_2 & b_2 & c_2 \\ a_3 & b_3 & c_3 \end{vmatrix}
+
\begin{vmatrix} mb_1 & b_1 & c_1 \\ mb_2 & b_2 & c_2 \\ mb_3 & b_3 & c_3 \end{vmatrix} .
$$

This last determinant may be written $m \begin{vmatrix} b_1 & b_1 & c_1 \\ b_2 & b_2 & c_2 \\ b_3 & b_3 & c_3 \end{vmatrix}$ which is zero by Property IV.

10. Show that $\begin{vmatrix} 3 & 2 & 2 & 1 \\ 6 & 5 & 4 & -2 \\ 9 & -3 & 6 & -5 \\ 12 & 2 & 8 & 7 \end{vmatrix} = 0.$

The number 3 may be factored from each element in the first column and 2 may be factored from each element in the third column to yield

$$
(3)(2) \begin{vmatrix} 1 & 2 & 1 & 1 \\ 2 & 5 & 2 & -2 \\ 3 & -3 & 3 & -5 \\ 4 & 2 & 4 & 7 \end{vmatrix}
$$

which equals zero since the first and third columns are identical.

11. Use Property VII to transform the determinant $\begin{vmatrix} 1 & -2 & 3 \\ 2 & -1 & 4 \\ -2 & 3 & 1 \end{vmatrix}$ into a determinant of equal value with zeros in the first row, second and third columns.

Multiply each element in the first column by 2 and add to the corresponding elements in the second column, thus obtaining

$$
\begin{vmatrix} 1 & (2)(1)-2 & 3 \\ 2 & (2)(2)-1 & 4 \\ -2 & (2)(-2)+3 & 1 \end{vmatrix}
=
\begin{vmatrix} 1 & 0 & 3 \\ 2 & 3 & 4 \\ -2 & -1 & 1 \end{vmatrix} .
$$

Multiply each element in the first column of the new determinant by −3 and add to the corresponding elements in the third column to obtain

$$
\begin{vmatrix} 1 & 0 & (-3)(1)+3 \\ 2 & 3 & (-3)(2)+4 \\ -2 & -1 & (-3)(-2)+1 \end{vmatrix}
=
\begin{vmatrix} 1 & 0 & 0 \\ 2 & 3 & -2 \\ -2 & -1 & 7 \end{vmatrix} .
$$

The result could have been obtained in one step by writing

$$
\begin{vmatrix} 1 & (2)(1)-2 & (-3)(1)+3 \\ 2 & (2)(2)-1 & (-3)(2)+4 \\ -2 & (2)(-2)+3 & (-3)(-2)+1 \end{vmatrix}
=
\begin{vmatrix} 1 & 0 & 0 \\ 2 & 3 & -2 \\ -2 & -1 & 7 \end{vmatrix} .
$$

The choice of the numbers 2 and −3 was made in order to obtain zeros in the desired places.

12. Use Property VII to transform the determinant $\begin{vmatrix} 3 & 6 & 2 & 3 \\ -2 & 1 & -2 & 2 \\ 4 & -5 & 1 & 4 \\ 1 & 3 & 4 & -2 \end{vmatrix}$ into an equal determinant having three zeros in the 4th row.

Multiply each element in the 1st column (the *basic* column shown shaded) by $-3, -4, +2$ and add respectively to the corresponding elements in the 2nd, 3rd, 4th columns. The result is

$$\begin{vmatrix} 3 & (-3)(3)+6 & (-4)(3)+2 & (2)(3)+3 \\ -2 & (-3)(-2)+1 & (-4)(-2)-2 & (2)(-2)+2 \\ 4 & (-3)(4)-5 & (-4)(4)+1 & (2)(4)+4 \\ 1 & (-3)(1)+3 & (-4)(1)+4 & (2)(1)-2 \end{vmatrix} = \begin{vmatrix} 3 & -3 & -10 & 9 \\ -2 & 7 & 6 & -2 \\ 4 & -17 & -15 & 12 \\ 1 & 0 & 0 & 0 \end{vmatrix}$$

Note that it is useful to choose a basic row or column containing the element 1.

13. Obtain 4 zeros in a row or column of the 5th order determinant $\begin{vmatrix} 3 & 5 & 4 & 6 & 2 \\ -2 & 3 & 2 & 3 & 4 \\ 4 & 1 & 3 & -2 & -3 \\ 6 & -3 & 2 & 4 & 3 \\ 2 & 2 & 5 & 3 & -2 \end{vmatrix}$.

We shall produce zeros in the 2nd column by use of the basic row shown shaded. Multiply the elements in this basic row by $-5, -3, 3, -2$ and add respectively to the corresponding elements in the 1st, 2nd, 4th, 5th rows to obtain

$$\begin{vmatrix} -17 & 0 & -11 & 16 & 17 \\ -14 & 0 & -7 & 9 & 13 \\ 4 & 1 & 3 & -2 & -3 \\ 18 & 0 & 11 & -2 & -6 \\ -6 & 0 & -1 & 7 & 4 \end{vmatrix}$$

14. Obtain 3 zeros in a row or column of the determinant $\begin{vmatrix} 3 & 4 & 2 & 3 \\ -2 & 2 & 3 & -2 \\ 2 & -3 & 3 & 4 \\ 4 & 5 & -2 & -2 \end{vmatrix}$ without changing its value.

It is convenient to use Property VII to obtain an element 1 in a row or column. For example, by multiplying each of the elements in column 2 by -1 and adding to the corresponding elements in column 3, we obtain

$$\begin{vmatrix} 3 & 4 & -2 & 3 \\ -2 & 2 & 1 & -2 \\ 2 & -3 & 6 & 4 \\ 4 & 5 & -7 & -2 \end{vmatrix} .$$

Using the 3rd column as basic column, multiply its elements by $2, -2, 2$ and add respectively to the 1st, 2nd, 4th columns to obtain

$$\begin{vmatrix} -1 & 8 & -2 & -1 \\ 0 & 0 & 1 & 0 \\ 14 & -15 & 6 & 16 \\ -10 & 19 & -7 & -16 \end{vmatrix}$$ which equals the given determinant.

EVALUATION OF DETERMINANTS BY MINORS (OR COFACTORS).

15. Write the minor and corresponding cofactor of the element in the second row, third column for the determinant

$$\begin{vmatrix} 2 & -2 & 3 & 1 \\ 1 & 3 & 2 & 5 \\ 1 & -2 & 5 & -1 \\ 2 & 1 & 3 & -2 \end{vmatrix} .$$

Crossing out the row and column containing the element, the minor is given by

$$\begin{vmatrix} 2 & -2 & 1 \\ 1 & -2 & -1 \\ 2 & 1 & -2 \end{vmatrix} .$$

Since the element is in the 2nd row, 3rd column and $2+3 = 5$ is an odd number, the associated sign is minus. Thus the cofactor corresponding to the given element is

$$-\begin{vmatrix} 2 & -2 & 1 \\ 1 & -2 & -1 \\ 2 & 1 & -2 \end{vmatrix} .$$

16. Write the minors and cofactors of the elements in the 4th row of the determinant

$$\begin{vmatrix} 3 & -2 & 4 & 2 \\ 2 & 1 & 5 & -3 \\ 1 & 5 & -2 & 2 \\ -3 & -2 & -4 & 1 \end{vmatrix} .$$

The elements in the 4th row are $-3, -2, -4, 1$.

Minor of element -3 = $\begin{vmatrix} -2 & 4 & 2 \\ 1 & 5 & -3 \\ 5 & -2 & 2 \end{vmatrix}$ Cofactor = $-$ Minor

Minor of element -2 = $\begin{vmatrix} 3 & 4 & 2 \\ 2 & 5 & -3 \\ 1 & -2 & 2 \end{vmatrix}$ Cofactor = $+$ Minor

Minor of element -4 = $\begin{vmatrix} 3 & -2 & 2 \\ 2 & 1 & -3 \\ 1 & 5 & 2 \end{vmatrix}$ Cofactor = $-$ Minor

Minor of element 1 = $\begin{vmatrix} 3 & -2 & 4 \\ 2 & 1 & 5 \\ 1 & 5 & -2 \end{vmatrix}$ Cofactor = $+$ Minor

17. Express the value of the determinant of Problem 16 in terms of minors or cofactors.

Value of determinant = sum of elements each multiplied by associated cofactor

$$= (-3)\left\{ -\begin{vmatrix} -2 & 4 & 2 \\ 1 & 5 & -3 \\ 5 & -2 & 2 \end{vmatrix} \right\} + (-2)\left\{ +\begin{vmatrix} 3 & 4 & 2 \\ 2 & 5 & -3 \\ 1 & -2 & 2 \end{vmatrix} \right\}$$

$$+ (-4)\left\{ -\begin{vmatrix} 3 & -2 & 2 \\ 2 & 1 & -3 \\ 1 & 5 & 2 \end{vmatrix} \right\} + (1)\left\{ +\begin{vmatrix} 3 & -2 & 4 \\ 2 & 1 & 5 \\ 1 & 5 & -2 \end{vmatrix} \right\}$$

Upon evaluating each of the 3rd order determinants the result -53 is obtained.

The method of evaluation here indicated is tedious. However, the labor involved may be considerably reduced by first transforming a given determinant into an equivalent one having zeros in a row or column by use of Property VII as shown in the following problem.

18. Evaluate the determinant in Problem 16 by first transforming it into one having three zeros in a row or column and then expanding by minors.

Choosing the basic column indicated, $\begin{vmatrix} 3 & -2 & 4 & 2 \\ 2 & 1 & 5 & -3 \\ 1 & 5 & -2 & 2 \\ -3 & -2 & -4 & 1 \end{vmatrix}$, multiply its elements by $-2, -5, 3$

and add respectively to the corresponding elements of the 1st, 3rd, 4th columns to obtain

$$\begin{vmatrix} 7 & -2 & 14 & -4 \\ 0 & 1 & 0 & 0 \\ -9 & 5 & -27 & 17 \\ 1 & -2 & 6 & -5 \end{vmatrix}.$$

Expand according to the cofactors of the elements in the second row and obtain

(0)(its cofactor) + (1)(its cofactor) + (0)(its cofactor) + (0)(its cofactor)

$$= \ (1)(\text{its cofactor}) \ = \ 1 \left\{ + \begin{vmatrix} 7 & 14 & -4 \\ -9 & -27 & 17 \\ 1 & 6 & -5 \end{vmatrix} \right\}.$$

Expanding this determinant, we obtain the value -53 which agrees with the result of Prob. 16.

Note that the method of this problem may be employed to evaluate 3rd order determinants in terms of 2nd order determinants.

19. Evaluate each of the following determinants.

a) $\begin{vmatrix} 4 & 1 & -2 & 3 \\ -1 & 2 & 1 & 4 \\ 3 & -1 & 3 & 4 \\ 2 & 3 & -3 & 2 \end{vmatrix}$

Multiply the elements in the indicated basic row by $-2, 1, -3$ and add respectively to the corresponding elements in the 2nd, 3rd, 4th rows to obtain

$$\begin{vmatrix} 4 & 1 & -2 & 3 \\ -9 & 0 & 5 & -2 \\ 7 & 0 & 1 & 7 \\ -10 & 0 & 3 & -7 \end{vmatrix} = \ 1 \left\{ - \begin{vmatrix} -9 & 5 & -2 \\ 7 & 1 & 7 \\ -10 & 3 & -7 \end{vmatrix} \right\}$$

$$= - \begin{vmatrix} -9 & 5 & -2 \\ 7 & 1 & 7 \\ -10 & 3 & -7 \end{vmatrix}.$$

Multiply the elements in the indicated basic column by -7 and add to the corresponding elements in the 1st and 3rd columns to obtain

$$- \begin{vmatrix} -44 & 5 & -37 \\ 0 & 1 & 0 \\ -31 & 3 & -28 \end{vmatrix} = \ -(1) \left\{ + \begin{vmatrix} -44 & -37 \\ -31 & -28 \end{vmatrix} \right\} = \ -85.$$

b) $\begin{vmatrix} 1 & -3 & 2 & -3 & 1 \\ -1 & 2 & 1 & 2 & -3 \\ -3 & 1 & -2 & -1 & 4 \\ 2 & -3 & 3 & 4 & -1 \\ 3 & -2 & -4 & 2 & 1 \end{vmatrix}$

Multiply the elements in the indicated basic column by $3, 2, 1, -4$ respectively and add to the corresponding elements in the 1st, 3rd, 4th, 5th columns to obtain

$$\begin{vmatrix} -8 & -3 & -4 & -6 & 13 \\ 5 & 2 & 5 & 4 & -11 \\ 0 & 1 & 0 & 0 & 0 \\ -7 & -3 & -3 & 1 & 11 \\ -3 & -2 & -8 & 0 & 9 \end{vmatrix} = \ 1 \left\{ - \begin{vmatrix} -8 & -4 & -6 & 13 \\ 5 & 5 & 4 & -11 \\ -7 & -3 & 1 & 11 \\ -3 & -8 & 0 & 9 \end{vmatrix} \right\} = \ - \begin{vmatrix} -8 & -4 & -6 & 13 \\ 5 & 5 & 4 & -11 \\ -7 & -3 & 1 & 11 \\ -3 & -8 & 0 & 9 \end{vmatrix}.$$

In the last determinant, multiply the elements in the indicated basic row by $6, -4$ and add respectively to the elements in the 1st and 2nd rows to obtain

$$- \begin{vmatrix} -50 & -22 & 0 & 79 \\ 33 & 17 & 0 & -55 \\ -7 & -3 & 1 & 11 \\ -3 & -8 & 0 & 9 \end{vmatrix} = -(1) \left\{ + \begin{vmatrix} -50 & -22 & 79 \\ 33 & 17 & -55 \\ -3 & -8 & 9 \end{vmatrix} \right\} = - \begin{vmatrix} -50 & -22 & 79 \\ 33 & 17 & -55 \\ -3 & -8 & 9 \end{vmatrix} .$$

Multiply the elements in the indicated row of the last determinant by 2 and add to the 2nd row to obtain

$$- \begin{vmatrix} -50 & -22 & 79 \\ 27 & 1 & -37 \\ -3 & -8 & 9 \end{vmatrix} .$$

Multiply the elements in the indicated row of the last determinant by $22, 8$ and add respectively to the elements in the 1st and 3rd rows to obtain

$$- \begin{vmatrix} 544 & 0 & -735 \\ 27 & 1 & -37 \\ 213 & 0 & -287 \end{vmatrix} = -(1) \left\{ + \begin{vmatrix} 544 & -735 \\ 213 & -287 \end{vmatrix} \right\} = -427.$$

20. Factor the following determinant.

$$\begin{vmatrix} x & y & 1 \\ x^2 & y^2 & 1 \\ x^3 & y^3 & 1 \end{vmatrix} = xy \begin{vmatrix} 1 & 1 & 1 \\ x & y & 1 \\ x^2 & y^2 & 1 \end{vmatrix}$$

Removing factors x and y from 1st and 2nd columns respectively.

$$= xy \begin{vmatrix} 0 & 0 & 1 \\ x-1 & y-1 & 1 \\ x^2-1 & y^2-1 & 1 \end{vmatrix}$$

Adding -1 times elements in 3rd column to the corresponding elements in 1st and 2nd columns.

$$= xy \begin{vmatrix} x-1 & y-1 \\ x^2-1 & y^2-1 \end{vmatrix}$$

$$= xy(x-1)(y-1) \begin{vmatrix} 1 & 1 \\ x+1 & y+1 \end{vmatrix}$$

Removing factors $(x-1)$ and $(y-1)$ from 1st and 2nd columns respectively.

$$= xy(x-1)(y-1)(y-x).$$

SIMULTANEOUS LINEAR EQUATIONS AND CRAMER'S RULE.

21. Solve the system
$$\begin{aligned} 2x + y - z + w &= -4 \\ x + 2y + 2z - 3w &= 6 \\ 3x - y - z + 2w &= 0 \\ 2x + 3y + z + 4w &= -5 \end{aligned}$$

$$\Delta = \begin{vmatrix} 2 & 1 & -1 & 1 \\ 1 & 2 & 2 & -3 \\ 3 & -1 & -1 & 2 \\ 2 & 3 & 1 & 4 \end{vmatrix} = 86$$

$$\Delta_1 = \begin{vmatrix} -4 & 1 & -1 & 1 \\ 6 & 2 & 2 & -3 \\ 0 & -1 & -1 & 2 \\ -5 & 3 & 1 & 4 \end{vmatrix} = 86 \qquad \Delta_2 = \begin{vmatrix} 2 & -4 & -1 & 1 \\ 1 & 6 & 2 & -3 \\ 3 & 0 & -1 & 2 \\ 2 & -5 & 1 & 4 \end{vmatrix} = -172$$

$$\Delta_3 = \begin{vmatrix} 2 & 1 & -4 & 1 \\ 1 & 2 & 6 & -3 \\ 3 & -1 & 0 & 2 \\ 2 & 3 & -5 & 4 \end{vmatrix} = 258 \qquad \Delta_4 = \begin{vmatrix} 2 & 1 & -1 & -4 \\ 1 & 2 & 2 & 6 \\ 3 & -1 & -1 & 0 \\ 2 & 3 & 1 & -5 \end{vmatrix} = -86$$

Then $\quad x = \dfrac{\Delta_1}{\Delta} = 1, \quad y = \dfrac{\Delta_2}{\Delta} = -2, \quad z = \dfrac{\Delta_3}{\Delta} = 3, \quad w = \dfrac{\Delta_4}{\Delta} = -1.$

22. The currents i_1, i_2, i_3, i_4, i_5 (measured in amperes) can be determined from the following set of equations. Find i_3.

$$i_1 - 2i_2 + i_3 = 3$$
$$i_2 + 3i_4 - i_5 = -5$$
$$i_1 + i_2 + i_3 - i_5 = 1$$
$$2i_2 + i_3 - 2i_4 - 2i_5 = 0$$
$$i_1 + i_3 + 2i_4 + i_5 = 3$$

$$\Delta_3 = \begin{vmatrix} 1 & -2 & 3 & 0 & 0 \\ 0 & 1 & -5 & 3 & -1 \\ 1 & 1 & 1 & 0 & -1 \\ 0 & 2 & 0 & -2 & -2 \\ 1 & 0 & 3 & 2 & 1 \end{vmatrix} = 38, \qquad \Delta = \begin{vmatrix} 1 & -2 & 1 & 0 & 0 \\ 0 & 1 & 0 & 3 & -1 \\ 1 & 1 & 1 & 0 & -1 \\ 0 & 2 & 1 & -2 & -2 \\ 1 & 0 & 1 & 2 & 1 \end{vmatrix} = 19, \qquad i_3 = \frac{\Delta_3}{\Delta} = 2 \text{ amp.}$$

CONSISTENT AND INCONSISTENT LINEAR EQUATIONS.

23. Determine whether the system $\begin{aligned} x - 3y + 2z &= 4 \\ 2x + y - 3z &= -2 \\ 4x - 5y + z &= 5 \end{aligned}$ is consistent.

$$\Delta = \begin{vmatrix} 1 & -3 & 2 \\ 2 & 1 & -3 \\ 4 & -5 & 1 \end{vmatrix} = 0. \qquad \text{However, } \Delta_1 = \begin{vmatrix} 4 & -3 & 2 \\ -2 & 1 & -3 \\ 5 & -5 & 1 \end{vmatrix} = -7.$$

Hence at least one of the determinants $\Delta_1, \Delta_2, \Delta_3 \neq 0$ so that the equations are inconsistent.

This could be seen in another way by multiplying the first equation by 2 and adding to the second equation to obtain $4x - 5y + z = 6$ which is not consistent with the last equation.

24. Determine whether the system $\begin{aligned} 4x - 2y + 6z &= 8 \\ 2x - y + 3z &= 5 \\ 2x - y + 3z &= 4 \end{aligned}$ is consistent.

$$\Delta = \begin{vmatrix} 4 & -2 & 6 \\ 2 & -1 & 3 \\ 2 & -1 & 3 \end{vmatrix} = 0 \qquad \Delta_1 = \begin{vmatrix} 8 & -2 & 6 \\ 5 & -1 & 3 \\ 4 & -1 & 3 \end{vmatrix} = 0$$

$$\Delta_2 = \begin{vmatrix} 4 & 8 & 6 \\ 2 & 5 & 3 \\ 2 & 4 & 3 \end{vmatrix} = 0 \qquad \Delta_3 = \begin{vmatrix} 4 & -2 & 8 \\ 2 & -1 & 5 \\ 2 & -1 & 4 \end{vmatrix} = 0$$

Nothing can be said about the consistency from these facts. On closer examination of the system it is noticed that the second and third equations are inconsistent. Hence the system is inconsistent.

25. Determine whether the system $\begin{aligned} 2x + y - 2z &= 4 \\ x - 2y + z &= -2 \\ 5x - 5y + z &= -2 \end{aligned}$ is consistent.

$\Delta = \Delta_1 = \Delta_2 = \Delta_3 = 0.$ Hence nothing can be concluded from these facts.

Solving the first two equations for x and y (in terms of z), $x = \frac{3}{5}(z+2)$, $y = \frac{4}{5}(z+2)$.
These values are found by substitution to satisfy the third equation. (If they did not satisfy the third equation the system would be inconsistent.)

Hence the values $x = \frac{3}{5}(z+2)$, $y = \frac{4}{5}(z+2)$ satisfy the system and there are infinite sets of solutions, obtained by assigning various values to z. Thus if $z = 3$, then $x = 3$, $y = 4$; if $z = -2$, then $x = 0$, $y = 0$; etc.

It follows that the given equations are *dependent*. This may be seen in another way by multiplying the second equation by 3 and adding to the first equation to obtain $5x - 5y + z = -2$ which is the third equation.

HOMOGENEOUS LINEAR EQUATIONS.

26. Does the system $\begin{aligned} 2x - 3y + 4z &= 0 \\ x + y - 2z &= 0 \\ 3x + 2y - 3z &= 0 \end{aligned}$ possess only the trivial solution $x = y = z = 0$?

$$\Delta = \begin{vmatrix} 2 & -3 & 4 \\ 1 & 1 & -2 \\ 3 & 2 & -3 \end{vmatrix} = -17 \qquad \Delta_1 = \Delta_2 = \Delta_3 = 0$$

Since $\Delta \neq 0$ and $\Delta_1 = \Delta_2 = \Delta_3 = 0$, the system has only the trivial solution.

27. Find non-trivial solutions for the system $\begin{aligned} x + 3y - 2z &= 0 \\ 2x - 4y + z &= 0 \\ x + y - z &= 0 \end{aligned}$ if they exist.

$$\Delta = \begin{vmatrix} 1 & 3 & -2 \\ 2 & -4 & 1 \\ 1 & 1 & -1 \end{vmatrix} = 0 \qquad \Delta_1 = \Delta_2 = \Delta_3 = 0$$

Hence there are non-trivial solutions.

To determine these non-trivial solutions solve for x and y (in terms of z) from the first two equations (this may not always be possible). We find $x = z/2$, $y = z/2$. These satisfy the third equation. An infinite number of solutions is obtained by assigning various values to z. For example, if $z = 6$, then $x = 3$, $y = 3$; if $z = -4$, then $x = -2$, $y = -2$; etc.

28. For what values of k will the system $\begin{aligned} x + 2y + kz &= 0 \\ 2x + ky + 2z &= 0 \\ 3x + y + z &= 0 \end{aligned}$ have non-trivial solutions?

Non-trivial solutions are obtained when $\Delta = \begin{vmatrix} 1 & 2 & k \\ 2 & k & 2 \\ 3 & 1 & 1 \end{vmatrix} = 0.$

Hence $\Delta = -3k^2 + 3k + 6 = 0$ or $k = -1, 2$.

SUPPLEMENTARY PROBLEMS

29. Determine the number of inversions in each grouping.

  a) $4, 3, 1, 2$ b) $3, 1, 5, 4, 2$ c) c, a, d, b, e

30. For the grouping d_3, b_4, c_1, e_2, a_5 determine

  a) the number of inversions in the subscripts when the letters are in alphabetical order,

  b) the number of inversions in the letters when the subscripts are in natural order.

31. In the expansion of $\begin{vmatrix} a_1 & b_1 & c_1 & d_1 \\ a_2 & b_2 & c_2 & d_2 \\ a_3 & b_3 & c_3 & d_3 \\ a_4 & b_4 & c_4 & d_4 \end{vmatrix}$ determine the signs associated with the terms

  $d_2 b_4 c_3 a_1$ and $b_3 c_2 a_4 d_1$.

32. a) Prove Property I: If the rows and columns of a determinant are interchanged, the value of the determinant is the same.

  b) Prove Property II: If each element in a row (or column) is zero, the value of the determinant is zero.

33. Show that the determinant $\begin{vmatrix} 1 & 2 & 3 & 4 \\ 2 & 4 & 6 & 3 \\ 3 & 8 & 12 & 2 \\ 4 & 16 & 24 & 1 \end{vmatrix}$ equals zero.

34. Transform the determinant $\begin{vmatrix} -2 & 4 & 1 & 3 \\ 1 & -2 & 2 & 4 \\ 3 & 1 & -3 & 2 \\ 4 & 3 & -2 & -1 \end{vmatrix}$ into an equal determinant having three zeros

  in the 3rd column.

35. Without changing the value of the determinant $\begin{vmatrix} 4 & -2 & 1 & 3 & 1 \\ -2 & 1 & -3 & -2 & -2 \\ 3 & 4 & 2 & 1 & 3 \\ 1 & -3 & 4 & -1 & -1 \\ 2 & -1 & 2 & 4 & 2 \end{vmatrix}$ obtain four zeros in

  the 4th column.

36. For the determinant $\begin{vmatrix} -1 & 2 & 3 & -2 \\ 4 & -1 & -2 & 2 \\ -3 & 1 & 2 & -1 \\ 2 & 4 & -1 & 3 \end{vmatrix}$

  a) write the minors and cofactors of the elements in the 3rd row,

  b) express the value of the determinant in terms of minors or cofactors,

  c) find the value of the determinant.

37. Transform the determinant $\begin{vmatrix} -2 & 1 & 2 & 3 \\ 3 & -2 & -3 & 2 \\ 1 & 2 & 1 & 2 \\ 4 & 3 & -1 & -3 \end{vmatrix}$ into a determinant having three zeros in a row

  and then evaluate the determinant by use of expansion by minors.

38. Evaluate each determinant.

  a) $\begin{vmatrix} 2 & -1 & 3 & 2 \\ -3 & 1 & 2 & 4 \\ 1 & -3 & -1 & 3 \\ -1 & 2 & -2 & -3 \end{vmatrix}$ b) $\begin{vmatrix} 3 & -1 & 2 & 1 \\ 4 & 2 & 0 & -3 \\ -2 & 1 & -3 & 2 \\ 1 & 3 & -1 & 4 \end{vmatrix}$ c) $\begin{vmatrix} 1 & 2 & -1 & 1 \\ -2 & 3 & 2 & -1 \\ 3 & -1 & 1 & -4 \\ -1 & 4 & -3 & 2 \end{vmatrix}$ d) $\begin{vmatrix} 3 & 2 & -1 & 3 & 2 \\ -2 & 0 & 3 & 4 & 3 \\ 1 & -3 & -2 & 1 & 0 \\ 2 & 4 & 1 & 0 & 1 \\ -1 & -1 & 2 & 1 & 0 \end{vmatrix}$

39. Factor each determinant: a) $\begin{vmatrix} a & b & c \\ a^2 & b^2 & c^2 \\ a^3 & b^3 & c^3 \end{vmatrix}$ b) $\begin{vmatrix} 1 & 1 & 1 & 1 \\ 1 & x & y & z \\ 1 & x^2 & y^2 & z^2 \\ 1 & x^3 & y^3 & z^3 \end{vmatrix}$

40. Solve each system: a) $\begin{cases} x - 2y + z - 3w = 4 \\ 2x + 3y - z - 2w = -4 \\ 3x - 4y + 2z - 4w = 12 \\ 2x - y - 3z + 2w = -2 \end{cases}$ b) $\begin{cases} 2x + y - 3z = -5 \\ 3y + 4z + w = 5 \\ 2z - w - 4x = 0 \\ w + 3x - y = 4 \end{cases}$

41. Find i_1 and i_4 for the system $\begin{cases} 2i_1 - 3i_3 - i_4 = -4 \\ 3i_1 + i_2 - 2i_3 + 2i_4 + 2i_5 = 0 \\ -i_1 - 3i_2 + 2i_4 + 3i_5 = 2 \\ i_1 + 2i_3 - i_5 = 9 \\ 2i_1 + i_2 = 5 \end{cases}$

42. Determine whether each system is consistent.

    a) $\begin{cases} 2x - 3y + z = 1 \\ x + 2y - z = 1 \\ 3x - y + 2z = 6 \end{cases}$ b) $\begin{cases} 2x - y + z = 2 \\ 3x + 2y - 4z = 1 \\ x - 4y + 6z = 3 \end{cases}$ c) $\begin{cases} x + 3y - 2z = 2 \\ 3x - y - z = 1 \\ 2x + 6y - 4z = 3 \end{cases}$ d) $\begin{cases} 2u + v - 3w = 1 \\ u - 2v - w = 2 \\ u + 3v - 2w = -2 \end{cases}$

43. Find non-trivial solutions, if they exist, for the system $\begin{cases} 3x - 2y + 4z = 0 \\ 2x + y - 3z = 0 \\ x + 3y - 2z = 0 \end{cases}$

44. For what value of k will the system $\begin{cases} 2x + ky + z + w = 0 \\ 3x + (k-1)y - 2z - w = 0 \\ x - 2y + 4z + 2w = 0 \\ 2x + y + z + 2w = 0 \end{cases}$ possess non-trivial

    solutions?

ANSWERS TO SUPPLEMENTARY PROBLEMS.

29. a) 5   b) 5   c) 3

30. a) 8   b) 8

31. − and + respectively

36. c) −38

37. 28

38. a) 38   b) −143   c) −108   d) 88

39. a) $abc(a-b)(b-c)(c-a)$ b) $(x-1)(y-1)(z-1)(x-y)(y-z)(z-x)$

40. a) $x = 2,\ y = -1,\ z = 3,\ w = 1$ b) $x = 1,\ y = -1,\ z = 2,\ w = 0$

41. $i_1 = 3,\ \ i_4 = -2$

42. a) consistent   b) dependent   c) inconsistent   d) inconsistent

43. only trivial solution $x = y = z = 0$

44. $k = -1$

CHAPTER 28

Partial Fractions

A RATIONAL FRACTION in x is the quotient $\dfrac{P(x)}{Q(x)}$ of two polynomials in x.

Thus $\dfrac{3x^2 - 1}{x^3 + 7x^2 - 4}$ is a rational fraction.

A PROPER FRACTION is one in which the degree of the numerator is less than the degree of the denominator.

Thus $\dfrac{2x - 3}{x^2 + 5x + 4}$ and $\dfrac{4x^2 + 1}{x^4 - 3x}$ are proper fractions.

An improper fraction is one in which the degree of the numerator is greater than or equal to the degree of the denominator.

Thus $\dfrac{2x^3 + 6x^2 - 9}{x^2 - 3x + 2}$ is an improper fraction.

By division, an improper fraction may always be written as the sum of a polynomial and a proper fraction.

Thus $\dfrac{2x^3 + 6x^2 - 9}{x^2 - 3x + 2} = 2x + 12 + \dfrac{32x - 33}{x^2 - 3x + 2}$.

PARTIAL FRACTIONS. A given proper fraction may often be written as the sum of other fractions (called partial fractions) whose denominators are of lower degree than the denominator of the given fraction.

For example, $\dfrac{3x - 5}{x^2 - 3x + 2} = \dfrac{3x - 5}{(x - 1)(x - 2)} = \dfrac{2}{x - 1} + \dfrac{1}{x - 2}$.

FUNDAMENTAL THEOREM. A proper fraction may be written as the sum of partial fractions according to the following rules.

1) LINEAR FACTORS NONE OF WHICH ARE REPEATED.

If a linear factor $ax + b$ occurs once as a factor of the denominator of the given fraction, then corresponding to this factor associate a partial fraction $\dfrac{A}{ax + b}$ where A is a constant $\neq 0$.

Example. $\dfrac{x + 4}{(x + 7)(2x - 1)} = \dfrac{A}{x + 7} + \dfrac{B}{2x - 1}$

275

2) LINEAR FACTORS SOME OF WHICH ARE REPEATED.

If a linear factor $ax + b$ occurs p times as a factor of the denominator of the given fraction, then corresponding to this factor associate the p partial fractions

$$\frac{A_1}{ax + b} + \frac{A_2}{(ax + b)^2} + \ldots + \frac{A_p}{(ax + b)^p}$$

where $A_1, A_2, \ldots, A_p$ are constants and $A_p \neq 0$.

Examples. $\quad \dfrac{3x - 1}{(x + 4)^2} = \dfrac{A}{x + 4} + \dfrac{B}{(x + 4)^2}$

$$\frac{5x^2 - 2}{x^3 (x + 1)^2} = \frac{A}{x^3} + \frac{B}{x^2} + \frac{C}{x} + \frac{D}{(x + 1)^2} + \frac{E}{x + 1}$$

3) QUADRATIC FACTORS NONE OF WHICH ARE REPEATED.

If a quadratic factor $ax^2 + bx + c$ occurs once as a factor of the denominator of the given fraction, then corresponding to this factor associate a partial fraction

$$\frac{Ax + B}{ax^2 + bx + c}$$

where A and B are constants which are not both zero.

Note. It is assumed that $ax^2 + bx + c$ cannot be factored into two real linear factors with integer coefficients.

Examples. $\quad \dfrac{x^2 - 3}{(x - 2)(x^2 + 4)} = \dfrac{A}{x - 2} + \dfrac{Bx + C}{x^2 + 4}$

$$\frac{2x^3 - 6}{x(2x^2 + 3x + 8)(x^2 + x + 1)} = \frac{A}{x} + \frac{Bx + C}{2x^2 + 3x + 8} + \frac{Dx + E}{x^2 + x + 1}$$

4) QUADRATIC FACTORS SOME OF WHICH ARE REPEATED.

If a quadratic factor $ax^2 + bx + c$ occurs p times as a factor of the denominator of the given fraction, then corresponding to this factor associate the p partial fractions

$$\frac{A_1 x + B_1}{ax^2 + bx + c} + \frac{A_2 x + B_2}{(ax^2 + bx + c)^2} + \ldots + \frac{A_p x + B_p}{(ax^2 + bx + c)^p}$$

where $A_1, B_1, A_2, B_2, \ldots, A_p, B_p$ are constants and A_p, B_p are not both zero.

Example. $\quad \dfrac{x^2 - 4x + 1}{(x^2 + 1)^2 (x^2 + x + 1)} = \dfrac{Ax + B}{x^2 + 1} + \dfrac{Cx + D}{(x^2 + 1)^2} + \dfrac{Ex + F}{x^2 + x + 1}$

SOLVED PROBLEMS

1. Resolve into partial fractions $\dfrac{x+2}{2x^2-7x-15}$ or $\dfrac{x+2}{(2x+3)(x-5)}$.

Let $\dfrac{x+2}{(2x+3)(x-5)} = \dfrac{A}{2x+3} + \dfrac{B}{x-5} = \dfrac{A(x-5)+B(2x+3)}{(2x+3)(x-5)} = \dfrac{(A+2B)x+3B-5A}{(2x+3)(x-5)}$.

We must find the constants A and B such that

$$\dfrac{x+2}{(2x+3)(x-5)} = \dfrac{(A+2B)x+3B-5A}{(2x+3)(x-5)} \quad \text{identically}$$

or $\qquad\qquad\qquad x + 2 = (A+2B)x + 3B - 5A.$

Equating coefficients of like powers of x, we have $1 = A+2B$ and $2 = 3B-5A$ which when solved simultaneously give $A = -1/13$, $B = 7/13$.

Hence $\dfrac{x+2}{2x^2-7x-15} = \dfrac{-1/13}{2x+3} + \dfrac{7/13}{x-5} = \dfrac{-1}{13(2x+3)} + \dfrac{7}{13(x-5)}$.

Another Method. $\quad x + 2 = A(x-5) + B(2x+3)$

To find B, let $x = 5$: $\qquad 5 + 2 = A(0) + B(10+3)$, $\qquad 7 = 13B$, $\quad B = 7/13$.

To find A, let $x = -3/2$: $\quad -3/2 + 2 = A(-3/2 - 5) + B(0)$, $\quad 1/2 = -13A/2$, $\quad A = -1/13$.

2. $\dfrac{2x^2+10x-3}{(x+1)(x^2-9)} = \dfrac{A}{x+1} + \dfrac{B}{x+3} + \dfrac{C}{x-3}$

$$2x^2 + 10x - 3 = A(x^2-9) + B(x+1)(x-3) + C(x+1)(x+3)$$

To find A, let $x = -1$: $\quad 2 - 10 - 3 = A(1-9)$, $\qquad A = 11/8$.

To find B, let $x = -3$: $\quad 18 - 30 - 3 = B(-3+1)(-3-3)$, $\quad B = -5/4$.

To find C, let $x = 3$: $\quad 18 + 30 - 3 = C(3+1)(3+3)$, $\qquad C = 15/8$.

Hence $\dfrac{2x^2+10x-3}{(x+1)(x^2-9)} = \dfrac{11}{8(x+1)} - \dfrac{5}{4(x+3)} + \dfrac{15}{8(x-3)}$.

3. $\dfrac{2x^2+7x+23}{(x-1)(x+3)^2} = \dfrac{A}{x-1} + \dfrac{B}{(x+3)^2} + \dfrac{C}{x+3}$

$$\begin{aligned} 2x^2 + 7x + 23 &= A(x+3)^2 + B(x-1) + C(x-1)(x+3) \\ &= A(x^2+6x+9) + B(x-1) + C(x^2+2x-3) \\ &= Ax^2 + 6Ax + 9A + Bx - B + Cx^2 + 2Cx - 3C \\ &= (A+C)x^2 + (6A+B+2C)x + 9A - B - 3C \end{aligned}$$

Equating coefficients of like powers of x, $\quad A+C = 2$, $\quad 6A+B+2C = 7$ and $9A-B-3C = 23$.

Solving simultaneously, $\quad A = 2$, $B = -5$, $C = 0$. Hence $\dfrac{2x^2+7x+23}{(x-1)(x+3)^2} = \dfrac{2}{x-1} - \dfrac{5}{(x+3)^2}$.

Another Method. $\qquad 2x^2 + 7x + 23 \;=\; A(x+3)^2 + B(x-1) + C(x-1)(x+3)$

To find A, let $x = 1$: $\quad 2 + 7 + 23 \;=\; A(1+3)^2,$ $\qquad\qquad A = 2.$

To find B, let $x = -3$: $\quad 18 - 21 + 23 \;=\; B(-3-1),$ $\qquad\quad B = -5.$

To find C, let $x = 0$: $\quad 23 \;=\; 2(3)^2 - 5(-1) + C(-1)(3),\quad C = 0.$

4. $\dfrac{x^2 - 6x + 2}{x^2(x-2)^2} \;=\; \dfrac{A}{x^2} + \dfrac{B}{x} + \dfrac{C}{(x-2)^2} + \dfrac{D}{x-2}$

$$
\begin{aligned}
x^2 - 6x + 2 \;&=\; A(x-2)^2 + Bx(x-2)^2 + Cx^2 + Dx^2(x-2)\\
&=\; A(x^2 - 4x + 4) + Bx(x^2 - 4x + 4) + Cx^2 + Dx^2(x-2)\\
&=\; (B+D)x^3 + (A - 4B + C - 2D)x^2 + (-4A + 4B)x + 4A
\end{aligned}
$$

Equating coefficients of like powers of x, $\;B + D = 0,\; A - 4B + C - 2D = 1,\; -4A + 4B = -6,\; 4A = 2.$
The simultaneous solution of these four equations is $\;A = 1/2,\; B = -1,\; C = -3/2,\; D = 1.$

Hence $\quad \dfrac{x^2 - 6x + 2}{x^2(x-2)^2} \;=\; \dfrac{1}{2x^2} - \dfrac{1}{x} - \dfrac{3}{2(x-2)^2} + \dfrac{1}{x-2}.$

Another Method. $\quad x^2 - 6x + 2 \;=\; A(x-2)^2 + Bx(x-2)^2 + Cx^2 + Dx^2(x-2)$

To find A, let $x = 0$: $2 = 4A$, $A = 1/2$. To find C, let $x = 2$: $4 - 12 + 2 = 4C$, $C = -3/2$.

To find B and D, let $x =$ any values except 0 and 2 (for example, let $x = 1$, $x = -1$).

Let $x \doteq 1$: $\;\; 1 - 6 + 2 = A(1-2)^2 + B(1-2)^2 + C + D(1-2) \qquad$ and $\quad$ 1) $B - D = -2.$

Let $x = -1$: $\;\; 1 + 6 + 2 = A(-1-2)^2 - B(-1-2)^2 + C + D(-1-2) \quad$ and $\quad$ 2) $9B + 3D = -6.$

The simultaneous solution of equations 1) and 2) is $\;B = -1,\; D = 1.$

5. $\dfrac{x^2 - 4x - 15}{(x+2)^3}.\quad$ Let $y = x + 2$; then $x = y - 2$.

$$
\begin{aligned}
\frac{x^2 - 4x - 15}{(x+2)^3} \;&=\; \frac{(y-2)^2 - 4(y-2) - 15}{y^3} \;=\; \frac{y^2 - 8y - 3}{y^3}\\[2mm]
&=\; \frac{1}{y} - \frac{8}{y^2} - \frac{3}{y^3} \;=\; \frac{1}{x+2} - \frac{8}{(x+2)^2} - \frac{3}{(x+2)^3}
\end{aligned}
$$

6. $\dfrac{7x^2 - 25x + 6}{(x^2 - 2x - 1)(3x - 2)} \;=\; \dfrac{Ax + B}{x^2 - 2x - 1} + \dfrac{C}{3x - 2}$

$$
\begin{aligned}
7x^2 - 25x + 6 \;&=\; (Ax + B)(3x - 2) + C(x^2 - 2x - 1)\\
&=\; (3Ax^2 + 3Bx - 2Ax - 2B) + Cx^2 - 2Cx - C\\
&=\; (3A + C)x^2 + (3B - 2A - 2C)x + (-2B - C)
\end{aligned}
$$

Equating coefficients of like powers of x, $\;3A + C = 7,\; 3B - 2A - 2C = -25,\; -2B - C = 6.$
The simultaneous solution of these three equations is $\;A = 1,\; B = -5,\; C = 4.$

Hence $\quad \dfrac{7x^2 - 25x + 6}{(x^2 - 2x - 1)(3x - 2)} \;=\; \dfrac{x - 5}{x^2 - 2x - 1} + \dfrac{4}{3x - 2}.$

7. $\dfrac{4x^2 - 28}{x^4 + x^2 - 6} = \dfrac{4x^2 - 28}{(x^2+3)(x^2-2)} = \dfrac{Ax + B}{x^2 + 3} + \dfrac{Cx + D}{x^2 - 2}$

$$4x^2 - 28 = (Ax + B)(x^2 - 2) + (Cx + D)(x^2 + 3)$$
$$= (Ax^3 + Bx^2 - 2Ax - 2B) + (Cx^3 + Dx^2 + 3Cx + 3D)$$
$$= (A + C)x^3 + (B + D)x^2 + (3C - 2A)x - 2B + 3D$$

Equating coefficients of like powers of x,
$$A + C = 0, \quad B + D = 4, \quad 3C - 2A = 0, \quad -2B + 3D = -28.$$

Solving simultaneously, $A = 0$, $B = 8$, $C = 0$, $D = -4$. Hence $\dfrac{4x^2 - 28}{x^4 + x^2 - 6} = \dfrac{8}{x^2 + 3} - \dfrac{4}{x^2 - 2}$.

SUPPLEMENTARY PROBLEMS

Resolve into partial fractions.

8. $\dfrac{x + 2}{x^2 - 7x + 12}$

9. $\dfrac{12x + 11}{x^2 + x - 6}$

10. $\dfrac{8 - x}{2x^2 + 3x - 2}$

11. $\dfrac{5x + 4}{x^2 + 2x}$

12. $\dfrac{x}{x^2 - 3x - 18}$

13. $\dfrac{10x^2 + 9x - 7}{(x + 2)(x^2 - 1)}$

14. $\dfrac{x^2 - 9x - 6}{x^3 + x^2 - 6x}$

15. $\dfrac{x^3}{x^2 - 4}$

16. $\dfrac{3x^2 - 8x + 9}{(x - 2)^3}$

17. $\dfrac{3x^3 + 10x^2 + 27x + 27}{x^2(x + 3)^2}$

18. $\dfrac{5x^2 + 8x + 21}{(x^2 + x + 6)(x + 1)}$

19. $\dfrac{5x^3 + 4x^2 + 7x + 3}{(x^2 + 2x + 2)(x^2 - x - 1)}$

20. $\dfrac{3x}{x^3 - 1}$

21. $\dfrac{7x^3 + 16x^2 + 20x + 5}{(x^2 + 2x + 2)^2}$

ANSWERS TO SUPPLEMENTARY PROBLEMS.

8. $\dfrac{6}{x - 4} - \dfrac{5}{x - 3}$

9. $\dfrac{7}{x - 2} + \dfrac{5}{x + 3}$

10. $\dfrac{3}{2x - 1} - \dfrac{2}{x + 2}$

11. $\dfrac{2}{x} + \dfrac{3}{x + 2}$

12. $\dfrac{2/3}{x - 6} + \dfrac{1/3}{x + 3}$

13. $\dfrac{3}{x + 1} + \dfrac{2}{x - 1} + \dfrac{5}{x + 2}$

14. $\dfrac{1}{x} - \dfrac{2}{x - 2} + \dfrac{2}{x + 3}$

15. $x + \dfrac{2}{x - 2} + \dfrac{2}{x + 2}$

16. $\dfrac{3}{x - 2} + \dfrac{4}{(x - 2)^2} + \dfrac{5}{(x - 2)^3}$

17. $\dfrac{1}{x} + \dfrac{3}{x^2} + \dfrac{2}{x + 3} - \dfrac{5}{(x + 3)^2}$

18. $\dfrac{2x + 3}{x^2 + x + 6} + \dfrac{3}{x + 1}$

19. $\dfrac{2x - 1}{x^2 + 2x + 2} + \dfrac{3x + 1}{x^2 - x - 1}$

20. $\dfrac{1}{x - 1} + \dfrac{-x + 1}{x^2 + x + 1}$

21. $\dfrac{7x + 2}{x^2 + 2x + 2} + \dfrac{2x + 1}{(x^2 + 2x + 2)^2}$

CHAPTER 29

Infinite Series

SEQUENCES. A set of numbers $u_1, u_2, u_3, \ldots$ in a definite order of arrangement and formed according to a definite rule is called a *sequence*. Each number in the sequence is called a *term* of the sequence. If the number of terms is finite it is called a *finite sequence*, otherwise an *infinite sequence*.

For example, the set of numbers $2, 5, 8, \ldots, 20$ is a finite sequence while the set $1, 1/3, 1/5, 1/7, \ldots$ is an infinite sequence.

We shall be concerned here with infinite sequences unless otherwise stated.

THE nth TERM OF A SEQUENCE is a law of formation by which any term in the sequence may be obtained. This law is often given by means of a formula dependent on n such that the substitution of $n = 1, 2, 3, \ldots$ in the formula yields the first, second, third, $\ldots$ terms. The nth term is the *general term* of the sequence.

Thus if the nth term is given by $u_n = \dfrac{1}{n^2 + 1}$, the 1st, 2nd, and 5th terms are respectively $u_1 = \dfrac{1}{1^2 + 1} = \dfrac{1}{2}$, $u_2 = \dfrac{1}{2^2 + 1} = \dfrac{1}{5}$, $u_5 = \dfrac{1}{5^2 + 1} = \dfrac{1}{26}$.

A SERIES is an indicated sum $u_1 + u_2 + u_3 + \ldots$ of the numbers in a sequence. The numbers $u_1, u_2, u_3, \ldots$ are called the first, second, third, $\ldots$ terms of the series; u_n is the nth term of the series. If the series has a finite number of terms it is called a *finite series*, otherwise an *infinite series*.

For example, if the nth term of a series is $u_n = \dfrac{1}{n!}$, then the series is

$$\frac{1}{1!} + \frac{1}{2!} + \frac{1}{3!} + \frac{1}{4!} + \frac{1}{5!} + \ldots \, .$$

The notation $\displaystyle\sum_{n=1}^{p} u_n$ is used to represent $u_1 + u_2 + u_3 + \ldots + u_p$.

The notation $\displaystyle\sum_{n=1}^{\infty} u_n$ is used to represent $u_1 + u_2 + u_3 + \ldots$.

For example, $\displaystyle\sum_{n=1}^{\infty} \frac{1}{n^2 + 1} = \frac{1}{1^2 + 1} + \frac{1}{2^2 + 1} + \frac{1}{3^2 + 1} + \ldots$.

LIMIT OF A SEQUENCE. A number L is said to be the limit of an infinite sequence $u_1, u_2, u_3, \ldots$ if for every preassigned positive number ϵ (no matter how small) one can find a number N such that $|u_n - L| < \epsilon$ for all integers $n > N$.

For example, for the sequence with nth term $u_n = 2 + \dfrac{1}{n} = \dfrac{2n+1}{n}$, i.e. for the sequence $3, 5/2, 7/3, 9/4, \ldots,$ one may show that the limit L is 2.

If the limit L of a sequence $u_1, u_2, u_3, \ldots$ exists, we write $\lim\limits_{n \to \infty} u_n = L$ and read "the limit of u_n as n approaches infinity is equal to L". It is understood that "n approaches infinity" is synonomous with "n gets larger and larger" or "n increases without bound".

Thus for the sequence with nth term given by $u_n = \dfrac{2n+1}{n}$ we may write $\lim\limits_{n \to \infty} \dfrac{2n+1}{n} = 2$.

If there exists no number L such that $|u_n - L| < \epsilon$ for all $n > N$, the sequence is said to have no limit. If the successive terms of the sequence increase without bound, then $\lim\limits_{n \to \infty} u_n$ does not exist and we use the notation $\lim\limits_{n \to \infty} u_n = \infty$ to indicate this.

For example, if $u_n = 2n$ then $\lim\limits_{n \to \infty} 2n = \infty$, i.e. in the sequence $2, 4, 6, 8, 10, \ldots$ the terms increase without bound.

A more intuitive but less rigorous way of expressing the concept of limit than given above is to say that the sequence $u_1, u_2, u_3, \ldots$ has a limit L if the successive terms get "closer and closer" in value to L. This intuitive way is often used to provide a "guess" as to the value of L. However, in order to establish the correctness of the "guess" recourse must be made to the definition above, although often the proof is difficult.

For example, in the sequence with nth term $u_n = \dfrac{2n+1}{n+1}$ the successive terms $(\frac{3}{2}, \frac{5}{3}, \frac{7}{4}, \frac{9}{5}, \ldots)$ get "closer and closer" to 2 and the limit is thus suspected to be 2. This suspicion is confirmed in Problem 7.

THEOREMS ON LIMITS. If $\lim\limits_{n \to \infty} a_n$ and $\lim\limits_{n \to \infty} b_n$ exist, then:

1. $\lim\limits_{n \to \infty} (a_n \pm b_n) = \lim\limits_{n \to \infty} a_n \pm \lim\limits_{n \to \infty} b_n$

2. $\lim\limits_{n \to \infty} (a_n \cdot b_n) = \lim\limits_{n \to \infty} a_n \cdot \lim\limits_{n \to \infty} b_n$

3. $\lim\limits_{n \to \infty} \dfrac{a_n}{b_n} = \dfrac{\lim\limits_{n \to \infty} a_n}{\lim\limits_{n \to \infty} b_n}$ provided $\lim\limits_{n \to \infty} b_n \neq 0$

   If $\lim\limits_{n \to \infty} b_n = 0$ and $\lim\limits_{n \to \infty} a_n \neq 0$, $\lim\limits_{n \to \infty} \dfrac{a_n}{b_n}$ does not exist.

   If $\lim\limits_{n \to \infty} b_n = 0$ and $\lim\limits_{n \to \infty} a_n = 0$, $\lim\limits_{n \to \infty} \dfrac{a_n}{b_n}$ may or may not exist.

4. $\lim\limits_{n\to\infty} (a_n)^p = (\lim\limits_{n\to\infty} a_n)^p$, p = any real number

BOUNDED, MONOTONE SEQUENCES. A sequence is called *bounded* if there is a positive number M, not depending on n, such that $|u_n| \leq M$ for $n = 1, 2, 3, \dots$.

For example, $3, 5/2, 7/3, 9/4, \dots$ is a bounded sequence since the absolute value of each term never exceeds 3. The sequence $3/4, -4/5, 5/6, -6/7, 7/8, \dots$ is bounded since the absolute value of each term never exceeds 1.

The sequence $2, 4, 6, 8, \dots$ is unbounded.

A sequence is called *monotone increasing* if $u_{n+1} \geq u_n$ and *monotone decreasing* if $u_{n+1} \leq u_n$, where $n = 1, 2, 3, \dots$. Thus the sequence $1/2, 3/4, 4/5, 5/6, \dots$ is monotone increasing, and the sequence $3, 2, 1, 0, -1, -2, -3, \dots$ is monotone decreasing.

A bounded, monotone increasing or monotone decreasing sequence has a limit. For example, the sequences

$$1/2, 3/4, 4/5, 5/6, \dots \quad \text{and} \quad 1/4, 1/8, 1/16, 1/32, \dots$$

are bounded and respectively monotone increasing and monotone decreasing. Hence the sequences have limits.

However, a sequence need not be monotone increasing or decreasing in order to have a limit. For example, the sequence $2/3, 5/4, 3/4, 6/5, 4/5, 7/6, 5/6, \dots$ is bounded but not monotone increasing or decreasing. Nevertheless a limit exists, namely 1.

CONVERGENCE AND DIVERGENCE OF INFINITE SERIES.

Let $S_n = u_1 + u_2 + u_3 + \dots + u_n$ be the sum of the first n terms of the series $u_1 + u_2 + u_3 + \dots$. The terms of the sequence $S_1, S_2, S_3, \dots$ are called the *partial sums* of the series. If $\lim\limits_{n\to\infty} S_n = S$, a finite number, the series $u_1 + u_2 + u_3 + \dots$ is *convergent*, and S is called the *sum* of the infinite series. A series which is not convergent is called *divergent*.

For example, the sum of the first n terms of the series $\dfrac{1}{2} + \dfrac{1}{2^2} + \dfrac{1}{2^3} + \dfrac{1}{2^4}$ $+ \dots$ is the same as the sum of the first n terms of a geometric progression with first term $\dfrac{1}{2}$ and ratio $\dfrac{1}{2}$; this sum is $S_n = 1 - \dfrac{1}{2^n}$. Since $\lim\limits_{n\to\infty} 1 - \dfrac{1}{2^n}$ $= 1$, the series is convergent and has sum $S = 1$.

In the series $1 - 1 + 1 - 1 + \dots$ the sum of the first n terms is 0 or 1, depending on whether the number of terms taken is even or odd. Hence $\lim\limits_{n\to\infty} S_n$ does not exist and the series is divergent.

If a series is convergent, its nth term *must* have limit zero as $n \to \infty$. If, however, the nth term does have limit zero as $n \to \infty$, the series may or may not converge. A series is divergent if its nth term *does not* have limit zero as $n \to \infty$.

COMPARISON TEST FOR CONVERGENCE of series of positive terms.

If, from some term on, each term of a given series of positive terms is less than or equal to the corresponding term of a known convergent series, the given series is convergent.

If, from some term on, each term of a given series of positive terms is greater than or equal to the corresponding term of a known divergent series, the given series is divergent.

The following two series are useful in the comparison test.

1) The Geometric Series, $a + ar + ar^2 + \ldots + ar^{n-1} + \ldots$ where a and r are given constants, is convergent if $|r| < 1$ and divergent if $|r| \geq 1$.

2) The p-Series, $\dfrac{1}{1^p} + \dfrac{1}{2^p} + \dfrac{1}{3^p} + \ldots + \dfrac{1}{n^p} + \ldots$ where p is a given constant, is convergent if $p > 1$ and divergent if $p \leq 1$. If $p = 1$ we have the divergent *harmonic* series $1 + \dfrac{1}{2} + \dfrac{1}{3} + \dfrac{1}{4} + \ldots$.

RATIO TEST FOR CONVERGENCE of series with like or mixed signs.

For the series $u_1 + u_2 + u_3 + \ldots$ with like or mixed signs, let

$$\lim_{n \to \infty} \left| \frac{u_{n+1}}{u_n} \right| = R.$$

Then the series

a) converges if $R < 1$, b) diverges if $R > 1$. c) If $R = 1$ the test fails.

AN ALTERNATING SERIES is a series whose terms are alternately positive and negative. For example, $1 - \dfrac{1}{2} + \dfrac{1}{3} - \dfrac{1}{4} + \ldots$.

An alternating series is convergent if

a) after a certain number of terms, the absolute value of each term is less than that of the preceding term, i.e. if $|u_{n+1}| < |u_n|$, and if

b) the nth term has limit zero as n increases without limit, i.e. if $\lim\limits_{n \to \infty} u_n = 0$.

ABSOLUTE AND CONDITIONAL CONVERGENCE.

A series is said to be *absolutely convergent* if the series formed by making all the signs positive converges. A convergent series which is not absolutely convergent is *conditionally convergent*. For example:

1) $1 - \dfrac{1}{2} + \dfrac{1}{2^2} - \dfrac{1}{2^3} + \ldots$ is absolutely convergent since $1 + \dfrac{1}{2} + \dfrac{1}{2^2} + \dfrac{1}{2^3} + \ldots$ converges.

2) $1 - \dfrac{1}{2} + \dfrac{1}{3} - \dfrac{1}{4} + \ldots$ converges but is not absolutely convergent since $1 + \dfrac{1}{2} + \dfrac{1}{3} + \dfrac{1}{4} + \ldots$ diverges. Hence the alternating series is conditionally convergent.

The terms of an absolutely convergent series may be arranged in any order and not affect the convergence. However, if the terms of a conditionally convergent series are rearranged the resulting series may diverge or converge to any desired sum.

POWER SERIES. A series of the form $c_0 + c_1 x + c_2 x^2 + c_3 x^3 + \ldots + c_n x^n + \ldots$, where the coefficients $c_0, c_1, c_2, \ldots$ are constants, is called a power series in x.

Similarly, $c_0 + c_1(x-a) + c_2(x-a)^2 + \ldots + c_n(x-a)^n + \ldots$, where a is a constant, is a power series in $(x-a)$.

Thus $1 + x + \dfrac{x^2}{2} + \dfrac{x^3}{3} + \ldots$ is a power series in x.

The set of values of x for which a power series converges is called its *interval of convergence*. This interval may be found by using the ratio test supplemented by other tests applied at the endpoints of the interval.

SOLVED PROBLEMS

THE GENERAL TERM.

1. Write the first four terms of the sequence having the indicated general term.

a) $\dfrac{n}{2n+1}$. The terms are $\dfrac{1}{2(1)+1}, \dfrac{2}{2(2)+1}, \dfrac{3}{2(3)+1}, \dfrac{4}{2(4)+1}$ or $\dfrac{1}{3}, \dfrac{2}{5}, \dfrac{3}{7}, \dfrac{4}{9}$.

b) $\dfrac{2n-1}{(n+1)^2}$. The terms are $\dfrac{2(1)-1}{(1+1)^2}, \dfrac{2(2)-1}{(2+1)^2}, \dfrac{2(3)-1}{(3+1)^2}, \dfrac{2(4)-1}{(4+1)^2}$ or $\dfrac{1}{4}, \dfrac{3}{9}, \dfrac{5}{16}, \dfrac{7}{25}$.

c) $\dfrac{2^{n-1}}{n^2+1}$. The terms are $\dfrac{2^0}{1^2+1}, \dfrac{2^1}{2^2+1}, \dfrac{2^2}{3^2+1}, \dfrac{2^3}{4^2+1}$ or $\dfrac{1}{2}, \dfrac{2}{5}, \dfrac{4}{10}, \dfrac{8}{17}$.

d) $\dfrac{(-1)^n}{n(n+1)}$. The terms are $\dfrac{(-1)^1}{1(1+1)}, \dfrac{(-1)^2}{2(2+1)}, \dfrac{(-1)^3}{3(3+1)}, \dfrac{(-1)^4}{4(4+1)}$ or $-\dfrac{1}{2}, \dfrac{1}{6}, -\dfrac{1}{12}, \dfrac{1}{20}$.

e) $\dfrac{x^{n+1}}{(n+1)!}$. The terms are $\dfrac{x^2}{2!}, \dfrac{x^3}{3!}, \dfrac{x^4}{4!}, \dfrac{x^5}{5!}$ or $\dfrac{x^2}{2}, \dfrac{x^3}{6}, \dfrac{x^4}{24}, \dfrac{x^5}{120}$.

2. Write the first four terms and the $(n+1)$st term of the infinite series whose nth term is given.

a) $\dfrac{n}{3^{n-1}}$. $1 + \dfrac{2}{3} + \dfrac{1}{3} + \dfrac{4}{27}$ $(n+1)$st term: $\dfrac{n+1}{3^n}$

b) $\dfrac{2n+1}{4n-2}$. $\dfrac{3}{2} + \dfrac{5}{6} + \dfrac{7}{10} + \dfrac{9}{14}$ $(n+1)$st term: $\dfrac{2(n+1)+1}{4(n+1)-2} = \dfrac{2n+3}{4n+2}$

c) $\dfrac{(-1)^{n-1}\sqrt{n}}{n+1}$ $\dfrac{1}{2} - \dfrac{\sqrt{2}}{3} + \dfrac{\sqrt{3}}{4} - \dfrac{\sqrt{4}}{5}$ $(n+1)$st term: $\dfrac{(-1)^n \sqrt{n+1}}{n+2}$

3. Write an nth term for each of the following sequences.

a) $\dfrac{1}{2}, \dfrac{2}{3}, \dfrac{3}{4}, \dfrac{4}{5}, \ldots \qquad \dfrac{n}{n+1}$

d) $\dfrac{4}{3\cdot5}, \dfrac{5}{4\cdot6}, \dfrac{6}{5\cdot7}, \dfrac{7}{6\cdot8}, \ldots \qquad \dfrac{n+3}{(n+2)(n+4)}$

b) $\dfrac{1}{2}, \dfrac{3}{4}, \dfrac{5}{6}, \dfrac{7}{8}, \ldots \qquad \dfrac{2n-1}{2n}$

e) $\dfrac{2}{1+3}, \dfrac{4}{2+3}, \dfrac{6}{3+3}, \dfrac{8}{4+3}, \ldots \qquad \dfrac{2n}{n+3}$

c) $\dfrac{2}{3}, \dfrac{4}{5}, \dfrac{8}{7}, \dfrac{16}{9}, \ldots \qquad \dfrac{2^n}{2n+1}$

4. Find an nth term for each series and also write the $(n+1)$st term.

a) $\dfrac{1}{3} + \dfrac{1}{5} + \dfrac{1}{7} + \dfrac{1}{9} + \cdots \qquad \dfrac{1}{2n+1}$; $\dfrac{1}{2(n+1)+1} = \dfrac{1}{2n+3}$

b) $\dfrac{1}{3} + \dfrac{1}{6} + \dfrac{1}{9} + \dfrac{1}{12} + \cdots \qquad \dfrac{1}{3n}$; $\dfrac{1}{3(n+1)} = \dfrac{1}{3n+3}$

c) $\dfrac{1}{1!} - \dfrac{1}{2!} + \dfrac{1}{3!} - \dfrac{1}{4!} + \cdots \qquad \dfrac{(-1)^{n-1}}{n!}$; $\dfrac{(-1)^n}{(n+1)!}$

d) $1 - \sqrt[3]{4} + \sqrt[3]{9} - \sqrt[3]{16} + \cdots \qquad (-1)^{n-1}\sqrt[3]{n^2}$; $(-1)^n\sqrt[3]{(n+1)^2}$

e) $\dfrac{3\cdot4\cdot5}{1!} + \dfrac{4\cdot5\cdot6}{3!} + \dfrac{5\cdot6\cdot7}{5!} + \dfrac{6\cdot7\cdot8}{7!} + \cdots \qquad \dfrac{(n+2)(n+3)(n+4)}{(2n-1)!}$; $\dfrac{(n+3)(n+4)(n+5)}{(2n+1)!}$

f) $\dfrac{x}{1\cdot3} + \dfrac{x^3}{3\cdot5} + \dfrac{x^5}{5\cdot7} + \dfrac{x^7}{7\cdot9} + \cdots \qquad \dfrac{x^{2n-1}}{(2n-1)(2n+1)}$; $\dfrac{x^{2n+1}}{(2n+1)(2n+3)}$

g) $1 - \dfrac{x}{2\cdot2^2} + \dfrac{x^2}{2^2\cdot3^2} - \dfrac{x^3}{2^3\cdot4^2} + \cdots \qquad \dfrac{(-1)^{n-1}x^{n-1}}{2^{n-1}n^2}$; $\dfrac{(-1)^n x^n}{2^n(n+1)^2}$

5. Two students were asked to write an nth term for the sequence $1, 16, 81, 256, \ldots$ and to write the 5th term of the sequence. One student gave the nth term as $u_n = n^4$. The other student, who did not recognize this simple law of formation, wrote $u_n = 10n^3 - 35n^2 + 50n - 24$. Which student gave the correct 5th term?

If $u_n = n^4$, then $u_1 = 1$, $u_2 = 2^4 = 16$, $u_3 = 3^4 = 81$ and $u_4 = 4^4 = 256$, which agrees with the sequence. Hence the first student gave the 5th term as $u_5 = 5^4 = 625$.

If $u_n = 10n^3 - 35n^2 + 50n - 24$, then $u_1 = 1$, $u_2 = 16$, $u_3 = 81$ and $u_4 = 256$, which also agrees with the sequence. Hence the second student gave the 5th term as $u_5 = 601$.

Both students were correct. Merely giving a finite number of terms of a sequence or series does not define a unique nth term. In fact, an infinite number of nth terms is possible.

6. Write the series represented by each of the following.

a) $\displaystyle\sum_{n=1}^{4} \dfrac{1}{n^4} = \dfrac{1}{1^4} + \dfrac{1}{2^4} + \dfrac{1}{3^4} + \dfrac{1}{4^4}$

b) $\displaystyle\sum_{n=1}^{6} \frac{\sqrt{n}}{n+2} = \frac{\sqrt{1}}{3} + \frac{\sqrt{2}}{4} + \frac{\sqrt{3}}{5} + \frac{\sqrt{4}}{6} + \frac{\sqrt{5}}{7} + \frac{\sqrt{6}}{8}$.

c) $\displaystyle\sum_{n=1}^{\infty} \frac{n^3}{n^4+1} = \frac{1^3}{1^4+1} + \frac{2^3}{2^4+1} + \frac{3^3}{3^4+1} + \cdots$

d) $\displaystyle\sum_{n=4}^{7} \frac{1}{\sqrt{n}} = \frac{1}{\sqrt{4}} + \frac{1}{\sqrt{5}} + \frac{1}{\sqrt{6}} + \frac{1}{\sqrt{7}}$

e) $\displaystyle\sum_{n=2}^{\infty} \frac{x^n}{n^2} = \frac{x^2}{2^2} + \frac{x^3}{3^2} + \frac{x^4}{4^2} + \frac{x^5}{5^2} + \cdots$

f) $\displaystyle\sum_{n=0}^{\infty} \frac{(-1)^n (x+2)^{n+1}}{n+1} = (x+2) - \frac{(x+2)^2}{2} + \frac{(x+2)^3}{3} - \frac{(x+2)^4}{4} + \cdots$

LIMIT OF A SEQUENCE.

7. A sequence has its *n*th term given by $u_n = \frac{2n+1}{n+1}$.

a) Write the 1st, 2nd, 5th, 10th, 100th, 1000th and 10,000th terms of the sequence in decimal form. Make a *guess* as to the limit of this sequence as $n \to \infty$.

b) Using the definition of limit, verify that the *guess* in *a)* is actually correct.

c) Evaluate the limit by using the theorems on limits.

a)

$n=1$	$n=2$	$n=5$	$n=10$	$n=100$	$n=1000$	$n=10,000$
1.50000	1.66667	1.83333	1.90909	1.99010	1.99900	1.99990

A good *guess* is that the limit is 2.

b) We must show that for a given positive number ϵ (no matter how small) there will be a value of *n*, say *N* (depending on ϵ), such that $\left| u_n - 2 \right| < \epsilon$ for all $n > N$.

Now $\left| \frac{2n+1}{n+1} - 2 \right| < \epsilon$ if $\left| -\frac{1}{n+1} \right| < \epsilon$, i.e. if $\frac{1}{n+1} < \epsilon$, $n+1 > \frac{1}{\epsilon}$ or $n > \frac{1}{\epsilon} - 1$.

For example, if $\epsilon = .01$, $n > 99$. This means that all the terms of the sequence beyond the 99th differ from 2 by less than .01.

If $\epsilon = .0001$, $n > 9999$. Thus all the terms of the sequence beyond the 9999th term differ from 2 by less than .0001.

c) Write $u_n = \frac{2n+1}{n+1} = \frac{2 + \dfrac{1}{n}}{1 + \dfrac{1}{n}}$, dividing numerator and denominator by *n*.

Then $\displaystyle\lim_{n\to\infty} u_n = \lim_{n\to\infty} \frac{2 + \dfrac{1}{n}}{1 + \dfrac{1}{n}} = \frac{\displaystyle\lim_{n\to\infty}\left(2 + \dfrac{1}{n}\right)}{\displaystyle\lim_{n\to\infty}\left(1 + \dfrac{1}{n}\right)} = \frac{2 + \displaystyle\lim_{n\to\infty}\dfrac{1}{n}}{1 + \displaystyle\lim_{n\to\infty}\dfrac{1}{n}} = \frac{2+0}{1+0} = 2$, since $\displaystyle\lim_{n\to\infty}\frac{1}{n} = 0$.

8. Evaluate each of the following.

a) $\lim\limits_{n\to\infty} \dfrac{2n^2+n}{5n^2-1} = \lim\limits_{n\to\infty} \dfrac{2+\dfrac{1}{n}}{5-\dfrac{1}{n^2}} = \dfrac{\lim\limits_{n\to\infty}(2+\dfrac{1}{n})}{\lim\limits_{n\to\infty}(5-\dfrac{1}{n^2})} = \dfrac{\lim\limits_{n\to\infty}2+\lim\limits_{n\to\infty}\dfrac{1}{n}}{\lim\limits_{n\to\infty}5-\lim\dfrac{1}{n^2}} = \dfrac{2+0}{5-0} = \dfrac{2}{5}$

b) $\lim\limits_{n\to\infty} \dfrac{3n^2+4n+5}{7n^2-4} = \lim\limits_{n\to\infty} \dfrac{3+\dfrac{4}{n}+\dfrac{5}{n^2}}{7-\dfrac{4}{n^2}} = \dfrac{3+0+0}{7-0} = \dfrac{3}{7}$

c) $\lim\limits_{n\to\infty} \dfrac{2n^2+3}{4n^3-1} = \lim\limits_{n\to\infty} \dfrac{\dfrac{2}{n}+\dfrac{3}{n^3}}{4-\dfrac{1}{n^3}} = \dfrac{0+0}{4-0} = 0$

d) $\lim\limits_{n\to\infty} \dfrac{n^2+2}{3n+2} = \lim\limits_{n\to\infty} \dfrac{1+\dfrac{2}{n^2}}{\dfrac{3}{n}+\dfrac{2}{n^2}} = \dfrac{1}{0} = \infty$, i.e. a limit does not exist

e) $\lim\limits_{n\to\infty} \dfrac{(n-2)!}{n!} = \lim\limits_{n\to\infty} \dfrac{(n-2)!}{n(n-1)(n-2)!} = \lim\limits_{n\to\infty} \dfrac{1}{n(n-1)} = 0$

f) $\lim\limits_{n\to\infty} (\dfrac{4n-2}{2n+3})^4 = (\lim\limits_{n\to\infty} \dfrac{4n-2}{2n+3})^4 = (\lim\limits_{n\to\infty} \dfrac{4-2/n}{2+3/n})^4 = (\dfrac{4}{2})^4 = 16$

g) $\lim\limits_{n\to\infty} (\sqrt{n+1}-\sqrt{n}) = \lim\limits_{n\to\infty} (\sqrt{n+1}-\sqrt{n})\dfrac{(\sqrt{n+1}+\sqrt{n})}{(\sqrt{n+1}+\sqrt{n})} = \lim\limits_{n\to\infty} \dfrac{1}{\sqrt{n+1}+\sqrt{n}} = 0$

BOUNDED, MONOTONE SEQUENCES.

9. Examine the following table.

Sequence	Bounded	Monotone Increasing	Monotone Decreasing	Limit Exists
2, 2.5, 3, 3.5, 4, 4.5, ...	No	Yes	No	No
1, −1, 1, −1, 1, −1, ...	Yes	No	No	No
1, 1.1, 1.11, 1.111, 1.1111, ...	Yes	Yes	No	Yes
1/10, 1/11, 1/12, 1/13, 1/14, ...	Yes	No	Yes	Yes
1, 3/4, 1, 4/5, 1, 5/6, 1, 6/7, ...	Yes	No	No	Yes

10. Prove that the sequence given by $u_n = \dfrac{3n+1}{n+2}$ is a) bounded and b) monotone increasing and therefore has a limit.

a) The sequence is bounded by 3 (or any number larger than 3) since $\dfrac{3n+1}{n+2} \leqq 3$ if and only if $3n+1 \leqq 3n+6$ or $1 \leqq 6$, which is true for all values of n.

b) The sequence is monotone increasing if $u_{n+1} \geqq u_n$. Now $\dfrac{3n+4}{n+3} \geqq \dfrac{3n+1}{n+2}$ if and only if

$(3n+4)(n+2) \geqq (3n+1)(n+3)$, $3n^2+10n+8 \geqq 3n^2+10n+3$ or $8 \geqq 3$, which is true for all values of n.

CONVERGENCE AND DIVERGENCE OF INFINITE SERIES.

11. A series has its nth term given by $u_n = \dfrac{1}{3^{n-1}}$.

   a) Write the first four terms of the series.

   b) If S_n denotes the sum of the first n terms of the series, find S_1, S_2, S_3, ..., S_8 in decimal form. Make a *guess* as to the limit of the sequence S_n as $n \to \infty$.

   c) By obtaining the sum of the first n terms of the series, verify that the guess in b) is correct.

   a) $\dfrac{1}{3^0} + \dfrac{1}{3^1} + \dfrac{1}{3^2} + \dfrac{1}{3^3} + \cdots$ or $1 + \dfrac{1}{3} + \dfrac{1}{9} + \dfrac{1}{27} + \cdots$

   b) $u_1 = 1.000000$ $S_1 = u_1 = 1.000000$
   $u_2 = .333333$ $S_2 = u_1 + u_2 = 1.333333$
   $u_3 = .111111$ $S_3 = u_1 + u_2 + u_3 = 1.444444$
   $u_4 = .037037$ $S_4 = u_1 + u_2 + u_3 + u_4 = 1.481481$
   $u_5 = .012346$ $S_5 = 1.493827$
   $u_6 = .004115$ $S_6 = 1.497942$
   $u_7 = .001372$ $S_7 = 1.499314$
   $u_8 = .000457$ $S_8 = 1.499771$

   A good *guess* is that $\lim\limits_{n \to \infty} S_n = 1.500000 = 3/2$.

   c) The series $1 + \dfrac{1}{3} + \dfrac{1}{9} + \dfrac{1}{27} + \cdots$ is a geometric series which has first term $a = 1$ and

   ratio $r = 1/3$. The sum of the first n terms is

$$S_n = \frac{a(1-r^n)}{1-r} = \frac{1-(1/3)^n}{1-1/3} = \frac{3}{2}\left[1 - \left(\frac{1}{3}\right)^n\right] \quad \text{and} \quad \lim_{n \to \infty} S_n = \frac{3}{2}.$$

   Hence the *guess* in b) is correct.

   Thus the series $1 + \dfrac{1}{3} + \dfrac{1}{9} + \dfrac{1}{27} + \cdots$ is convergent and has sum $= \dfrac{3}{2}$.

12. Given the series $\dfrac{1}{3} - \dfrac{1}{3} + \dfrac{1}{3} - \dfrac{1}{3} + \cdots$

   a) Compute S_n, the sum of the first n terms of the series.

   b) What is $\lim\limits_{n \to \infty} S_n$?

   c) Is the series convergent?

   a) $S_1 = \dfrac{1}{3}$, $S_2 = \dfrac{1}{3} - \dfrac{1}{3} = 0$, $S_3 = \dfrac{1}{3} - \dfrac{1}{3} + \dfrac{1}{3} = \dfrac{1}{3}$, $S_4 = \dfrac{1}{3} - \dfrac{1}{3} + \dfrac{1}{3} - \dfrac{1}{3} = 0$, etc.

   Hence $S_n = \dfrac{1}{3}$ if n is odd and $S_n = 0$ if n is even.

   b) The sequence S_1, S_2, S_3, ...   or $\dfrac{1}{3}$, 0, $\dfrac{1}{3}$, 0, $\dfrac{1}{3}$, $\cdots$ does not have a limit.

   c) The series is divergent since $\lim\limits_{n \to \infty} S_n$ does not exist.

   This is also evident from the fact that the nth term does not approach zero as $n \to \infty$.

13. Establish the convergence or divergence of each of the following series.

a) $1 + \dfrac{2}{3} + \dfrac{3}{5} + \dfrac{4}{7} + \cdots + \dfrac{n}{2n-1} + \cdots$ The nth term has limit given by $\displaystyle\lim_{n\to\infty} \dfrac{n}{2n-1} = \dfrac{1}{2}$.

Since the nth term does not approach zero as $n \to \infty$, the series diverges.

b) $1 + \dfrac{1}{2} + \dfrac{1}{3} + \dfrac{1}{4} + \cdots + \dfrac{1}{n} + \cdots$

The nth term has limit given by $\displaystyle\lim_{n\to\infty} \dfrac{1}{n} = 0$; however, this does not prove convergence. Actually, the series diverges as shown in Problem 14d).

c) $\dfrac{1}{2} - \dfrac{2}{3} + \dfrac{3}{4} - \dfrac{4}{5} + \cdots \dfrac{(-1)^{n-1}n}{n+1} + \cdots$

The nth term does not approach zero as $n \to \infty$; hence the series diverges.

COMPARISON TEST.

14. Examine the following series for convergence.

a) $1 + \dfrac{1}{4} + \dfrac{1}{16} + \dfrac{1}{64} + \cdots$ c) $1 + \dfrac{1}{2^3} + \dfrac{1}{3^3} + \dfrac{1}{4^3} + \cdots$ e) $1 + \dfrac{1}{2^{\frac{1}{2}}} + \dfrac{1}{3^{\frac{1}{2}}} + \dfrac{1}{4^{\frac{1}{2}}} + \cdots$

b) $1 + \dfrac{3}{2} + \dfrac{9}{4} + \dfrac{27}{8} + \cdots$ d) $1 + \dfrac{1}{2} + \dfrac{1}{3} + \dfrac{1}{4} + \cdots$ f) $5 + \dfrac{5}{2^2} + \dfrac{5}{3^2} + \dfrac{5}{4^2} + \cdots$

a) This is an infinite geometric series with $r = 1/4 < 1$ and is convergent.
b) This is an infinite geometric series with $r = 3/2 > 1$ and is divergent.
c) This is a p-series with $p = 3 > 1$ and is convergent.
d) This is a p-series with $p = 1$ (or a harmonic series) and is divergent.
e) This is a p-series with $p = 1/2 < 1$ and is divergent.
f) This is term by term equal to five times the convergent p-series with $p = 2 > 1$ and hence is convergent.

15. Investigate the convergence of the following infinite series, using the comparison test.

a) $\dfrac{1}{1^2 + 1} + \dfrac{1}{2^2 + 1} + \dfrac{1}{3^2 + 1} + \dfrac{1}{4^2 + 1} + \cdots$ The general term is $u_n = \dfrac{1}{n^2 + 1} < \dfrac{1}{n^2}$.

The given series is convergent since it is term by term less than the convergent p-series $(p = 2 > 1)$ $\quad 1 + \dfrac{1}{2^2} + \dfrac{1}{3^2} + \cdots$.

b) $\dfrac{3}{2\cdot4} + \dfrac{4}{3\cdot5} + \dfrac{5}{4\cdot6} + \dfrac{6}{5\cdot7} + \cdots$ The general term is $\dfrac{n+2}{(n+1)(n+3)} > \dfrac{1}{n+3}$.

The given series is divergent since it is term by term greater than the divergent series $\dfrac{1}{4} + \dfrac{1}{5} + \dfrac{1}{6} + \cdots$,

c) $\dfrac{1}{1\cdot2} + \dfrac{1}{2\cdot2^2} + \dfrac{1}{3\cdot2^3} + \dfrac{1}{4\cdot2^4} + \cdots$ The general term is $u_n = \dfrac{1}{n\,2^n} \leq \dfrac{1}{2^n}$.

The given series is term by term less than or equal to the convergent geometric series $(r = 1/2 < 1)$ $\frac{1}{2} + \frac{1}{4} + \frac{1}{8} + \cdots$ and hence is convergent.

d) $\frac{\sqrt{2}}{1} + \frac{\sqrt{3}}{2} + \frac{\sqrt{4}}{3} + \frac{\sqrt{5}}{4} + \cdots$ The general term is $u_n = \frac{\sqrt{n+1}}{n} > \frac{\sqrt{n}}{n} = \frac{1}{n^{1/2}}$.

The given series is term by term greater than the divergent p-series $(p = 1/2 < 1)$ $1 + \frac{1}{2^{1/2}} + \frac{1}{3^{1/2}} + \cdots$ and hence is divergent.

e) $1 + \frac{1}{4} + \frac{1}{7} + \frac{1}{10} + \cdots$ The general term is $u_n = \frac{1}{3n-2} > \frac{1}{3n}$.

The given series is term by term greater than one-third the divergent harmonic series (p-series with $p = 1$) $1 + \frac{1}{2} + \frac{1}{3} + \cdots$ and hence is divergent.

f) $\frac{2}{3 \cdot 1^3} + \frac{3}{4 \cdot 2^3} + \frac{4}{5 \cdot 3^3} + \frac{5}{6 \cdot 4^3} + \cdots$ The general term is $u_n = \frac{n+1}{(n+2)n^3} < \frac{1}{n^3}$.

The series is term by term less than the convergent p-series ($p = 3 > 1$) $1 + \frac{1}{2^3} + \frac{1}{3^3} + \cdots$ and hence is convergent.

RATIO TEST.

16. Apply the ratio test to investigate the convergence of the following series. If the ratio test fails, use the comparison test.

a) $\frac{3}{2^2} + \frac{4}{2^3} + \frac{5}{2^4} + \frac{6}{2^5} + \cdots$

$u_n = \frac{n+2}{2^{n+1}}, \quad u_{n+1} = \frac{n+3}{2^{n+2}}$ and $\frac{u_{n+1}}{u_n} = \frac{n+3}{2^{n+2}} \cdot \frac{2^{n+1}}{n+2} = \frac{n+3}{2(n+2)}$.

Then $\lim_{n \to \infty} \left| \frac{u_{n+1}}{u_n} \right| = \frac{1}{2} \lim_{n \to \infty} \frac{n+3}{n+2} = \frac{1}{2} = R < 1$ and the series converges.

b) $\frac{3}{1^4} + \frac{3^2}{2^4} + \frac{3^3}{3^4} + \frac{3^4}{4^4} + \cdots$

$u_n = \frac{3^n}{n^4}, \quad u_{n+1} = \frac{3^{n+1}}{(n+1)^4}$ and $\frac{u_{n+1}}{u_n} = \frac{3^{n+1}}{(n+1)^4} \cdot \frac{n^4}{3^n} = \frac{3n^4}{(n+1)^4}$.

Then $\lim_{n \to \infty} \left| \frac{u_{n+1}}{u_n} \right| = 3 \lim_{n \to \infty} \left(\frac{n}{n+1} \right)^4 = 3 = R > 1$ and the series diverges.

c) $\frac{1 \cdot 2}{3^2} + \frac{2 \cdot 3}{3^3} + \frac{3 \cdot 4}{3^4} + \frac{4 \cdot 5}{3^5} + \cdots$ $u_n = \frac{n(n+1)}{3^{n+1}}, \quad u_{n+1} = \frac{(n+1)(n+2)}{3^{n+2}}$ and $\frac{u_{n+1}}{u_n} = \frac{n+2}{3n}$.

Then $\lim_{n \to \infty} \left| \frac{u_{n+1}}{u_n} \right| = \frac{1}{3} \lim_{n \to \infty} \frac{n+2}{n} = \frac{1}{3} = R < 1$ and the series converges.

$d)$ $\dfrac{2^2}{3^3} + \dfrac{2^4}{3^5} + \dfrac{2^6}{3^7} + \dfrac{2^8}{3^9} + \cdots$ $\qquad u_n = \dfrac{2^{2n}}{3^{2n+1}}, \quad u_{n+1} = \dfrac{2^{2n+2}}{3^{2n+3}} \quad$ and $\quad \dfrac{u_{n+1}}{u_n} = \dfrac{2^2}{3^2} = \dfrac{4}{9}.$

Then $\quad \lim\limits_{n\to\infty} \left| \dfrac{u_{n+1}}{u_n} \right| = \dfrac{4}{9} = R < 1 \quad$ and the series converges.

$e)$ $\dfrac{1}{1\cdot 2^2} + \dfrac{1}{2\cdot 3^2} + \dfrac{1}{3\cdot 4^2} + \cdots$ $\qquad u_n = \dfrac{1}{n(n+1)^2}, \quad u_{n+1} = \dfrac{1}{(n+1)(n+2)^2}, \quad \dfrac{u_{n+1}}{u_n} = \dfrac{n(n+1)}{(n+2)^2}.$

Then $\quad \lim\limits_{n\to\infty} \dfrac{n(n+1)}{(n+2)^2} = \lim\limits_{n\to\infty} \dfrac{n^2+n}{n^2+4n+4} = \lim\limits_{n\to\infty} \dfrac{1+1/n}{1+4/n+4/n^2} = 1 \quad$ and the ratio test fails.

Now $\quad \dfrac{1}{n(n+1)^2} \leqq \dfrac{1}{(n+1)^2} \quad$ for all $n \geqq 1$. But $\quad \dfrac{1}{2^2} + \dfrac{1}{3^2} + \dfrac{1}{4^2} + \cdots$ converges; therefore the

given series converges.

$f)$ $1 + \dfrac{1}{1!} + \dfrac{1}{2!} + \dfrac{1}{3!} + \cdots$

$u_n = \dfrac{1}{n!}$ for all $n \geqq 0 \quad (0! = 1), \quad u_{n+1} = \dfrac{1}{(n+1)!} = \dfrac{1}{n!(n+1)} \quad$ and $\quad \dfrac{u_{n+1}}{u_n} = \dfrac{1}{n+1}.$

Then $\quad \lim\limits_{n\to\infty} \dfrac{1}{n+1} = 0 \quad$ and the series converges.

The sum of the given series is designated by e, approximately equal to 2.71828, and is the base of the natural or Napierian logarithms.

ALTERNATING SERIES.

17. Establish the convergence or divergence of each of the following series.

$a)$ $1 - \dfrac{1}{2^2} + \dfrac{1}{3^2} - \dfrac{1}{4^2} + \cdots$ $\qquad$ Here $|u_n| = \dfrac{1}{n^2} \quad$ and $\quad |u_{n+1}| = \dfrac{1}{(n+1)^2}.$

Since $\quad |u_{n+1}| < |u_n| \quad$ and $\quad \lim\limits_{n\to\infty} u_n = 0, \quad$ the series converges.

$b)$ $\dfrac{1}{2\cdot 3} - \dfrac{1}{3\cdot 4} + \dfrac{1}{4\cdot 5} - \dfrac{1}{5\cdot 6} + \cdots$ $\qquad$ Here $|u_n| = \dfrac{1}{(n+1)(n+2)} \quad$ and $\quad |u_{n+1}| = \dfrac{1}{(n+2)(n+3)}$

Since $\quad |u_{n+1}| < |u_n| \quad$ and $\quad \lim\limits_{n\to\infty} u_n = 0, \quad$ the series converges.

$c)$ $\dfrac{1}{3} - \dfrac{2}{5} + \dfrac{3}{7} - \dfrac{4}{9} + \cdots$ $\qquad$ Here $|u_n| = \dfrac{n}{2n+1} \quad$ and $\quad |u_{n+1}| = \dfrac{n+1}{2n+3}.$

Since $\quad \lim\limits_{n\to\infty} \dfrac{n}{2n+1} = \dfrac{1}{2} \neq 0, \quad$ the series is divergent.

ABSOLUTE AND CONDITIONAL CONVERGENCE.

18. Examine the following series for absolute or conditional convergence.

a) $1 - \dfrac{1}{\sqrt{2}} + \dfrac{1}{\sqrt{3}} - \dfrac{1}{\sqrt{4}} + \cdots$ $\qquad$ Here $\quad |u_n| = \dfrac{1}{n^{\frac{1}{2}}}\quad$ and $\quad |u_{n+1}| = \dfrac{1}{(n+1)^{\frac{1}{2}}}$.

The series of absolute values $\quad 1 + \dfrac{1}{\sqrt{2}} + \dfrac{1}{\sqrt{3}} + \dfrac{1}{\sqrt{4}} + \cdots\quad$ is divergent, being a p-series with $p = \frac{1}{2} < 1$.

The given series is convergent since $\quad |u_{n+1}| < |u_n|\quad$ and $\quad \lim\limits_{n \to \infty} u_n = 0$.
Thus the given series is conditionally convergent.

b) $1 - \dfrac{1}{2^3} + \dfrac{1}{3^3} - \dfrac{1}{4^3} + \cdots$

The series of absolute values $\quad 1 + \dfrac{1}{2^3} + \dfrac{1}{3^3} + \dfrac{1}{4^3} + \cdots\quad$ is convergent, being a p-series with $p = 3$. Thus the given series is absolutely convergent (and hence convergent).

c) $\sqrt{3/2} - \sqrt{4/3} + \sqrt{5/4} - \sqrt{6/5} + \cdots$ $\qquad$ Here $\quad |u_n| = \left(\dfrac{n+2}{n+1}\right)^{\frac{1}{2}}$.

Since $\quad \lim\limits_{n \to \infty} \left(\dfrac{n+2}{n+1}\right)^{\frac{1}{2}} = 1 \ne 0$, the series diverges.

d) $\dfrac{1}{2} - \dfrac{1}{4} + \dfrac{1}{6} - \dfrac{1}{8} + \cdots$ $\qquad$ Here $\quad |u_n| = \dfrac{1}{2n}\quad$ and $\quad |u_{n+1}| = \dfrac{1}{2n+2}$.

The series of absolute values $\quad \dfrac{1}{2} + \dfrac{1}{4} + \dfrac{1}{6} + \dfrac{1}{8} + \cdots = \dfrac{1}{2}\left(1 + \dfrac{1}{2} + \dfrac{1}{3} + \dfrac{1}{4} + \cdots\right)\quad$ is half the harmonic series (or half the p-series with $p = 1$) and is divergent.

The given series converges since $\quad \dfrac{1}{2n+2} < \dfrac{1}{2n}\quad$ and $\quad \lim\limits_{n \to \infty} \dfrac{1}{2n} = 0$.
Thus the given series is conditionally convergent.

19. Prove that the series $\quad \dfrac{1}{1^2} + \dfrac{1}{2^2} - \dfrac{1}{3^2} - \dfrac{1}{4^2} + \dfrac{1}{5^2} + \dfrac{1}{6^2} - \cdots\quad$ is convergent.

The series of absolute values $\quad \dfrac{1}{1^2} + \dfrac{1}{2^2} + \dfrac{1}{3^2} + \dfrac{1}{4^2} + \cdots\quad$ is convergent, being a p-series with $p = 2 > 1$.

Thus the given series is absolutely convergent and hence convergent.

20. Use the ratio test to investigate the convergence of the following series.

a) $\dfrac{1}{2} - \dfrac{2}{4} + \dfrac{3}{8} - \dfrac{4}{16} + \cdots$ $\qquad$ Here $\quad |u_n| = \dfrac{n}{2^n}\quad$ and $\quad |u_{n+1}| = \dfrac{n+1}{2^{n+1}}$.

$$\lim_{n\to\infty}\left|\frac{u_{n+1}}{u_n}\right| \;=\; \lim_{n\to\infty}\frac{n+1}{2^{n+1}}\cdot\frac{2^n}{n} \;=\; \lim_{n\to\infty}\frac{n+1}{2n} \;=\; \frac{1}{2} \;=\; R < 1 \quad\text{and the alternating series is}$$

convergent.

Since the ratio test also shows that the series of absolute values $\frac{1}{2}+\frac{2}{4}+\frac{3}{8}+\frac{4}{16}+\cdots$ is convergent, the given series is absolutely convergent.

b) $\dfrac{3}{2!}-\dfrac{3^2}{4!}+\dfrac{3^3}{6!}-\dfrac{3^4}{8!}+\cdots$ \qquad Here $|u_n|=\dfrac{3^n}{(2n)!}$ and $|u_{n+1}|=\dfrac{3^{n+1}}{(2n+2)!}$.

$$\lim_{n\to\infty}\left|\frac{u_{n+1}}{u_n}\right| \;=\; \lim_{n\to\infty}\frac{3(2n)!}{(2n+2)!} \;=\; \lim_{n\to\infty}\frac{3(2n)!}{(2n+2)(2n+1)(2n)!} \;=\; \lim_{n\to\infty}\frac{3}{4n^2+6n+2} \;=\; 0 = R < 1.$$

Thus the series is absolutely convergent.

c) $1-\dfrac{1}{2^{2/3}}+\dfrac{1}{3^{2/3}}-\dfrac{1}{4^{2/3}}+\cdots$ \qquad Here $|u_n|=\dfrac{1}{n^{2/3}}$ and $|u_{n+1}|=\dfrac{1}{(n+1)^{2/3}}$.

$$\lim_{n\to\infty}\left|\frac{u_{n+1}}{u_n}\right| \;=\; \lim_{n\to\infty}\left(\frac{n}{n+1}\right)^{2/3} \;=\; 1 \quad\text{and the ratio test fails.}$$

But, since $|u_{n+1}| < |u_n|$ and $\lim\limits_{n\to\infty} u_n = 0$, the given alternating series is conditionally convergent.

POWER SERIES.

21. Find the interval of convergence of the following series.

a) $x+\dfrac{x^2}{2}+\dfrac{x^3}{3}+\dfrac{x^4}{4}+\cdots$ \qquad Here $u_n=\dfrac{x^n}{n}$ and $u_{n+1}=\dfrac{x^{n+1}}{n+1}$.

$$\lim_{n\to\infty}\left|\frac{u_{n+1}}{u_n}\right| \;=\; \lim_{n\to\infty}\left|\frac{x^{n+1}}{n+1}\cdot\frac{n}{x^n}\right| \;=\; |x|\lim_{n\to\infty}\frac{n}{n+1} \;=\; |x|.$$

The series converges for $|x| < 1$ or $-1 < x < 1$, and diverges for $|x| > 1$ or $x > 1$ and $x < -1$. The test fails for $|x| = 1$ or $x = 1$ and $x = -1$.

For $x = 1$, the series is $1+\dfrac{1}{2}+\dfrac{1}{3}+\dfrac{1}{4}+\cdots$ which diverges.

For $x = -1$, the series is $-1+\dfrac{1}{2}-\dfrac{1}{3}+\dfrac{1}{4}-\cdots$ which converges.

Thus the interval of convergence of the given series is $-1 \leqq x < 1$. This is represented graphically by

where the thick line represents the interval of convergence, and the thin line the intervals on which the series diverges. The solid circle indicates that the series converges at end-point -1, and the open circle shows that it diverges at end-point 1.

b) $\dfrac{x}{2} + \dfrac{x^2}{2^2} + \dfrac{x^3}{2^3} + \dfrac{x^4}{2^4} + \cdots$ Here $u_n = \dfrac{x^n}{2^n}$ and $u_{n+1} = \dfrac{x^{n+1}}{2^{n+1}}$.

$$\lim_{n\to\infty} \left|\frac{u_{n+1}}{u_n}\right| = \lim_{n\to\infty} \left|\frac{x^{n+1}}{2^{n+1}} \cdot \frac{2^n}{x^n}\right| = \lim_{n\to\infty} \left|\frac{x}{2}\right| = \frac{1}{2}|x|.$$

The series converges for $\frac{1}{2}|x| < 1$ or $-2 < x < 2$.

For $x = -2$, the series is $-1 + 1 - 1 + 1 - \cdots$ which diverges.
For $x = 2$, the series is $1 + 1 + 1 + 1 + \cdots$ which diverges.

Thus the given series is convergent on the interval $-2 < x < 2$.

c) $1 + x + \dfrac{x^2}{2!} + \dfrac{x^3}{3!} + \cdots$ Here $u_n = \dfrac{x^{n-1}}{(n-1)!}$ and $u_{n+1} = \dfrac{x^n}{n!}$.

$$\lim_{n\to\infty} \left|\frac{x^n}{n!} \cdot \frac{(n-1)!}{x^{n-1}}\right| = \lim_{n\to\infty} \left|\frac{x(n-1)!}{n(n-1)!}\right| = |x| \lim_{n\to\infty} \frac{1}{n} = 0 \quad \text{for all } x.$$

Thus the series converges for all values of x, written $-\infty < x < \infty$.

d) $x - \dfrac{x^2}{\sqrt{2}} + \dfrac{x^3}{\sqrt{3}} - \dfrac{x^4}{\sqrt{4}} + \cdots$ Here $|u_n| = \left|\dfrac{x^n}{n^{1/2}}\right|$ and $|u_{n+1}| = \left|\dfrac{x^{n+1}}{(n+1)^{1/2}}\right|$.

$$\lim_{n\to\infty} \left|\frac{x^{n+1}}{(n+1)^{1/2}} \cdot \frac{n^{1/2}}{x^n}\right| = |x| \lim_{n\to\infty} \left(\frac{n}{n+1}\right)^{1/2} = |x|.$$

The series converges for $|x| < 1$ or $-1 < x < 1$.

For $x = 1$, the series is $1 - \dfrac{1}{\sqrt{2}} + \dfrac{1}{\sqrt{3}} - \dfrac{1}{\sqrt{4}} + \cdots$ which converges.

For $x = -1$, the series is $-1 - \dfrac{1}{\sqrt{2}} - \dfrac{1}{\sqrt{3}} - \dfrac{1}{\sqrt{4}} - \cdots$ which diverges.

Thus the required interval of convergence is $-1 < x \leqq 1$.

e) $x + 2!x^2 + 3!x^3 + 4!x^4 + \cdots$ Here $|u_n| = |n!x^n|$ and $|u_{n+1}| = |(n+1)!x^{n+1}|$.

$$\lim_{n\to\infty} \left|\frac{(n+1)!\,x^{n+1}}{n!\,x^n}\right| = |x| \lim_{n\to\infty} (n+1) = \infty \text{ unless } x = 0.$$

The series converges only for $x = 0$.

f) $(x-3) - \dfrac{(x-3)^2}{2} + \dfrac{(x-3)^3}{3} - \dfrac{(x-3)^4}{4} + \cdots$ $|u_n| = \left|\dfrac{(x-3)^n}{n}\right|$ and $|u_{n+1}| = \left|\dfrac{(x-3)^{n+1}}{n+1}\right|$.

$$\lim_{n\to\infty} \left|\frac{(x-3)^{n+1}}{n+1} \cdot \frac{n}{(x-3)^n}\right| = |x-3| \lim_{n\to\infty} \frac{n}{n+1} = |x-3|.$$

The series converges for $|x-3| < 1$ or $2 < x < 4$.

For $x = 2$, the series is $-1 - 1/2 - 1/3 - 1/4 - \cdots$ which diverges.
For $x = 4$, the series is $1 - 1/2 + 1/3 - 1/4 + \cdots$ which converges.

Thus the given series is convergent on the interval $2 < x \leqq 4$.

SUPPLEMENTARY PROBLEMS

22. Write the first four terms of the sequence with the indicated nth term.

a) $\dfrac{2n}{n+2}$ b) $\dfrac{3n-2}{2^n}$ c) $\dfrac{(-1)^n}{n!}$ d) $\dfrac{n}{2n^2-1}$ e) $\dfrac{x^{2n}}{2n+1}$

23. Write the first four terms and the $(n+1)$st term of the infinite series whose nth term is given.

a) $\dfrac{n}{3^n+1}$ b) $\dfrac{\sqrt{n}}{n+1}$ c) $\dfrac{(-1)^{n-1}}{(2n)!}$ d) $\dfrac{2n-1}{2n+1}$ e) $\dfrac{x^{n-1}}{n(n+1)(n+2)}$

24. Find an nth term for each sequence.

a) $\dfrac{1}{3}, \dfrac{2}{5}, \dfrac{3}{7}, \dfrac{4}{9}, \ldots$

c) $\dfrac{3}{4\cdot5}, \dfrac{4}{5\cdot6}, \dfrac{5}{6\cdot7}, \dfrac{6}{7\cdot8}, \ldots$

b) $\dfrac{1}{2+1}, \dfrac{1}{2^2+1}, \dfrac{1}{2^3+1}, \dfrac{1}{2^4+1}, \ldots$

d) $\dfrac{2}{2^2-1}, \dfrac{4}{2^3-1}, \dfrac{6}{2^4-1}, \dfrac{8}{2^5-1}, \ldots$

25. Find an nth term for each series and also write the $(n+1)$st term.

a) $1 + \dfrac{1}{2} + \dfrac{1}{4} + \dfrac{1}{8} + \cdots$

d) $1 - \dfrac{1}{3} + \dfrac{1}{5} - \dfrac{1}{7} + \cdots$

b) $\dfrac{1}{2} + \dfrac{3}{4} + \dfrac{5}{8} + \dfrac{7}{16} + \cdots$

e) $x - \dfrac{x^2}{2} + \dfrac{x^3}{3} - \dfrac{x^4}{4} + \cdots$

c) $\dfrac{2}{1\cdot3} + \dfrac{4}{3\cdot5} + \dfrac{6}{5\cdot7} + \dfrac{8}{7\cdot9} + \cdots$

f) $1 + \dfrac{x}{2\cdot3!} + \dfrac{x^2}{2^2\cdot5!} + \dfrac{x^3}{2^3\cdot7!} + \cdots$

26. Write the series represented by each of the following.

a) $\displaystyle\sum_{n=1}^{5} \dfrac{1}{2n+2}$

c) $\displaystyle\sum_{n=1}^{\infty} \dfrac{1}{(2n+1)^2}$

e) $\displaystyle\sum_{n=1}^{\infty} \dfrac{(-1)^{n-1}(x-1)^n}{n!}$

b) $\displaystyle\sum_{n=1}^{3} \dfrac{(-1)^n}{n}$

d) $\displaystyle\sum_{n=2}^{\infty} \dfrac{x^{n-1}}{(n-1)^3}$

27. The sequence $\dfrac{2}{3}, \dfrac{5}{4}, \dfrac{8}{5}, \dfrac{11}{6}, \ldots$ has nth term given by $u_n = \dfrac{3n-1}{n+2}$.

a) Write the 10th, 20th, 100th, 1000th, 10,000th terms of the sequence. Make a guess as to the limit of this sequence as $n \to \infty$.

b) Using the definition of limit, verify that the guess in a) is actually correct.

c) Evaluate the limit by using the theorems on limits.

28. Find the limit as $n \to \infty$ of each of the following sequences having the indicated nth term (or general term).

a) $\dfrac{5n-2}{3n+1}$ b) $\dfrac{4n^2}{(n+2)^2}$ c) $\dfrac{n}{n^2+1}$ d) $\sqrt[3]{\dfrac{2n}{n+1}}$ e) $\dfrac{n^3+n}{2n^3+1}$

29. Determine whether the sequence is bounded, is monotone increasing or decreasing, and whether a limit exists.

a) 3, 3.3, 3.33, 3.333, ...

b) 1.9, 2.1, 1.99, 2.01, 1.999, 2.001, ...

c) 1, -2, 3, -4, 5, -6, ...

d) 1, 2, 1, 3, 1, 4, 1, 5, ...

e) $\frac{2}{3}, \frac{3}{5}, \frac{4}{7}, \frac{5}{9}, \ldots$

30. A series has its nth term given by $u_n = \dfrac{2^{n-1}}{3^n}$.

a) Write the first four terms of the series.

b) If S_n denotes the sum of the first n terms of the series, find $S_1, S_2, S_3, \ldots, S_{10}$ in decimal form. Make a guess as to the limit of the sequence S_n as $n \to \infty$.

c) By obtaining the sum of the first n terms of the series, verify that the guess in b) is correct. Is the series convergent?

31. Given the series $1 - 3 + 5 - 7 + \ldots$

a) Compute S_n, the sum of the first n terms of the series.

b) What is $\lim_{n \to \infty} S_n$?

c) Is the series convergent?

32. State whether the following conclusions are valid on the basis of reasons given.

a) The series $\dfrac{1}{3} + \dfrac{1}{5} + \dfrac{1}{7} + \dfrac{1}{9} + \cdots$ is convergent since the successive terms are smaller and smaller.

b) The series $\dfrac{1}{3} + \dfrac{2}{5} + \dfrac{3}{7} + \dfrac{4}{9} + \cdots$ is divergent since the nth term does not approach zero.

c) The series $\dfrac{1}{1^2} + \dfrac{1}{2^2} + \dfrac{1}{3^2} + \dfrac{1}{4^2} + \cdots$ is convergent because the nth term approaches zero.

33. Examine each series for convergence.

a) $2 + \dfrac{2}{3} + \dfrac{2}{9} + \dfrac{2}{27} + \cdots$

b) $1 + \dfrac{4}{3} + (\dfrac{4}{3})^2 + (\dfrac{4}{3})^3 + \cdots$

c) $\dfrac{3}{2} + \dfrac{3}{2^2} + \dfrac{3}{2^3} + \dfrac{3}{2^4} + \cdots$

d) $\dfrac{1}{1^2} + \dfrac{1}{2^2} + \dfrac{1}{3^2} + \dfrac{1}{4^2} + \cdots$

e) $\dfrac{1}{\sqrt[3]{1}} + \dfrac{1}{\sqrt[3]{2}} \quad \dfrac{1}{\sqrt[3]{3}} + \dfrac{1}{\sqrt[3]{4}} + \cdots$

f) $\dfrac{1}{10} + \dfrac{1}{20} + \dfrac{1}{30} + \dfrac{1}{40} + \cdots$

34. Employ the comparison test to investigate the convergence of each series.

a) $\dfrac{1}{2+1} + \dfrac{1}{2^2+1} + \dfrac{1}{2^3+1} + \dfrac{1}{2^4+1} + \cdots$

b) $\dfrac{1}{1} + \dfrac{1}{3} + \dfrac{1}{5} + \dfrac{1}{7} + \cdots$

c) $\dfrac{3}{3^3+1} + \dfrac{4}{4^3+1} + \dfrac{5}{5^3+1} + \dfrac{6}{6^3+1} + \cdots$

d) $\dfrac{1}{2 \cdot 1^4} + \dfrac{1}{4 \cdot 2^4} + \dfrac{1}{6 \cdot 3^4} + \dfrac{1}{8 \cdot 4^4} + \cdots$

e) $\dfrac{2}{2^2-1} + \dfrac{3}{3^2-1} + \dfrac{4}{4^2-1} + \dfrac{5}{5^2-1} + \cdots$

f) $\dfrac{1}{\sqrt{1 \cdot 2}} + \dfrac{1}{\sqrt{2 \cdot 3}} + \dfrac{1}{\sqrt{3 \cdot 4}} + \dfrac{1}{\sqrt{4 \cdot 5}} + \cdots$

35. Apply the ratio test to investigate the convergence of each series. If the ratio test fails, use the comparison test.

a) $\dfrac{1}{2 \cdot 3^2} + \dfrac{1}{3 \cdot 3^3} + \dfrac{1}{4 \cdot 3^4} + \dfrac{1}{5 \cdot 3^5} + \cdots$

b) $\dfrac{2^2}{1} + \dfrac{2^3}{2} + \dfrac{2^4}{3} + \dfrac{2^5}{4} + \cdots$

c) $\dfrac{3}{2^2} + \dfrac{4}{3^2} + \dfrac{5}{4^2} + \dfrac{6}{5^2} + \cdots$

e) $\dfrac{1}{3\cdot 4} + \dfrac{1}{4\cdot 5} + \dfrac{1}{5\cdot 6} + \dfrac{1}{6\cdot 7} + \cdots$

d) $\dfrac{2}{1!} + \dfrac{3}{2!} + \dfrac{4}{3!} + \dfrac{5}{4!} + \cdots$

f) $\dfrac{1}{2}\left(\dfrac{3}{2}\right) + \dfrac{1}{4}\left(\dfrac{3}{2}\right)^2 + \dfrac{1}{6}\left(\dfrac{3}{2}\right)^3 + \dfrac{1}{8}\left(\dfrac{3}{2}\right)^4 + \cdots$

36. Establish the convergence or divergence of each series.

a) $1 - \dfrac{1}{3} + \dfrac{1}{5} - \dfrac{1}{7} + \cdots$

c) $\dfrac{1}{2} - \dfrac{2}{3} + \dfrac{3}{4} - \dfrac{4}{5} + \cdots$

b) $\dfrac{1}{3} - \dfrac{1}{9} + \dfrac{1}{27} - \dfrac{1}{81} + \cdots$

d) $\dfrac{2}{3^2} - \dfrac{3}{4^2} + \dfrac{4}{5^2} - \dfrac{5}{6^2} + \cdots$

37. Examine each series for absolute or conditional convergence.

a) $\dfrac{1}{3} - \dfrac{1}{3^2} + \dfrac{1}{3^3} - \dfrac{1}{3^4} + \cdots$

d) $\dfrac{2}{1\cdot 4} - \dfrac{3}{2\cdot 5} + \dfrac{4}{3\cdot 6} - \dfrac{5}{4\cdot 7} + \cdots$

b) $\dfrac{1}{4} - \dfrac{1}{8} + \dfrac{1}{12} - \dfrac{1}{16} + \cdots$

e) $\dfrac{1}{3\cdot 2^2} - \dfrac{1}{4\cdot 3^2} + \dfrac{1}{5\cdot 4^2} - \dfrac{1}{6\cdot 5^2} + \cdots$

c) $\dfrac{1}{2+1} - \dfrac{1}{2^2+1} + \dfrac{1}{2^3+1} - \dfrac{1}{2^4+1} + \cdots$

f) $\dfrac{1}{1!} - \dfrac{1}{3!} + \dfrac{1}{5!} - \dfrac{1}{7!} + \cdots$

38. Examine the validity of each of the following statements.
    a) An alternating series in which each term is less than the preceding term in absolute value must be convergent.
    b) An alternating series which is absolutely convergent is convergent.

39. Find the interval of convergence of each series.

a) $\dfrac{x}{1^2} + \dfrac{x^2}{2^2} + \dfrac{x^3}{3^2} + \dfrac{x^4}{4^2} + \cdots$

d) $2x - 4x^2 + 8x^3 - 16x^4 + \cdots$

b) $\dfrac{x}{3\cdot 1} + \dfrac{x^2}{3^2\cdot 2} + \dfrac{x^3}{3^3\cdot 3} + \dfrac{x^4}{3^4\cdot 4} + \cdots$

e) $\dfrac{x+2}{\sqrt{1}} + \dfrac{(x+2)^2}{\sqrt{2}} + \dfrac{(x+2)^3}{\sqrt{3}} + \dfrac{(x+2)^4}{\sqrt{4}} + \cdots$

c) $x - \dfrac{x^3}{3} + \dfrac{x^5}{5} - \dfrac{x^7}{7} + \cdots$

f) $x - \dfrac{x^3}{3!} + \dfrac{x^5}{5!} - \dfrac{x^7}{7!} + \cdots$

ANSWERS TO SUPPLEMENTARY PROBLEMS.

22. a) $\dfrac{2}{3}, 1, \dfrac{6}{5}, \dfrac{4}{3}$ b) $\dfrac{1}{2}, 1, \dfrac{7}{8}, \dfrac{5}{8}$ c) $-1, \dfrac{1}{2}, -\dfrac{1}{6}, \dfrac{1}{24}$ d) $1, \dfrac{2}{7}, \dfrac{3}{17}, \dfrac{4}{31}$ e) $\dfrac{x^2}{3}, \dfrac{x^4}{5}, \dfrac{x^6}{7}, \dfrac{x^8}{9}$

23. a) $\dfrac{1}{4} + \dfrac{2}{10} + \dfrac{3}{28} + \dfrac{4}{82}$; $\dfrac{n+1}{3^{n+1}+1}$

c) $\dfrac{1}{2!} - \dfrac{1}{4!} + \dfrac{1}{6!} - \dfrac{1}{8!}$; $\dfrac{(-1)^n}{(2n+2)!}$

b) $\dfrac{\sqrt{1}}{2} + \dfrac{\sqrt{2}}{3} + \dfrac{\sqrt{3}}{4} + \dfrac{\sqrt{4}}{5}$; $\dfrac{\sqrt{n+1}}{n+2}$

d) $\dfrac{1}{3} + \dfrac{3}{5} + \dfrac{5}{7} + \dfrac{7}{9}$; $\dfrac{2n+1}{2n+3}$

e) $\dfrac{1}{1\cdot 2\cdot 3} + \dfrac{x}{2\cdot 3\cdot 4} + \dfrac{x^2}{3\cdot 4\cdot 5} + \dfrac{x^3}{4\cdot 5\cdot 6}$; $\dfrac{x^n}{(n+1)(n+2)(n+3)}$

24. a) $\dfrac{n}{2n+1}$ b) $\dfrac{1}{2^n+1}$ c) $\dfrac{n+2}{(n+3)(n+4)}$ d) $\dfrac{2n}{2^{n+1}-1}$

25. *a)* $u_n = \dfrac{1}{2^{n-1}}$; $u_{n+1} = \dfrac{1}{2^n}$ *d)* $u_n = \dfrac{(-1)^{n-1}}{2n-1}$; $u_{n+1} = \dfrac{(-1)^n}{2n+1}$

b) $u_n = \dfrac{2n-1}{2^n}$; $u_{n+1} = \dfrac{2n+1}{2^{n+1}}$ *e)* $u_n = \dfrac{(-1)^{n-1}x^n}{n}$; $u_{n+1} = \dfrac{(-1)^n x^{n+1}}{n+1}$

c) $u_n = \dfrac{2n}{(2n-1)(2n+1)}$; $u_{n+1} = \dfrac{2n+2}{(2n+1)(2n+3)}$ *f)* $u_n = \dfrac{x^{n-1}}{2^{n-1}(2n-1)!}$; $u_{n+1} = \dfrac{x^n}{2^n(2n+1)!}$

26. *a)* $\dfrac{1}{4} + \dfrac{1}{6} + \dfrac{1}{8} + \dfrac{1}{10} + \dfrac{1}{12}$

d) $\dfrac{x}{1^3} + \dfrac{x^2}{2^3} + \dfrac{x^3}{3^3} + \dfrac{x^4}{4^3} + \cdots$

b) $-1 + \dfrac{1}{2} - \dfrac{1}{3}$

c) $\dfrac{1}{3^2} + \dfrac{1}{5^2} + \dfrac{1}{7^2} + \dfrac{1}{9^2} + \cdots$

e) $\dfrac{(x-1)}{1!} - \dfrac{(x-1)^2}{2!} + \dfrac{(x-1)^3}{3!} - \dfrac{(x-1)^4}{4!} + \cdots$

27. *a)* $u_{10} = 29/12 = 2.4166\ldots,$ $u_{20} = 59/22 = 2.6818\ldots,$ $u_{100} = 299/102 = 2.9313\ldots,$ $u_{1000} = 2.9930\ldots,$ $u_{10000} = 2.9993\ldots$ Our guess is 3.

28. *a)* 5/3   *b)* 4   *c)* 0   *d)* $\sqrt[3]{2}$ *e)* 1/2

29. *a)* bounded, monotone increasing, limit exists
    b) bounded, not monotone increasing or decreasing, limit exists
    c), *d)* not bounded, not monotone increasing or decreasing, limit does not exist
    e) bounded, monotone decreasing, limit exists

30. *a)* $\dfrac{1}{3} + \dfrac{2}{9} + \dfrac{4}{27} + \dfrac{8}{81}$

b) $S_1 = .33333\ldots,$ $S_2 = .55555\ldots,$ $S_3 = .70370\ldots,$ $S_4 = .80246\ldots,$ $S_5 = .86831\ldots$
    $S_6 = .91220\ldots,$ $S_7 = .94147\ldots,$ $S_8 = .96098\ldots,$ $S_9 = .97398\ldots,$ $S_{10} = .98265\ldots$
    Our guess is 1.

c) Sum of first *n* terms, $S_n = 1 - (2/3)^n$. $\lim\limits_{n \to \infty} S_n = 1$, series is convergent.

31. *a)* $S_1 = 1,\ S_2 = -2,\ S_3 = 3,\ S_4 = -4,\ \ldots$ *b)* $\lim\limits_{n \to \infty} S_n$ does not exist. *c)* Series is divergent.

32. *a)* No   *b)* Yes   *c)* No

33. *a)* convergent   *b)* divergent   *c)* convergent   *d)* convergent   *e)* divergent   *f)* divergent

34. *a)* convergent   *b)* divergent   *c)* convergent   *d)* convergent   *e)* divergent   *f)* divergent

35. *a)* convergent   *b)* divergent   *c)* divergent   *d)* convergent   *e)* convergent   *f)* divergent

36. *a)* convergent   *b)* convergent   *c)* divergent   *d)* convergent

37. *a)* absolutely convergent   *d)* conditionally convergent
    b) conditionally convergent   *e)* absolutely convergent
    c) absolutely convergent   *f)* absolutely convergent

38. *a)* false   *b)* true

39. *a)* $-1 \leqq x \leqq 1$ *c)* $-1 \leqq x \leqq 1$ *e)* $-3 \leqq x < -1$

    b) $-3 \leqq x < 3$ *d)* $-\dfrac{1}{2} < x < \dfrac{1}{2}$ *f)* $-\infty < x < \infty$

APPENDIX

Four-Place Common Logarithms

N	0	1	2	3	4	5	6	7	8	9
10	0000	0043	0086	0128	0170	0212	0253	0294	0334	0374
11	0414	0453	0492	0531	0569	0607	0645	0682	0719	0755
12	0792	0828	0864	0899	0934	0969	1004	1038	1072	1106
13	1139	1173	1206	1239	1271	1303	1335	1367	1399	1430
14	1461	1492	1523	1553	1584	1614	1644	1673	1703	1732
15	1761	1790	1818	1847	1875	1903	1931	1959	1987	2014
16	2041	2068	2095	2122	2148	2175	2201	2227	2253	2279
17	2304	2330	2355	2380	2405	2430	2455	2480	2504	2529
18	2553	2577	2601	2625	2648	2672	2695	2718	2742	2765
19	2788	2810	2833	2856	2878	2900	2923	2945	2967	2989
20	3010	3032	3054	3075	3096	3118	3139	3160	3181	3201
21	3222	3243	3263	3284	3304	3324	3345	3365	3385	3404
22	3424	3444	3464	3483	3502	3522	3541	3560	3579	3598
23	3617	3636	3655	3674	3692	3711	3729	3747	3766	3784
24	3802	3820	3838	3856	3874	3892	3909	3927	3945	3962
25	3979	3997	4014	4031	4048	4065	4082	4099	4116	4133
26	4150	4166	4183	4200	4216	4232	4249	4265	4281	4298
27	4314	4330	4346	4362	4378	4393	4409	4425	4440	4456
28	4472	4487	4502	4518	4533	4548	4564	4579	4594	4609
29	4624	4639	4654	4669	4683	4698	4713	4728	4742	4757
30	4771	4786	4800	4814	4829	4843	4857	4871	4886	4900
31	4914	4928	4942	4955	4969	4983	4997	5011	5024	5038
32	5051	5065	5079	5092	5105	5119	5132	5145	5159	5172
33	5185	5198	5211	5224	5237	5250	5263	5276	5289	5302
34	5315	5328	5340	5353	5366	5378	5391	5403	5416	5428
35	5441	5453	5465	5478	5490	5502	5514	5527	5539	5551
36	5563	5575	5587	5599	5611	5623	5635	5647	5658	5670
37	5682	5694	5705	5717	5729	5740	5752	5763	5775	5786
38	5798	5809	5821	5832	5843	5855	5866	5877	5888	5899
39	5911	5922	5933	5944	5955	5966	5977	5988	5999	6010
40	6021	6031	6042	6053	6064	6075	6085	6096	6107	6117
41	6128	6138	6149	6160	6170	6180	6191	6201	6212	6222
42	6232	6243	6253	6263	6274	6284	6294	6304	6314	6325
43	6335	6345	6355	6365	6375	6385	6395	6405	6415	6425
44	6435	6444	6454	6464	6474	6484	6493	6503	6513	6522
45	6532	6542	6551	6561	6571	6580	6590	6599	6609	6618
46	6628	6637	6646	6656	6665	6675	6684	6693	6702	6712
47	6721	6730	6739	6749	6758	6767	6776	6785	6794	6803
48	6812	6821	6830	6839	6848	6857	6866	6875	6884	6893
49	6902	6911	6920	6928	6937	6946	6955	6964	6972	6981
50	6990	6998	7007	7016	7024	7033	7042	7050	7059	7067
51	7076	7084	7093	7101	7110	7118	7126	7135	7143	7152
52	7160	7168	7177	7185	7193	7202	7210	7218	7226	7235
53	7243	7251	7259	7267	7275	7284	7292	7300	7308	7316
54	7324	7332	7340	7348	7356	7364	7372	7380	7388	7396
N	0	1	2	3	4	5	6	7	8	9

Four-Place Common Logarithms

N	0	1	2	3	4	5	6	7	8	9
55	7404	7412	7419	7427	7435	7443	7451	7459	7466	7474
56	7482	7490	7497	7505	7513	7520	7528	7536	7543	7551
57	7559	7566	7574	7582	7589	7597	7604	7612	7619	7627
58	7634	7642	7649	7657	7664	7672	7679	7686	7694	7701
59	7709	7716	7723	7731	7738	7745	7752	7760	7767	7774
60	7782	7789	7796	7803	7810	7818	7825	7832	7839	7846
61	7853	7860	7868	7875	7882	7889	7896	7903	7910	7917
62	7924	7931	7938	7945	7952	7959	7966	7973	7980	7987
63	7993	8000	8007	8014	8021	8028	8035	8041	8048	8055
64	8062	8069	8075	8082	8089	8096	8102	8109	8116	8122
65	8129	8136	8142	8149	8156	8162	8169	8176	8182	8189
66	8195	8202	8209	8215	8222	8228	8235	8241	8248	8254
67	8261	8267	8274	8280	8287	8293	8299	8306	8312	8319
68	8325	8331	8338	8344	8351	8357	8363	8370	8376	8382
69	8388	8395	8401	8407	8414	8420	8426	8432	8439	8445
70	8451	8457	8463	8470	8476	8482	8488	8494	8500	8506
71	8513	8519	8525	8531	8537	8543	8549	8555	8561	8567
72	8573	8579	8585	8591	8597	8603	8609	8615	8621	8627
73	8633	8639	8645	8651	8657	8663	8669	8675	8681	8686
74	8692	8698	8704	8710	8716	8722	8727	8733	8739	8745
75	8751	8756	8762	8768	8774	8779	8785	8791	8797	8802
76	8808	8814	8820	8825	8831	8837	8842	8848	8854	8859
77	8865	8871	8876	8882	8887	8893	8899	8904	8910	8915
78	8921	8927	8932	8938	8943	8949	8954	8960	8965	8971
79	8976	8982	8987	8993	8998	9004	9009	9015	9020	9025
80	9031	9036	9042	9047	9053	9058	9063	9069	9074	9079
81	9085	9090	9096	9101	9106	9112	9117	9122	9128	9133
82	9138	9143	9149	9154	9159	9165	9170	9175	9180	9186
83	9191	9196	9201	9206	9212	9217	9222	9227	9232	9238
84	9243	9248	9253	9258	9263	9269	9274	9279	9284	9289
85	9294	9299	9304	9309	9315	9320	9325	9330	9335	9340
86	9345	9350	9355	9360	9365	9370	9375	9380	9385	9390
87	9395	9400	9405	9410	9415	9420	9425	9430	9435	9440
88	9445	9450	9455	9460	9465	9469	9474	9479	9484	9489
89	9494	9499	9504	9509	9513	9518	9523	9528	9533	9538
90	9542	9547	9552	9557	9562	9566	9571	9576	9581	9586
91	9590	9595	9600	9605	9609	9614	9619	9624	9628	9633
92	9638	9643	9647	9652	9657	9661	9666	9671	9675	9680
93	9685	9689	9694	9699	9703	9708	9713	9717	9722	9727
94	9731	9736	9741	9745	9750	9754	9759	9763	9768	9773
95	9777	9782	9786	9791	9795	9800	9805	9809	9814	9818
96	9823	9827	9832	9836	9841	9845	9850	9854	9859	9863
97	9868	9872	9877	9881	9886	9890	9894	9899	9903	9908
98	9912	9917	9921	9926	9930	9934	9939	9943	9948	9952
99	9956	9961	9965	9969	9974	9978	9983	9987	9991	9996
N	0	1	2	3	4	5	6	7	8	9

Compound Amount: $(1 + i)^n$

n \ i	1%	$1\frac{1}{4}\%$	$1\frac{1}{2}\%$	2%	$2\frac{1}{2}\%$	3%	4%	5%	6%
1	1.0100	1.0125	1.0150	1.0200	1.0250	1.0300	1.0400	1.0500	1.0600
2	1.0201	1.0252	1.0302	1.0404	1.0506	1.0609	1.0816	1.1025	1.1236
3	1.0303	1.0380	1.0457	1.0612	1.0769	1.0927	1.1249	1.1576	1.1910
4	1.0406	1.0509	1.0614	1.0824	1.1038	1.1255	1.1699	1.2155	1.2625
5	1.0510	1.0641	1.0773	1.1041	1.1314	1.1593	1.2167	1.2763	1.3382
6	1.0615	1.0774	1.0934	1.1262	1.1597	1.1941	1.2653	1.3401	1.4185
7	1.0721	1.0909	1.1098	1.1487	1.1887	1.2299	1.3159	1.4071	1.5036
8	1.0829	1.1045	1.1265	1.1717	1.2184	1.2668	1.3688	1.4775	1.5938
9	1.0937	1.1183	1.1434	1.1951	1.2489	1.3048	1.4233	1.5513	1.6895
10	1.1046	1.1323	1.1605	1.2190	1.2801	1.3439	1.4802	1.6289	1.7908
11	1.1157	1.1464	1.1779	1.2434	1.3121	1.3842	1.5395	1.7103	1.8983
12	1.1268	1.1608	1.1956	1.2682	1.3449	1.4258	1.6010	1.7959	2.0122
13	1.1381	1.1753	1.2136	1.2936	1.3785	1.4685	1.6651	1.8856	2.1329
14	1.1495	1.1900	1.2318	1.3195	1.4130	1.5126	1.7317	1.9799	2.2609
15	1.1610	1.2048	1.2502	1.3459	1.4483	1.5580	1.8009	2.0789	2.3966
16	1.1726	1.2199	1.2690	1.3728	1.4845	1.6047	1.8730	2.1829	2.5404
17	1.1843	1.2351	1.2880	1.4002	1.5216	1.6528	1.9479	2.2920	2.6928
18	1.1961	1.2506	1.3073	1.4282	1.5597	1.7024	2.0258	2.4066	2.8543
19	1.2081	1.2662	1.3270	1.4568	1.5987	1.7535	2.1068	2.5270	3.0256
20	1.2202	1.2820	1.3469	1.4859	1.6386	1.8061	2.1911	2.6533	3.2071
21	1.2324	1.2981	1.3671	1.5157	1.6796	1.8603	2.2788	2.7860	3.3996
22	1.2447	1.3143	1.3876	1.5460	1.7216	1.9161	2.3699	2.9253	3.6035
23	1.2572	1.3307	1.4084	1.5769	1.7646	1.9736	2.4647	3.0715	3.8197
24	1.2697	1.3474	1.4295	1.6084	1.8087	2.0328	2.5633	3.2251	4.0489
25	1.2824	1.3642	1.4509	1.6406	1.8539	2.0938	2.6658	3.3864	4.2919
26	1.2953	1.3812	1.4727	1.6734	1.9003	2.1566	2.7725	3.5557	4.5494
27	1.3082	1.3985	1.4948	1.7069	1.9478	2.2213	2.8834	3.7335	4.8223
28	1.3213	1.4160	1.5172	1.7410	1.9965	2.2879	2.9987	3.9201	5.1117
29	1.3345	1.4337	1.5400	1.7758	2.0464	2.3566	3.1187	4.1161	5.4184
30	1.3478	1.4516	1.5631	1.8114	2.0976	2.4273	3.2434	4.3219	5.7435
31	1.3613	1.4698	1.5865	1.8476	2.1500	2.5001	3.3731	4.5380	6.0881
32	1.3749	1.4881	1.6103	1.8845	2.2038	2.5751	3.5081	4.7649	6.4534
33	1.3887	1.5067	1.6345	1.9222	2.2589	2.6523	3.6484	5.0032	6.8406
34	1.4026	1.5256	1.6590	1.9607	2.3153	2.7319	3.7943	5.2533	7.2510
35	1.4166	1.5446	1.6839	1.9999	2.3732	2.8139	3.9461	5.5160	7.6861
36	1.4308	1.5639	1.7091	2.0399	2.4325	2.8983	4.1039	5.7918	8.1473
37	1.4451	1.5835	1.7348	2.0807	2.4933	2.9852	4.2681	6.0814	8.6361
38	1.4595	1.6033	1.7608	2.1223	2.5557	3.0748	4.4388	6.3855	9.1543
39	1.4741	1.6233	1.7872	2.1647	2.6196	3.1670	4.6164	6.7048	9.7035
40	1.4889	1.6436	1.8140	2.2080	2.6851	3.2620	4.8010	7.0400	10.2857
41	1.5038	1.6642	1.8412	2.2522	2.7522	3.3599	4.9931	7.3920	10.9029
42	1.5188	1.6850	1.8688	2.2972	2.8210	3.4607	5.1928	7.7616	11.5570
43	1.5340	1.7060	1.8969	2.3432	2.8915	3.5645	5.4005	8.1497	12.2505
44	1.5493	1.7274	1.9253	2.3901	2.9638	3.6715	5.6165	8.5572	12.9855
45	1.5648	1.7489	1.9542	2.4379	3.0379	3.7816	5.8412	8.9850	13.7646
46	1.5805	1.7708	1.9835	2.4866	3.1139	3.8950	6.0748	9.4343	14.5905
47	1.5963	1.7929	2.0133	2.5363	3.1917	4.0119	6.3178	9.9060	15.4659
48	1.6122	1.8154	2.0435	2.5871	3.2715	4.1323	6.5705	10.4013	16.3939
49	1.6283	1.8380	2.0741	2.6388	3.3533	4.2562	6.8333	10.9213	17.3775
50	1.6446	1.8610	2.1052	2.6916	3.4371	4.3839	7.1067	11.4674	18.4202

Present Value after n Periods: $(1 + i)^{-n}$

n \ i	1%	$1\frac{1}{4}$%	$1\frac{1}{2}$%	2%	$2\frac{1}{2}$%	3%	4%	5%	6%
1	.99010	.98765	.98522	.98039	.97561	.97087	.96154	.95238	.94340
2	.98030	.97546	.97066	.96117	.95181	.94260	.92456	.90703	.89000
3	.97059	.96342	.95632	.94232	.92860	.91514	.88900	.86384	.83962
4	.96098	.95152	.94218	.92385	.90595	.88849	.85480	.82270	.79209
5	.95147	.93978	.92826	.90573	.88385	.86261	.82193	.78353	.74726
6	.94205	.92817	.91454	.88797	.86230	.83748	.79031	.74622	.70496
7	.93272	.91672	.90103	.87056	.84127	.81309	.75992	.71068	.66506
8	.92348	.90540	.88771	.85349	.82075	.78941	.73069	.67684	.62741
9	.91434	.89422	.87459	.83676	.80073	.76642	.70259	.64461	.59190
10	.90529	.88318	.86167	.82035	.78120	.74409	.67556	.61391	.55839
11	.89632	.87228	.84893	.80426	.76214	.72242	.64958	.58468	.52679
12	.88745	.86151	.83639	.78849	.74356	.70138	.62460	.55684	.49697
13	.87866	.85087	.82403	.77303	.72542	.68095	.60057	.53032	.46884
14	.86996	.84037	.81185	.75788	.70773	.66112	.57748	.50507	.44230
15	.86135	.82999	.79985	.74301	.69047	.64186	.55526	.48102	.41727
16	.85282	.81975	.78803	.72845	.67362	.62317	.53391	.45811	.39365
17	.84438	.80963	.77639	.71416	.65720	.60502	.51337	.43630	.37136
18	.83602	.79963	.76491	.70016	.64117	.58739	.49363	.41552	.35034
19	.82774	.78976	.75361	.68643	.62553	.57029	.47464	.39573	.33051
20	.81954	.78001	.74247	.67297	.61027	.55368	.45639	.37689	.31180
21	.81143	.77038	.73150	.65978	.59539	.53755	.43883	.35894	.29416
22	.80340	.76087	.72069	.64684	.58086	.52189	.42196	.34185	.27751
23	.79544	.75147	.71004	.63416	.56670	.50669	.40573	.32557	.26180
24	.78757	.74220	.69954	.62172	.55288	.49193	.39012	.31007	.24698
25	.77977	.73303	.68921	.60953	.53939	.47761	.37512	.29530	.23300
26	.77205	.72398	.67902	.59758	.52623	.46369	.36069	.28124	.21981
27	.76440	.71505	.66899	.58586	.51340	.45019	.34682	.26785	.20737
28	.75684	.70622	.65910	.57437	.50088	.43708	.33348	.25509	.19563
29	.74934	.69750	.64936	.56311	.48866	.42435	.32065	.24295	.18456
30	.74192	.68889	.63976	.55207	.47674	.41199	.30832	.23138	.17411
31	.73458	.68038	.63031	.54125	.46511	.39999	.29646	.22036	.16425
32	.72730	.67198	.62099	.53063	.45377	.38834	.28506	.20987	.15496
33	.72010	.66369	.61182	.52023	.44270	.37703	.27409	.19987	.14619
34	.71297	.65549	.60277	.51003	.43191	.36604	.26355	.19035	.13791
35	.70591	.64740	.59387	.50003	.42137	.35538	.25342	.18129	.13011
36	.69892	.63941	.58509	.49022	.41109	.34503	.24367	.17266	.12274
37	.69200	.63152	.57644	.48061	.40107	.33498	.23430	.16444	.11579
38	.68515	.62372	.56792	.47119	.39128	.32523	.22529	.15661	.10924
39	.67837	.61602	.55953	.46195	.38174	.31575	.21662	.14915	.10306
40	.67165	.60841	.55126	.45289	.37243	.30656	.20829	.14205	.09722
41	.66500	.60090	.54312	.44401	.36335	.29763	.20028	.13528	.09172
42	.65842	.59348	.53509	.43530	.35448	.28896	.19257	.12884	.08653
43	.65190	.58616	.52718	.42677	.34584	.28054	.18517	.12270	.08163
44	.64545	.57892	.51939	.41840	.33740	.27237	.17805	.11686	.07701
45	.63905	.57177	.51171	.41020	.32917	.26444	.17120	.11130	.07265
46	.63273	.56471	.50415	.40215	.32115	.25674	.16461	.10600	.06854
47	.62646	.55774	.49670	.39427	.31331	.24926	.15828	.10095	.06466
48	.62026	.55086	.48936	.38654	.30567	.24200	.15219	.09614	.06100
49	.61412	.54406	.48213	.37896	.29822	.23495	.14634	.09156	.05755
50	.60804	.53734	.47500	.37153	.29094	.22811	.14071	.08720	.05429

Amount of an Annuity: $s_{\overline{n}|i} = \dfrac{(1+i)^n - 1}{i}$

n \ i	1%	$1\frac{1}{4}$%	$1\frac{1}{2}$%	2%	$2\frac{1}{2}$%	3%	4%	5%	6%
1	1.0000	1.0000	1.0000	1.0000	1.0000	1.0000	1.0000	1.0000	1.0000
2	2.0100	2.0125	2.0150	2.0200	2.0250	2.0300	2.0400	2.0500	2.0600
3	3.0301	3.0377	3.0452	3.0604	3.0756	3.0909	3.1216	3.1525	3.1836
4	4.0604	4.0756	4.0909	4.1216	4.1525	4.1836	4.2465	4.3101	4.3746
5	5.1010	5.1266	5.1523	5.2040	5.2563	5.3091	5.4163	5.5256	5.6371
6	6.1520	6.1907	6.2296	6.3081	6.3877	6.4684	6.6330	6.8019	6.9753
7	7.2135	7.2680	7.3230	7.4343	7.5474	7.6625	7.8983	8.1420	8.3938
8	8.2857	8.3589	8.4328	8.5830	8.7361	8.8923	9.2142	9.5491	9.8975
9	9.3685	9.4634	9.5593	9.7546	9.9545	10.1591	10.5828	11.0266	11.4913
10	10.4622	10.5817	10.7027	10.9497	11.2034	11.4639	12.0061	12.5779	13.1808
11	11.5668	11.7139	11.8633	12.1687	12.4835	12.8078	13.4864	14.2068	14.9716
12	12.6825	12.8604	13.0412	13.4121	13.7956	14.1920	15.0258	15.9171	16.8699
13	13.8093	14.0211	14.2368	14.6803	15.1404	15.6178	16.6268	17.7130	18.8821
14	14.9474	15.1964	15.4504	15.9739	16.5190	17.0863	18.2919	19.5986	21.0151
15	16.0969	16.3863	16.6821	17.2934	17.9319	18.5989	20.0236	21.5786	23.2760
16	17.2579	17.5912	17.9324	18.6393	19.3802	20.1569	21.8245	23.6575	25.6725
17	18.4304	18.8111	19.2014	20.0121	20.8647	21.7616	23.6975	25.8404	28.2129
18	19.6147	20.0462	20.4894	21.4123	22.3863	23.4144	25.6454	28.1324	30.9057
19	20.8109	21.2968	21.7967	22.8406	23.9460	25.1169	27.6712	30.5390	33.7600
20	22.0190	22.5630	23.1237	24.2974	25.5447	26.8704	29.7781	33.0660	36.7856
21	23.2392	23.8450	24.4705	25.7833	27.1833	28.6765	31.9692	35.7193	39.9927
22	24.4716	25.1431	25.8376	27.2990	28.8629	30.5368	34.2480	38.5052	43.3923
23	25.7163	26.4574	27.2251	28.8450	30.5844	32.4529	36.6179	41.4305	46.9958
24	26.9735	27.7881	28.6335	30.4219	32.3490	34.4265	39.0826	44.5020	50.8156
25	28.2432	29.1354	30.0630	32.0303	34.1578	36.4593	41.6459	47.7271	54.8645
26	29.5256	30.4996	31.5140	33.6709	36.0117	38.5530	44.3117	51.1135	59.1564
27	30.8209	31.8809	32.9867	35.3443	37.9120	40.7096	47.0842	54.6691	63.7058
28	32.1291	33.2794	34.4815	37.0512	39.8598	42.9309	49.9676	58.4026	68.5281
29	33.4504	34.6954	35.9987	38.7922	41.8563	45.2189	52.9663	62.3227	73.6398
30	34.7849	36.1291	37.5387	40.5681	43.9027	47.5754	56.0849	66.4388	79.0582
31	36.1327	37.5807	39.1018	42.3794	46.0003	50.0027	59.3283	70.7608	84.8017
32	37.4941	39.0504	40.6883	44.2270	48.1503	52.5028	62.7015	75.2988	90.8898
33	38.8690	40.5386	42.2986	46.1116	50.3540	55.0778	66.2095	80.0638	97.3432
34	40.2577	42.0453	43.9331	48.0338	52.6129	57.7302	69.8579	85.0670	104.1838
35	41.6603	43.5709	45.5921	49.9945	54.9282	60.4621	73.6522	90.3203	111.4348
36	43.0769	45.1155	47.2760	51.9944	57.3014	63.2759	77.5983	95.8363	119.1209
37	44.5076	46.6794	48.9851	54.0343	59.7339	66.1742	81.7022	101.6281	127.2681
38	45.9527	48.2629	50.7199	56.1149	62.2273	69.1594	85.9703	107.7095	135.9042
39	47.4123	49.8662	52.4807	58.2372	64.7830	72.2342	90.4091	114.0950	145.0585
40	48.8864	51.4896	54.2679	60.4020	67.4026	75.4013	95.0255	120.7998	154.7620
41	50.3752	53.1332	56.0819	62.6100	70.0876	78.6633	99.8265	127.8398	165.0477
42	51.8790	54.7973	57.9231	64.8622	72.8398	82.0232	104.8196	135.2318	175.9505
43	53.3978	56.4823	59.7920	67.1595	75.6608	85.4839	110.0124	142.9933	187.5076
44	54.9318	58.1883	61.6889	69.5027	78.5523	89.0484	115.4129	151.1430	199.7580
45	56.4811	59.9157	63.6142	71.8927	81.5161	92.7199	121.0294	159.7002	212.7435
46	58.0459	61.6646	65.5684	74.3306	84.5540	96.5015	126.8706	168.6852	226.5081
47	59.6263	63.4354	67.5519	76.8172	87.6679	100.3965	132.9454	178.1194	241.0986
48	61.2226	65.2284	69.5652	79.3535	90.8596	104.4084	139.2632	188.0254	256.5645
49	62.8348	67.0437	71.6087	81.9406	94.1311	108.5406	145.8337	198.4267	272.9584
50	64.4632	68.8818	73.6828	84.5794	97.4843	112.7969	152.6671	209.3480	290.3359

Present Value of an Annuity: $a_{\overline{n}|i} = \dfrac{1-(1+i)^{-n}}{i}$

n $\diagdown$ i	1%	1¼%	1½%	2%	2½%	3%	4%	5%	6%
1	0.9901	0.9877	0.9852	0.9804	0.9756	0.9709	0.9615	0.9524	0.9434
2	1.9704	1.9631	1.9559	1.9416	1.9274	1.9135	1.8861	1.8594	1.8334
3	2.9410	2.9265	2.9122	2.8839	2.8560	2.8286	2.7751	2.7232	2.6730
4	3.9020	3.8781	3.8544	3.8077	3.7620	3.7171	3.6299	3.5460	3.4651
5	4.8534	4.8178	4.7826	4.7135	4.6458	4.5797	4.4518	4.3295	4.2124
6	5.7955	5.7460	5.6972	5.6014	5.5081	5.4172	5.2421	5.0757	4.9173
7	6.7282	6.6627	6.5982	6.4720	6.3494	6.2303	6.0021	5.7864	5.5824
8	7.6517	7.5681	7.4859	7.3255	7.1701	7.0197	6.7327	6.4632	6.2098
9	8.5660	8.4623	8.3605	8.1622	7.9709	7.7861	7.4353	7.1078	6.8017
10	9.4713	9.3455	9.2222	8.9826	8.7521	8.5302	8.1109	7.7217	7.3601
11	10.3676	10.2178	10.0711	9.7868	9.5142	9.2526	8.7605	8.3064	7.8869
12	11.2551	11.0793	10.9075	10.5753	10.2578	9.9540	9.3851	8.8633	8.3838
13	12.1337	11.9302	11.7315	11.3484	10.9832	10.6350	9.9856	9.3936	8.8527
14	13.0037	12.7706	12.5434	12.1062	11.6909	11.2961	10.5631	9.8986	9.2950
15	13.8651	13.6005	13.3432	12.8493	12.3814	11.9379	11.1184	10.3797	9.7122
16	14.7179	14.4203	14.1313	13.5777	13.0550	12.5611	11.6523	10.8378	10.1059
17	15.5623	15.2299	14.9076	14.2919	13.7122	13.1661	12.1657	11.2741	10.4773
18	16.3983	16.0295	15.6726	14.9920	14.3534	13.7535	12.6593	11.6896	10.8276
19	17.2260	16.8193	16.4262	15.6785	14.9789	14.3238	13.1339	12.0853	11.1581
20	18.0456	17.5993	17.1686	16.3514	15.5892	14.8775	13.5903	12.4622	11.4699
21	18.8570	18.3697	17.9001	17.0112	16.1845	15.4150	14.0292	12.8212	11.7641
22	19.6604	19.1306	18.6208	17.6580	16.7654	15.9369	14.4511	13.1630	12.0416
23	20.4558	19.8820	19.3309	18.2922	17.3321	16.4436	14.8568	13.4886	12.3034
24	21.2434	20.6242	20.0304	18.9139	17.8850	16.9355	15.2470	13.7986	12.5504
25	22.0232	21.3573	20.7196	19.5235	18.4244	17.4131	15.6221	14.0939	12.7834
26	22.7952	22.0813	21.3986	20.1210	18.9506	17.8768	15.9828	14.3752	13.0032
27	23.5596	22.7963	22.0676	20.7069	19.4640	18.3270	16.3296	14.6430	13.2105
28	24.3164	23.5025	22.7267	21.2813	19.9649	18.7641	16.6631	14.8981	13.4062
29	25.0658	24.2000	23.3761	21.8444	20.4535	19.1885	16.9837	15.1411	13.5907
30	25.8077	24.8889	24.0158	22.3965	20.9303	19.6004	17.2920	15.3725	13.7648
31	26.5423	25.5693	24.6461	22.9377	21.3954	20.0004	17.5885	15.5928	13.9291
32	27.2696	26.2413	25.2671	23.4683	21.8492	20.3888	17.8736	15.8027	14.0840
33	27.9897	26.9050	25.8790	23.9886	22.2919	20.7658	18.1476	16.0025	14.2302
34	28.7027	27.5605	26.4817	24.4986	22.7238	21.1318	18.4112	16.1929	14.3681
35	29.4086	28.2079	27.0756	24.9986	23.1452	21.4872	18.6646	16.3742	14.4982
36	30.1075	28.8473	27.6607	25.4888	23.5563	21.8323	18.9083	16.5469	14.6210
37	30.7995	29.4788	28.2371	25.9695	23.9573	22.1672	19.1426	16.7113	14.7368
38	31.4847	30.1025	28.8051	26.4406	24.3486	22.4925	19.3679	16.8679	14.8460
39	32.1630	30.7185	29.3646	26.9026	24.7303	22.8082	19.5845	17.0170	14.9491
40	32.8347	31.3269	29.9158	27.3555	25.1028	23.1148	19.7928	17.1591	15.0463
41	33.4997	31.9278	30.4590	27.7995	25.4661	23.4124	19.9931	17.2944	15.1380
42	34.1581	32.5213	30.9941	28.2348	25.8206	23.7014	20.1856	17.4232	15.2245
43	34.8100	33.1075	31.5212	28.6616	26.1664	23.9819	20.3708	17.5459	15.3062
44	35.4555	33.6864	32.0406	29.0800	26.5038	24.2543	20.5488	17.6628	15.3832
45	36.0945	34.2582	32.5523	29.4902	26.8330	24.5187	20.7200	17.7741	15.4558
46	36.7272	34.8229	33.0565	29.8923	27.1542	24.7754	20.8847	17.8801	15.5244
47	37.3537	35.3806	33.5532	30.2866	27.4675	25.0247	21.0429	17.9810	15.5890
48	37.9740	35.9315	34.0426	30.6731	27.7732	25.2667	21.1951	18.0772	15.6500
49	38.5881	36.4755	34.5247	31.0521	28.0714	25.5017	21.3415	18.1687	15.7076
50	39.1961	37.0129	34.9997	31.4236	28.3623	25.7298	21.4822	18.2559	15.7619

INDEX